The Changing Ocean Carbon Cycle

A midterm synthesis of the Joint Global Ocean Flux Study

The world's oceans act as a reservoir, with the capacity to absorb and retain carbon dioxide. The air–sea exchange of carbon is driven by physico-chemical forces, photosynthesis and respiration, and has an important influence on atmospheric composition. Variability in the ocean carbon cycle could therefore exert significant feedback effects during conditions of climate change. The Joint Global Ocean Flux Study (JGOFS) is the first multidisciplinary programme to directly address the interactions between the biology, chemistry and physics of marine systems, with emphasis on the transport and transformations of carbon within the ocean and across its boundaries. This unique volume, written by an international panel of scientists, provides a synthesis of JGOFS science and its achievements to date. It will therefore appeal to all those seeking a recent overview of the role of ocean processes in Earth system science and their wider implications on climate change.

Roger Hanson is the Executive Officer of the JGOFS at the International Project Office in Bergen, Norway. He was previously the Associated Program Director in Geosciences and Office of Polar Programs at the US National Science Foundation.

Hugh Ducklow is Professor of Marine Science at the School of Marine Science at the College of William and Mary in Virginia and is currently Chair of the US Joint Global Ocean Flux Study.

John Field is Professor of Zoology at the University of Cape Town. He is currently SCOR President and has served both as a member of the IGBP Scientific Committee and as Chair of the JGOFS Scientific Steering Committee.

The International Geosphere–Biosphere Programme was estabished in 1986 by the International Council of Scientific Unions, with the stated aim

To describe and understand the interactive physical, chemical and biological processes that regulate the total Earth system, the unique environment that it provides for life, the changes that are occurring in this system, and the manner in which they are influenced by human activities.

A wide-ranging and multi-disciplinary project of this kind is unlikely to be effective unless it identifies priorities and goals, and the IGBP defined six key questions that it seeks to answer. These are:

- How is the chemistry of the global atmosphere regulated, and what is the role of biological processes in producing and consuming trace gases?
- How will global changes affect terrestrial ecosystems?
- How does vegetation interact with physical processes of the hydrological cycle?
- How will changes in land-use, sea level and climate alter coastal ecosystems, and what are the wider consequences?
- How do ocean biogeochemical processes influence and respond to climate change?
- What significant climatic and environmental changes occurred in the past, and what were their causes?

The **International Geosphere–Biosphere Programme Book Series** bring new work on topics within these themes to the attention of the wider scientific audience.

Titles in the series

1. **Plant Functional Types**
   T. M. Smith, H. H. Shugart & F. I. Woodward (Editors)
   0 521 48231 3 (hardback)
   0 521 56643 6 (paperback)

2. **Global Change and Terrestrial Ecosystems**
   B. H. Walker & W. L. Steffen (Editors)
   0 521 57094 8 (hardback)
   0 521 57810 8 (paperback)

3. **Asian Change in the Context of Global Change**
   J. N. Galloway & J. M. Melillo (Editors)
   0 521 62343 X (hardback)
   0 521 63888 7 (paperback)

4. **The Terrestrial Biosphere and Global Change**
   B. Walker, J. Ingram, W. Steffen & J. Canadell (Editors)
   0 521 62429 0 (hardback)
   0 521 63888 7 (paperback)

5. **The Changing Ocean Carbon Cycle**
   R. B. Hanson, H. W. Ducklow & J. G. Field (Editors)
   0 521 65199 9 (hardback)
   0 521 65603 6 (paperback)

INTERNATIONAL GEOSPHERE–BIOSPHERE PROGRAMME BOOK SERIES

The Changing Ocean Carbon Cycle

A midterm synthesis of the Joint Global Ocean Flux Study

Edited by

Roger B. Hanson
University of Bergen

Hugh W. Ducklow
Virginia Institute of Marine Science
The College of William and Mary

John G. Field
University of Cape Town

CAMBRIDGE
UNIVERSITY PRESS

PUBLISHED BY THE PRESS SYNDICATE OF THE UNIVERSITY OF CAMBRIDGE
The Pitt Building, Trumpington Street, Cambridge, United Kingdom

CAMBRIDGE UNIVERSITY PRESS
The Edinburgh Building, Cambridge CB2 2RU, UK www.cup.cam.ac.uk
40 West 20th Street, New York, NY 10011–4211, USA www.cup.org
10 Stamford Road, Oakleigh, Melbourne 3166, Australia
Ruiz de Alarcón 13, 28014 Madrid, Spain

First published 2000

Printed in the United Kingdom at the University Press, Cambridge

Typeface Ehrhardt 11/14pt [VN]

A cataogue record for this book is available from the British Library

ISBN 0 521 65199 9 hardback
ISBN 0 521 65603 6 paperback

Contents

PART THREE REGIONAL-SCALE ANALYSIS AND INTEGRATION

PART FOUR GLOBAL-SCALE ANALYSIS AND INTEGRATION

PART FIVE FUTURE CHALLENGES

PART SIX CONCLUSION

Contributors

R. T. Barber
Duke University Nicholas School of the Environment Marine Laboratory, 135 Duke Marine Lab Road, Beaufort, NC 28516–9721, USA

U. Bathmann
Alfred-Wegener Institute for Polar and Marine Research, D-27515 Bremerhaven, Germany

P. W. Boyd
NIWA Centre for Chemical and Physical Oceanography, Department of Chemistry, University of Otago, Dunedin, New Zealand

F. Chai
School of Marine Sciences, University of Maine, Orono, ME 04469–5741, USA

S.-Y. Chao
Horn Point Laboratory, Center for Environmental Science, University of Maryland, Cambridge, MD 21613–0775, USA

H. J. W. de Baar
Netherlands Institute for Sea Research, P.O. Box 59, 1790 AB Den Burg, Texel, The Netherlands

K. L. Denman
Institute of Ocean Sciences, 9860 West Saanich Road, P.O. Box 6000, Sidney, British Columbia V8L 4B2, Canada

S. C. Doney
National Center for Atmospheric Research, P.O. Box 3000, Boulder, CO 80307–3000, USA

H. W. Ducklow
Virginia Institute of Marine Science, The College of William and Mary, PO Box 1346, Gloucester Point, VA 23062–1346, USA

G. T. Evans
Department of Fisheries and Oceans, Northwest Atlantic Fisheries Centre, PO Box 5667, St. John's, Newfoundland A1C 5X1, Canada

M. J. R. Fasham
Southampton Oceanography Centre, European Way, Empress Dock, Southampton SO14 3ZH, UK

J. G. Field
Zoology Department, University of Cape Town, 7700 Rondebosch, Cape Town, South Africa

W. D. Gardner
Department of Oceanography, Texas A&M University, College Station, TX 77843, USA

J. Hall
NIWA-Ecosystems, National Institute of Water and Atmospheric Research, 100 Aurora Terrace, PO Box 11–115, Hamilton, New Zealand

R. B. Hanson
JGOFS International Project Office, Center for the Study of Environment and Resources, University of Bergen, High Technology Center, N-5020 Bergen, Norway

K. Iseki
National Research Institute of Fisheries Science, 12–4, Fukuura 2–chome, Kanazawa-ku, Yokohama 236, Japan

D. M. Karl
Department of Oceanography, University of Hawaii, Honolulu, HA 96822, USA

A. H. Knap
Bermuda Biological Station for Research, Inc., 17 Biological Station Lane, Ferry Reach, St. George's, Bermuda GE 01

S. T. Lindley
Southwest Fisheries Science Center, 3150 Paradise Drive, Tiburon, CA 94920, USA

P. S. Liss
School of Environmental Sciences, University of East Anglia, Norwich NR4 7TJ, UK

K.-K. Liu
Institute of Oceanography, National Taiwan University, P.O. Box 23–13, Taipei, Taiwan, Republic of China

A. Longhurst
Department of Oceanography, Dalhousie University, Halifax, Nova Scotia B3H 4J1, Canada

M. Lucas
Zoology Department, University of Cape Town, 7700 Rondebosch, Cape Town, South Africa

J. J. McCarthy
Museum of Comparative Zoology, 26 Oxford Street, Harvard University, Cambridge, MA 02138, USA

A. F. Michaels
Wrigley Institute for Environmental Studies, University of Southern California, Los Angeles, CA 90089–0371, USA

A. Morel
Laboratoire de Physique et Chimie Marines, Université Pierre et Marie Curie et CNRS, F-06238 Villefranche-sur-Mer, Cédex, France

J. Parslow
Antarctic Program, CSIRO Division of Fisheries, GPO Box 1538, Hobart, Tasmania 7001, Australia

M. A. Peña
Institute of Ocean Sciences, 9860 West Saanich Road, P.O. Box 6000, Sidney, British Columbia, Canada V8L 4B2 (present address: Virginia Institute of Marine Science, The College of William and Mary, PO Box 1346, Gloucester Point, VA 23062–1346, USA)

T. Platt
Biological Oceanography Division, Bedford Institute of Oceanography, Box 1006, Dartmouth, Nova Scotia B2Y 4A2, Canada

J. Priddle
British Antarctic Survey, NERC, Madingley Road, Cambridge, Cambridge CB3 0ET, UK

E. Sakshaug
Trondheim Biological Station, Norwegian University of Science and Technology, N-7034 Trondheim, Norway

S. Sathyendranath
Department of Oceanography, Dalhousie University, Halifax, Nova Scotia B3H 4J1, Canada; Biological Oceanography Division, Bedford Institute of Oceanography, Dartmouth, Nova Scotia B2Y 4A2, Canada

D. Slagstad
SINTEF Civil and Environmental, Engineering, N-7034 Trondheim, Norway

K. Tangen
OCEANOR ASA, Pirsentret, N-7005 Trondheim, Norway

J. R. Toggweiler
GFDL/NOAA, Princeton University, P.O. Box 308, Princeton, NJ 08542, USA

P. Tréguer
URA CNRS 1513, Institut Universitaire Européen de la Mer, Université de Bretagne Occidental, BP 809, F-29285 Brest, Cédex, France

S. M. Turner
School of Environmental Sciences, University of East Anglia, Norwich NR4 7TJ, UK

D. W. R. Wallace
Oceanographic & Atmospheric Sciences Division, Brookhaven National Laboratory, Building 318, Upton, NY 11973, USA

P. J. le B. Williams
School of Ocean Sciences, University of Wales-Bangor, Menai Bridge, Gwynedd LL59 5EY, UK

Preface

International studies are essential to obtain global-scale understanding of ocean processes. The Joint Global Ocean Flux Study (JGOFS, jointly sponsored by the International Geosphere–Biosphere Programme, IGBP, and the Scientific Committee on Oceanic Research, SCOR) is the first international ocean project to directly address the interactions between the biology, chemistry and physics of marine systems. Its emphasis is on carbon exchange, cycling and export within the ocean and across the ocean's boundaries, with the atmosphere, the sea floor and coastal waters. The air–sea exchange of carbon is driven by physico-chemical forces, photosynthesis and respiration, and has an important influence on atmospheric composition. The temporal and spatial variability of these biogeochemical processes in the ocean carbon cycle could therefore exert significant feedback effects during conditions of climate change.

The first JGOFS Scientific Symposium held at Villefranche-sur-Mer, France, in May 1995, in which some 150 ocean scientists participated, inspired this book. The editors selected and re-structured some of the invited plenary presentations from the Symposium to provide a balanced synthesis of our understanding of ocean biogeochemistry towards the end of the twentieth century, as advanced by JGOFS since its inception.

Many organisations and people have worked together to make this publication possible. The Symposium was organised by the French National JGOFS Committee and financed under SCOR, IGBP, US National Science Foundation, the International Council of Scientific Unions (ICSU), Observatoire Oceanologique de Villefranche-sur-Mer, and several JGOFS National Committees. The Organising Committee comprised: Liliane Merlivat (Chair), Arthur C. T. Chen, Hugh Ducklow, John Field, Elizabeth Gross, Guy Jacques, André Morel, Paul Nival, Trevor Platt, Jarl-Ove Strömberg, and Neil Swanberg.

Roger B. Hanson
Hugh W. Ducklow
John G. Field

Bergen, June 1998

PART ONE

INTRODUCTION

1 The evolution of the Joint Global Ocean Flux Study project

J. J. McCarthy

Keywords: ocean carbon cycle, carbon dioxide exchange, biogeochemical processes, greenhouse effect, primary production, SCOR, IGBP

Introduction

It is always interesting to note that numerous ancestors appear swelling with pride when the offspring is a great success. The Joint Global Ocean Flux Study (JGOFS) project is certainly such a success, and, not surprisingly, many recognise within JGOFS sequences of the genome of a favourite ancestral project or committee report. Since the editors of this volume have given me the liberty to inject personal views in this chapter, I should like to begin arranging the stage upon which JGOFS science has evolved, set as it was in the mid-1960s. (I will use the acronym JGOFS to refer to immediate precursors of this project and various national elements of it, many of which have other names.)

The ocean carbon cycle

Students beginning graduate studies in ocean sciences at the Scripps Institution of Oceanography three decades ago, as I did, received their first introduction to an oceanographic perspective of the carbon cycle with books such as Dietrich's 1963 text entitled *General Oceanography* (originally published in German in 1957). The schematic of the carbon cycle depicted by Dietrich is given in Fig. 1.1. It is essentially the same figure used earlier by Borchert (1951; in German as cited by Dietrich 1963) and even earlier by Kalle (1945; in German as cited by Dietrich 1963). By today's conventions, the dimensions for reservoirs and fluxes (mass C per unit area, and mass C per unit area × time) are both unusual and only awkwardly translated into dimensions that are more modern. One can see, however, that much of the cycle portrayed is in order. Dietrich summarises the wisdom of the time, that the terrestrial biological cycle is 'practically closed in itself' and the 'great geological cycle . . . is also balanced'. It is particularly

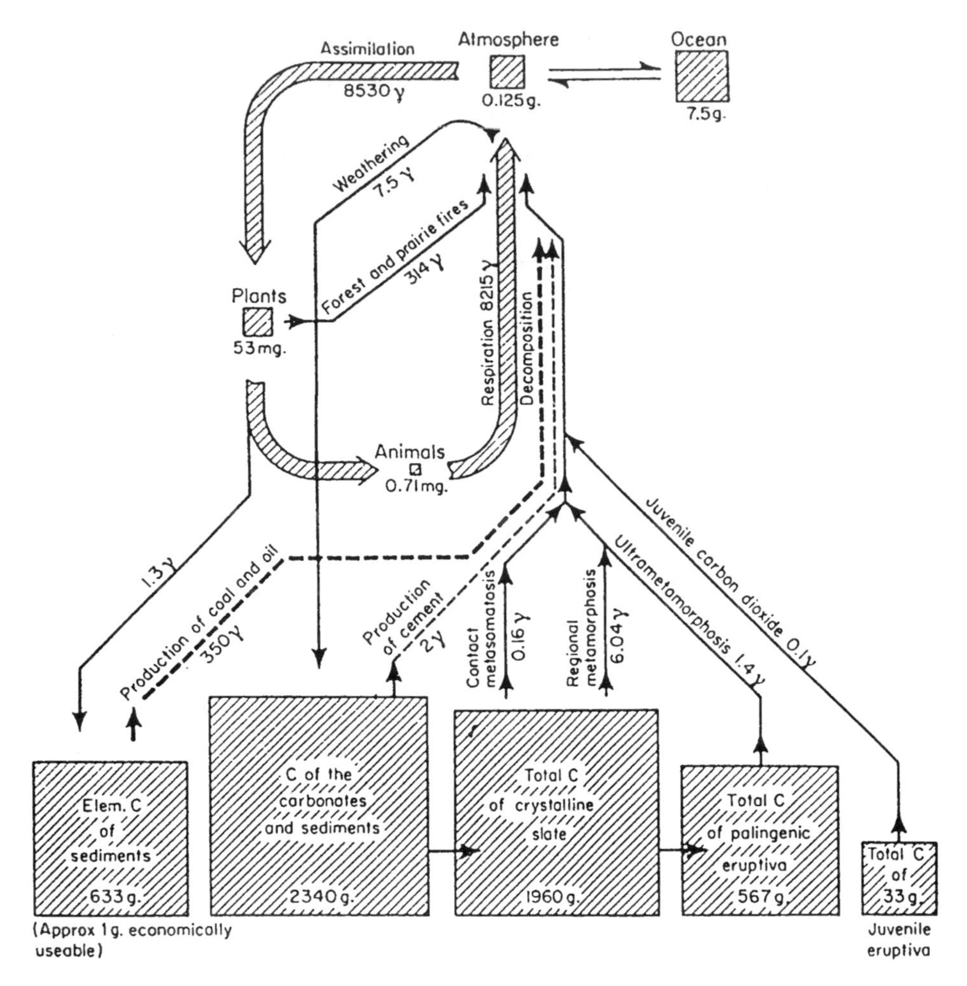

Figure 1.1 *Carbon cycle in nature (after K. Kalle, 1945; improved after H. Borchert, 1951). Dietrich, Copyright © (1963 Dietrich). Reprinted by permission of John Wiley & Sons, Inc.*

interesting to note how little was then known about the role of the ocean in the carbon cycle. Although the ocean reservoir of inorganic carbon is correctly indicated as being about sixty times larger than the atmospheric reservoir, there is no flux estimate for, or even an indication of, an exchange between the atmospheric and oceanic reservoirs. Implicit in this omission was an assumption that any such fluxes were either negligible or offsetting so as to yield zero net flux.

The understanding reflected in this diagram never could have been used to justify a project such as JGOFS. The ocean biota and associated sedimentation of organic and inorganic carbon are not even alluded to. Accompanying text acknowledges, however, that 'Although carbon stored in living substances . . .

represents a minute part of the total carbon content in the Earth's crust . . . it must be considered the actual motor which puts the geochemical carbon cycles in motion'. The importance of sedimentary burial of organic carbon in the evolution of Earth's atmosphere, and the stochiometric equivalence between this burial and Earth's atmospheric O_2 content, was not to be fully realised for another two decades (see Kasting, 1993).

Speculations regarding a potential increase in the so-called greenhouse effect of the Earth's atmosphere arising from release of CO_2 with fossil-fuel combustion were well known from the work of Callendar (1938) two decades before Dietrich, and even earlier, around the turn of the century, from the works of Arrhenius (1896) and Chamberlin (1899). Dietrich comments on the possibility of such a consequence of industrialisation by noting that the 'artificial supply of . . . carbon – (when) compared with the natural carbon dioxide production . . . is completely negligible – (and it could) have led to a doubling of the atmospheric carbon dioxide content since the start of industrialisation'. He then adds, 'This probably would have caused serious climatic consequences. It is doubtless a result of the great buffer capacity of the ocean that this increase in the carbon dioxide content of the atmosphere has remained so small as to be undetectable so far' (Dietrich 1963).

Skirrow, in the first volume of Riley & Skirrow's *Chemical Oceanography*, is himself the author of the chapter on carbon dioxide (1965). He uses exactly the carbon cycle schematic presented earlier by Dietrich, and in 105 pages on carbon dioxide in seawater, only with six lines does he treat the carbon cycle *per se*. He gives no more attention to the role of marine biogeochemical processes in this cycle than did Dietrich (1963).

The year 1957 is remembered by many in conjunction with the International Geophysical Year. This year was profoundly important for many aspects of Earth sciences, and the carbon cycle is no exception. In 1957, Roger Revelle and Charles David Keeling were preparing to implement the longest and most highly resolved time series that exists for global anthropogenic influence on atmospheric composition. Roger used to love to tell how difficult it was to sustain early funding for the Mauna Loa time series that Keeling began. Having demonstrated the seasonal cycle and slight upward trend at the end of the first few years was not, in the minds of some, sufficiently interesting to warrant additional support.

Another important piece of the ocean carbon cycle was also coming into focus at this same time. Sufficient data had been collected with the ^{14}C technique for assessing rates of primary productivity to make new global estimates of this process. Prior assessments based upon the oxygen light–dark bottle technique and various proxies had yielded estimates of global production as high as 130 Gt C yr^{-1} (Table 1.1). Steeman-Nielsen & Jensen (1957) presented the first

Table 1.1. *Estimates of global plankton production*

Year	Gt C yr^{-1}	Reference
1942	50–130	Sverdrup *et al.*
1946	126 ± 82	Riley
1957	20–25	Steeman-Nielsen & Jensen
1970	23	Koblentz-Mishke *et al.*
1971	44	Bruevich & Ivanenkov
1975	31	Platt & Subba Rao
1979	44	de Vooys
1987	42	Martin *et al.*
1995	45–50	Longhurst *et al.*

global composite of the ^{14}C data, and estimated a dramatically lower value of 20 Gt C yr^{-1}.

Estimates of global marine primary production of about 20 Gt C yr^{-1} prevailed for the next dozen years, culminating with the schematic for geographic distribution of primary production published by Koblentz-Mishke *et al.* (1970) (Fig. 1.2). Regardless of subsequent upward revisions of this estimate by 150–200% through the 1970s and 1980s, the original estimate and accompanying schematic of Koblentz-Mishke *et al.* remain popular choices in many overview and synthesis chapters and books, probably because of the power conveyed with the single global map. Interestingly, though, although derivative versions of this map are widely used, most delete detail in the original legend that indicated that, with the exception of the Indian Ocean, virtually no direct measurements were available to compute annual rates of primary production for the southern-hemisphere oceans.

The map of Koblentz-Mishke *et al.* helped to perpetuate the notion that offshore regions are without strong annual cycles in primary production. Indeed, in 1971, little was known about temporal variability in timing and magnitude of seasonal blooms. The three-year time series of Menzel & Ryther at Station 'S' southeast of Bermuda (Menzel & Ryther, 1961), which was truly exceptional documentation of such variability when published, remained exceptional in its detail and duration for oceanic waters until the advent of the JGOFS time-series stations near Bermuda and Hawaii in 1987. In fact the extent of temporal and spatial variability in primary productivity in many offshore as well as near-shore regions was only first known with the Earth orbiting Coastal Zone Color Scanner observations in the late 1970s.

By the early 1980s the well-established annually averaged secular trend in atmospheric CO_2 concentration permitted the determination that about half of human society's fossil-fuel CO_2 emissions currently remain in the atmosphere. Results of GEOSECS (Geochemical Ocean Sections) and other studies led to

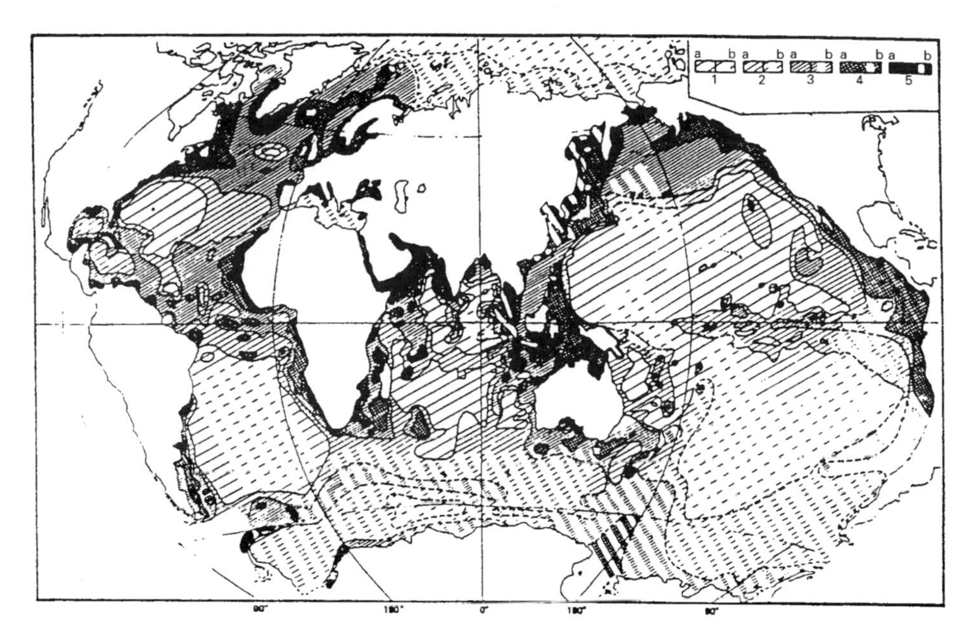

Figure 1.2 *Distribution of primary production in the World Ocean. Units are mg C m^{-2} d^{-1}: 1, less than 100; 2, 100–150; 3, 150–250; 4, 250–500, 5, over 500;* a, *data from direct ^{14}C measurements;* b, *data from phytoplankton biomass, hydrogen, or oxygen saturation. Reprinted with permission from* Scientific Exploration of the Southern Pacific. *Copyright © (1970 Koblentz-Mishke* et al.*) by the National Academy of Sciences. Courtesy of the National Academy Press, Washington, D.C.*

the strong arguments that the remaining portion, the so-called missing half, was being absorbed by the ocean. Then in 1982 came a most startling observation, the fact that polar ice samples contained evidence for a previously unknown oscillation in atmospheric CO_2, one of 80–100 ppm amplitude, which was in phase with temperature during the last glacial cycle (Neftel *et al.*, 1982) (Fig. 1.3). This finding stimulated much discussion and speculation as to which ocean processes could have contributed to such dramatic changes in the atmosphere. Certain marine biogeochemical responses and feedbacks were considered, and productivity and storage of organic carbon in the Southern Ocean were invoked by several groups (Knox & McElroy, 1984; Sarmiento & Toggweiler, 1984; Siegenthaler & Wenk, 1984).

At about the same time another discovery shattered a long-standing paradigm in ocean science. It had long been assumed that the deep ocean floor was without seasonality. The term 'rain of detritus' was coined by Alexander Agassiz (1888) to describe the slow continuous flux of fine particulate material from the upper ocean to deep benthic habitats. However, in the late 1960s it was noted that at depths of a few thousand metres certain marine invertebrates experience annual cycles in reproductive state (Schoener, 1968). Thus, in a habitat without

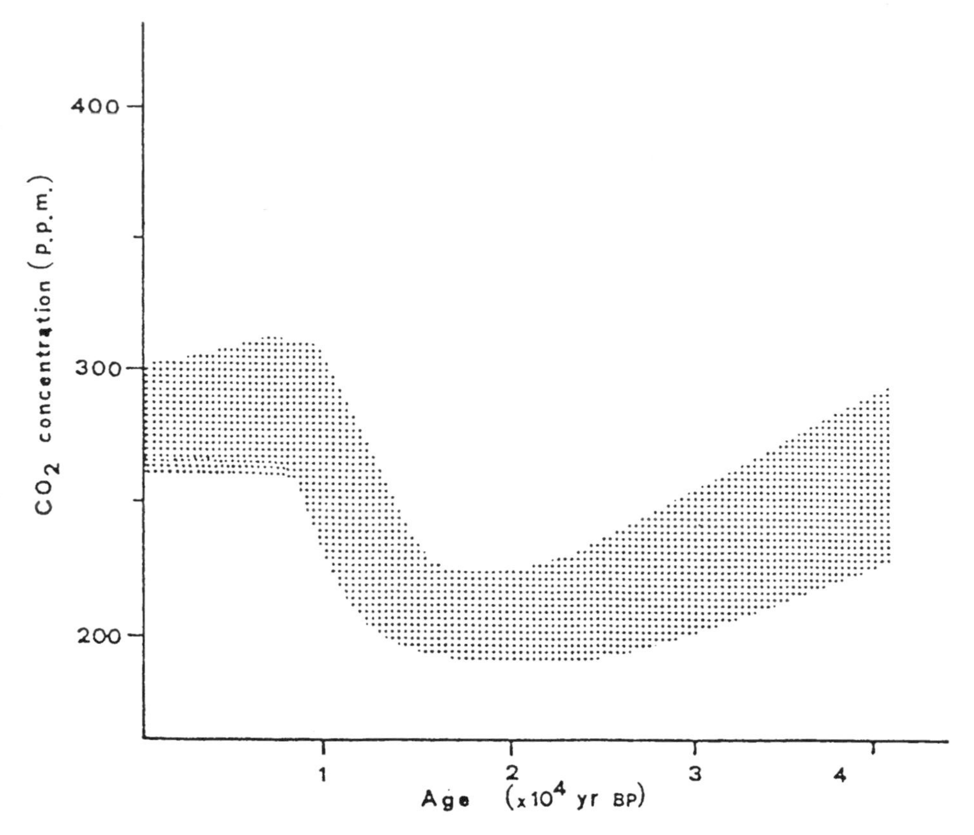

Figure 1.3 *Estimate range of atmospheric CO_2 during the past 40 000 yr. Reprinted with permission from* Nature *and the author. Copyright © (1982 Neftel* et al.*) by Macmillan Magazines Ltd.*

seasonal signals in either light or temperature, seasonality in the supply of sinking particulate material was a likely source of this effect. In the late 1970s, Werner Deuser began his time-series sediment-trap work southeast of Bermuda, and established (Deuser & Ross, 1980) that the seasonal cycle of organic flux to the deep ocean mimics the seasonal cycle of primary production in the overlying water (Fig. 1.4). Shortly thereafter, Honjo (1984), with higher temporal resolution, demonstrated pronounced interannual flux in organic matter to the deep ocean at Station 'P' in the North Pacific. The concept of seasonal and annual variability in vertical flux to great depth is now so well established that we tend to forget how recent these findings are.

The Earth system context

Another important development, and I shall argue a critical one, was to set the significance of these marine processes in a larger context. Although linkages

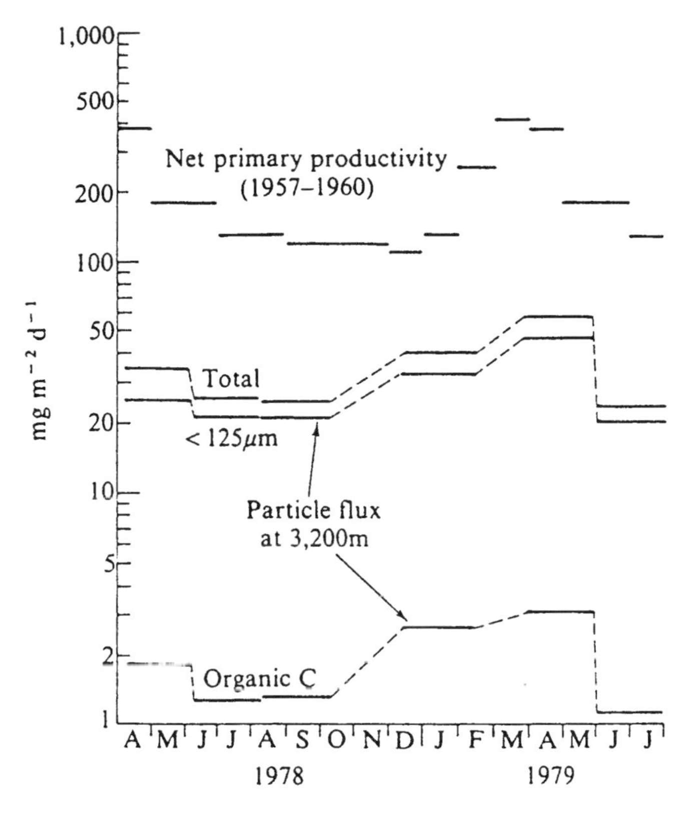

Figure 1.4 *Comparisons of monthly averages of daily net primary productivity (in terms of carbon) for the years 1957–60 with mean daily yields of total sediment (dry mass) and particulate organic carbon during six two-month sampling periods in 1978–79. Reprinted with permission from* Nature *and author. Copyright © (1980 Deuser & Ross) by Macmillan Magazines Ltd.*

between biogeochemical processes and climate had been inferred from polar ice-core data, a scientific plan that would permit the quantitative assessment of these and other relationships within the climate system was lacking

The stage was set to do just this. The early 1980s were years of expansive and optimistic mood in ocean science. New satellite technology was providing information on ocean processes at temporal and spatial scales we had never before imagined possible, and priorities in ocean science shifted accordingly. Many moderate-scale oceanographic projects had been successfully completed, and several of these had been interdisciplinary. Within and between the marine, terrestrial, and atmospheric domains, research, and even new journals, were addressing questions relating to biogeochemical cycles. On a more general front, scientists in many disciplines were pondering potential consequences of the inexorable rise in atmospheric CO_2, as the words of Revelle & Suess (1957) that 'human beings are now carrying out a large scale experiment of a kind that could

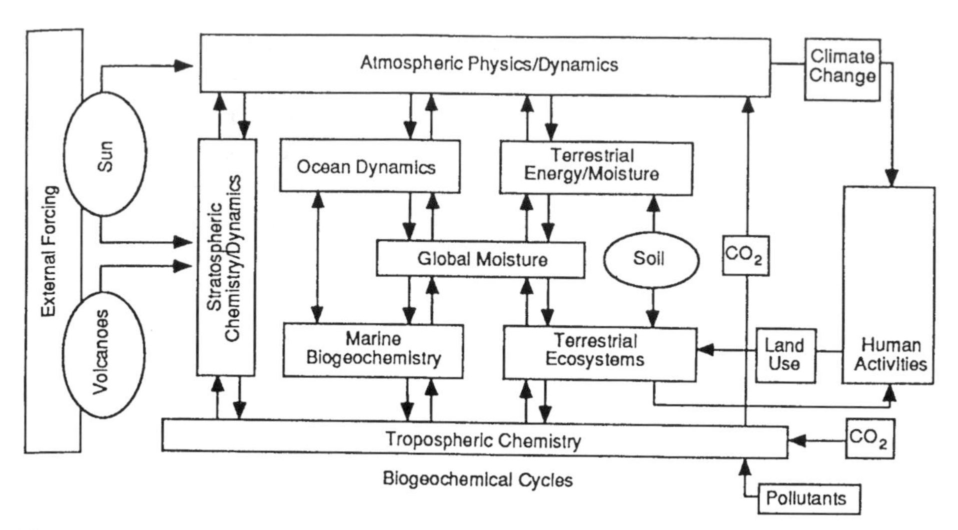

Figure 1.5 *Earth's climate system (ESSC, 1988).*

not have happened in the past nor be reproduced in the future' began to strike home.

One example of the larger context that was needed is evident in the schematic produced in the mid-1980s by the Earth System Science Committee in the USA. This committee began with a charge to report to the National Aeronautics and Space Administration (NASA) on research directions in Earth science, but its final report was embraced by the National Science Foundation (NSF) and the National Oceanic and Atmospheric Administration (NOAA) as well. This schematic (Fig. 1.5) (ESSC, 1988) was developed in various levels of detail and allowed scientists across the community of Earth sciences, broadly defined, to envision couplings among the various disciplinary segments in an overall union of physical dynamics and biology via biogeochemical cycles. In the USA, at least, the setting of any aspect of Earth science research in this larger context helped both to build the case for and allow access to new funds for large co-ordinated studies.

An important aspect of this approach was the design of effective observing capabilities and experiments, the results of which could be used to initialise and refine models of interactive ocean processes for the purpose of extrapolating to larger temporal and spatial scales, and to develop scenarios that could be tested by further observations. Needless to say, the role of models was and still is debated in this science. In early 1985 a distinguished geochemist involved in helping to organise US contributions to JGOFS stated that modelling is a 'buzz word', and argued that, although it is important, it must be kept in its proper place, which is in the design of data-gathering strategies. I believe this helped to set a tone that has been evident, at least in US JGOFS, until very recently.

By the early to mid-1980s, many pieces of the marine carbon cycle were falling into place. Considerable progress had been made in refining methodologies for measuring rates of primary production, new understanding had emerged regarding the role of the so-called 'biological pump' in the global carbon cycle, and strong inferences were being drawn about the interactions among ocean biogeochemical cycles and climate. The power of satellite observations for extending spatial distributions of certain physical and biological properties that historically were only measured *in situ*, and for inferring, with caution, rates of key biological processes on larger space and time scales, made a truly global study feasible. Products of the Coastal Zone Color Scanner enabled visualisation of spatial and temporal dimensions of plankton bloom cycles and episodes that were previously unknown owing to under-sampling with conventional methods. Moreover, as plans for the World Ocean Circulation Experiment (WOCE) were evolving, it was apparent that appending a JGOFS CO_2 sampling component to the World Hydrographic Program of WOCE would result in a very efficient use of resources.

Were there tensions within the scientific community regarding JGOFS objectives and plans for implementation? Of course there were, both within the community of supporting scientists and in adjacent fields as well. To accomplish the objectives of JGOFS, unprecedented international co-operation within the biological and chemical ocean science communities was required. No one nation had the resources necessary to undertake this task. Moreover, it was believed by some that individual national efforts designed to address JGOFS-related scientific questions would be difficult or impossible to co-ordinate within the structure of a single international project. In part, this perception was based upon understandable national differences in interest and programmatic emphasis, between coastal – open ocean, atmosphere–ocean, upper ocean – deep ocean, and deep ocean – sediment fluxes.

In addition to truly scientific differences, there were also issues relating to resources that attracted the attention of other segments of the ocean science community. Just as some scientists within JGOFS argued until recently that modelling should be given lower priority than observations and experiments, certain physical oceanographers argued in national and international meetings that JGOFS should be postponed until WOCE was completed. Nevertheless, momentum for a JGOFS study grew rapidly through the latter half of 1984. A pivotal meeting for US support was the National Research Council summer study convened by Kenneth Bruland in Woods Hole. About 60 scientists from 7 nations attended. The report from this meeting (NAS, 1984) provided a summary of the state of the science, and argued for a focused effort to understand biogeochemical cycles of the ocean sufficiently well to predict the interaction between the oceanic, atmospheric and sedimentary cycles of biologically active elements.

In an effort to capitalise on this momentum and to effect the requisite international support for such a project, Roger Revelle thought there was a good chance of convincing the now defunct Committee on Climate Change and the Ocean (CCCO), which had been instrumental in launching the Tropical Ocean and Global Atmosphere programme (TOGA) and WOCE under the auspices of the World Climate Research Programme, to endorse this programmatic concept. Roger took Peter Brewer, Richard Gammon and me to CCCO VI in November 1984 to make this case. It was proposed that this new study would have two aspects: (1) measurement over time of the constituents of the carbon dioxide system in surface and subsurface ocean waters down to great depths, and the rates and locations of carbon dioxide exchange between the sea and the air; and (2) studies of the interaction between biological activity in the ocean and atmospheric and ocean carbon dioxide. This recommendation included determination of the flux of organic particles from the euphotic zone into deeper waters or onto the bottom and their chemical transformation, as well as estimates of the effects of biological production in surface waters on the air–sea carbon dioxide exchange.

Many participants in the CCCO meeting made complimentary remarks regarding the science, but several argued that there was no point in engaging in such a study until the physics of the ocean was better understood. In addition, some participants argued ardently that the ocean science community could never convince governments to support an altimeter and a new colour sensor at the same time. In the end, the CCCO rejected Roger's proposal, but the biogeochemical community did not heed their warnings.

National efforts related to JGOFS continued to emerge and flourish, and in early 1987 the Scientific Committee on Oceanic Research (SCOR) came to the rescue, by convening a meeting in Paris under the chairmanship of D. James Baker, Jr. The report of this meeting encouraged SCOR to provide the international home this project needed, and SCOR took this action. Subsequently, through an agreement of joint sponsorship with the International Geosphere–Biosphere Programme (IGBP), JGOFS became the first marine science project of this programme.

Much of the science that is revealed in the following chapters of this book would not have been possible without the architecture and infrastructure of JGOFS. The interdisciplinary and collaborative aspects of this project have been immensely productive, and for a young scientist just now entering this field it is probably difficult to imagine how tenuous the prospects for this type of research were only a decade ago. Those of us who have had the opportunity to pursue science of fundamental interest to ourselves in the context of this large project are indeed fortunate. For many of us in ocean science, the problems at

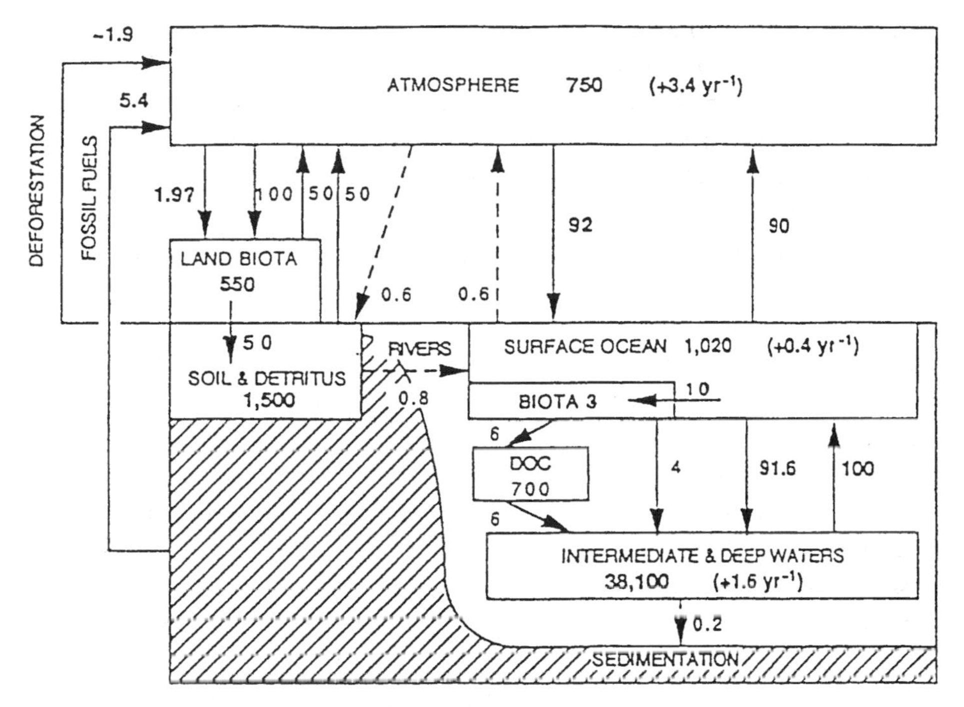

Figure 1.6 *Carbon cycle 1980–89. Units are Gt C and Gt C yr*$^{-1}$. *Reprinted with permission from* Nature *and author. Copyright © (1993 Siegenthaler & Sarmiento) by Macmillan Magazines Ltd.*

the interfaces of the traditional scientific disciplines have always had particular appeal. However, in this past decade the sense of societal relevance for aspects of these problems that relate to global and climate change has given both new purpose and a particular urgency to this science.

The future

Over the past three decades, knowledge and understanding of Earth's carbon cycle and the ocean's role in regulating atmospheric CO_2 content has grown substantially. This is abundantly evident in research papers (see, for example, Siegenthaler & Sarmiento, 1993) (Fig. 1.6) and in recent assessments of the Intergovernmental Panel on Climate Change (IPCC). It is necessary for us to improve the precision of estimates for essential reservoirs and fluxes. However, questions regarding the capacity of the ocean to sequester carbon in the future loom large. Model scenarios for future climate states typically presume that the ocean component of the carbon cycle will continue to function as it does at present. If under altered climate conditions, ocean circulation differs substantially from its present state, then the ocean's capacity to absorb and retain CO_2 will certainly change as well.

History tells us that there were a couple of important periods in the evolution of the science that led to JGOFS, one in the late 1950s and another in the early 1980s. In both instances leaps in scientific understanding raised expectations and mobilised the ocean science community. I predict that the late 1990s will be another such moment, as the product of JGOFS propels a new wave of understanding as to how interaction among biogeochemical and physical processes determines Earth's climate.

References

Agassiz, A. (1888). *Three Cruises of the United States Coast and Geodetic Survey Steamer 'Blake'*. Boston: Houghton, Mifflin and Company.

Arrhenius, S. (1896). On the influence of carbonic acid in the air upon the temperature of the ground. *Philosophical Magazine*, **41**, 237.

Bruevich, S. V. & Ivanenkov, V. N. (1971). Problems of the chemical balance of the world ocean. *Okeanologiya*, **11**, 835–41.

Callendar, G. S. (1938). The artificial production of carbon dioxide and its influence on temperature. *Quarterly Journal of the Royal Meteorological Society*, **64**, 223.

Chamberlin, T. C. (1899). An attempt to frame a working hypothesis of the cause of glacial periods on an atmospheric basis. *Journal of Geology* 7, **575**, 667–51.

De Vooys, C. G. N. (1979). Primary production in aquatic environments. In *The Global Carbon Cycle*, ed. B. Bolin, E. T. Degens, S. Kempe & P. Ketner, pp. 259–92. Chichester: John Wiley & Sons.

Deuser, W. G. & Ross, E. H. (1980). Seasonal change in the flux of organic carbon to the deep Sargasso Sea. *Nature*, **283**, 364–5.

Dietrich, G. (1963). *General Oceanography: An Introduction*. New York: John Wiley & Sons.

Earth System Science Committee (ESSC). 1988. *Earth System Science: A Closer View*. Washington, D.C.: National Aeronautics and Space Administration.

Honjo, S. (1984). In *Study of ocean fluxes in time and space by bottom-tethered sediment trap arrays: a recommendation*, pp. 305–24. Washington, D.C.: National Academy Press.

Kasting, J. F. (1993). Earth's early atmosphere. *Science*, **259**, 920–6.

Knox, F. & McElroy, M. (1984). Changes in atmospheric CO_2: Influence of the marine biota at high latitudes. *Journal of Geophysical Research*, **89**, 4629–37.

Koblentz-Mishke, O. J., Volkovinsky, V. V. & Kabanova, J. G. (1970). Plankton primary production of the world ocean. In *Scientific Exploration of the Southern Pacific*, ed. W. S. Wooster, pp. 183–93. Washington, D.C.: National Academy of Sciences.

Longhurst, A., Sathyendranath, S., Platt, T. & Caverhill, C. (1995). An estimate of global primary production in the ocean from satellite radiometer data. *Journal of Plankton Research*, **17**, 1245–71.

Martin, J. H., Knauer, G. A., Karl, D. M. & Broenkow, W. W. (1987). VERTEX: Carbon cycling in the northeast Pacific. *Deep-Sea Research*, **34**, 267–85.

Menzel, D. W. & Ryther, J. H. (1961). Annual variations in primary production of the Sargasso Sea off Bermuda. *Deep-Sea Research*, **7**, 282–8.

National Academy of Science. (1984). *Study of ocean fluxes in time and space by bottom-tethered sediment trap arrays: a recommendation*, Washington, D.C.: National Academy Press.

Neftel, A., Oeschger, H., Schwander, J.,

Stauffer, B. & Zumbrunn, R. (1982). Ice core sample measurements give atmospheric CO_2 content during the past 40,000 yr. *Nature*, **295**, 220–3.
Platt, T. & Subba Rao, D. V. (1975). Primary production of marine microphytes. In *Photosynthesis and Productivity in Different Environments, International Biological Programme*, ed. J. P. Cooper, pp. 249–80. Cambridge University Press.
Revelle, R. & Suess, H. E. (1957). Carbon dioxide exchange between atmosphere and ocean and the question of an increase of atmospheric CO_2 during the past decades. *Tellus*, **9**, 18–27.
Riley, G. A. (1946). Factors controlling phytoplankton populations on Georges Bank. *Journal of Marine Research*, **6**, 54–73.
Sarmiento, J. R. & Toggweiler, J. R. (1984). A new model for the role of the oceans in determining atmospheric pCO_2. *Nature*, **308**, 621–4.
Schoener, A. (1968). Evidence for reproductive periodicity in the deep sea. *Ecology*, **49**, 81–7.
Siegenthaler, U. & Sarmiento, J. L. (1993). Atmospheric carbon dioxide and the ocean. *Nature*, **365**, 119–25.
Siegenthaler, U. & Wenk, T. (1984). Rapid atmospheric CO_2 variations and ocean circulation. *Nature*, **308**, 624–6.
Skirrow, G. (1965). The dissolved gases – carbon dioxide. In *Chemical Oceanography*, ed. J. P. Riley & G. Skirrow, pp. 227–322. London: Academic Press.
Steeman-Nielsen, E. & Jensen, E.A. (1957). The autotrophic production of organic matter in the oceans. *Galathea Report*, **1**, 49–124.
Sverdrup, H. U., Johnson, M. W. & Fleming, R. H. (1942). *The Oceans: Their Physics, Chemistry, and General Biology*. Englewood Cliffs: Prentice-Hall.

PART TWO

CARBON EXCHANGE PROCESSES AND THEIR VARIABILITY

2 Marine primary production and the effects of wind

E. Sakshaug, K. Tangen and D. Slagstad

Keywords: wind, vertical mixing, 'new' primary productivity, polar phytoplankton

Introduction

Marine pelagic primary productivity is very unevenly distributed in the oceans. Only when and where the difference between the growth and loss rates of phytoplankton is significant in a positive sense, over some time, can phytoplankton blooms form and the primary productivity rate become very high. According to chlorophyll *a* maps made by NASA on the basis of ocean colour data from the CZCS sensor, large phytoplankton standing crops (measured as chlorophyll *a* concentration) are restricted to continental shelves and the Subarctic–Arctic circumpolar zone (the Greenland–Iceland–Norwegian–Barents Seas and the Bering Sea).

Areas of high productivity are characterised by adequate irradiance for at least half the year and high rates of supply of 'new' nutrients. At moderate to high latitudes, the vertical mixing that takes place during winter supplies the 'new' nutrients that form the basis for the spring bloom, which is initiated when the insolation becomes adequate. Secondary peaks in algal biomass may form later in the growth season; such peaks, however, are caused by the immediate supply of nutrients, alternatively a relaxation of the grazing pressure (Longhurst, 1995).

In shelf regions, vertical mixing created by the bottom topography (for instance banks or islands) and current shears are important in creating vertical mixing that brings deep waters rich in plant nutrients to the upper illuminated layers. Where such continual features are absent, wind may be the primary or only important factor that generates upward transport of plant nutrients. The 'classical' cases in this respect are the medium- to low-latitude eastern seaboards of the Pacific and Atlantic Oceans. Here, persistent longshore winds blow the surface waters offshore, making the Peruvian and West African coasts among the most productive in the world. Total primary productivity (new + regenerative) reaches 300–1000 g C m^{-2} yr^{-1} (Dugdale & Wilkerson, 1991). In oligotrophic ocean basins, annual total primary productivity may be as low as around 130 g C m^{-2} yr^{-1} (Ducklow & Carlson, 1992).

In addition, in the northern belt of Westerlies, where low-pressure systems (or depressions) traverse the North Atlantic from Labrador and Iceland to Scandinavia and, usually, into the Norwegian and Barents Seas, wind is important in promoting primary productivity (Sakshaug & Slagstad, 1992). In the Antarctic Ocean, a production-enhancing impact of the Westerlies may be less obvious: mixing may frequently reach too deep, creating strong light-limitation (Sakshaug & Holm-Hansen, 1984). In addition, the ecosystem in the open 'blue' waters in this region may be deficient in iron (Martin *et al.*, 1990; de Baar & Boyd, Chapter 4, this volume).

Primary productivity in the Subarctic and the Antarctic has been extensively studied (see Smith & Sakshaug (1990) for references), However, studies from the point of view that the northern belt of Westerlies may constitute an important factor in the regulation of subarctic primary production are sparse. This chapter gives a short overview of some numerical modelling studies carried out using data collected by marine environmental data buoys.

The northern Westerlies

The northern belt of Westerlies is situated at 50–75°N. The trajectory of the individual depressions is described by the jet stream, typically forming a lobed circumpolar feature (Rossby waves), that marks the border between Arctic air masses and warmer air masses to the south (Fig. 2.1). A corresponding belt of Westerlies, which is characterised by generally stronger winds than in the northern belt, surrounds the Antarctic continent ('the roaring forties').

The individual depressions move eastwards along the Polar Front and may, on average, cross a given site inside the belt of depressions from once or twice per week to once every second week. This creates a local periodicity in the weather in the sense that it appears to repeat itself in cycles of 1–2 weeks (Sakshaug & Slagstad, 1992). This periodicity may, however, not last for more than 1–2 months in one stretch, because the Rossby waves may not stay in the same configuration longer than that.

Reconfiguration of the jet-stream lobes, for instance in conjunction with a seasonal change in the area covered by arctic air masses, may take place rather abruptly (in a day or two). A relocation of a lobe involving just a few hundred kilometres may create a weather pattern that is entirely different from the previous one, changing, for instance, from a low-pressure weather type (repeatedly windy and cloudy) to a high-pressure type (predominantly calm and

Figure 2.1 *Schematic illustration of the jet stream and low-pressure systems that constitute the belt of Westerlies in the northern hemisphere. Single depressions may actually be over 1000 km in diameter. From Sakshaug* et al. *(1995).*

sunny) or vice versa. Thus, variations in the distribution of the Rossby waves may render some summers wet and cold, and others dry and warm (or wet and mild vs. dry and cold in winter).

Atmospheric depressions are anticyclonic. In the northern hemisphere, winds blow anti-clockwise, whereas they blow clockwise in the southern hemisphere. This and the eastward movement of the centre makes the southeast side particularly windy in the northern hemisphere (or, correspondingly, the northeast side in the southern hemisphere). The high resultant wind speed is the sum of the rotational wind velocity and the eastward speed of the whole depression, a fact well known to experienced seafarers. On the northern side (southern in the southern hemisphere), the resultant wind speed corresponds to the difference between the two, resulting in calmer weather.

The distribution of winds associated with a depression implies that the front passes a location a little south of an east-moving depression centre (north in the southern hemisphere). A depression carries strong winds that induce strong vertical mixing in the water column. A calmer weather type may follow when the centre passes, weakening the vertical mixing. Subsequently, another depression will form and the weather pattern described above will repeat itself as it passes. Thus the strong winds, and the mixing of the water column, will be highly episodic, as opposed to the more continual winds in the trade-wind belt.

Wind and vertical mixing

Vertical mixing is one of the decisive factors in the distribution of phytoplankton biomass and productivity. It controls the nutrient supply from deeper waters, and too-deep mixing may cause light limitation. Sverdrup (1953) elegantly formulated this in a model that introduced the concept of critical depth.

Vertical, turbulent mixing is generated by internal friction between water masses. The friction force is proportional to the velocity gradient (dv/dz) and inversely proportional to the coefficient of friction. At the surface, mechanical action from waves, which is generated by winds, also causes mixing. Vertical mixing due to cooling of the surface layer may be important, particularly in the absence of wind.

In the models dealt with in this chapter, the vertical mixing is a function of the Richardson number and the wave height. Empirically, if the Richardson number becomes less than 0.65, laminar flows are transformed into turbulent flows (Price & Weller, 1986). Dense water that overlies less dense water produces a negative Richardson number and thus increases the vertical mixing by convection. The effect of waves on the vertical mixing depends on the significant wave height, the average wave period and the wave number, according to a simple model by Ichiye (1967). A significant wave height is proportional to the square of the wind speed for storms in Norwegian waters (Hasselmann *et al.*, 1973), a realistic assumption for the areas in question.

'New' compared with 'regenerative' production

The positive impact of wind on primary production is a consequence of the lifting of deep nutrient-rich waters towards the surface, by upwelling or vertical mixing (the negative impact being light limitation, caused by the transport of phytoplankton downwards). Thus wind contributes significantly to the regulation of 'new' production (the open-system mode), as opposed to the 'regenerative' production that is based on recycling of nutrients already present within the euphotic zone (the closed-system mode); total production is defined as 'new' + 'regenerative' (see Thingstad & Sakshaug, 1990).

It is the 'new' rather than the total primary production that determines the harvestability of a pelagic ecosystem in terms of biological resources by non-inhabitants such as fishermen and hunters, as well as the sedimentation rate of a pelagic ecosystem (Sakshaug & Slagstad, 1992). It is therefore important to distinguish 'new' from 'regenerative' and total primary production in fisheries- and carbon-flux-related contexts.

Spring blooms of phytoplankton, being based on winter nutrients, may represent about 50% 'new' production or even more whereas a tropical rainforest, although it is very productive, is close to 100% regenerative and thus not harvestable by non-inhabitants in any sustainable fashion; this has, regrettably, been demonstrated in several industrial-scale 'experiments'.

Whereas the central oligotrophic ocean basins are typically characterised by a high incidence of regenerative production and a low total, the marine high-productive areas, such as continental shelves and the Subarctic, exhibit a high incidence of 'new' production. In the Barents Sea, the production is about 60 of a total of 110 g C m^{-2} annually (Kristiansen & Farbrot, 1991; Sakshaug & Slagstad, 1992). This high level of 'new' production is in part related to a high incidence of wind-induced mixing. In an average year, it is expected that about 50 g C m^{-2} is grazed and 45 g C m^{-2} leaves the euphotic zone directly as ungrazed phytoplankton biomass (75% of the 'new' and 41% of the total production). About 15 g C m^{-2} represents production of dissolved organic carbon (DOC) (Sakshaug *et al.*, 1994). Considering that the average depth of the Barents Sea is as little as 230 m, presumably as much as 10% of the matter that falls out of the surface water may reach the seabed (Wassmann, 1990).

Wind-related annual primary production

The 'classical' cases of high wind-driven primary production are the ecosystems in the zones of coastal upwelling, for example the coastal waters of Peru and West Africa. Such systems may exhibit annual total primary productivity of 300–1000 g C m^{-2} (Dugdale & Wilkerson, 1991).

Primary productivity in the belt of northern Westerlies hardly reaches that of upwelling areas in low latitudes, except in landlocked and polluted areas. As mentioned above, the average total productivity of the Barents Sea is not more than one third of that in an upwelling area. Yet in the southwestern Barents Sea, where wind may mix the column to 40–80 m depth in the growth season, the total annual productivity may reach as high as 165 g C m^{-2} yr^{-1}, of which about

90 g C m^{-2} is 'new' production. In the northern parts of the Barents Sea, where water from melting sea ice contributes to a stable water column and limits wind-mixing deeper than 15–40 m during the growth season, 'new' production is about 40–70 g C m^{-2} yr^{-1} (Sakshaug & Slagstad, 1992). In the Bering Sea, the horizontal gradients are also pronounced. The annual total productivity may range from 30–50 g C m^{-2} on the Alaskan side to 285 g C m^{-2} in Anadyr waters on the western side (Walsh *et al.*, 1989).

There are two main reasons for the lower 'new' productivity in the northern belt of Westerlies relative to the upwelling zones in the trade-wind belt. The main reason is the short growth season at high latitudes: the spring bloom may begin as late as mid-May or early June owing to deep mixing earlier in the season in waters not affected by meltwater from sea ice (see Sverdrup, 1953). For a bloom in Atlantic waters to begin earlier than May at latitude 60–75° N, the depth of mixing should not be deeper than 80 m, assuming average atmospheric irradiance in the belt of Westerlies (30% of the irradiance during clear-sky conditions; Sakshaug & Slagstad, 1991). The other reason is the episodic nature of wind-generated nutrient supply following the spring bloom in this belt: presumably there is not more than one significant nutrient spike per one or two weeks, as opposed to the continual wind-driven supplies in the trade-wind belt.

Data and models

One reason for the scarcity of studies of the effects of wind on algal biomass and primary production in the oceanic stretches of the northern belt of Westerlies is that the weather is often characterised by dense cloud cover, making frequent observation of ocean colour from satellites unachievable. Further, frequent periods of notoriously rough weather conditions restrict detailed studies of the water column from research vessels when the vertical mixing presumably is strongest. For this reason, three-dimensional (3-D) models are particularly helpful tools for understanding the dynamics of primary production in the belt of Westerlies. The realism of the models, however, depends ultimately on good experimental studies to obtain model coefficients and field data for validation. In rough-weather zones, telemetering buoys with strings of sensors are potentially of great value.

It is believed that models used in the present study for the Barents Sea are realistic enough for the present purposes. Validation of the models has been carried out by comparison of model runs and data collected during the Norwegian research program Pro Mare in 1984–89 (Sakshaug *et al.*, 1995) and by Russian researchers (Titov, 1995).

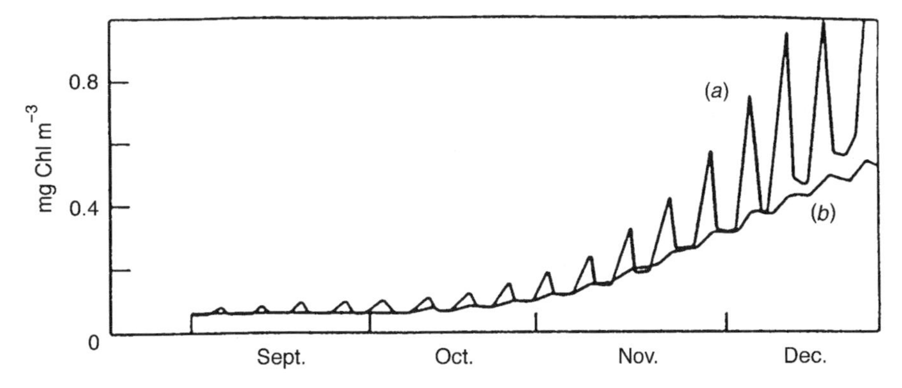

Figure 2.2 *Simulated progress of the spring phytoplankton bloom ('new' biomass only) in Antarctic waters as a function of:* (a) *alternating (1–week cycle) depths of mixing (60 and 200 m) and irradiance (80 and 20% of clear days), and* (b) *constant depth of mixing (200 m; variations in irradiance included). Data from Sakshaug* et al. *(1991).*

Our first attempt at a wind-driven simulation model for primary productivity was a one-dimensional (1-D) model that was derived from a plankton model for the Barents Sea (Slagstad & Støle-Hansen, 1991). In the modified model, the passage of atmospheric depressions was idealised. The depth of the mixed layer was alternated between 30 and 100 m or between 60 and 200 m on a weekly cycle; the irradiance hitting the surface was alternated concomitantly from 80 to 20% of irradiance during clear sky conditions (simulating dense cloud cover accompanying the strongest winds and vice versa). Applied to Antarctic waters (Sakshaug *et al.*, 1991), this model suggested a repeated pattern of growth and dilution of algal biomass, yielding a serrated biomass curve (Fig. 2.2) giving a slow increase in algal biomass towards midsummer. Moreover, when the model accounted algal self-shading, the model indicated that extremely high nutrient concentrations in Antarctic waters were unlikely to be completely depleted unless the depth of mixing was less than 10 m. This result is in accordance with field observations (Holm-Hansen *et al.*, 1989; Kocmur *et al.*, 1990) and it emphasises that complete depletion of nutrients in Antarctic waters may be a rather scarce event except in fairly sheltered ice-filled waters. These same model also indicated that removing iron deficiency in Antarctic deep-sea waters might enhance primary productivity only to a small extent because light limitation introduced by deep vertical mixing and/or self-shading by the phytoplankton might be about as limiting as iron.

Our first wind-driven model for the Barents Sea (Sakshaug & Slagstad, 1992) was based on a 3-D model of the hydrographic properties of the Barents Sea. Applying real wind data for 1983, this model clearly demonstrated the effect of wind on algal biomass and primary production, as opposed to a situation

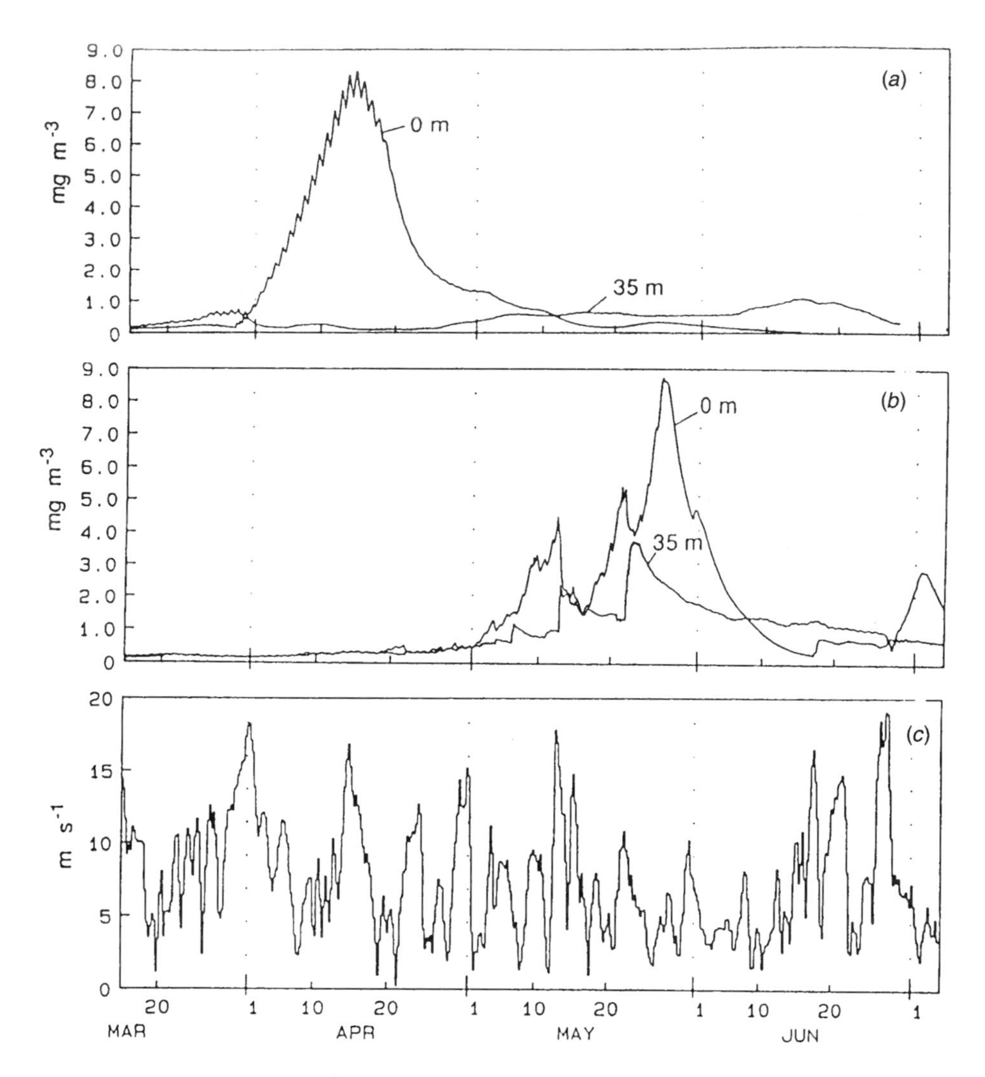

Figure 2.3 *Simulated progress of chlorophyll* a *concentration at 0 and 35 m depth in spring ('new' biomass only) in Atlantic waters of the Barents Sea.* (a) *Simulation without wind,* (b) *simulation using wind data for 1983, and* (c) *wind speeds modelled for 1983. Wind data have been obtained from the Hindcast model of the Norwegian Meteorological Institute. From Sakshaug & Slagstad (1992).*

without wind (compare Fig. 2.3*b* with 2.3*a*). A weekly periodicity in the wind speed is evident (Fig. 2.3*c*), indicating regular passage of low-pressure systems. Because the model, when run without wind, leaves cooling of the surface waters as the only factor generating vertical mixing, the depth of mixing is restricted to only 5–10 m in the growth season. This, however, permits culmination of the spring bloom as early as mid-April, as opposed to the model that includes wind and predicts a bloom as late as late May, because the wind-induced vertical mixing until then reaches deeper than 40–80 m.

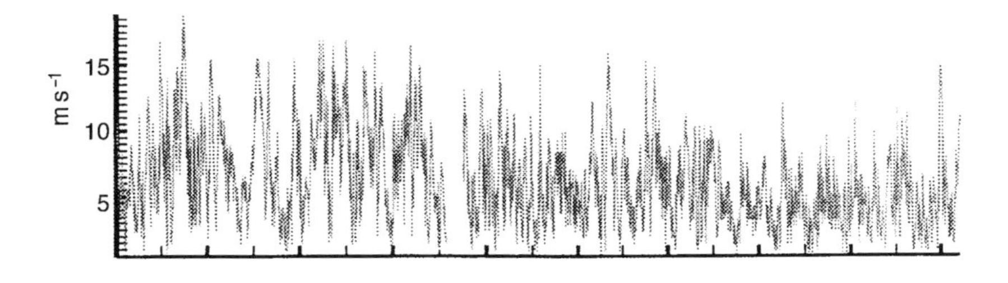

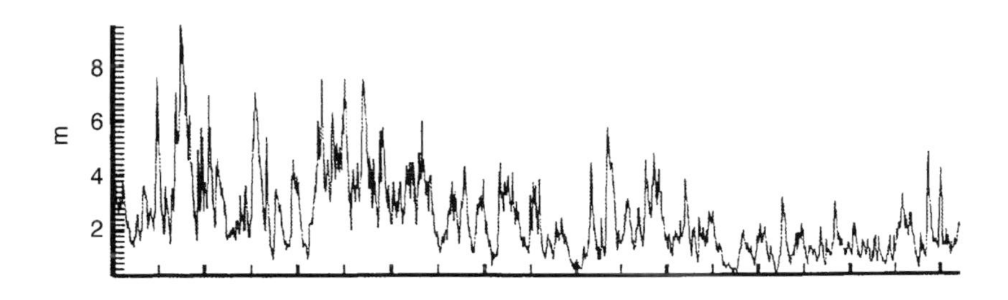

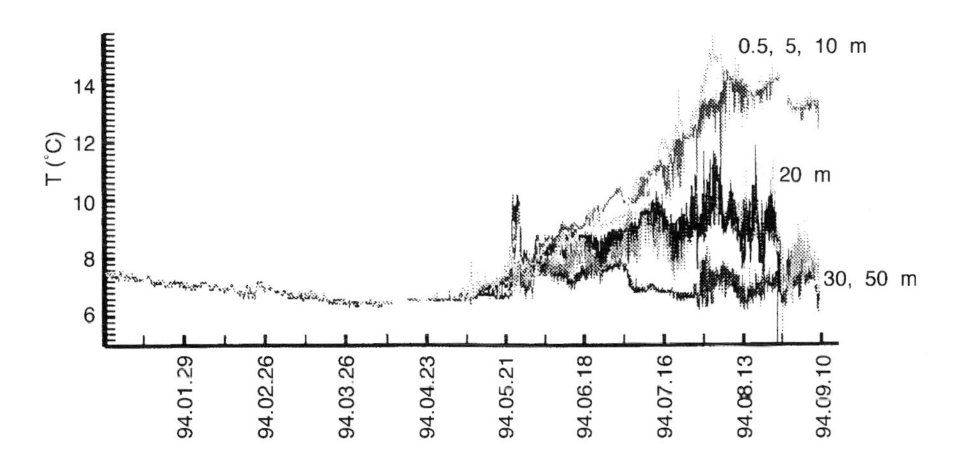

Figure 2.4 *Seasonal progress of wind speed (m s^{-1}), significant wave height (m) and hourly recorded temperatures at the Halten Bank Central Norway (65°5.0′N, 34.0°E; depth 255 m), in January–September 1994. Seawatch Buoy data at Oceanor AS, Trondheim.*

Late stabilisation of the upper layers is typical also in the Norwegian Sea and Norwegian coastal waters, leading to blooms beginning as late as late April or even early June (the Norwegian Sea). An example of late stabilisation is provided by hourly temperature recordings from a Seawatch telemetering buoy placed at the Halten Bank off Central Norway, where the waters are characteristically coastal, with salinities around 34–34.8 (Fig. 2.4). This record, which includes observations in the 0–50 m range, shows that the water column is homogeneous with respect to temperature down to a depth of over 50 m until 7 May. After then, a temperature gradient builds up in July and reaches as much as 6.4 °C at 50 m to 15.5 °C at the surface. In July to early August, only the waters in the upper 10 m appear to be wind-mixed. From late August onwards, the temperature gradient in the water column decreases markedly in response to cooling and the first strong gale in autumn. Although a wind-driven periodicity may be evident in the summer season, the winds are not strong enough to mix the waters appreciably in these 'low-salinity' coastal waters, as opposed to Atlantic waters in the Barents Sea.

In the same study it was assumed that annual 'new' productivity in the northern part of the Barents Sea (characterised as Polar water) would be less than half of that in the southern part (characterised as Atlantic water). This characterisation corresponds to a model run without wind in terms of productivity, owing to the high stability of the water column in the northern part (Fig. 2.3*a*). Our most recent wind-driven 3-D model of the Barents Sea, which is a study of a *Phaeocystis* bloom observed in 1987 (Sakshaug *et al.*, 1995), indicates that this may indeed be likely. Interestingly, applying a 'diatom' parameter set instead of a '*Phaeocystis*' set in this model made the predictions unrealistically high with respect to chlorophyll *a* distribution owing to the higher Chl *a* : C ratio of diatoms. The predictions of this model imply that a reduction of the salinity of the surface layer by only 0.3 parts per thousand in mid-May is sufficient to reduce the depth of mixing to half its value. That is, the mixed layer shifts from a depth of 80 m at a station characterised by Atlantic water (left half of Fig. 2.5) to depth of about 40 m in a more northerly station that is influenced by lower-salinity Polar water (right half of Fig. 2.5).

Moreover, regardless of the three gale-force wind peaks, which occurred from mid-May to late June, the resulting depth of the mixed layer was roughly halved in the same period at both stations. At the Atlantic and Polar stations, the mixed-layer depth was about 40 and 25 m, respectively (Fig. 2.5). As a consequence of a decrease in the heat flux from the water to the air in summer, sea temperatures increased slightly. The depth of mixing is, accordingly, most sensitive to the density gradient in the water column. The stronger the gradient, the more momentum is needed to mix the column to a given depth.

Considering that salinities of the Polar surface waters that overlie the Atlantic waters in the northern half of the Barents Sea are as low as 31–34.6, in contrast to over 34.9 in the Atlantic surface layer of the southern half, one cannot expect low-pressure systems of plausible magnitude to generate winds strong enough to mix the water column deeply. In fact, the depth of the wind-mixed layer increases only slowly with time in the northern half of the Barents Sea, from 20–25 m to a maximum of 35–40 m in the course of the growth season (Loeng,

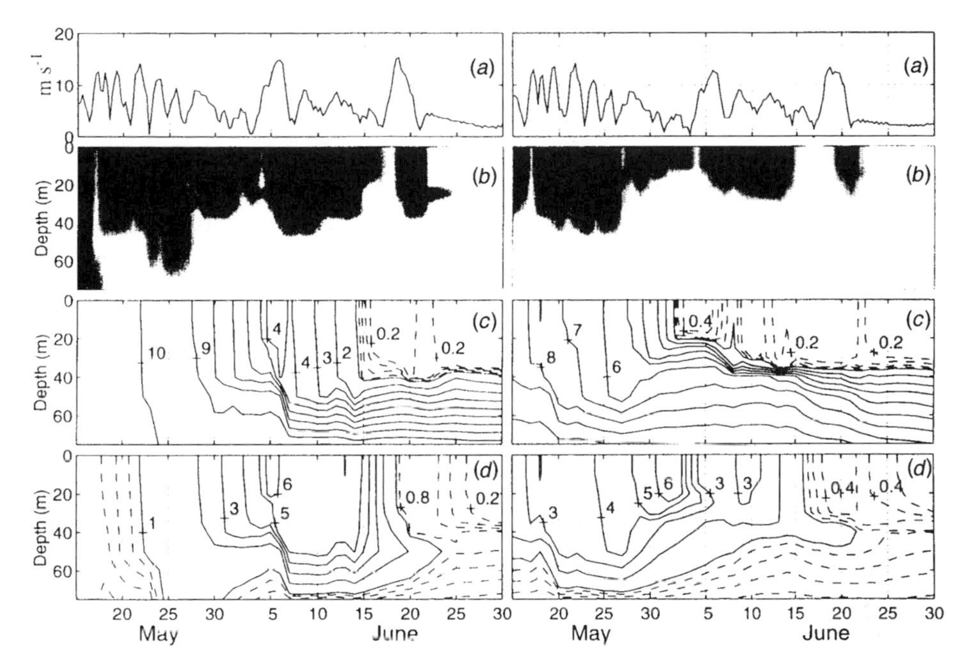

Figure 2.5 *Simulated progress of (a) wind speed, (b) depth of the mixed layer (defined as the lower boundary for waters with diffusion coefficients greater than 0.001 m^2 s^{-1}), (c) nitrate concentration, and (d) chlorophyll* a *concentration in Atlantic (left) and Polar (right) waters of the Barents Sea. Data from Sakshaug* et al. *(1995).*

1991; Slagstad & Støle-Hansen, 1991). Thus, there is little 'new' production, save some in the chlorophyll maximum layer near the pycnocline, in the Polar waters after the winter nutrients have been depleted by the so-called ice-edge bloom.

Ice-edge phytoplankton blooms, first predicted by Gran (1931) and later observed in polar seas with seasonal ice cover (see Sakshaug & Holm-Hansen (1984) for review), form a belt 20–50 km wide along the ice edge in the Barents Sea. Temporally, the bloom sweeps across the northern half of the Barents Sea towards summer, following the retreating ice edge (Rey *et al.*, 1987). The ice-edge bloom creates a very rich biological zone that moves with the ice edge. However, the average annual productivity for the whole northern half of the Barents Sea is low because there is almost no 'new' production except in the ice-edge bloom (Sakshaug *et al.*, 1994).

Secondary peaks in the primary productivity following the spring (or ice-edge) blooms are restricted to the permanently ice-free waters where the vertical salinity gradient is so small that winds can mix the water column appreciably. Because the 'new' productivity in May–June includes the spring blooms, they may be of fairly equal size in Atlantic waters and the southernmost

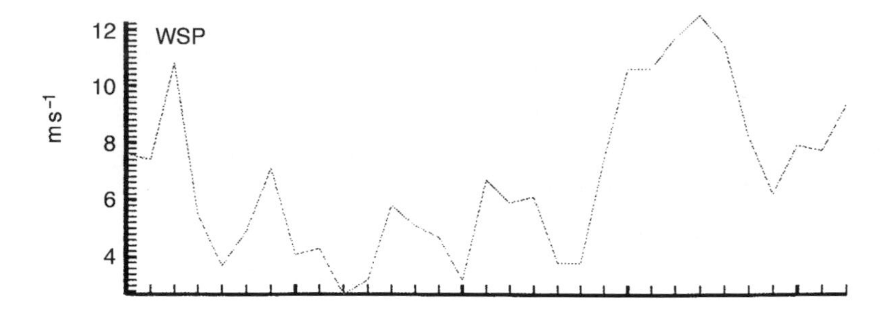

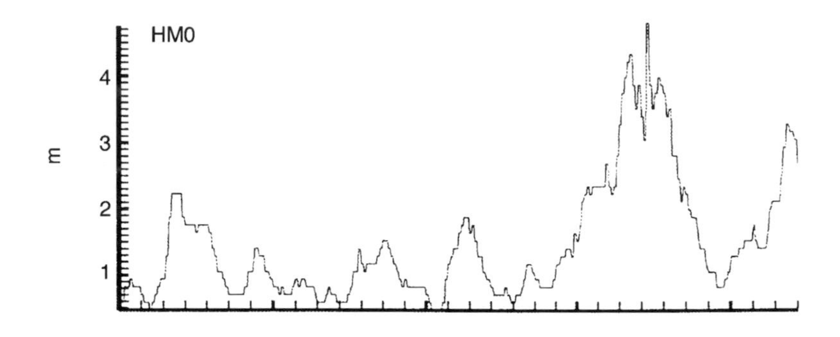

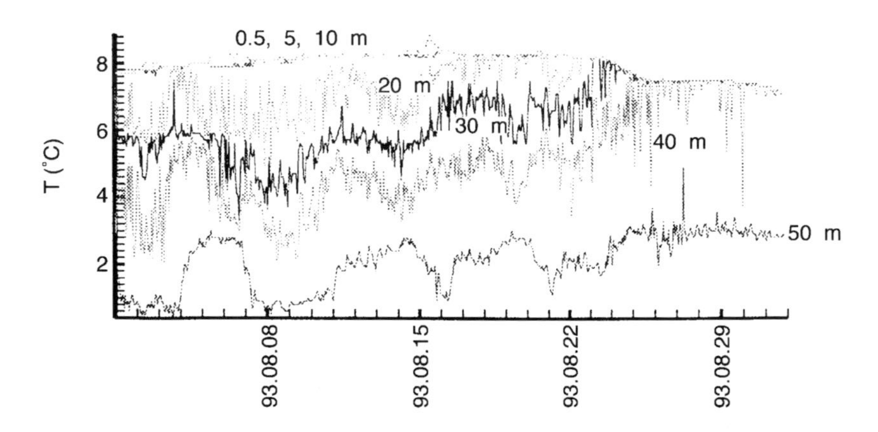

Figure 2.6 *Response of significant wave height (HM0) and temperature in the water column to wind speed (WSP) at Shtockmanovskoye, west of Novaya Zemlya (73°0.5′N, 43°50.5′E; depth 326 m), in August 1993. Seawatch Buoy data at Oceanor AS, Trondheim. Wind speeds have been obtained from the Hindcast model of the Norwegian Meteorological Institute.*

part of the Polar water in the Barents Sea (40–50 g C m^{-2}; Sakshaug *et al.*, 1995). Thus, the difference in annual productivity between Polar and Atlantic waters in the Barents Sea is a consequence mainly of the number and size of wind-generated nutrient spikes in the surface layers in the Atlantic waters during July–September. Although wind fields are not very different in the northern and southern halves of the Barents Sea, mixing in the northern half is restricted by the lower salinity of the surface layer. This is obviously more important in terms of annual productivity than differences between the two areas regarding the length of the growth season. In fact, both Sakshaug *et al.* (1995) and models (Figs. 2.3*a*, 2.5) indicate that the growth season is shorter in the Atlantic water than in the southernmost ranges of the Polar water. This is a consequence of the earlier stabilisation of the water masses in the latter region (Fig. 2.5; Sakshaug & Skjoldal, 1989).

Data from a Seawatch buoy at Shtockmanovskoye, west of Novaya Zemlya, indicate, as predicted by the models, that wind-generated episodes of vertical mixing do occur (Fig. 2.6). In this particular case, the water column was stable, as one might expect for August. There was a marked temperature gradient (0.5–8 °C) from the surface to 50 m depth. Then, on 24–25 August, the upper 40 m suddenly became mixed with a temperature around 7 °C, while at 50 m depth the temperature remained about 3 °C. The upper layer was obviously mixed as a result of a sudden onset of winds of more than 12 m s^{-1}, which also created significant waves (heights up to 5 m). Like the models, the ocean observations in the Barents Sea showed that summer mixing of the water column (more than 20 m) occurs when the weak density gradients are broken down with strong winds (over 12 m s^{-1}) and high waves (over 4 m).

Wind and the whole ecosystem

We have presented evidence that the northern Westerlies exhibit a marked positive effect on 'new' productivity. In the southern part of the Barents Sea, where the surface waters are Atlantic and the vertical density gradient of the water column is relatively weak, the episodic vertical mixing generated by the Westerlies implies a doubling of the annual 'new' primary productivity over calm periods. The latter may not be an utterly unrealistic case in productivity terms. It does, however, with respect to vertical mixing, correspond to the seasonally ice-filled northern half of the Barents Sea. At this time of the year, the density gradient of the water column is too steep for mixing deeper than 20–40 m. Thus, high 'new' primary productivity is restricted to the narrow ice-edge zone that sweeps through this area, following the ice edge when it retreats northwards during late spring and summer.

The distribution of zooplankton and fish productivity also indicates that the southern, Atlantic part of the Barents Sea is the biologically richer one. Admittedly, the ice-edge zone in the northern half is teeming with life. However, this counts for less when averaged for the whole northern Barents Sea. Thus, the average annual productivity of *Calanus* may be twice as large in the southern part (*C. finmarchicus*: 10 g C m^{-2}) as in the northern part (*C. glacialis*: 4.5 g C m^{-2}) (Tande & Slagstad, 1990, 1992). Besides, virtually the whole krill stock, responsible for an additional productivity of 1.5 g C m^{-2} annually, is restricted to the Atlantic part (Dalpadado & Skjoldal, 1991). Moreover, the commercial fish species, primarily capelin and cod, but also herring and haddock, are distributed mainly in the southern half of the Barents Sea, except that part of the capelin stock temporarily feeds near the ice edge in summer (Skjoldal *et al.*, 1992).

The regional differences described above do not imply that wind-related differences in the primary production alone regulate the productivity of the higher trophic levels. In fact, each trophic level is also affected directly by environmental variation. Again, the belt of Westerlies may be essential. The course of the Westerlies and the intensity of the individual depressions determine (together with the deep-water formation in the Greenland and Norwegian Seas), the magnitude of the North Atlantic Current. The Westerlies are presumably decisive in determining the flow of Atlantic water that is deflected into the Barents Sea. This flow may, as a monthly average, range from 1 to 4 Sverdrup (1 Sverdrup = 1 million cubic metres per second), the highest inflows occurring during low-pressure periods. Typically, there are marked peaks in November–January; the highest peaks recorded so far are those in 1991, 1992 and 1993 (Loeng *et al.*, 1995). Variation in this inflow causes 'good' and 'bad' years in the Barents Sea in terms of biological production. In the Bering Sea, too, the distribution of wind is known to create 'good' and 'bad' years in biological productivity terms (Walsh *et al.*, 1989).

The good years are the years that are 'warm', that is, years with a relatively high temperature in the upper 50–70 m, demonstrating a large transport of heat with the Atlantic inflow. The 'Kola section' data collected by PINRO, Murmansk (see Loeng, 1991), shows an evident periodicity in the temperature of this layer (Fig. 2.7). Yearly temperature cycles have been identified over 3–5-year, 10–11-year and 18-year periods. The latter two correspond to the sunspot and lunar node cycles, respectively. Such cycles are known from many climatically related variables dating back to the Middle Ages: the onset of the wine harvests in France and Germany, the cod fisheries off France and England, and the sea level at Stockholm (Currie *et al.*, 1994). Moreover, this information emphasises the role of the North Atlantic Current as a carrier of climate-related messages, of astronomical origin or otherwise.

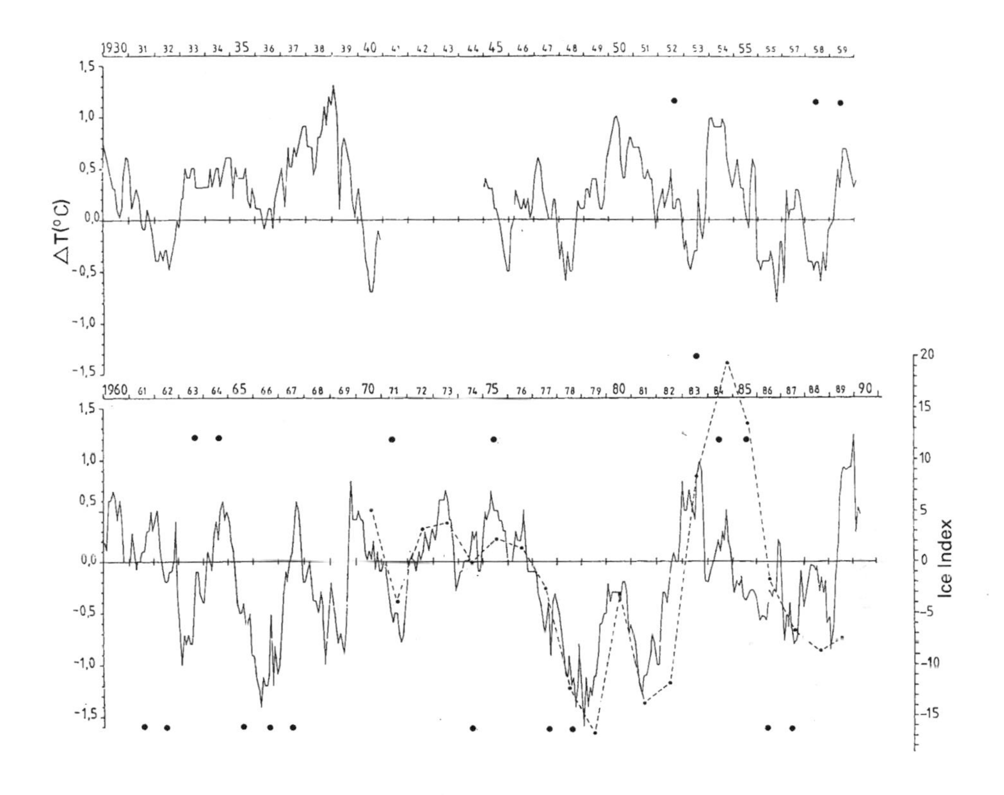

Figure 2.7 *Recruitment successes and failures of cod (upper and lower circles, respectively) plotted against temperature anomalies in the Kola section, 1930–90. The broken line is the ice index (high values indicate little ice). Hydrographical data from PINRO, Murmansk ('Kola section'), in Loeng (1991). Cod recruitment data from Tretyak* et al. *(1995).*

The 'good' (warm) and 'bad' (cold) years differ in several respects, however, above all in the distribution of sea ice. In a warm year, the maximum ice cover may be as little as 0.7 million km² whereas in a cold year it may cover up to 1.1 million km² (Vinje & Kvambekk, 1991). Consequently, primary production in the Barents Sea as a whole may be 30% higher in a warm year; for the northern half alone there may be a fourfold higher productivity (Slagstad & Stokke, 1994). Second, extended and late-melting ice covers have a catastrophic impact on the zooplankton productivity in the northern part of the Barents Sea; at the same time the import of zooplankton with the North Atlantic Current from the south is at a minimum (Tande & Slagstad, 1990, 1992). This, and the low temperature in itself, in turn have grave implications for the recruitment of fish stocks in the Barents Sea: data indicate that only the warm years are 'good' in terms of cod recruitment and it is just as clear that the cold years are failures

(Fig. 2.7; Tretyak *et al.*, 1995). The size of the capelin and cod stocks respond strongly. Year-to-year variations are in the range 30–700 and 150–700 kg C km^{-2}, respectively (Sakshaug *et al.*, 1994). The most critical stage is the transition from a sequence of good years to one of bad years. Then all stocks suddenly become very unbalanced relative to each other from a prey–predator point of view. This is when capelin stocks may collapse, mass mortality of capelin-eating birds such as the common guillemot may occur, and harp seals may flee south along the Norwegian Coast. Extreme imbalances ('capelin catastrophes') occurred in 1988–89 and happened in 1901, before there was any high-technology fisheries in the open seas.

The positive effect of the Westerlies on 'new' primary production in Atlantic waters of the north represents but one aspect of the impact of wind on the pelagic ecosystem. In the partly ice-filled Barents Sea, the role of the Westerlies in determining the influx of Atlantic waters and the area of ice cover may be just as important, if not more so, considering the ecosystem as a whole.

Whether one studies weather in Northern Europe or fisheries in its coastal waters, the essential variables are the same: the magnitude of the North Atlantic Current, and the course and intensity of the Westerlies. Both may be strongly affected by global warming: the North Atlantic Drift may be reduced in spring and summer (Sakshaug, 1996), and the Westerlies may intensify (Hall *et al.*, 1994). Thus, the Barents Sea is coupled to the global heat transport system. After all, the North Atlantic Current, together with the East Greenland Current, form a loop that prevents the lower-latitude areas from overheating, by exporting salt and heat.

Acknowledgements

Thanks are due to Dr Stephen Barstow for critically reading the manuscript, and Sivert Moen and Kjersti Andresen for assistance with the figures.

References

Ådlandsvik, B. & Loeng, H. (1991). A study of the climatic system in the Barents Sea. *Polar Research*, **10**, 45–50.

Currie, R. G., Wyatt, T. & O'Brien, D. P. (1994). Deterministic signals in European fish catches, wine harvests, and sea-level, and further experiments. *International Journal of Climatology*, **13**, 665–87.

Dalpadado, P. & Skjoldal, H. R. (1991). Distribution and life history of krill from the Barents Sea. *Polar Research*, **10**, 443–60.

Ducklow, H. W. & Carlson, C. A. (1992). Oceanic bacterial production. *Advances in Microbial Ecology*, **12**, 113–81.

Dugdale, R. C. & Wilkerson, F. P. (1991). Low specific nitrate uptake rate: A common feature of high-nutrient, low-chlorophyll marine ecosystems. *Limnology and Oceanography*, **36**, 1678–88.

Gran, H. H. (1931). On the conditions for

the production of plankton in the sea. *Rapports et Procés Verbaux des Réunions, Conseil Permanent International de l'Exploration de Mer*, **75**, 37–46.

Hall, N. M. J., Hoskins, B. J., Valdes, P. J. & Senior, C. A. (1994). Storm tracks in a high-resolution GCM with doubled carbon dioxide. *Quarterly Journal of the Royal Meteorological Society*, **120**, 1209–30.

Hasselmann, K., Barnett, T. P., Bouws, E., Carlson, H., Cartwright, D. E., Enke, K., Ewing, J. A., Gienapp, H., Hasselmann, D. E., Kruseman, P., Meerburg, A., Müller, P., Olbers, D. J., Richter, K., Sell, W. & Walden, H. (1973). Measurements of wind-wave growth and swell decay during the Joint North Sea Wave Project (JONSWAP). *Ergänzungsheft zur Deutschen Hydrographischen Zeitschrift, Reihe A (8°)*, **12**, 95 pp.

Holm-Hansen, O., Mitchell, B. G., Hewes, C. D. & Karl, D. M. (1989). Phytoplankton blooms near Palmer Station, Antarctica. *Polar Biology*, **10**, 49–57.

Ichiye, T. (1967). Upper ocean boundary-layer flow determined by dye diffusion. *Physics of Fluids*, **10S**, 270–7.

Kocmur, S. F., Vernet, M. & Holm-Hansen, O. (1990). RACER: Nutrient depletion by phytoplankton during the 1989 austral spring bloom. *Antarctic Journal of the US*, **25**, 138–41.

Kristiansen, S. & Farbrot, T. (1991). Nitrogen uptake rates in the phytoplankton and in the ice algae in the Barents Sea. *Polar Research*, **10**, 187–92.

Loeng, H. (1991). Features of the oceanographic conditions of the Barents Sea. *Polar Research*, **10**, 5–18.

Loeng, H., Ozhigin, V. & Adlandsvik, B. (1995). Water fluxes through the Barents Sea. *ICES C.M. 1995/Mini*: **10**, 12 pp.

Longhurst, A. (1995). Seasonal cycles of pelagic production and consumption. *Progress in Oceanography*, **36**, 77–167.

Martin, J. H., Gordon, R. M. & Fitzwater, S. E. (1990). Iron in Antarctic waters. *Nature*, **345**, 156–8.

Price, J. F. & Weller, A. (1986). Diurnal cycling: Observations and models of the upper ocean response to diurnal heating, cooling, and wind mixing. *Journal of Geophysical Research*, **91**, 8411–27.

Rey, F., Skjoldal, H. R. & Slagstad, D. (1987). Primary production in relation to climatic changes in the Barents Sea. In *Proceedings of the 3d Soviet-Norwegian Symposium, Murmansk 1986*, ed. H. Loeng, pp. 29–46. Bergen, Norway: Institute of Marine Research.

Sakshaug, E. (1996). Weather and climate: effects on life in the sea. *Kongelige Norske Videnskabers Selskabs Forhandlinger 1995*, pp. 103–23.

Sakshaug, E., Bjorge, A., Gulliksen, B., Loeng, H. & Mehlum, F. (1994). Structure, biomass distribution, and energetics of the pelagic ecosystem in the Barents Sea: A synopsis. *Polar Biology*, **14**, 405–11.

Sakshaug, E. & Holm-Hansen, O. (1984). Factors governing pelagic production in polar waters. In *Marine Phytoplankton and Productivity*, ed. O. Holm-Hansen, L. Bolis & R. Gilles, pp. 1–17. (Lecture notes on coastal and estuarine studies, 8.) Berlin: Springer.

Sakshaug, E., Rey, F. & Slagstad, D. (1995). Wind forcing of marine primary production in the northern atmospheric low-pressure belt. In *Ecology of Fjords and Coastal Waters*, ed. H. R. Skjoldal, C. Hopkins, K. E. Erikstad & H. P. Leinaas, pp. 15–25. Amsterdam: Elsevier Science.

Sakshaug, E. & Skjoldal, H. R. (1989). Life at the ice edge. *Ambio*, **8**, 60–7.

Sakshaug, E. & Slagstad, D. (1991). Light and productivity of phytoplankton in polar marine ecosystems: a physiological view. *Polar Research*, **10**, 69–85.

Sakshaug, E. & Slagstad, D. (1992). Sea ice and wind: Effects on primary productivity in the Barents Sea. *Atmosphere and Ocean*, **30**, 579–91.

Sakshaug, E., Slagstad, D. & Holm-Hansen, O. (1991). Factors controlling the development of phytoplankton blooms in the Antarctic Ocean – a mathematical model. *Marine Chemistry*, **35**, 259–71.

Skjoldal, H. R., Gjøsæter, H. & Loeng, H. (1992). The Barents Sea ecosystem in the

1980s: ocean climate, plankton and capelin growth. *ICES Marine Science Symposia*, **195**, 278–90.

Slagstad, D. & Stokke, S. (1994). Simulering av strømfelt, hydrografi, isdekke og primærproduksjon i Det nordlige Barentshavet (Simulation of current fields, hydrography, ice cover, and primary production in the northern Barents Sea). *Fisken og Havet*, **9**, 1–46 (English summary).

Slagstad, D. & Støle-Hansen, K. (1991). Dynamics of plankton growth in the Barents Sea: model studies. *Polar Research*, **10**, 173–86.

Smith, W. O. Jr & Sakshaug, E. (1990). Polar phytoplankton. In *Polar Oceanography*, part B: *Chemistry, biology and geology*, ed. W. O. Smith, Jr, pp. 477–525. New York: Academic Press.

Sverdrup, H. U. (1953). On conditions for the vernal blooming of phytoplankton. *Journal du Conseil permanent international de l'Exploration de Mer*, **18**, 287–95.

Tande, K. S. & Slagstad, D. (1990). Growth and production of the herbivorous copepod, Calanus glacialis, in the arctic waters of the Barents Sea. *Marine Ecology Progress Series*, **63**, 189–99.

Tande, K. S. & Slagstad, D. (1992). Regional and interannual variation in biomass and productivity of *Calanus finmarchicus*, in subarctic environments. *Oceanologica Acta*, **15**, 309–21.

Thingstad, T. F. & Sakshaug, E. (1990). Control of phytoplankton growth in nutrient recycling ecosystems. Theory and terminology. *Marine Ecology Progress Series*, **63**, 261–72.

Titov, O. V. (1995). Seasonal dynamics of primary production in the Barents Sea. *ICES CM 1995/Mini*: **16**, 25 pp.

Tretyak, V. L., Ozhigin, V. K., Yaragina, N. A. & Ivshin, V. A. (1995). Role of oceanographic conditions in Arcto-Norwegian cod recruitment dynamics. *ICES CM 1995/Mini*: **15**, 20 pp.

Vinje, T. & Kvambekk, Å. S. (1991). Barents Sea drift ice characteristics. *Polar Research*, **10**, 59–68.

Walsh, J. J., McRoy, C. P., Blackburn, T. H., Coachman, L. K., Goering, J. J., Henriksen, K., Andersen, P., Nihoul, J. J., Parker, P. L., Springer, A. M., Tripp, R. B., Whitledge, T. E. & Wirick, C. D. (1989). The role of Bering Strait in the carbon/nitrogen fluxes of polar marine ecosystems. In *Proceedings of the 6th Conference of the Committee, Arctique International*, ed. L. Rey & V. Alexander, pp. 90–120. Brill, Leiden, the Netherlands.

Wassmann, P. (1990). Dynamics of primary production and sedimentation in shallow fjords and polls of western Norway. *Oceanography and Marine Biology Annual Review*, **91**, 87–154.

3 Net production, gross production and respiration: what are the interconnections and what controls what?

P. J. le B. Williams

Keywords: marine plankton, photosynthesis, respiration, autotrophs, heterotrophs, net production

Introduction

The oceanic biogeochemical cycle of elements is driven by the two primary physiological processes, photosynthesis and respiration. The balance between these two processes is net community production (NCP). Over the past four decades we have amassed an impressive data set on the primary photosynthetic fixation of carbon (gross primary production; GPP) in the sea and are able to model the process over large tracts of the oceans with good apparent success (Platt *et al.*, 1995). By contrast, our knowledge of the rates of the other two (respiration and NCP) in the sea is disturbingly poor. Williams (1984) noted the paucity of the database for respiration well over a decade ago, and yet there is no sign of any substantial improvement. The situation is even worse for NCP. NCP represents the potential of the plankton for organic export and as such its assessment is central to the aims of JGOFS. The lack of data in many respects is odd. It is not a difficult measurement to make and it can be measured with the least uncertainty of interpretation of most, if not all, major planktonic processes (e.g. ^{14}C-determined primary production, ^{15}N-determined recycled production and O_2-determined respiration). It is notable that, within the JGOFS protocols, none exists for the measurement of net community production of carbon, despite its clear relevance to the aims of JGOFS.

The purposes of this chapter are to review the combined data sets available for respiration and NCP, primarily from the UK BOFS programme. Moreover, the author will explore whether we are yet able to determine what factors control the rates of these processes and how they relate to the process of gross production. Embedded in this general question is a further and fundamental one. Net community production, arithmetically, is the balance between gross primary production and respiration (either the sum or the difference, depending upon whether respiration is given a positive or negative sign). If one starts with the simplifying assumption that photosynthesis is determined by the autotrophic photosynthetic capacity and the availability of photons, then the question arises whether net production is controlled by respiration (i.e. does GPP − Resp → NCP) or that it is a process in its own right and respiration determined by net production (i.e. GPP − NCP → Resp). It is obvious that the question is critical when exploring controls on and constructing models of NCP and respiration. If the first circumstance prevails (i.e. GPP − Resp → NCP) then obviously to model NCP we need to develop models of both GPP and respiration; the former has been largely achieved (see Platt *et al.*, 1995), but by contrast models of respiration are at a very early stage of development. Consequently, our capability to model NCP by this route is very uncertain. Conversely, if the second circumstance prevails (i.e. GPP − NCP → Resp) then to model NCP we shall need to establish what ecological factors control NCP and how to incorporate them into models. Here we are in a worse state. This chapter is an attempt to wrestle with the above issues.

Analysis and discussion

Field observations of net community production

I shall draw upon seven sets of observations derived by our laboratory and collaborators over the past decade and a half. Most of the data are now in the public domain. All rates reported are derived from oxygen flux measurement, chosen because of the greater sensitivity of this technique and its longer availability in comparison with total CO_2 measurements. I have only used rates determined *in vitro*; the advantage of these data sets is that the precision of the rate measurement is easy to establish, furthermore they represent by far the major part of the information available to us and presently have greater consistency than *in situ* rate measurements.

In chemical measurement of dark respiration and the calculated rates of chemically determined gross production, we can eliminate insignificant rates statistically by comparing the signal with the precision of its measurement. Data where the signal is less than twice its standard error have not been used in the following analysis. Similarly, 'impossible' values, i.e. negative values of GPP and respiration, have been eliminated. About 5–10% of the data set is shed in this manner. No equivalent filter can be used for NCP, because by the nature of the process the range of true values will straddle zero.

The following data sets have been processed.

The BOFS North Atlantic Bloom Experiment (NABE) data set

The data set derives from a series of stations at approximately 47° and 60°N along the 20°W meridian over the period 13 May to 4 June 1989. These and the following two sets of data are now available on compact disk from the British Oceanographic Data Centre, Bidston. After filtering (see above) the data, 40 sets of observations remain.

The BOFS Lagrangian data set

This data set comprises two consecutive time series in the northeast Atlantic, separated by eight days (sampling days 1–19 May and 28 May – 14 June) centred round a drifting drogue in the vicinity of 55°N, 20°W. The filtered data set contains 50 sets of observations.

The BOFS coccolithophorid data set

A set of observations taken in the vicinity of 60°N and 20°W in the NE Atlantic over the period 16–29 June 1991 in a coccolithophorid bloom. The filtered data set contains 26 sets of observations.

The EROS NW Mediterranean data set

This comprises a series of cruises over the years 1990–93. The details of the cruises are given in Lefèvre *et al.* (1997); the numerical data will be found in the papers of Williams & Robinson (1990, 1991) and Lefèvre *et al.* (1995). The filtered data set contains 82 sets of observations.

The Menai Strait time series

The data set comprises a time series made at a fixed site in the Menai Strait (North Wales, UK) over the period January 1992 – December 1994. Details of sampling, location, etc. are given in Blight *et al.* (1995). The data are only in graphical form in this publication, the numerical data can be provided on disk by the author. The filtered data set contains 62 sets of observations.

The PRPOOS data set

A series of *in vitro* observations made in the region of 28–29°N and 154–155°W in the N Central Pacific Gyre over the period 23 August – 4 September 1984. The numerical data are available in the paper of Williams & Purdie (1991). The filtered data set contains 12 sets of observations.

The ARABESQUE Arabian Sea data set

This comprised a set of data obtained during the post-monsoon period during September 1994. The individual data are given in graphical form in Robinson & Williams (1999). The full numerical data set has been deposited in the British Oceanographic Data Centre. The filtered data set contains 21 sets of observations.

Relations between concurrent rates

Before discussing the analyses it is important to note that there are certain formal boundaries that should constrain the data: (1) in the positive sector of the plot of net community production against gross production, NCP cannot exceed GPP, for GPP sets the upper boundary to NCP; (2) in the negative sector of the plot of NCP against respiration, NCP cannot exceed respiration, i.e. respiration sets the lower boundary to NCP (these well-defined formal boundaries have been plotted in Figs. 3.2 and 3.3 as solid lines); and (3) the NCP $\leq$ GPP constraint can also be incorporated into the plot of NCP against respiration by defining the boundary to NCP in relation to photosynthesis and respiration. As may be seen from the NCP–respiration plots (Fig. 3.3), this boundary is less distinct than the lower one. The upper boundary will consist of situations of maximum P: R ratio (because NCP $= P - R$). One may expect that these will be instances where respiration by organisms other than the algae is minimal. This being so, the slope of the upper boundary would be determined by the P_{max}:R_a ratio, being equal to $(P_{max} - R_a)/R_a$. The evidence is that P_{max}: R_a ratio is not fixed in algae, but is group-specific (Langdon, 1993). This, among other things, would explain the more diffuse nature of the upper compared with the lower boundary in the NCP–respiration plots. Langdon (1993) gives modal P_{max}: R_a ratios for diatoms, prymnesiophytes and chlorophytes of 6 (the first two groups) and 8 (for the third). These would give $(P_{max} - R_a)/R_a$ slopes of 5 and 7. Because of their less certain nature, these are given as dashed lines in Fig. 3.3. This plot by its nature gives a spatial separation of the time-dependent phases of the bloom (autotrophic, heterotrophic and equilibrium phases) as illustrated in Fig. 3.3*d*. These are discussed later.

Gross production and respiration relations

There are no formal constraints to the instantaneous relation between these two parameters. In a closed system, the time-dependent integral of the two must be equal, but the distribution of the ratio of these two processes in space and time will not be constant. Despite the lack of any fixed relation between their concurrent values, there are physiological and ecological controls that do not allow the relation to be wholly random. It may be seen from Fig. 3.1*a–g* that, although there is a great deal of scatter, in a number of cases there is an indication of a curvilinear relation. Assuming this, the data have been analysed to determine the extent of any relations on the expectation that it would be an asset to have this information when modelling or setting a general pattern to respiration. An objective fit to the simple (but arbitrary) hyperbolic expression

$$R = \frac{PR_{max}}{P + K}$$

to the data was made by searching for a local minimum of the residual sum of squares between the data and the line. Of six data sets analysed (the PRPOOS data set, Fig. 3.1*e*, contained insufficient data to merit analysis) a minimum was found in five (Table 3.1). In the BOFS NABE data set, the solutions gave values of K and R_{max} more than 10^{10}, i.e. essentially a linear fit, and it is the only instance where the data seem to approximate to a linear, rather than a curvilinear, relation. In Fig 3.1, the fitted curves are shown as continuous lines. It was interesting to see how well a single equation would fit all data. The dotted line in Fig. 3.1*a–g* shows the fit with an arbitrary set of constants ($R_{max} = 10$ mmol O_2 m^{-3} d^{-1} and $K = 7.5$ mmol O_2 m^{-3} d^{-1}). In most cases (see Table 3.1) the fit with this single set of constants was almost as good as that found by the minimisation procedure. It is recognised that the approach is crude and the equation arbitrary. None the less, considering the wide diversity of the environments included in the analysis, it is interesting to find how much of the variance can be accounted for in such a simple manner.

Gross production and net community production relations

The set of plots in Fig. 3.2*a–g* looks more organised than others. This better organisation is in part illusory because one is in fact plotting NCP against the sum of NCP and respiration, which sets constraints to the occupiable space. The plots consequently tell us more about the behaviour of respiration than about the properties plotted. As discussed above, the upper boundary has a slope of unity, set by the constraint that GPP $\geq$ NCP. The most striking observation is that negative rates of NCP (i.e. respiration $>$ GPP) mainly occur at the lower end of the span of gross production rates, i.e. the points are not randomly distributed either side of zero. This derives from the form of non-linearity of the relation between gross production and respiration and is seen in the gross production – respiration plots (Fig. 3.1*a–g*) where, at low production rates, respiration rates frequently exceed production, whereas at the high rates of production the reverse applies. An explanation for this behaviour is a partial displacement in the time and space between the production and consumption of organic material; this is examined in a later section.

It was pointed out earlier that the measurement of NCP enjoys the least uncertainty of interpretation of all *in vitro* methods. Thus, the general respect of GPP and respiration observations for the formal boundaries allow us to conclude that the data yield no evidence for systematic under-estimates of respiration (and, as a consequence of the calculation, also of GPP).

The relation between planktonic respiration and photosynthesis has recently been debated (del Giorgio *et al.*, 1997; Williams, 1998; Duarte & Agusti, 1998). The topic of the debate is adjacent to the main direction of this chapter, but the form of analysis, however, is relevant. In these recent papers log–log analyses

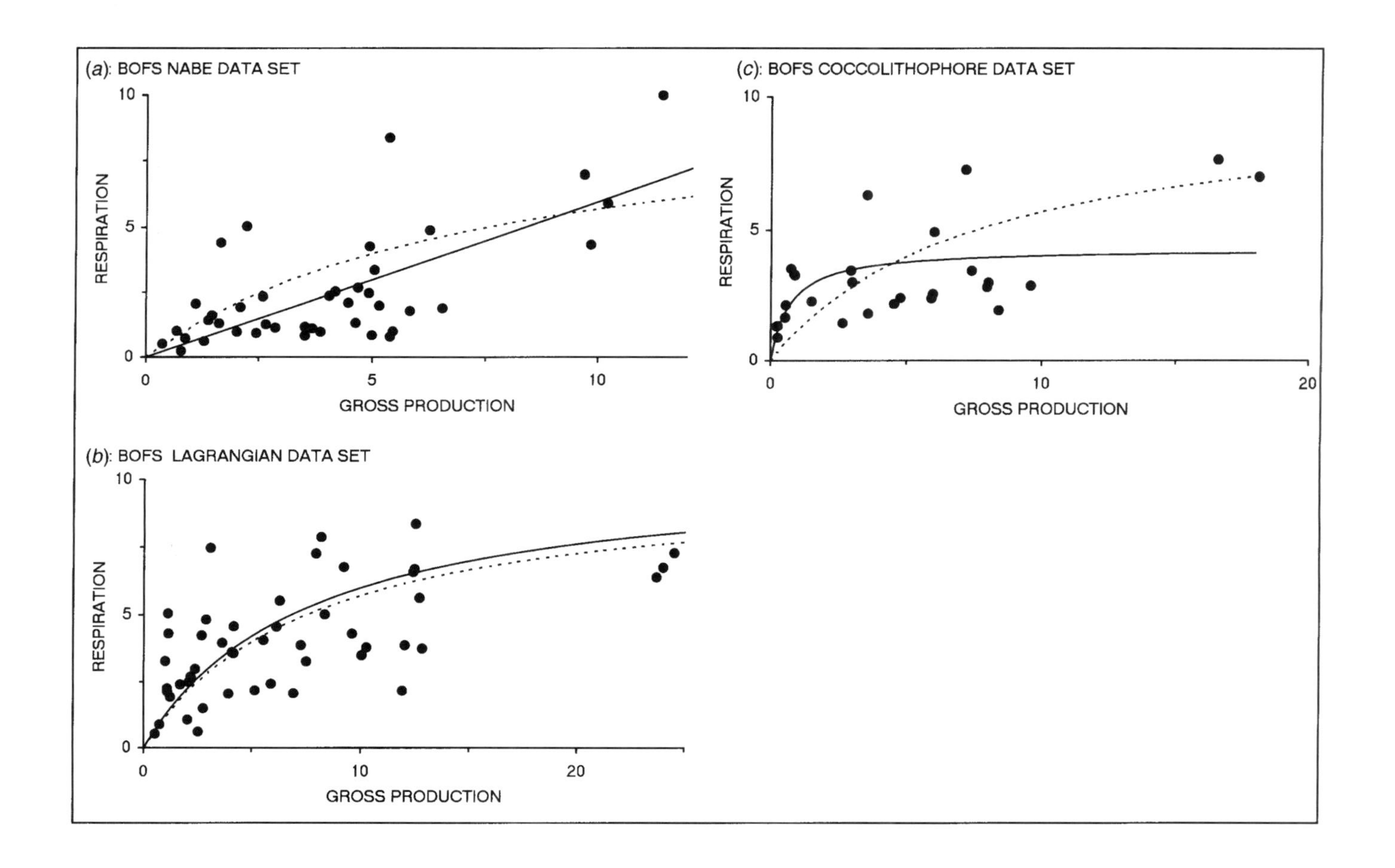
(a): BOFS NABE DATA SET
RESPIRATION
GROSS PRODUCTION
(b): BOFS LAGRANGIAN DATA SET
RESPIRATION
GROSS PRODUCTION
(c): BOFS COCCOLITHOPHORE DATA SET
RESPIRATION
GROSS PRODUCTION

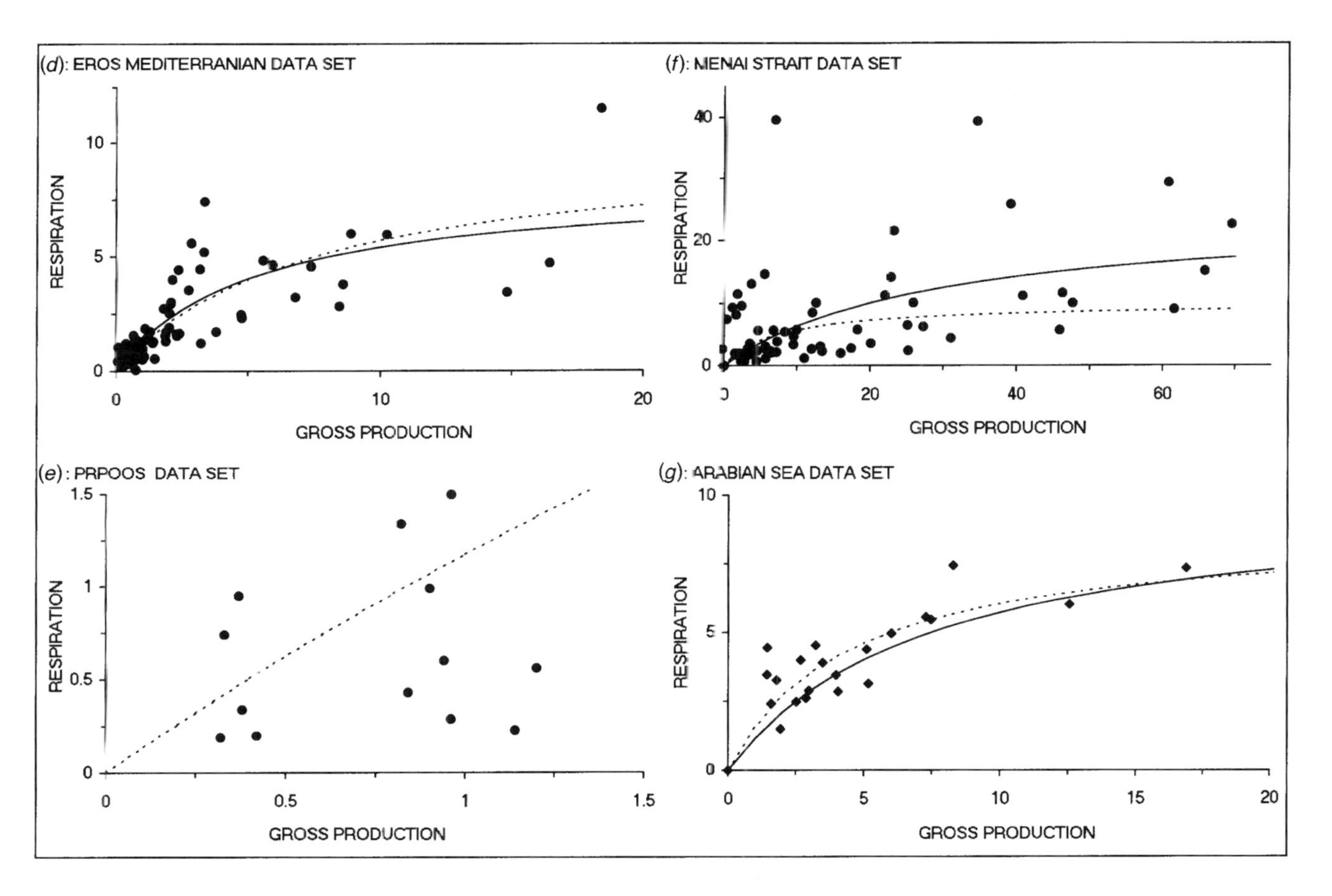

Figure 3.1 *Respiration and gross production. All rates are as mmol $O_2\,m^{-3}\,d^{-1}$. The solid curve is derived from a least-squares fit procedure (see text and table 3.1); the dotted line is that with K and R_{max} set, respectively, to 7.5 and 10 mmol $O_2\,m^{-3}\,d^{-1}$.*

Table 3.1. *Summary of results of least-squares derivation of constants for K and* R_{max}

Values in square brackets are the residual sum of squares as a percentage of the original. Units as mmol O_2 m^{-3} d^{-1}.

Figure number	Data set	Found values for K and R_{max}	Total sum of squares	Residual sum of squares using: Found values for K and R_{max}	Residual sum of squares using: $K = 7.5$, $R_{max} = 10$
3.1*a*	BOFS NABE	no local minimum, $K > 10^{10}$, $R_{max} > 10^{11}$ ($n = 40$)	451	107 [24]	148 [33]
3.1*b*	BOFS Lagrangian study	$R_{max} = 10.5$, $K = 7.5$ ($n = 50$)	1457	254 [17]	257 [18]
3.1*c*	BOFS coccolithophorid cruise	$R_{max} = 4.3$, $K = 0.69$ ($n = 26$)	367	67 [18]	86 [23]
3.1*d*	EROS Mediterranean data	$R_{max} = 8.2$, $K = 5.2$ ($n = 82$)	665	112 [17]	115 [17]
3.1*e*	PRPOOS N Pacific gyre study	(data not analysed) ($n = 12$)	7.82	—	2.75 [35]
3.1*f*	Menai Strait data	$R_{max} = 24.4$, $K = 28.7$ ($n = 62$)	8215	3631 [44]	4189 [51]
3.1*g*	Arabian Sea data	$R_{max} = 8.78$, $K = 4.53$ ($n = 21$)	403	20 [5]	26 [6]

have been used to derive allometric relations between respiration and photosynthesis; they appear to be as successful as, if not more so than, the rectangular hyperbolae used in the present study. Neither has any theoretical basis but of the two the allometric expressions are easier to handle mathematically and probably give a better account of respiration at low photosynthetic rates.

Net community production and respiration

The relation between NCP and respiration is interesting and appears to be the most variable of the three plots. Although these are analytically independent measurements, there are formal constraints to the immediate relation. The boundaries have been discussed above and for the most part are respected. If a degree of analytical variance in the measurement of the rates is acknowledged, then very few points fall outside the specified boundaries. The notable exception to this is the Arabian Sea data. Robinson & Williams (1999) have noted inexplicable anomalies in the respiration data from this study.

It is the pattern within these boundaries that is most interesting to explore, particularly the distribution in relation to the upper boundary, which may be able to give insight into the status of the community. Three general sectors may be envisaged: (1) an area adjacent to the upper ($(P_{max} - R_a)/R_a$) boundary (Sector 1 in Fig. 3.3*d*, characteristic of an autotroph-dominated community), (2) an area adjacent to the lower (i.e. community respiration 1 : − 1 line) boundary, i.e. a heterotroph-dominated community (Sector 2), and (3) an area along the NPC = 0 axis, i.e. an equilibrium community (Sector 3). This type of plankton succession was discussed by Margalef (1958) and observed by Blight *et al.* (1995) and comprises a sequence of an initial purely autotrophic phase, followed by a net heterotrophic one, leading to an equilibrium phase where respiration and photosynthesis are tightly coupled and balanced. The sequence is illustrated in Fig. 3.3*d*.

Given this rationalisation, the differences in the pattern of distribution, seen in the various plots, would reflect where the environment lay in this succession at the time of study. The BOFS NABE and the Menai Strait observations (Fig. 3.3*a*,*f*) appear to cover all three sectors, implying that all three phases are represented in the observations; in the Menai Strait study the time sequence has been observed (Blight *et al.*, 1995). The BOFS Lagrangian and coccolithophorid studies (Fig. 3.3*b*,*c*) show fewer observations in Sector 1, suggesting that the main part of the first phase has been missed. The EROS Mediterranean data set is interesting in that most of the data lies in the Sector 3, suggesting that, even though measurements were made in all seasons, the heterotrophic and autotrophic communities were always tightly coupled and the autotrophic and heterotrophic phases not represented. The Arabian Sea and the

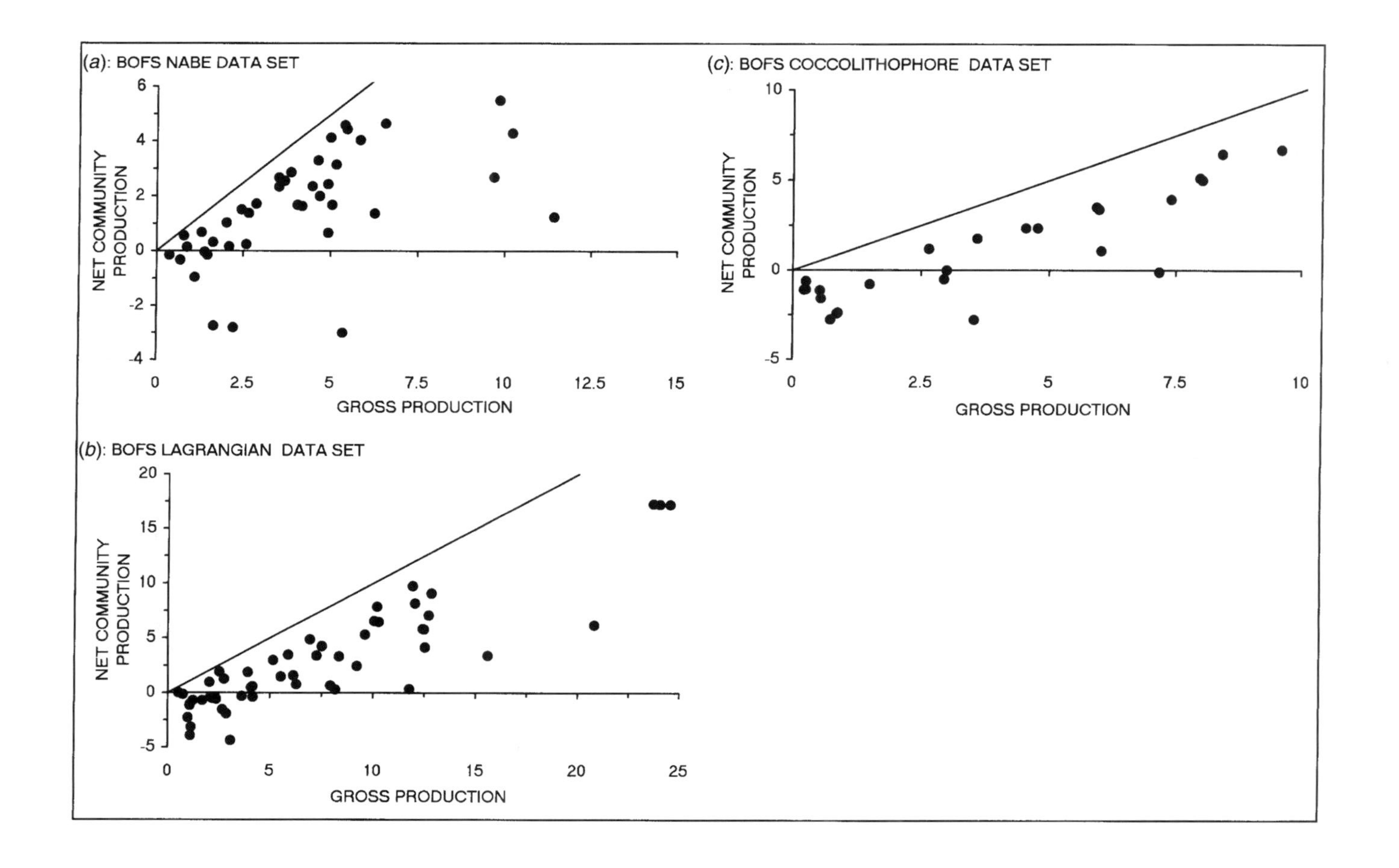

(a): BOFS NABE DATA SET
NET COMMUNITY PRODUCTION
GROSS PRODUCTION
(b): BOFS LAGRANGIAN DATA SET
NET COMMUNITY PRODUCTION
GROSS PRODUCTION
(c): BOFS COCCOLITHOPHORE DATA SET
NET COMMUNITY PRODUCTION
GROSS PRODUCTION

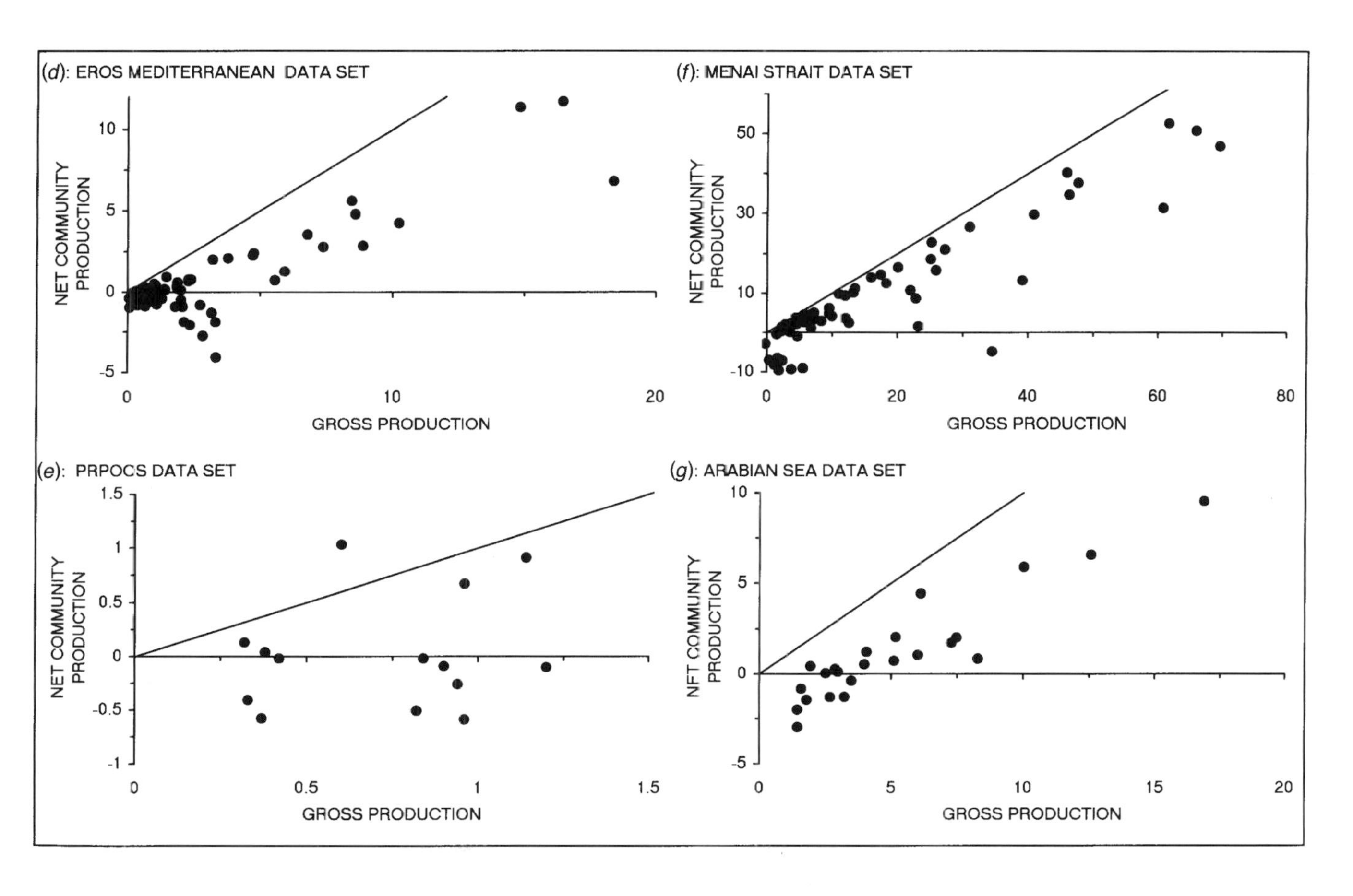

Figure 3.2 *Net community production and gross production. The solid line has a slope of unity and represents the boundary for the NCP–GPP relation. Units as Fig. 3.1.*

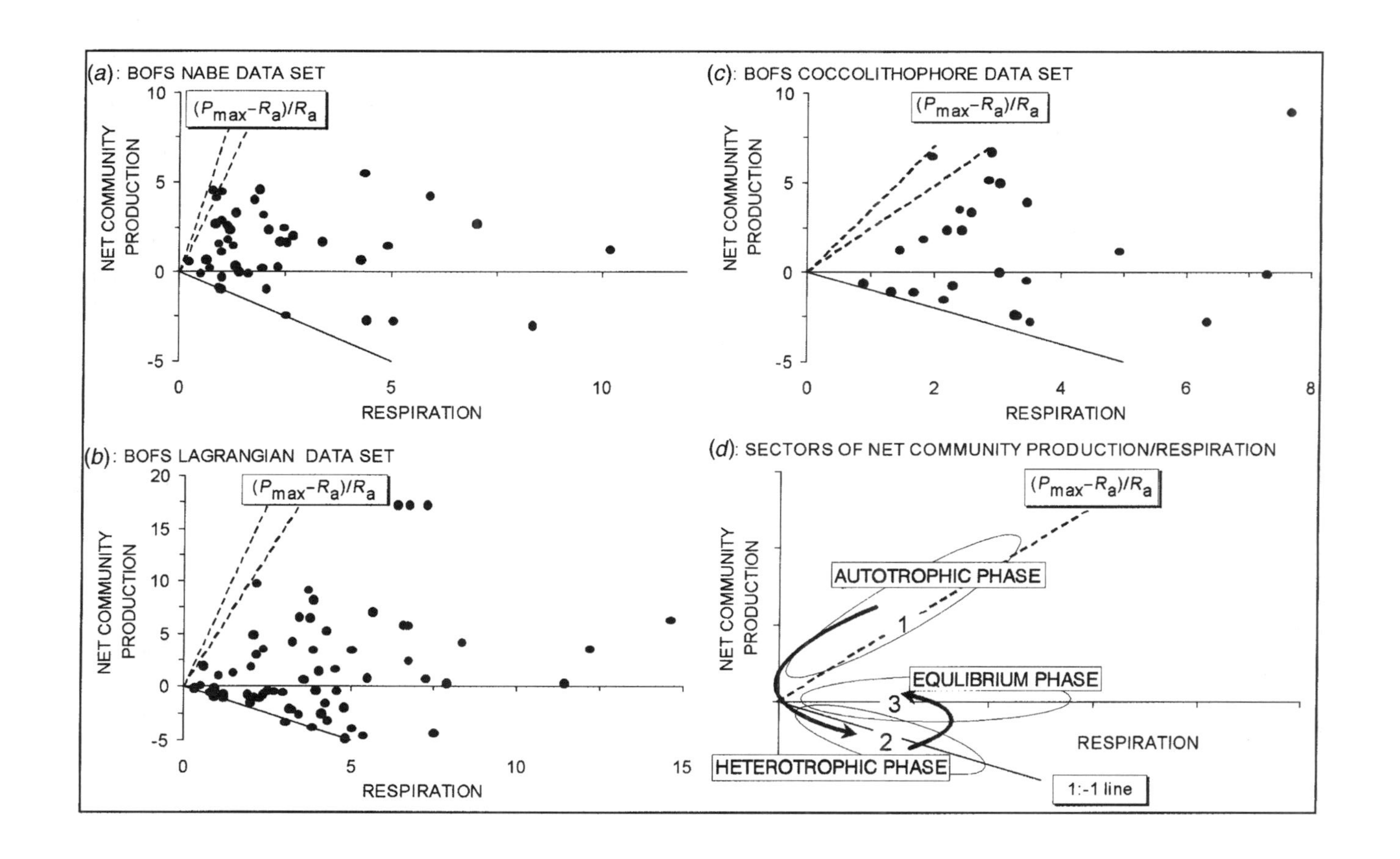

(a): BOFS NABE DATA SET
(c): BOFS COCCOLITHOPHORE DATA SET
(b): BOFS LAGRANGIAN DATA SET
(d): SECTORS OF NET COMMUNITY PRODUCTION/RESPIRATION
NET COMMUNITY PRODUCTION
RESPIRATION
$(P_{max}-R_a)/R_a$
AUTOTROPHIC PHASE
EQULIBRIUM PHASE
HETEROTROPHIC PHASE
1:-1 line
1
2
3

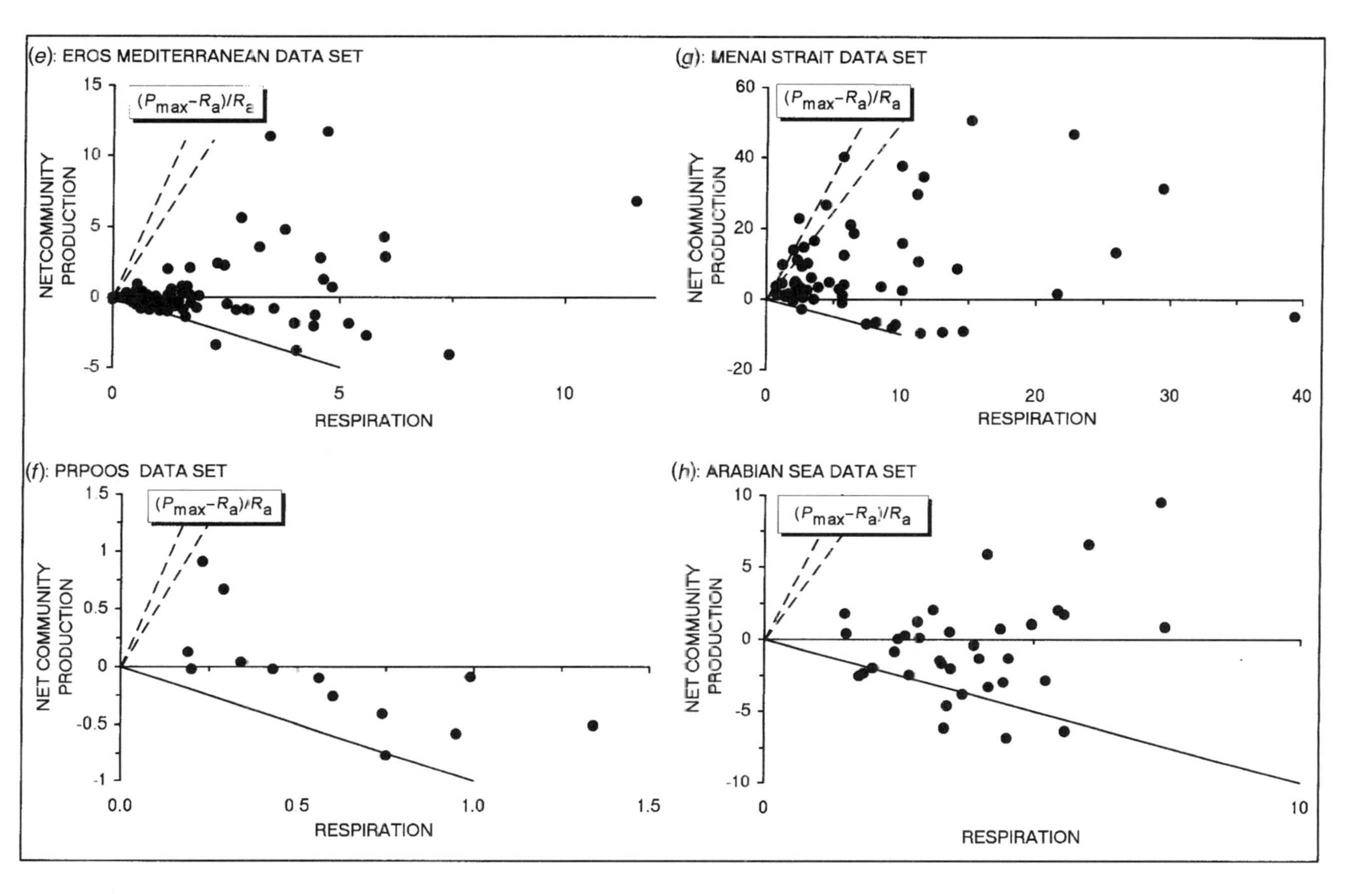

Figure 3.3 *Net community production and respiration. The solid line has a slope of −1 and represents the lower boundary of the relation. The upper dashed lines have slopes of 5 and 7 and represent the upper boundary of the relation (see text for details). Fig. 3.3*d *is a conceptual diagram of the sectors of the space occupied by particular phases of the plankton succession. Units as Fig. 3.1.*

PRPOOS data occupy shorted time spans; the data again imply a tightly coupled system.

We could use this compliance as evidence that the dark-bottle procedure used for the measurement of respiration does not incur serious and universal under-estimates. However, this conclusion is not as general as one might wish and does not resolve the uncertainties over the question of the importance of light respiration, for the present observations only apply when photosynthetic rates approach zero. In such situations, light respiration would be of little quantitative importance and so in this respect the test is weak.

Time and space separation of primary production and community respiration

The classical study of spatio-temporal displacement of production and consumption was probably that of Odum (1956), where the distribution of these two properties was measured along the length of a river bed and where a displacement in space (and thus time) was observed. Time and space are interlinked; separating out the spatial and temporal components is difficult or requires knowledge of the relative time scales of the physical and biological processes. With a benthic community, one may use a simple Eulerian sampling programme to examine the combined space and time displacements. In the case of planktonic populations, aliasing due to advection and diffusion makes this a profoundly difficult task. The ideal circumstances to study space–time connections in the plankton is in a Lagrangian sampling programme; however, the 1990 BOFS field programme and the first IronEx study (Martin *et al.*, 1994) revealed that open-ocean Lagrangian experiments are not easy studies to manage.

Evidence for separations in time have been found in Eulerian studies. The observations of Blight *et al.* (1995) are thought to reveal an instance of the classical autotrophic–heterotrophic temporal sequence and were analysed by the authors in relation to the routes and passage time of organic material between the autotrophs and the heterotrophs. The BOFS 1990 Lagrangian experiment was designed to follow events in time during a bloom. The observations (Fig. 3.5*a*) show no evidence for a temporal separation between respiration and photosynthesis; the temporal patterns of the two are close to mirror images of one another, the amplitude of respiration being perhaps one half that of production. The lack of any evident displacement in time may have been due to limitations of sampling, or because the study was not truly Lagrangian (see Savidge *et al.*, 1992), or because ship logistics, which gave rise to a seven-day hiatus in the middle of the sampling programme, could have

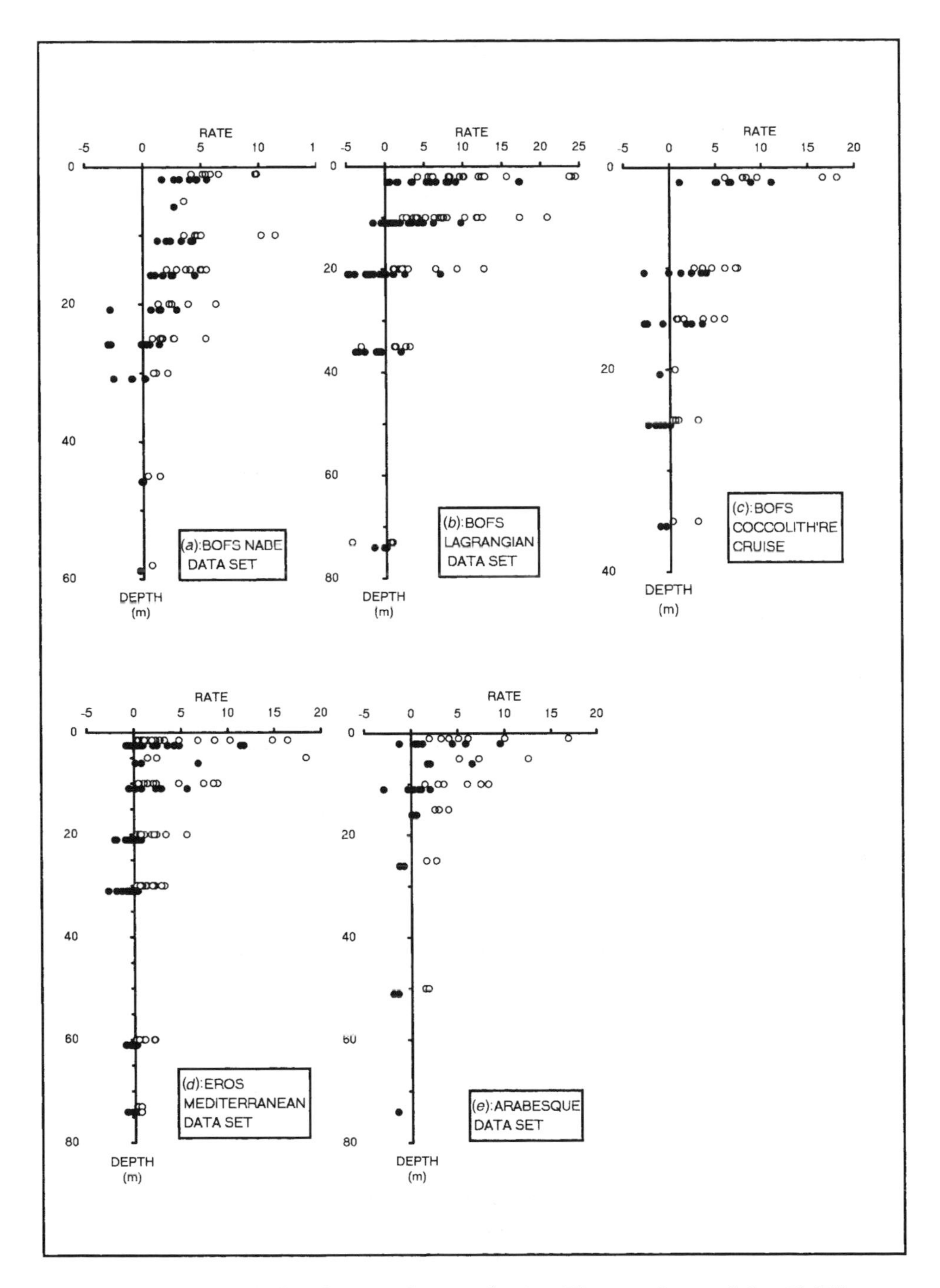

Figure 3.4 *Depth profiles of gross and net production. The rates (as mmol $O_2\ m^{-3}\ d^{-1}$) shown are: filled circles, net community production; open circles, gross primary production. The two rates are offset for clarity.*

resulted in missing the relapse into negative NCP. It is thought, however, that it may reflect a more fundamental ecological matter, i.e. that the time scale of the separation of autotrophic and subsequent heterotrophic processes will be dependent on the form of the organic coupling between the autotrophic and heterotrophic populations (see discussion by Blight *et al.*, 1995). If the coupling takes the form of low molecular-mass, readily assimilated organic excretions, it would be so tight as not to have a resolvable time separation with conventional ecological field approaches. This would not be the case if the link involved the formation and decomposition of organic detritus.

Although short-term temporal phasing of processes is hard to tease out of field observations of planktonic populations, simple Eulerian sampling procedures have the potential to reveal separations in space. This has been examined in Fig. 3.4*a–e* by plotting both NCP and GPP with depth. *A priori*, one would expect a downward displacement of respiration due to particle settling. In a number of data sets, a displacement between respiration and production is revealed as a zone of negative community production in the lower half of the euphotic zone. In the N Atlantic data set, where the maximum photosynthetic rates are seen at or near the surface, a NCP minimum (i.e. a negative maximum) is seen at around 20 m. If we assume a settling rate for particles of 10–20 m d^{-1} (Bodungden *et al.*, 1981), this implies a time displacement of about a day or so between the two processes. If this is the case, then it would be very difficult to observe a displacement of this order from conventional time-series field measurements of autotrophic and heterotrophic processes.

In summary, in the data sets analysed, the field observations show the closest relation between NCP and GPP, the relations between GPP and respiration and between NCP and respiration show much more scatter. Negative rates of NCP are mainly encountered at low production rates: this is also reflected in the form of the GPP–respiration plots. It has been possible to account for a substantial amount (characteristically 70–80%) of the variance of the GPP–respiration relation in plots with a simple hyperbolic curve and a single set of constants. Formal boundaries have been set for the NCP–respiration and NCP–GPP observations. The fact that these are generally respected by the data is evidence, albeit weak, for the absence of a systematic under-estimation of respiration by the dark-bottle measurements. Three sectors have been delimited in the NCP–GPP plots, representing the autotrophic, heterotrophic and balance phases of plankton community succession, and the data from the various study areas have been analysed in this context. Whereas a time sequence may be resolved in favourable conditions, it is not seen in oceanic studies. By contrast, a displacement in space between the autotrophic and heterotrophic situations is seen in a number of offshore data sets: a minimum in NCP (i.e. a negative maximum) is seen some 20 m below the GPP maximum.

Interactions between net community production and respiration

The question was raised in the Introduction as to what extent respiration controls net community production or *vice versa*. Almost certainly, the question is too simply posed to have a simple answer, but it is instructive to see how much progress can be made.

As a first step, we may recognise that net production (or better, net accumulation) may arise in three general compartments: the net production of the autotrophs (after respiration exudation and grazing) and that of the heterotrophs (after respiration and the formation of detritus) and that of organic detritus (both particulate and dissolved, after organic grazing and bacterial decomposition). There are potentially complicating matters of time and size scales as discussed by Williams (1993), but they will be ignored in the following discussion as they are not seen to be matters of major consequence. Algebraically, the various plankton rate processes may be expressed as follows:

$$\text{NCP} = NP_a + NP_h + NP_d;$$
$$\text{NPP} = NP_a + NP_h + NP_d + R_h;$$
$$\text{GPP} = NP_a + NP_h + NP_d + R_h + R_a,$$

where NP_a, NP_h and NP_d are, respectively, the net production (accumulation) in the categories autotrophs, heterotrophs and detritus, and R_a and R_h are the respective respiration rates of the autotrophs and the heterotrophs (the respiration rates are given a positive sign to allow simple summation).

NCP has been argued (Platt *et al.*, 1989) to be that part of gross production equivalent to new production, and as such, sustained by nitrate assimilation. Thus the remaining part of gross production (i.e. GPP – NCP = $R_a + R_h$) would then be equivalent to regenerated production (i.e. that part of production sustained by ammonia). Platt *et al.* (1989) made the argument on the basis that in practice (and new and regenerated production are largely empirical properties) production measured by the ^{14}C technique gives rates closer to NPP than to GPP. The signs are that this is correct (Williams *et al.*, 1996). In this case, regenerated production would be equal to R_h. The argument obviously collapses at the base of the euphotic zone, where NPP becomes negative and the equivalence between NPP and ^{14}C-determined production can no longer be sustained. This problem is seen to be a local one and for the purpose of the present discussion can probably be ignored. If we make the assumption that, in measuring new and regenerated production, there is no short-term recycling of ammonia by nitrification, we can come to the conclusion that heterotrophic respiration should have no immediate effect on NCP, because the latter is nitrate-driven. Thus nitrate utilisation should control NCP; the difference between NPP and NCP is sustained by (and if nitrogen is limiting, controlled by) heterotroph respiration via ammonia production. In the situation discussed above, heterotrophic respiration would affect gross production rather than NCP rates, i.e. *NCP rather than GPP would be the fixed property*, thus, one should not start from GPP to predict NCP, but rather the reverse (i.e. NCP + Resp → GPP). Although respiration does not seem to be the factor controlling NCP, it has been argued earlier that the various forms of respiration are involved in setting the formal boundaries of NCP, i.e. $(P_{max} - R_a) \geq \text{NCP} \geq -(R_a + R_h)$. The above discussion leads to the conclusion that the poising of NCP within these boundaries is determined by nitrate assimilation.

Clearly respiration is ultimately dependent upon production in a closed system but the connections between their two processes vary for the different forms of respiration. Photosynthesis and algal respiration are obviously connected intracellularly, so the connection will be very much a physiological phenomenon with molecular diffusion and the permeability of intracellular membranes playing key roles. This contrasts with the relation between NPP and heterotroph respiration, which requires some form of trophic transfer and so the relation will be more strongly ecologically controlled. Algal respiration may be subdivided into short-term photosynthetically related forms of respiration and a more long-term biomass-related form (see reviews by Beardall & Raven, 1990; Langdon, 1993). The former types of respiration are so intimately connected with photosynthesis that they are difficult to separate from it, making accurate measurements of GPP very difficult.

Although the routes of supply of the organic material supporting algal biomass-related respiration and heterotrophic respiration are very different, the processes themselves have strong physiological similarities. For example, the enzyme mechanisms are much the same, and the overall processes are probably concerned with cell maintenance and so operate over time scales of days or longer and are dependent on earlier NCP. The disconnections in time between gross production (and concurrent NCP) and dark respiration (presumably mainly algal biomass-related and heterotroph respiration) were discussed earlier. These disconnections make it very difficult to explore the time constants and causal relations. Separations in space present the conceptual problems of (1) whether one considers Eulerian or Lagrangian space, and (2) in the latter case, what reference should be used for the co-ordinates (it is not obvious whether it should be the water, or the algal cell, or the organic material itself). Clearly, if the water is used as the space reference, then a time sequence of events occurring within a sinking algal cell (e.g. the production and subsequent consumption of organic material) would be considered to be occurring in different spaces but yet the same metabolic entity.

In summary, conventional theory leads to the conclusion that nitrate assimilation rather than respiration controls the rates of net community production during its positive phase, with respiration (presumably mainly heterotroph respiration) setting the lower boundary, and photosynthesis (P_{max}) and algal respiration combining to set the upper boundary. In a closed system, NCP must set the eventual limit to the integrated autotrophic biomass-related and heterotrophic respirations: the photosynthetic-dependent forms of algal respiration are independent of NCP.

Net production and critical depth

Critical depth (Sverdrup, 1953) is an important concept in explaining and modelling the development of a plankton bloom (Platt *et al.*, 1991). Expressed simply (and naively) it is based on the balance between photosynthesis and respiration (i.e. NCP). Smetacek & Passow (1990) point out that there are uncertainties over what group of organisms are considered responsible for the respiration. They point out that 'modern renditions of Sverdrup's model, however, generally ignore zooplankton and consider only phytoplankton respiration'. These authors are of the view that zooplankton respiration (because it is a sink for carbon) should be included. Neither they nor Sverdrup invoke any discussion of microbial heterotrophic respiration, even though there is evidence that it can be a major component of overall respiration (see, for example, Williams, 1981, 1984; Blight *et al.*, 1995; Boyd *et al.*, 1995). Because, as with zooplankton respiration, it is a carbon sink, it should be included in the mass balance equation. Taking this to be so, the integral of net community production to the depth at which it reaches zero should give an estimate of critical depth.

Very few data sets for marine environments allow a calculation of critical depth to be made that includes the contribution to respiration by the microbial fraction of the community. Boyd *et al.* (1995) made a calculation for a single station. The BOFS 1990 Lagrangian cruises were set up to allow a time series of the depth distribution of production parameters to be determined. Wood (1992) analysed these. Although a simple integral of net production could have been used, the logic she adopted was that photosynthesis and, to a somewhat lesser extent, respiration have more readily predictable depth relations, so it was regarded as more reliable to fit these two processes separately. Thus, the procedure Wood adopted was to fit log-linear curves to the GPP and respiration rates against depth profiles to derive the slope and intercept parameters, and then the coefficients from these were used in simple integrals to calculate critical depth. The calculated values (Z_{crit} in Fig. 3.5*b*) were compared with the estimates of mixed depth (taken from the BODC BOFS database) and the difference between the two ($Z_{crit} - Z_{mixed}$) gives an indication of the scope for growth of the community. The observations analysed by Wood do provide what

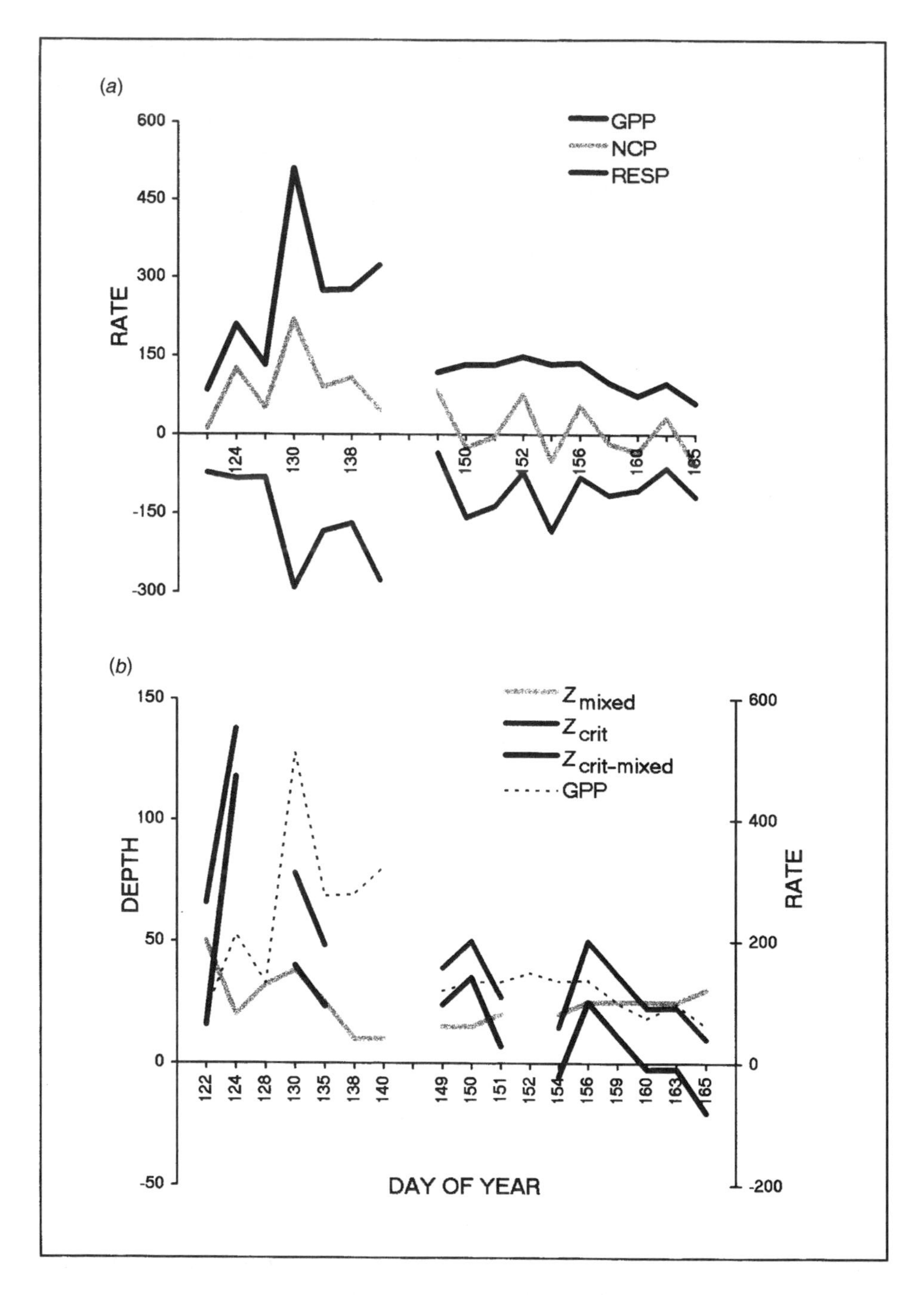

Figure 3.5 *The BOFS Lagrangian data set.* (a) *Time series of gross and net production and respiration, integrated to 35 m, calculated from observations made during the 1990 BOFS Lagrangian cruises. Rates as mmol* $O_2\,m^{-2}\,d^{-1}$ (b) *Time series of calculated mixed depth* (Z_{mixed}), *critical depth* (Z_{crit}), *and the difference between the two* ($Z_{crit-mixed}$). *Also plotted is integrated GPP (units as in Fig. 3.5*a).

seem to be reasonable estimates of critical depth, when compared with mixed depth. Shown also in Fig. 3.5*a–b* are the integrated values of GPP, NCP and respiration down to 35 m (the maximum depth regularly sampled during the study). The observations in Fig. 3.5*a* have been discussed earlier. A full analysis of the relationship between NCP and $Z_{crit} - Z_{mixed}$ would need to take into careful account the time response of the population to avoid the obvious circularity that, since the correlation between GPP and NCP is close, there will be a second-order relation between concomitant values of GPP and Z_{crit}.

The main feature revealed by Fig. 3.5*b* is that, over the period of the study, critical depth was more variable than mixed depth. One would expect this to be the case because the biology (notably photosynthesis) has the potential to respond much more rapidly and extensively to changes in external forcing, such as incoming radiation, than the physics. There were no storm events during the period of the study to give rise to rapid changes in mixed depth. Because the biology itself was the main determinant of $Z_{crit} - Z_{mixed}$, the system had the potential for positive feedback, although this did not seem to be expressed. What is interesting, and not explained, is that after the initial peak in GPP, following a high value for $Z_{crit} - Z_{mixed}$, the integrated photosynthetic rates are very constant, but NCP by contrast seems to oscillate either side of zero with a frequency of about 3–5 d. Whether this is a freak observation or a fundamental oscillation in the nature of the respiring community is not known.

The modelling of net production

The JGOFS objectives effectively call for models to describe carbon export; both organic and inorganic exports require predictions of net community production. The discussion above leads to the conclusion that, in the foreseeable future, NCP will need to be modelled as the difference between photosynthesis (since our models are invariably parameterised on ^{14}C observations, the modelled production is most probably NPP rather than GPP) and respiration. If NPP is indeed the form of production modelled, then only heterotrophic respiration should be included in these models. The value of the field observations of NCP is that, because they are not involved in the parameterisation of the models, they provide a wholly independent test. To take full advantage of this, the models need to have the capability to resolve not only time but space, because the spatial separation of primary production and net community production is more readily resolved than the temporal. Thus, depth-integrated models will not be able to take advantage of the observations.

Summary

This chapter has considered the processes of plankton net community production and respiration, with particular attention to the former. It has been possible to set boundaries for the space occupied by these processes in plots of the rates of net production against respiration and net production against gross production. There are no formal constraints to the gross production – respiration plots, other than that neither process can be negative (in contrast to net production). Although there is a great deal of scatter in the GPP–respiration plots, a substantial fraction of the variance may be accounted for by a simple hyperbolic relation. A single set of constants gave a surprisingly good fit to data from three cruises in the N Atlantic and a composite of four cruises in the Mediterranean. The fit weakens in extreme oligotrophic environments (the N Central Pacific Gyre) and inshore eutrophic systems (the Menai Strait). The residual variance will arise from measurement error and environmental variability stemming from the different phases of the cycle of production and consumption. The fact that, with few exceptions, the formal boundaries are respected, confirms our view that measurement error is not a major source of the observed variance. The ecological factors that poise respiration and NCP are difficult to determine from the data sets, in part owing to temporal and spatial displacements between the autotrophic and heterotrophic phases. If, or where, one can equate new and net production, then nitrate assimilation would be the determinant of NCP; however, during periods of time and in parts of the water column where NCP is negative, new and net production cannot be equated and the above relation collapses.

Acknowledgements

The author is pleased to acknowledge the support of the Natural Environment Research Council as part of the BOFS programme (Grant No GST/02/304); the MAST II project MAS2–CT92–0031 (MEICE); and the EC STEP Programme (CT.90.0080 DSCN). He has greatly valued stimulating discussions, comment and help from his colleagues Drs Carol Robinson and Dominique Lefèvre and Stephen Blight. During the period the paper was prepared he was recipient of a Higher Educational Funding Council for Wales Professorship. The author is grateful for the thoughtful comments of three anonymous referees.

References

Beardall, J. & Raven, J. A. (1990). Pathways and mechanisms of respiration in microalgae. *Marine Microbial Food Webs*, **4**, 7–30.

Blight, S. P., Bentley, T. L., Lefèvre, D., Robinson, C., Rodrigues, R., Rowlands, J. & Williams, P. J. le B. (1995). The phasing of autotrophic and heterotrophic plankton metabolism in a temperate coastal ecosystem. *Marine Ecology Progress Series*, **128**, 61–75.

Bodungden, B. von, Brökel, K. von, Smetacek, V. & Zeitzschel, B. (1981). Growth and sedimentation of the phytoplankton spring bloom in the Bornholm Sea (Baltic Sea). *Kieler Meeresforschungen* (Special Issue), **5**, 49–60.

Boyd, P., Robinson, C., Savidge, G. & Williams, P. J. le B. (1995). Water column and sea ice primary production during austral spring in the Bellingshausen Sea. *Deep-Sea Research*, **42**, 1177–200.

del Giorgio, P. A., Cole, J. J. & Cimbleris, A. (1997). Respiration rates in bacteria exceed plankton production in unproductive aquatic systems. *Nature*, **385**, 148–51.

Duarte, C. M. & Agusti, S. (1998). The CO_2 balance of unproductive aquatic ecosystems. *Science*, **281**, 234–6.

Langdon, C. (1993). The significance of respiration in production measurements based on oxygen. *ICES Marine Science Symposium*, **197**, 67–78.

Lefèvre, D., Minas, H. J., Minas, M., Robinson, C., Williams, P. J. le B. & Woodward, E. M. S. (1997). Review of gross community production, primary production, net community production and dark community respiration in the Gulf of Lions. *Deep-Sea Research*, part II, **44**, 801–32.

Lefèvre, D., Williams, P. J. le B. & Bentley, T. L. (1995). Oxygen based production measurements in the Gulf of Lions. In *Water Pollution Research Reports, EROS 2000 (European River Ocean System)*, ed. J. M. Martin & H. Barth, pp. 125–32. (Fifth workshop on the northwest Mediterranean Sea.) Brussels: Commission of the European Communities.

Margalef, R. (1958). Temporal succession and spatial heterogeneity in phytoplankton. In *Perspectives in Marine Biology*, ed. A. A. Buzzati-Traverso, pp. 323–49. Berkeley: University of California Press.

Martin, J., Coale, K. H. *et al.* (1994). Testing the iron hypothesis in ecosystems of the equatorial Pacific Ocean. *Nature*, **371**, 123–9.

Odum, H. T. (1956). Primary production in flowing waters. *Limnology and Oceanography*, **1**, 102–17.

Platt, T., Harrison, W. G., Lewis, M. R., Li, W. K., Sathyendranath, S., Smith, R. E. & Vezina, A. F. (1989). Biological production of the oceans: the case for a consensus. *Marine Ecology Progress Series*, **52**, 77–88.

Platt, T., Bird, D. F. & Sathyendranath, S. (1991). Critical depth and marine primary production. *Proceeding of the Royal Society of London* B **246**, 205–17.

Platt, T., Sathyendranath, S. & Longhurst, A. (1995). Remote sensing of primary production in the ocean: promise and fulfilment. *Philosophical Transactions of the Royal Society of London* B **1324**, 191–202.

Robinson, C. & Williams, P. J. le B. (1999). Plankton net production and dark respiration in the Arabian Sea during September 1994. *Deep-Sea Research*, part II (in press).

Savidge, G., Turner, D. R., Burkill, P. H., Watson, A. J., Angel, M. V., Pingreee, R. D., Leach, H. & Richards, K. J. (1992). The BOFS 1990 Spring Bloom Experiment: temporal evolution and spatial variability of the hydrographic field. *Progress in Oceanography*, **29**, 235–81.

Smetacek, V. & Passow, U. (1990). Spring bloom initiation and Sverdrup's mixed depth model. *Limnology and Oceanography*, **35**, 228–33.

Sverdrup, H. U. (1953). On conditions for the vernal blooming of phytoplankton. *Journal du Conseil*, 18, 287–95.
Williams, P. J. le B. (1981). Microbial contribution to overall plankton community respiration studies in enclosures. In *Marine Microcosms*, ed. G. D. Grice & M. R. Reeves, pp. 305–21. Berlin: Springer-Verlag.
Williams, P. J. le B. (1984). A review of measurements of respiration rates of marine plankton populations. In *Heterotrophy in the Sea*, ed. J. E. Hobbie & P. J. le B. Williams, pp. 357–89. New York: Plenum Press.
Williams, P. J. le B. (1993). Chemical and tracer methods of measuring plankton production: what do they in fact measure? In *The ^{14}C technique considered. ICES Marine Science Symposium*, **197**, 20–36.
Williams, P. J. le B. (1998). The balance of plankton respiration and photosynthesis in the open oceans. *Nature*, **394**, 55–7.
Williams, P. J. le B. & Purdie, D. A. (1991). *In vitro* and *in situ* derived rates of gross production, net community production and respiration of oxygen in the oligotrophic subtropical gyre of the North Pacific Ocean. *Deep-Sea Research*, 38, 891–910.
Williams, P. J. le B. & Robinson, C. (1990). Seasonal differences in the control of productivity in the Rhone outfall region of the Gulf of Lions. In *Water Pollution Research Reports 'EROS 2000' (European River Ocean System)*, ed. J. M. Martin & H. Barth, pp. 145–54. (Second workshop on the northwest Mediterranean Sea.) Brussels: Commission of the European Communities.
Williams, P. J. le B. & Robinson, C. (1991). Analysis of oxygen-based productivity measurements throughout phase I of the EROS 2000 programme. In *Water Pollution Research Reports 'EROS 2000' (European River Ocean System)*, ed. J. M. Martin & H. Barth, pp. 151–62. (Third workshop on the northwest Mediterranean Sea.) Brussels: Commission of the European Communities.
Williams, P. J. le B., Robinson, C., Jespersen, A.-M., Søndergaard, M., Bentley, T. L., Lefèvre, D., Richardson, K. & Riemann, B. (1996). Algal ^{14}C and total carbon metabolisms. 2. Experimental observations with the diatom *Skeletonema costatum*. *Journal of Plankton Research*, 18, 1959–74.
Wood, E. S. (1992). The community photosynthetic quotient and the assimilation of nitrogen by oceanic plankton. Ph.D. Dissertation, University of Wales, UK.

4 The role of iron in plankton ecology and carbon dioxide transfer of the global oceans

H. J. W. de Baar and P. W. Boyd

Keywords: iron, plankton, diatoms, carbon dioxide, oceans

Introduction

The biological production of the oceans by photosynthetic fixation of dissolved CO_2, as well as the related air–sea exchanges of CO_2, are well-recognised forcings of the abundance of CO_2 in the atmosphere (Denman *et al.*, 1996). Variations in atmospheric CO_2 affect the past, present and future of the global climate (Houghton *et al.*, 1996). The major objective of the Joint Global Ocean Flux Study (JGOFS) is to determine and understand the fluxes of carbon, and associated biogenic elements, involved in the biological cycle of the oceans (de Baar *et al.*, 1989*a*). For this, it is essential to know which factors are governing, or limiting, the biological production.

In the central gyres of temperate oceans, the growth rate and biomass of phytoplankton appear to be largely limited by the availability of major nutrients, such as nitrogen and phosphorus, as well as silicon for diatoms and other algae with opaline skeletal parts. This has been well known since about 1930 (Mills, 1989). On the other hand, little is known about the possible limiting role of essential trace nutrients Fe, Mn, Zn, Cu, Ni and Co in these central gyres.

Nowadays there is a revival of the awareness that in the polar and subpolar regions, as well as equatorial upwelling zones, the concentrations of major nutrients are high throughout the year, whereas phytoplankton stocks remain low (Chisholm & Morel, 1991). These vast regions are often referred to as high nutrient – low chlorophyll (HNLC; Minas *et al.*, 1986) areas, but in fact they comprise very different oceanic regimes. For example, in the Pacific Basin these regions together constitute about 40% of all surface waters (Reid, 1962). However, the subarctic (Fig. 4.1), the equatorial, and the Antarctic (Fig. 4.2)

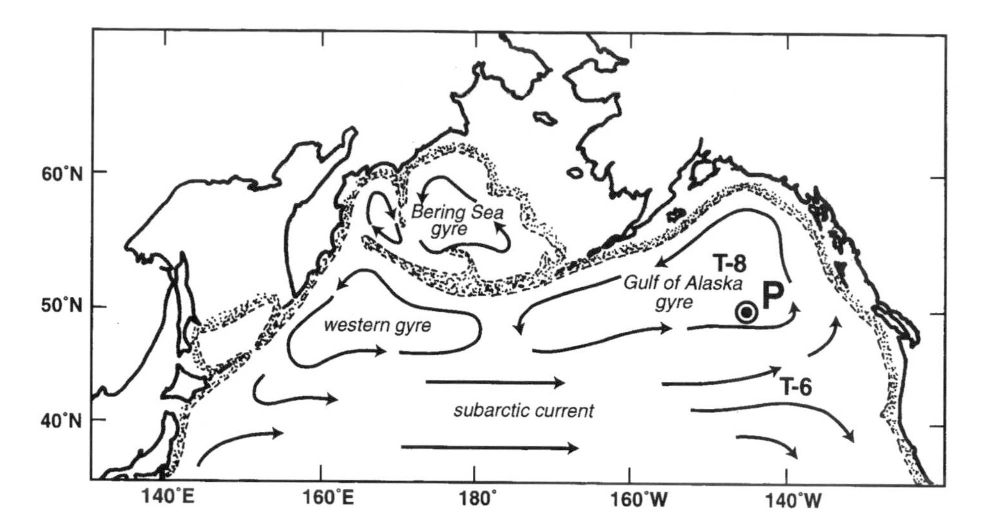

Figure 4.1 *Map of the Subarctic Pacific Ocean showing the main circulation patterns and the position of station P where studies were conducted by the Subarctic Pacific Ecosystems Research (SUPER) and VERTEX programmes during the 1980s and by the JGOFS Canada programme during the 1990s. Station P was referred to as station T-7 during the August 1987 VERTEX study (Martin & Fitzwater, 1988; Coale, 1991). Stations T-6 and T-8 denote two other VERTEX 1987 iron-enrichment stations (Martin* et al.*, 1989). Map reproduced from Miller* et al. *(1991).*

and Pacific Ocean are very different ecosystems. Similarly, in the Atlantic Ocean, at latitudes higher than about 45–55° in both hemispheres, the major nutrients are always present in surface waters. However, the subarctic and Arctic have less nutrients than the Antarctic Ocean (Smith, 1991), and the sea-ice regimes are also quite different.

For all these HNLC regions, limitation by iron supply (Fe) has been suggested (Martin & Fitzwater, 1988). In this chapter, we summarise what has been learned about Fe limitation in these HNLC regions. First, we define eight working hypotheses for Fe limitation and supply. This is done based on general concepts, the biological function of iron, and the various iron-limitation hypotheses in the existing literature. Secondly, the existing observational evidence is presented describing significant research efforts in the Antarctic, the subarctic Pacific and the equatorial Pacific Oceans. These studies, albeit impressive, have only scratched the surface; the seasonal and regional coverage of these oceans (Figs. 4.1, 4.2) remains modest. Moreover, there are no observations yet for the high Arctic and the subarctic Atlantic Oceans. Thirdly, with a brief discussion, we validate the eight working hypotheses, by either acceptance or falsification.

The iron hypothesis

Concepts for regulation of plankton production

In the oceans, the rate of growth of phytoplankton

$$\text{Net increase} = \text{Primary production} - \text{Loss terms} \qquad (4.1)$$

may be restricted either by the shortage of light and nutrients affecting primary productivity, or by marked losses due to intense grazing or sedimentation events. It is the variations in the rate of production and the rates of loss that determine the standing stock. The diurnal to seasonal variations of production and loss can be driven by biotic and abiotic factors. Changes in life cycle and strategy are taking place, not only in the zooplankton assemblage responsible for the grazing loss, but also in the phytoplankton, where formation of colonies (Lancelot, 1995; Lancelot *et al.*, 1998) and sedimentation in relation to sexual reproduction (Waite & Harrison, 1992; Crawford, 1995; Crawford *et al.*, 1997) and nutrient stress (Muggli *et al.*, 1996) have been reported. Riley (1946) and others have devised various formulations for the loss terms in Equation 4.1 above. The primary production term in Equation 4.1 is mostly affected by changes in abiotic factors, notably light and nutrients, but replenishment of nutrients in turn is largely composed of the biological processes of respiration and mineralisation (Fig. 4.3).

For the primary production of a single species of alga, and by approximation of an algal community, there exist optimum conditions for light and nutrients. For light the specific growth rate μ [s^{-1}] can be determined from the photosynthesis (P) versus irradiance (I) curves:

$$\mu/\mu_{max} = I - \exp(\alpha I/K_{max}))\exp(-\beta I/K_{max}), \qquad (4.2)$$

where, normalized to biomass, the α is the photosynthetic rate assuming no photoinhibition while the term $\exp(-\beta I/K_{max})$ is an expression for photoinhibition, all after Platt *et al.* (1980). Similarly for a given nutrient, for example nitrogen (N) the growth curves for nutrient saturation (Monod, 1950):

$$\mu/\mu_{max} = [N]/(K_N + [N]) \qquad (4.3)$$

describe the specific growth rate μ (s^{-1}), where K is the half-saturation constant. The Michaelis & Menten (1913) formulation typically used to describe nutrient uptake is the same, with μ / μ_{max} replaced by V/V_{max} (Dugdale *et al.*, 1995). More refined descriptions are often used for light as well as nutrients and their internal pools (Droop, 1974) but are not necessary for this chapter. In the field situation the irradiance, I, decreases with depth (Sakshaug *et al.*, 1991) and can be estimated quite accurately. When concentrations of all major nutrients (Fe, N, P and Si) are known, this may be combined into a large, general-community production term:

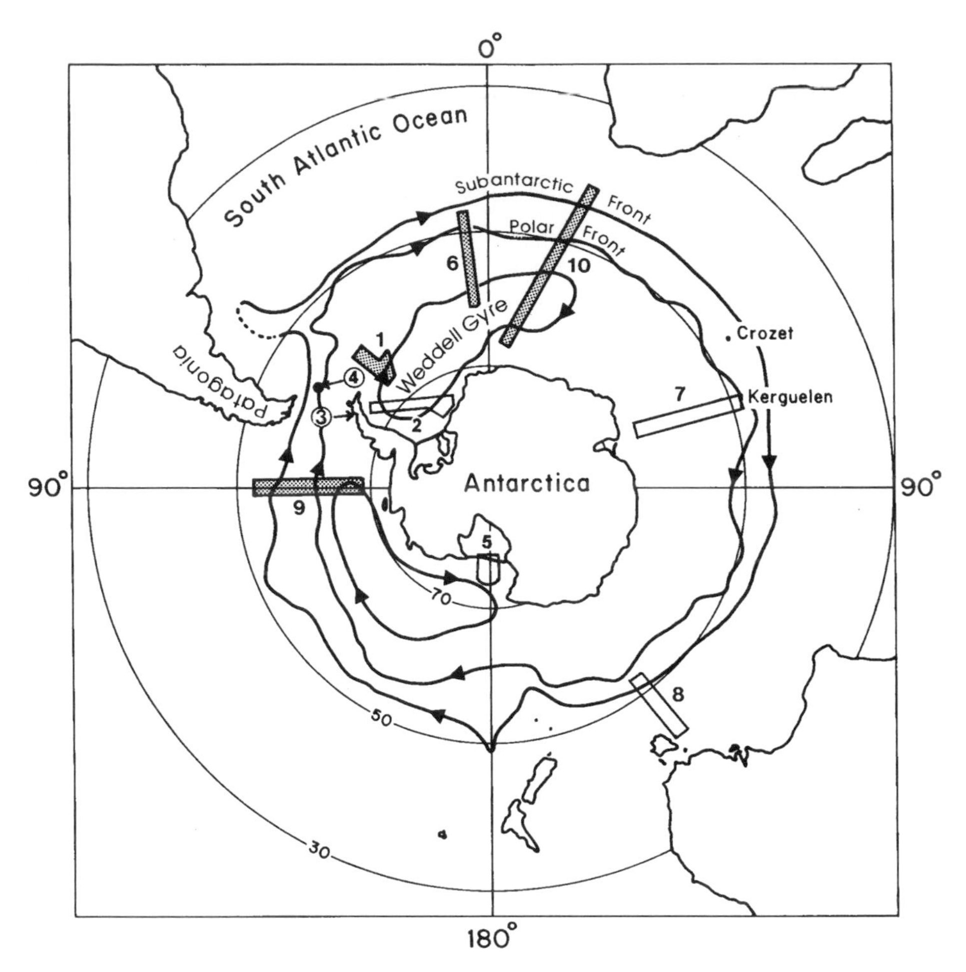

Figure 4.2 *The Southern Ocean defined (Deacon, 1984) as all the waters south of the Subtropical Front at about 40°S (not shown). Drafted after Whitworth (1988) and Peterson & Stramma (1991). Most dominant is the eastward-flowing Antarctic Circumpolar Current (ACC) and its concentric circumpolar frontal systems adjoining (or flowing through) the Indian, Pacific and Atlantic basins. Within the overall ACC the Subantarctic Front, at about 45°S, divides the Subantarctic Zone (SAZ over* c. *40–45°S region) and the Polar Frontal Zone (PFZ over* c. *45–50°S region). At about 50°S the Polar Front exists as a distinct jet stream and serves as the southern boundary of the PFZ. Thus, the Polar Front divides the Polar Frontal Zone part of the ACC from the broad, southern branch or sACC. All waters south of the Polar Front comprise the Antarctic Ocean proper (Deacon, 1984). More towards the Antarctic continent the sACC exhibits the less well-known southern Polar Front (sPF, not shown here; see Veth* et al., *1997) and then the boundaries with the major gyres of the Weddell and Ross Seas. Downstream of the Peninsula the ACC – Weddell Gyre Boundary (AWB) is also known as the Confluence region between the Scotia Sea, as part of the sACC, and the Weddell Sea proper. Also shown are the sites where iron determinations and/or iron-biota studies have taken place thus far. (Sites of studies by NIOZ–AWI–RUG, 1988–96, are shaded.)*

$$\mu = (1/C)(\mathrm{d}C/\mathrm{d}t) = \mu_{max}\{(1 - \exp(\alpha I/K_{max}))\exp(-\beta I/K_{max})\}\{[Fe]/(K_{Fe} + [Fe])\}\{[N]/(K_N + [N])\}\{[P]/(K_P + [P])\}\{[Si]/(K_{Si} + [Si]\}, \quad (4.4)$$

where C is the concentration of phytoplankton biomass, often expressed in units of chlorophyll *a* (Chl *a*) or particulate organic carbon, POC. Moreover, there is no *a priori* reason for other essential metals such as Mn, Zn, Co not to be limiting as well. Other complications arise owing to the fact that N can be supplied in three major forms, urea, ammonia and nitrate. Thus in the central gyres of the temperate ocean co-limitation by light and all major and trace nutrients may occur simultaneously. Equation 4.4 cannot be solved at present; commonly, just one, or at best two, of the multiple terms on the right-hand side are applied (see, for example, Taylor *et al.*, 1993; Townsend *et al.*, 1994). Sometimes this is justified by the assumption that, in general, just one factor is limiting the rate of growth at any one time (Droop, 1974). In fact multiple limitations, as in Equation 4.4, may just as well be the *a priori* expectation (O'Neill *et al.*, 1989), unless it is demonstrated otherwise in a particular case study.

In the HNLC regions the concentrations of the major nutrients nitrate, phosphate and silicate are high, and deemed to be not at all limiting. Moreover,

Caption for Figure 4.2 (*cont.*)

(1) EPOS in Weddell and Scotia Seas: European Polarstern *Study; 26.11.88–05.01.89; five suites of iron incubations (de Baar* et al., *1989b, 1990; Buma* et al., *1991); 52 dissolved Fe values at three stations (Nolting* et al., *1991); overall 64 stations (nos. 143–207) for plankton ecology; modelling by Lancelot* et al. *(1993).*
(2) Weddell Sea: Stena Arktika, *21.12.88–12.02.89; 239 total Fe values and 56 dissolved Fe values at 36 stations (Westerlund & Öhman, 1991).*
(3) Drake Passage and Gerlache Strait: 27.03.89–01.04.89; 14 dissolved Fe values at 3 stations (Martin et al., *1990*a*).*
(4) Drake Passage and near Seal Island: Surveyor, *Jan–Feb 1990 and Jan–March 1991; 5 experiments (Helbling* et al., *1991).*
(5) Ross Sea: Jan–Feb 1990; 4 experiments, see Martin et al. *(1990*b*); see also Dugdale & Wilkerson (1990); 8 particulate Fe values, see Martin* et al. *(1991*a*).*
(6) Southern Ocean JGOFS: Polarstern *ANT X/6 at 29.09.92–29.11.92 (Smetacek* et al., *1997; de Baar* et al., *1995, 1997; Löscher* et al., *1997; van Leeuwe* et al., *1997; Scharek* et al., *1997; Bakker* et al., *1997; Bathmann* et al., *1997; Quéguiner* et al., *1997; Rutgers van der Loeff* et al., *1997*a*).*
(7) Antares II of JGOFS-France: RV Marion Dufresne, *February 1994. Distributions of total Fe in seawater (Sarthou* et al., *1997).*
(8) Australian Subantarctic region: R.V. Southern Surveyor, *18.01.95–31.01.95. Dissolved and total dissolvable Fe and Mn values (Sedwick* et al., *1997).*
(9) Southern Ocean JGOFS: Polarstern *ANT XII/4 at 20.03.95–15.05.95 (de Baar* et al., *1999; van Leeuwe* et al., *1998a,b; Timmermans* et al., *1998; Nolting* et al., *1998).*
(10) Southern Ocean JGOFS: Polarstern *ANT XIII/2 at 04.12.95–22.01.96; de Jong* et al., *1998).*

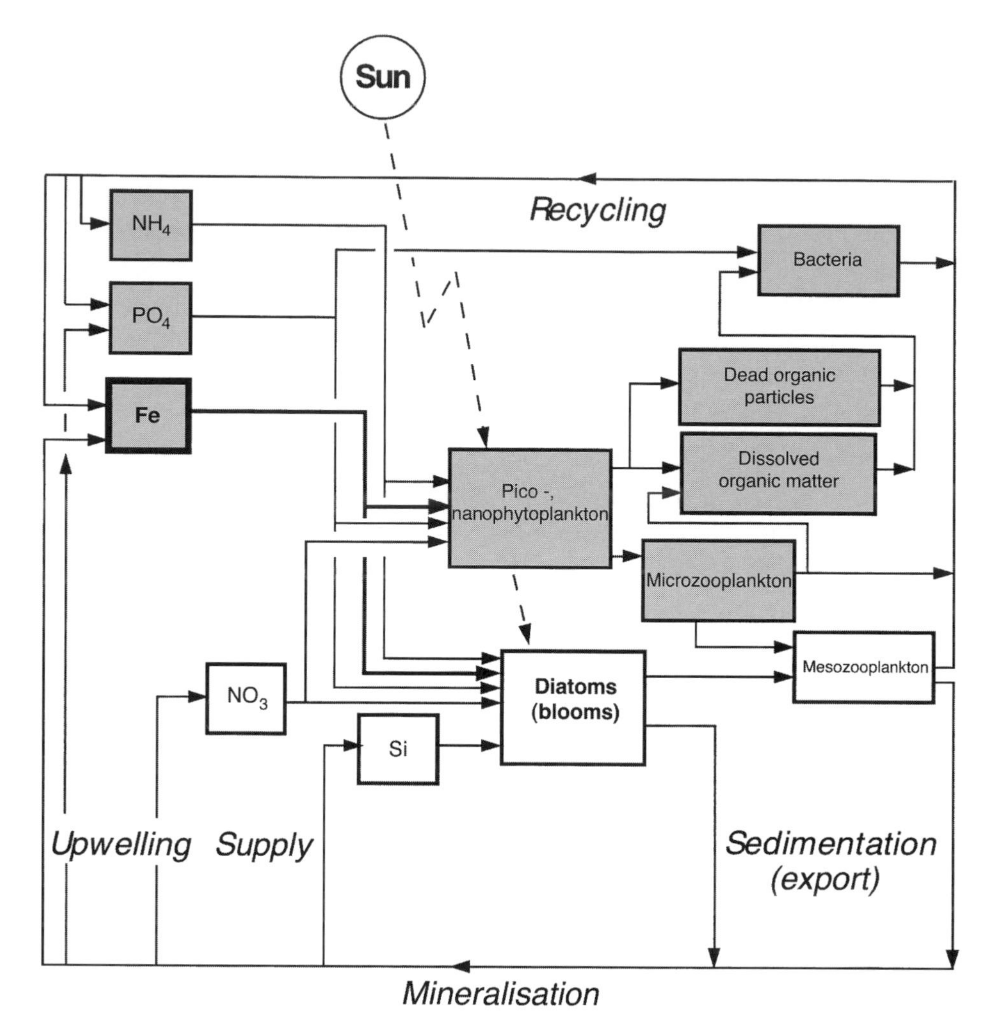

Figure 4.3 *Simple view of the plankton ecosystem in the euphotic zone of the upper ocean, re-drafted after de Baar (1994). The shaded boxes represent the background system of small planktonic organisms in functional compartments of phytoplankton, microzooplankton and bacteria. This is also known as the microbial food web or the microbial loop, i.e. the retention system with little or no exchange of matter with either the atmosphere above or the intermediate and deep waters below. Superimposed on this, there can occasionally be a transient bloom. In all the Fe studies thus far, this is invariably a bloom of diatoms, a remarkably coherent finding in all regions and experiments. Blooms other than diatoms do of course exist, but so far have not been linked to the Fe issue. Diatom growth and grazing by large zooplankton are not as closely coupled as the microbial food web, thus allowing blooms to flourish. The bloom and export system (open boxes) does exchange matter with above and below. It is therefore of major interest for global cycling of elements such as C. In the diagram not all relationships are depicted; for example, the consumption of bacteria by zooplankton is important but not shown. The cycling of urea is not shown either but is similar to the ammonium cycle. The figure can be seen as a simplified presentation of the mechanistic model (Lancelot* et al., *1993, 1996) of which some results are shown (Figs. 4.5, 4.6, 4.7 and 4.12) and discussed in the text.*

based on several experiments, the trace nutrients Mn, Zn and Co are deemed not to be limiting either (Buma *et al.*, 1991; Coale, 1991; Scharek *et al.*, 1997). Hence, in HNLC waters there exists the unique situation where a simple equation is actually justified (Lancelot *et al.*, 1996). Light and iron are the only factors regulating the production term:

$$(1/C)(\mathrm{d}C/\mathrm{d}t) = \mu_{\max}\{(1 - \exp(\alpha I/K_{\max}))\exp(-\beta I/K_{\max})\}\{[\mathrm{Fe}]/(K_{\mathrm{Fe}} + [\mathrm{Fe}])\}. \quad (4.5)$$

This combined control of production by light and iron (Equation 4.5), along with the loss by grazing (Equation 4.1) are the leading themes in our review. For Equation 4.5, two weaknesses must be realised. Firstly, the underlying Michaelis–Menten equation assumes a steady state, whereas in ocean modelling it is applied to non-steady-state conditions, notably the wax and wane of a plankton bloom. Another caveat is the implicit assumption (Equation 4.5) that the resource factors are treated as independent, whereas in fact they are not. For the HNLC situation it is known that light stress will alter cellular iron requirements (Raven, 1990), or that iron stress will affect nitrate assimilation (Timmermans *et al.*, 1994; Price *et al.*, 1994). The physiological responses of algal cells to combined iron and light deficiency are currently being investigated (Muggli & Harrison, 1996; Kudo & Harrison, 1997; van Leeuwe & Stefels, 1998; Stefels & van Leeuwe, 1998; van Leeuwe, 1997). The responses are complex. Eventually, this may yield more suitable equations. At the time of writing, the above Equation 4.5 is still the most useful parameterisation, in keeping with the state of the art. Finally, several lines of evidence suggest that high abundance of ammonium may inhibit the assimilation of nitrate (Wheeler & Kokkinakis, 1990). Here this mechanism will not be treated extensively, as excellent studies of the Southern Ocean (Goeyens *et al.*, 1995) and the equatorial Pacific Ocean (Price *et al.*, 1994) do exist.

Functions of iron in the cell

Within the algal cell, iron plays a major role, often in processes of electron transfer. Excellent overviews of the role of iron in plankton physiology and oceanography have recently become available (Geider & La Roche, 1994; Hutchins, 1995; van Leeuwe, 1997). Iron is essential in several steps of the photosynthetic route of CO_2 fixation. For example, several iron-containing enzymes that are involved in the Calvin cycle are essential for producing the substrate ribulose biphosphate, which is then converted by the enzyme Rubisco, and for the synthesis of the photon-harvesting molecule of chlorophyll *a*. Other major functions are in the nitrate and nitrite reductase enzymes necessary for reduction of nitrate and nitrite to ammonium for the synthesis of amino acids (Rueter & Ades, 1987; Timmermans *et al.*, 1994). Nitrogen can also be obtained by direct fixation of N_2. This requires much more Fe (Raven, 1988) and therefore has been reported more often in freshwater systems than in Fe-starved ocean waters. In the absence of Fe it would appear to be more efficient to utilise reduced N-moieties such as ammonia or urea. Several other functions and enzymes in the cell also require Fe (Geider & La Roche, 1994).

Iron limitation

The just-discovered (Ruud, 1930) high abundance of major nutrients in Antarctic surface waters led Gran (1931) to suggest, as well as to investigate, iron limitation. Hart (1934, 1942), in describing Antarctic plankton communities, hypothesised that the greater abundance of diatoms in nearshore, so-called *neritic*, waters may be favoured by coastal sources of, for example, Fe, Mn or organic compounds. However, it cannot be too strongly emphasised that in all probability phytoplankton production is always governed by a complex of inter-dependent factors, rather than by one or two that are clearly definable (Hart, 1934). The history of iron limitation and its investigation has been described elsewhere (de Baar, 1994). It took until August 1987 before the first realistic, i.e. uncontaminated, experimental tests of the iron limitation were reported, in this case for the subarctic Pacific (Martin & Fitzwater, 1988). Here Martin & Fitzwater (1988) suggested that this area could become a classical marine example of Liebig's law of the minimum, i.e. an iron hypothesis for limitation of all phytoplankton. However, Coale (1991) phrased more conservatively that his findings were contradictory to the hypothesis that iron is the *only* limiting nutrient, and thus Liebig's law of the minimum would not apply. De Baar *et al.* (1990) echoed the sentiments of Hart (1934) by realising that, whenever a single limiting factor has been hypothesised, field measurements and experiments have provided contradicting results. From their control experiments alone it became apparent that other factors such as light and grazing (Buma *et al.*, 1991) are also operative in the Antarctic Ocean, although the Fe enrichments also showed a distinct additional stimulation. Martin *et al.* (1990*a*) presumably were testing the hypothesis that Antarctic phytoplankton suffer from iron deficiency, which prevents them from blooming and using up the abundant supplies of major nutrients in vast areas of the Southern Ocean. Otherwise, the paper presented some dissolved iron observations, but no testing of the hypothesis. Martin *et al.* (1991*a*) believed that phytoplankton growth in major nutrient-rich waters is limited by iron deficiency, but then provided a more extensive working hypothesis.

The possible role of iron in plankton ecology was also related to cycling of CO_2 between the Southern Ocean and the atmosphere, in the present and the past (Martin & Gordon, 1988). Thus Fe may have played a role in modulating atmospheric CO_2 levels between glacial and interglacial times (Martin, 1990). Soon it was also suggested that oceanic iron fertilisation aimed at the enhancement of phytoplankton production may turn out to be the most feasible method of stimulating active removal of the greenhouse gas CO_2 from the atmosphere (Martin *et al.*, 1990*a*).

Cullen (1991) made a strong case for the hypothesis that there is not enough available iron in high-nutrient waters to support the net community production necessary for depletion of the major nutrients. This also appears to have been the major hypothesis to be tested in the IronEx *in situ* enrichment experiment (see also Liss & Turner, Chapter 5, this volume), although this was not clearly stated (Cullen, 1995) by lead author Coale and co-authors (Martin *et al.*, 1994).

Where initially the debate among some authors focused on either iron limitation or grazing as opposing mechanisms, there evolved a refinement. Briefly, it was realised that at low ambient Fe concentrations the very small cells would suffer less from limitation owing to their favourable surface: volume ratio for diffusive transport and uptake into the cell (Hudson & Morel, 1990, 1993). However, these small algae are kept at low numbers owing to intensive grazing by microzooplankton. Enrichment with Fe would allow larger algae such as diatoms to grow, and would yield blooms because the large cells are not easily grazed by the small microzooplankton (Buma *et al.*, 1991). Morel *et al.* (1991*b*) proposed an 'ecumenical' iron-limitation hypothesis. Briefly, the community is seen as consisting of small cells, under some degree of Fe stress and thus incapable of taking up nitrate efficiently. Instead the community is utilising ammonium, and it is heavily grazed upon by small microzooplankton. Upon Fe enrichment, the community would shift to larger cells using nitrate as their chief N source. This concept was largely confirmed for the equatorial Pacific, mostly based on size-fractionated nitrogen assimilation (Price *et al.*, 1994). Cullen (1995) views the HNLC condition as a grazer-controlled phytoplankton population in an iron-limited ecosystem (Price *et al.*, 1994) and refers to this as the ecumenical hypothesis. This would suggest that the interplay of factors, rather than a single factor, is the key to understanding this system (de Baar *et al.*, 1990). Otherwise, Cullen (1995) observed problems of testability of such an ecumenical hypothesis. The broader ecosystem viewpoint with several factors at play (Hart, 1934; Dugdale & Wilkerson, 1990; Buma *et al.*, 1991; Buma, 1992; Mitchell *et al.*, 1991; Nelson & Smith, 1991; de Baar, 1994), as well as the recent findings (de Baar *et al.*, 1995) led Cullen (1995) to realise an important shift in the iron hypothesis. Perhaps it is time to recast the hypothesis.

The plankton ecosystem

To try to define the various iron hypotheses more precisely, we have used a conceptual model of the pelagic microbial ecosystem characterised by a recycling system consisting of the small size classes of phyto-zoo-bacterioplankton (Fig. 4.3). Here organic intermediates such as ammonium and dissolved organic carbon (DOC) play an important role. Overall, such a system represents a low and more or less constant standing stock. The uptake of the abiotic substrates CO_2, NO_3 and PO_4 is balanced by recycling, such that overall the geochemical mass balance is virtually achieved.

Superimposed on this can be a regional (i.e. coastal, neritic, frontal) and/or seasonal (i.e. spring) expression of a plankton bloom, characterised in general by larger algal cell size such as diatoms, sometimes with nanoplankton as well (Becquevort *et al.*, 1993; Jochem *et al.*, 1995). The larger algae are being grazed upon by larger mesozooplankton, or may exhibit a massive sedimentation loss. This leads to a downward export production of settling larger biogenic particles, i.e. sinking intact diatom frustules (Crawford, 1995; Takahashi *et al.*, 1990) or faecal pellets (Smetacek, 1985; Fischer *et al.*, 1988; Wefer *et al.*, 1988; Wefer & Fischer, 1991). This is the classical foodweb as known to Gran, Hart and their contemporaries. The export of the biogenic elements C, N, P, Si and Fe in particles causes a deficiency of the same elements in the euphotic zone. Hence CO_2 in the surface waters tends to become undersaturated relative to the atmosphere; nitrate, phosphate, silicate and iron become depleted. Replenishment of these inorganic substrates may occur by an influx of CO_2 from the atmosphere (Bakker *et al.*, 1997), upwelling of nitrate (Eppley *et al.*, 1988) or aeolian inputs of N (Donaghay *et al.*, 1991), and similarly replenishment of iron may come from either upwelling or the atmosphere. It is these influxes into the euphotic zone, balancing the export flux into deeper waters, that are of interest for global geochemical budgets, notably that of CO_2 in the atmosphere. Whereas plankton ecologists are interested in both the recycling system and the bloom-and-export system, geochemists are mainly concerned with the latter.

Both systems are interrelated and by necessity can only be studied together. It should be noted that, at a given site, the importance of one system relative to that of the other will be different for the different chemical elements C, N, P, Si and Fe. In other words, the recycling efficiencies of these elements vary; hence, the elemental ratio of the export term may deviate very strongly from that in the standing stock. Deviations of the classical Redfield ratio have therefore been recognised as interesting (Cooper, 1937) and renewed attention is apparent for the major elements (Fanning, 1992; Sambrotto *et al.*, 1993; de Baar *et al.*, 1997). In addition, the recycling of iron has become the subject of investigations (Hutchins *et al.*, 1993, 1995 Hutchins & Bruland, 1994, 1995; Barbeau *et al.*, 1996). Here the elements are also interacting: iron is essential for the necessary enzymatic reduction of nitrate to ammonium inside the cell, whereas N_2 fixation by certain strains of phytoplankton requires a cellular Fe quota about an order of magnitude higher (Raven, 1988). Hence at very low ambient iron the impairment of nitrate uptake may yield deviations in the elemental N:P ratio (de Baar *et al.*, 1997), whereas a strong aeolian supply of iron may support massive N_2 fixation by surface-dwelling diazotrophs, also perturbing the

Table 4.1. *The iron hypotheses*

1. Inadequate external supply of dissolved iron limits the overall rate of primary productivity, hence phytoplankton blooms and export production, hence replenishment by, for example, CO_2 drawdown.
2. Inadequate external supply of dissolved iron limits the rate of growth of the larger, bloom-forming algae, hence phytoplankton blooms and export production, hence replenishment by, for example, CO_2 drawdown.
3. Inadequate external supply of dissolved iron limits the full utilisation until depletion of major nutrients N and P (and Si), and hence impairs maximum drawdown of CO_2 from the atmosphere.
4. Aeolian input of continental aerosols and their (partial) dissolution is the major supply of iron.
5. Input from below by vertical mixing or upwelling from intermediate or deep waters is the major supply of dissolved iron.
6. Lateral transport of dissolved iron from ocean-margin sediment sources is the major supply of dissolved iron.
7. What is the internal supply, or recycling efficiency within the euphotic zone, of dissolved Fe vis-à-vis that of dissolved N vis-à-vis that of dissolved C?
8. The chemical speciation and kinetics of iron is similar in different HNLC ecosystems such that availability for rate of biological uptake is largely proportional to the total concentration of dissolved iron.

classical Redfield ratio (Michaels & Knap, 1995). In summary, the Redfield-ratio stoichiometry, as for example,

$$C:N:P:Fe = 106:16:1:(10)^{-3.5} \quad (4.6)$$

does not necessarily apply to the input and export of the surface ocean ecosystem in large-scale budgets.

Defining hypotheses towards validation

The conceptual model of the plankton ecosystem (Fig. 4.3) is simple but serves for attempting to define the iron hypotheses more precisely for a given HNLC ecosystem, where three different hypotheses are initially apparent (Table 4.1, hypotheses 1–3).

It is clear that the external source of supply of dissolved iron to the surface waters is critical and this yields three additional hypotheses, which oppose one another (Table 4.1, hypotheses 4–6).

Over longer climatic or geological time scales, the importance of iron inputs may shift. For example, in the last glacial maximum, with arid periods and higher wind velocities, the aeolian input of continental dust may have been higher (Martin, 1990; Kumar *et al.*, 1995; Murray *et al.*, 1995*b*). Enhanced wind stress would on the other hand also enhance upwelling as well as vertical mixing, i.e. the input from below. Obviously, the latter three hypotheses (4–6) are closely related with what really is more of a question than a hypothesis (Table 4.1, hypothesis 7).

This relates to turnover rates and brings us to the essential issue of rate of growth under iron limitation being limited by the uptake kinetics of iron to the cells. These kinetics depend on the existence of different physico-chemical forms in seawater, i.e. dissolved (nominally less than 0.2 μm in diameter), colloidal and particulate forms, each group of which may be subdivided into various speciations, e.g. organically bound and inorganic dissolved iron, where the multiple valencies II and III further complicate the matter (Wells *et al.*, 1995). For example, it was recently found that apparently over 99% of dissolved iron is organically complexed (van den Berg, 1995; Gledhill & van den Berg, 1994, 1995; Rue & Bruland, 1995). This would leave only 1% or less of dissolved inorganic iron for direct uptake by the cells, unless the organic form either is taken up directly (for example, by involving siderophores; Trick *et al.*, 1983; Reid & Butler, 1991; Trick & Wilhelm, 1995; Soria-Deng & Horstmann, 1995), or dissociates very rapidly to replenish the inorganic pool at a sufficient rate. This complex of interactions at the molecular and cellular level is the topic of ongoing and several new investigations, but currently not resolved. Awaiting the outcome of such studies, in this chapter we mostly consider the rate of supply of total dissolved iron as a first-order assessment at the basin scale of various HNLC ecosystems. This leads to a final hypothesis as a provocative target for future falsification (Table 4.1, hypothesis 8).

In summary, the first three hypotheses (Table 4.1) more or less cover the various hypotheses in the existing literature. Gran (1931) and Hart (1934, 1942) studied the classical food chain (diatoms – zooplankton – higher predators) and presumably were not aware of the microbial food web. For them, hypothesis 2 would be no different from hypothesis 1, but modern investigators are aware of the distinction (Fig. 4.3). The law of the minimum (J. von Liebig) has occasionally been mentioned (Martin, 1991*b*) but in fact it is not relevant here (de Baar, 1994). Finally, the supply hypotheses 4–6 may be seen as multiple (competing) hypotheses (Chamberlain, 1890), but this does not appear to be the case for the first three hypotheses, as confirmation of one does not always rule out the others.

Methods for validation

Some general principles are to be considered before trying to synthesise recent research findings into an overall description.

Firstly, for a given HNLC ecosystem one seeks validation for all seasons, whereas in fact the collection of field observations has usually been done in one or two seasons only. Seasonality is relatively minor at the equator but becomes dominant towards the poles. Hence, among the three major HNLC regions it would affect the equatorial Pacific less than the subarctic Pacific. On the other hand, seasonality of incoming light, wind velocity and sea-ice cover is the overruling forcing factor in the ecology of the Southern Ocean. In addition to seasonal oscillations, there are interannual forcings. The El Niño – Southern Oscillation (ENSO) drastically shifts the equatorial Pacific ecosystem at intervals of 3–4 years. Recently, anomalies in sea-surface temperature, meridional wind and ice extent of the Southern Ocean have been shown to be part of a two-cycle Antarctic Circumpolar Wave creating local variability at 4–5 years, with a likely coupling to ENSO oscillations at lower latitudes (White & Peterson, 1996).

Secondly, when contemplating methodology, Lalli (1991) and Lalli & Parsons (1993) have suggested a combination of three lines of approach in resolving an issue in marine ecology: (1) observations in the field; (2) mesocosm experiments; and (3) controlled experiments in the laboratory.

No single approach in itself can provide the final and firm answer, but the combination may. For the issue of iron limitation affecting the rate of plankton growth this materialises as follows.

1. Measurement of state and rate variables in the *unperturbed* natural system (Bathmann *et al.*, 1994*b*; de Baar *et al.*, 1995; Smetacek *et al.*, 1997) and simulation modelling (Mitchell *et al.*, 1991; Lancelot *et al.*, 1993, 1996). More recently, this has been done also with physiological indicators diagnostic of iron stress (La Roche *et al.*, 1996).

2. Perturbation experiments of whole plankton assemblages. This has been done either by shipboard incubations in, for example, 20 l vessels (Martin & Fitzwater, 1988; Martin *et al.*, 1990*b*, 1991*a*, 1993; Coale, 1991; de Baar *et al.*, 1990; Buma *et al.*, 1991; Helbling *et al.*, 1991; Boyd *et al.*, 1996; van Leeuwe et al., 1997), or by *in situ* iron enrichment (Martin *et al.*, 1994; Coale *et al.*, 1996*b*).

3. Uptake and growth kinetics studies of monospecific algal cultures in the laboratory, and a comparison of findings for different species; see, for example, Andersen & Morel (1982), Brand (1991), Sunda *et al.* (1991), Hudson & Morel (1990), Sunda & Huntsman (1995), Muggli & Harrison, 1996 or Muggli *et al.* (1996).

From (1) to (3), observations become increasingly removed from reality. On the other hand, the cause–effect relationships become more directly proven. In this respect, the reproducibility of laboratory experiments (approach 3) is less hazardous than that of observations (approaches 1 and 2) between one cruise and season and another. Below we will find that researchers have chosen a combination of one or two of these approaches for a given project.

For the perturbation experiments (approach 2) a longer period of several days will often yield a more convincing response, but ironically the perturbed system deviates more and more from the natural situation. This is true for bottle incubations and *in situ* perturbations alike. In some respects short experiments of 24–48 h may be preferable, at intervals of the order of 12–24 h these would become synonymous with the assessment of rate variables (e.g. primary production, nitrate reductase activity) in the natural system (approach 1). For this reason the virtually immediate *in situ* detection of fluorescence response (Kolber *et al.*, 1994; Behrenfeld *et al.*, 1996) is most convincing. In addition, short-term incubations for assessment of physiological indicators of iron stress (nitrate reductase activity, ^{15}N-labelled nitrate assimilation rate) have recently been applied (Timmermans *et al.*, 1994, 1998; Price *et al.*, 1994). Moreover, bottle incubations tend to exclude some of the large mesozooplankton, unless these are intentionally added (Boyd *et al.*, 1996). Either way is valid as experimental methodology and in both ways useful information is obtained. For *in situ* perturbations, the mesozooplankton have the same opportunity as they do in the unperturbed system. Some shipboard bottle incubations have been done at optimal light so as to ensure that light is not at all a limiting factor; others have employed the simulated *in situ* light climate in order to mimic the natural ecosystem. Again both approaches are valid but the difference must be realised.

Finally, one seeks observations of rate variables (Equations 4.1 – 4.5), but the fact of the matter is that these are generally much more difficult to obtain and interpret than state variables. For example, the concentration of dissolved iron in seawater is already difficult to measure, but the kinetic rate of iron uptake by an algal assemblage is much more difficult to quantify in a reproducible manner; its actual value will also depend on experimental conditions. As a result, when trying to interpret in terms of rates of growth, one is often relying on a small data set of actual state variables.

The observations

Southern Ocean

In the southern hemisphere, circumpolar features (Fig. 4.2) characterise the Southern Ocean. It is defined (Deacon, 1984) as all the waters south of the Subtropical Front at about 40°S (not shown). Most dominant is the eastward-flowing Antarctic Circumpolar Current (ACC) and its circumpolar frontal systems adjoining (or flowing through) the Indian, Pacific and Atlantic basins (Fig. 4.2). Within the overall ACC, the Subantarctic Front, at about 45°S, divides the Subantarctic Zone (SAZ) and the Polar Frontal Zone (PFZ). At about 50°S the Polar Front exists as a distinct jet stream. All waters south of the Polar Front comprise the Antarctic Ocean proper (Deacon, 1984). The Polar Front divides the Polar Frontal Zone from the broad, southern branch or sACC. Closer to the continent the sACC extends to its boundaries with the major gyres of the Weddell and Ross Seas.

South of the Polar Front major nutrients are present all year; typical concentrations in surface waters are in the order of 1.6–2.0 μM phosphate, 20–30 μM nitrate and 30–100 μM silicate (Gordon *et al.*, 1986; Kamykowski & Zentara, 1989; Levitus *et al.*, 1993). In the Polar Frontal Zone, these values decrease towards the Subantarctic Front (Bathmann *et al.*, 1994*b*, their Fig. 6.5.5; de Baar *et al.*, 1997, their Fig. 3). This gradient is most pronounced for silicate (Plate 3 of Comiso *et al.*, 1993; Sullivan *et al.*, 1993, their Fig. 4.4B), and in the northerly PFZ silicate may become limiting for diatom growth. In the adjacent Subantarctic Zone, all major nutrients are commonly depleted. Hence, within the overall Southern Ocean, only the Antarctic Ocean proper, south of the Polar Front, is truly a region where HNLC conditions prevail for all major nutrients.

Antarctic investigations before JGOFS

WEDDELL AND SCOTIA SEAS (1988 EPOS) The hypothesis of iron limitation of the Southern Ocean was tested in 1988–89 (de Baar *et al.*, 1989*b*, 1990; Buma *et al.*, 1991). In all five experimental runs iron supply stimulated phytoplankton growth, most prominently in the more offshore Confluence (Fig. 4.4) and Scotia Sea, where ambient dissolved iron was low albeit still neritic at more than 1 nM (Nolting *et al.*, 1991), compared with higher iron concentrations in more nearshore waters of the Weddell Sea (Westerlund & Öhman, 1991; Nolting *et al.*, 1991) similar to those of the inshore Gerlache Strait at the Pacific side of the Peninsula (Martin *et al.*, 1990*a*). However, algal biomass in the controls were also consistently higher than the algal biomass *in situ*, conversely suggesting that factors other than iron were largely controlling the phytoplankton standing stocks *in situ*. Optimal light in the bottle experiment, and exclusion of mesozooplankton, suggest light and grazing as the major *in situ* controls on plankton stocks in these neritic waters (Buma *et al.*, 1991).

Close examination of the data from Fe enrichments over the first few days showed only minor iron-mediated increases in algal stocks, as also observed elsewhere (Martin & Fitzwater, 1988). In all incubations an eventual shift towards larger size classes, essentially of diatoms, was observed and ascribed to exclusion of mesozooplankton grazers. The shift was most pronounced in the iron-enriched bottles, suggesting a superimposed stimulation by iron of larger microplankton (Buma *et al.*, 1991). From the latter observation the daily rate of cell division of the dominant diatoms was calculated to be about 10% enhanced in the Fe-enriched bottles, suggesting suboptimal growth rates in the field situation. This difference is significant but too small to be discernible in models (see, for example, Lancelot *et al.*, 1993).

In this frontal region both dissolved Fe (Nolting *et al.*, 1991) and particulate

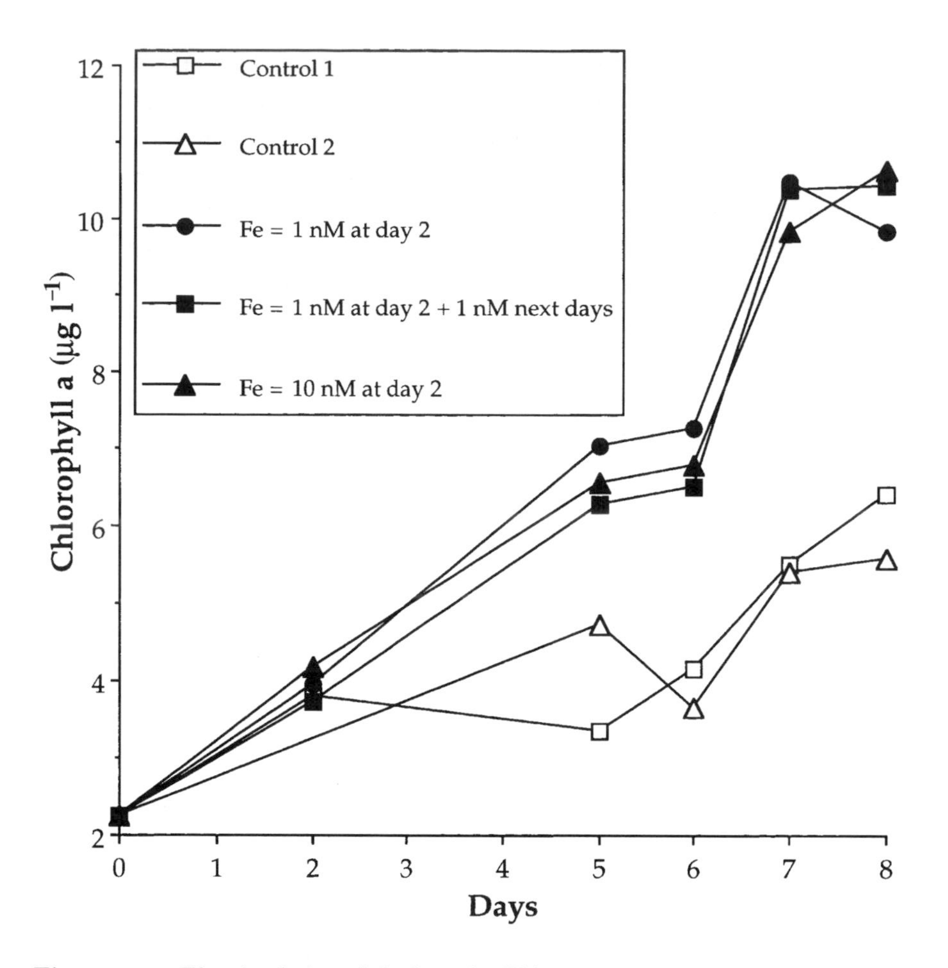

Figure 4.4 *The stimulation of plankton (as Chl* a*) on Fe addition (at day* t = *2) of 1 nM as well as 10 nM, and of 1 nM with additional 1 nM at days 3–7, compared with two untreated controls. EPOS station 158 (59°S, 49°W) at the Scotia Front near the Weddell–Scotia Confluence zone at 08.12.88 (after de Baar* et al., *1990). Notice the reproducible increase in biomass from t = 0 to t = 2 (before Fe was added), demonstrating the reliability of clean techniques. Similarly, at days 2–8 both the two controls and the three additions have excellent reproducibility. Apparently just 1 nM Fe addition is required for optimal growth; the extra Fe (9 nM as well as next day's 1 nM) had no extra effect on overall Chl* a, *but the 10 nM treatment did yield a distinct increase in the biovolume of the largest diatoms (*Corethron *sp.), as assessed from microscopy (Buma* et al., *1991).*

Al concentrations were found to be quite high (Dehairs *et al.*, 1992). At 49°W at the Scotia Front, then at *c.* 49°S, a sub-surface maximum of particulate Al as high as 30 nM was found at a depth of *c.* 200 m (Fig. 4.5) and ascribed to upstream shelf sediment sources of the Peninsula and Archipelago region. Assuming 10% dissolution and the given Fe: Al ratio, there is the potential to bring between 1.1 and 4.2 nM Fe into solution (Dehairs *et al.*, 1992). This is comparable to the dissolved Fe measured in the region (Nolting *et al.*, 1991). At this same site (59°S, 49°W) one of the Fe enrichment incubations had also been performed (Fig. 4.4). Moreover, the Scotia Front and Confluence region is largely coincident with the Marginal Ice Zone (MIZ), which is a major location of observed plankton blooms (MIZ is defined as less than 50% ice cover, as distinct from more than 50% in the Closed Pack-Ice Zone, CPIZ). The sea-ice melting at this MIZ region is an almost year-round phenomenon, as the CPIZ in the western Weddell Sea is continuously drifting in a northerly direction. The drift velocity of several kilometres per day provides a continuous supply of sea-ice, on average 2 years old (Eicken, 1992), with the occasional dirty (i.e. Fe-containing 'dirt'; de Baar *et al.*, 1990) iceberg. This supply allows the local MIZ to recede no more than about five degrees during summer. Upon melting this MIZ not only provides freshening favourable for stratification, but also supplies accumulated aeolian Fe. This, together with the iron derived from margin sediments as mentioned above, could stimulate phytoplankton blooms. Thus the combination of field and modelling observations and results from mesocosm experiments had shown that light limitation and grazing pressure, more than iron stress, were the dominant factors maintaining high macronutrient levels and low phytoplankton stocks in the more or less neritic waters of the Weddell and Scotia Seas.

DRAKE PASSAGE In the austral summers of 1990 and 1991, Helbling *et al.* (1991) found little or no effect of iron addition at their inshore sites A and F or nearshore sites B and E (water depth *c.* 1000 m), but significant effects of either 1 or 5 nM additions at offshore sites C and D. This is consistent with, respectively, the higher dissolved iron at *c.* 4.7 – 7.4 nM in inshore Gerlache Strait and lower dissolved Fe (*c.* 0.1–0.88 nM) in offshore Drake Passage waters (Martin *et al.*, 1990*a*) (when ignoring rejected values in the 0.52 – 1.55 nM range). At offshore site C, the Fe enrichment selectively stimulated the larger microplankton, as opposed to the control incubation. Hence, from this experiment alone, it would appear that in the field the microplankton are limited predominantly by iron stress, rather than selectively removed by mesozooplankton grazing. This would tentatively characterise the offshore Drake Passage as oligotrophic for iron.

ROSS SEA In the austral summer of 1990, a suite of four incubation experiments was done in the Ross Sea (Martin *et al.*, 1990*b*). The results were similar to the previous observations in the Weddell and Scotia Seas. Iron always stimulated plankton growth, most strongly at the offshore station, but the untreated controls also showed a steady increase of biomass exceeding *in situ* chlorophyll *a* levels. Dugdale & Wilkerson (1990) interpreted the findings in terms of grazing pressure and the changes thereof in the incubations, such that

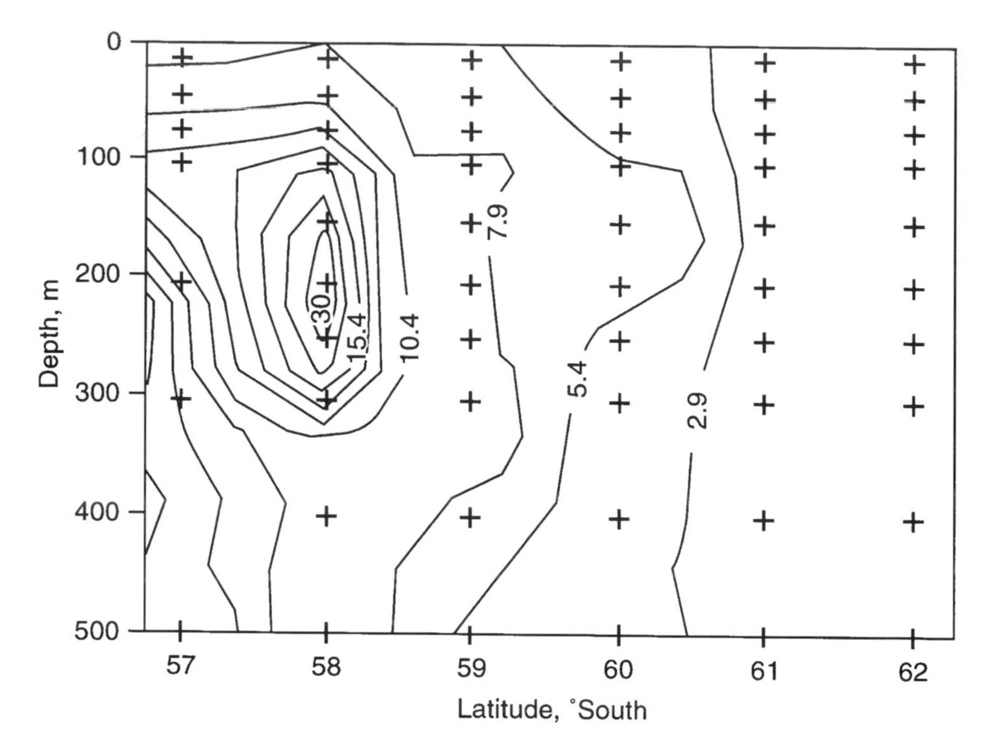

Figure 4.5 *Total particulate Al at a section at 49°W sampled during the 1988 EPOS programme. The maximum at* c. *58°S coincides with the position of the Scotia Front at that time, and is ascribed to lateral transport of mineral particles from upstream margin sediments (taken from Dehairs* et al., *1992).*

iron stress would appear less dominant in the field situation. Unfortunately, there is no information on ambient concentrations of dissolved iron. Particulate iron concentrations (Martin *et al.*, 1991*a*) are higher (*c.* 0.7 – 2.7 nM) inshore than at the offshore site (less than 0.5 nM), suggesting a decrease in dissolved iron with distance offshore, consistent with the more pronounced effect of Fe enrichment at the offshore site.

Atlantic Antarctic Circumpolar Current and Polar Front (1992 JGOFS)

The three studies thus far had demonstrated a modest role for iron stress in largely neritic nearshore waters, albeit with intriguing trends towards stronger iron stress in offshore directions. In the austral spring 1992, the first Southern Ocean JGOFS expeditions took place (Turner *et al.*, 1995; Bathmann *et al.*, 1994*b*; Smetacek *et al.*, 1997). For the Atlantic sector an integrated study of the plankton ecosystem along a truly offshore, large-scale, section (more than 1000 km length at *c.* 6°W; Fig. 4.6) was designed. Here the spatial distribution of ambient dissolved iron in the upper 300 m of the water column (Löscher *et al.*, 1997) was combined with the full suite of measurements of the SO-JGOFS

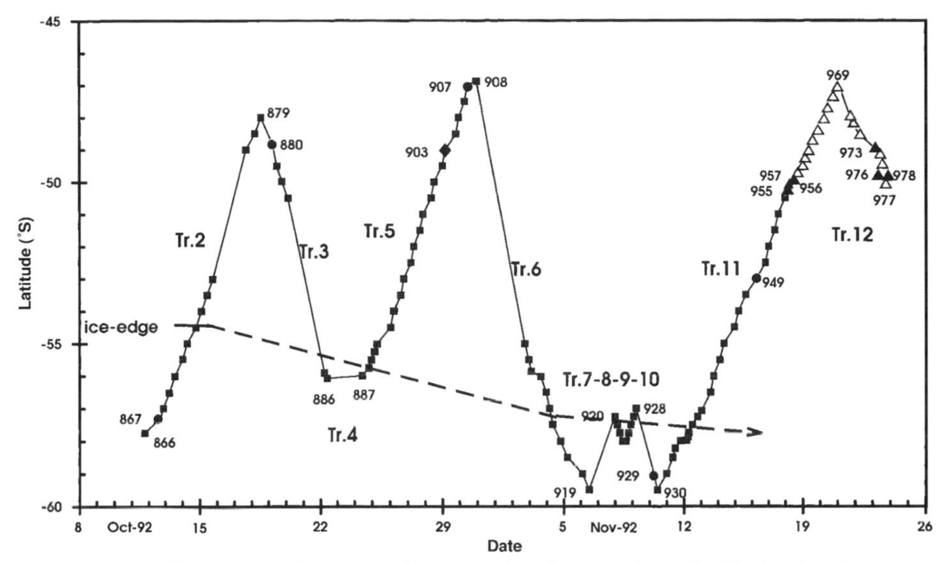

Figure 4.6 *Chronology of transects (Tr.) numbered 2–12, all at 6°W latitude, of Southern Ocean JGOFS ANT X/6, after Veth* et al. *(1997) and de Baar* et al. *(1997). Large filled diamond and filled triangles indicate where major blooms of the large diatom* Fragilariopsis kerguelensis *were encountered; open triangles, blooms of other large diatoms (*Corethron *spp.); small squares, remaining stations where no blooms were present; filled circles, Fe enrichment incubation experiments of van Leeuwe* et al. *(1997) and Scharek* et al. *(1997).*

programme (Smetacek *et al.*, 1997). Such observations in the unperturbed natural system are probably the most reliable approach for demonstrating iron deficiency. In order to understand and demonstrate the underlying causal relationships, a suite of shipboard mesocosm incubations was also undertaken (van Leeuwe *et al.*, 1997; Scharek *et al.*, 1997).

ATLANTIC ACC AND SOUTHERLY FRONTS In the southern branch of the ACC (*c.* 51–56°S) the dissolved Fe concentrations decreased seasonally from an average of 0.49 nM to an average of 0.31 nM some weeks later (Fig. 4.7, Table 4.2). The lowest observed concentration of dissolved Fe was 0.17 nM. The mean surface water levels, at *c.* 0.31 or *c.* 0.49 nM, were adequate to sustain a low and constant phytoplankton biomass (Chl *a* less than 0.25 $\mu g\,l^{-1}$), but were insufficient for bloom development (de Baar *et al.*, 1995). The vertical stability of the water column is the weakest in these central ACC waters, such that the light regime is also unfavourable for phytoplankton growth (Jochem *et al.*, 1995). The upwelling supply of dissolved iron was calculated to be about 1.68×10^{-12} mol m^{-2} s^{-1}. This would be adequate for supporting a primary productivity of about 170 mg C m^{-2} d^{-1}, consistent with observed productivities ranging from 80 to 300 mg C m^{-2} d^{-1} (Jochem *et al.*, 1995). The similar average upward flux of nitrate was 4.54×10^{-6} mol m^{-2} d^{-1} and would support a primary production of *c.* 720 mg C m^{-2} d^{-1} (de Baar *et al.*, 1995). The observed primary production is about half that value, i.e. only about half of the supply of nitrate is utilised. Thus other limiting factors, such as Fe, light and grazing, are controlling production rates. The majority of primary production was due to small algal cells (Jochem *et al.*, 1995), which presumably are more efficient at taking up iron owing to their more favourable surface: volume ratio (Hudson & Morel, 1990). The pico- and nanophytoplankton also constitute the production term of the small food web (Fig. 4.3) in the recycling ecosystem (Smetacek *et al.*, 1990), which has little impact on inventories of the dissolved gases CO_2 and O_2 or export of organic matter into deeper waters. Indeed the export production, assessed from $^{234}Th:^{238}U$ inventories, was negligible (Rutgers van der Loeff *et al.*, 1997*a*). The corresponding inventories of dissolved nutrients and total CO_2 remained virtually unchanged throughout the season. The concentration of undissociated $[CO_2]_{aq}$ in surface waters remained virtually constant as well (Bakker *et al.*, 1997). However, owing to seasonal warming its temperature-dependent solubility $K(S, T)$ versus the fugacity (f_{CO_2}) in relation

$$[CO_2]_{aq} = Kf_{CO_2} \tag{4.7}$$

decreased, where f_{CO_2} represents partial pressure p_{CO_2} after minor correction for non-ideal behaviour (Weiss, 1974). Therefore, the partial pressure of CO_2 in surface waters became oversaturated with respect to the atmosphere (Fig. 4.7).

At the site of the recently retreated ice edge there was an input of about 0.5 nM dissolved iron (latitude *c.* 54°S in section 11, Fig. 4.7). This is ascribed to instant release of the aeolian iron accumulated in the *c.* 1 m thick sea-ice over the winter period, and coincided with a maximum of the sea-ice-dwelling diatoms *Nitzschia prolongatoides* (Veth *et al.*, 1997). Budget calculations assuming *c.* 30 nM Fe in the sea-ice demonstrated the feasibility of an increase of *c.* 0.4 nM (de Baar *et al.*, 1995). The assumption of 30 nM Fe was based on sea-ice measurements in the range 10–60 nM (Löscher *et al.*, 1997), and in retrospect happens to be consistent with the 30 nM predicted independently for sea-ice dust (Martin & Gordon, 1988) on the basis of the mean Al concentrations in Holocene ice cores of the Antarctic continent (De Angelis *et al.*, 1987). However, at this site (54°S), the mixed layer was deep, and strong wind mixing episodes once or twice per week prevented Marginal Ice Zone blooming.

ATLANTIC POLAR FRONT In contrast with the southern branch of the ACC, the Polar Frontal jet at about 49–50°S, meandering eastward across the 6°W section, was characterised by elevated dissolved iron of *c.* 1.87 nM in section 5/6, decreasing to *c.* 1.14 nM two to three weeks later (Fig. 4.7). This decrease was consistent with a general increase of particulate plankton matter,

Table 4.2. *The average values of dissolved Fe, Chl* a *and* f_{CO_2} *in the southern branch of the Antarctic Circumpolar Current and in the Polar Front, consecutive sections 5/6 and 11 of JGOFS 1992 (Polarstern ANT X/6)*
See also Fig. 4.7*a,b*. The analytical methods and reproducibility of the Fe data are reported by Löscher *et al.* (1997). Similarly, the methods for CO_2 and Chl *a* are reported by Bakker *et al.* (1997) and Bathmann *et al.* (1997), respectively.

	ACC (southern branch)	Polar Front
Section 5		
Fe (nM)	0.49	1.87
Chl *a* ($\mu g\,kg^{-1}$)	0.25	0.74
f_{CO_2} (μatm)	356	339
Section 11		
Fe (nM)	0.31	1.14
Chl *a* ($\mu g\,kg^{-1}$)	0.24	0.33
f_{CO_2} (μatm)	358	321

measured as POC and PON (particulate organic nitrogen) (Bathmann *et al.*, 1997) or Chl *a* (Fig. 4.7). Simultaneously, the distributions of particulate and dissolved ^{234}Th had shown pronounced scavenging of ^{234}Th on the biogenic particles. This supports the contention of similar removal of dissolved iron, by uptake into or adsorption onto planktonic particles. This is partly confirmed by the large percentage of easily leachable particulate iron on suspended particles, compared with the more southerly oligotrophic ACC waters (Fig. 4.8). Otherwise, the total particulate Fe was significantly higher at the Polar Front, as was the total particulate Al determined in the same samples (Löscher *et al.*, 1997). The strong signal of particulate Al suggests a continental source term for this Al, as well as for the particulate and dissolved Fe. Similarly, higher concentrations of lithogenic silica were also found here (Quéguiner *et al.*, 1997). From the coincidence of these features with the hydrography of the frontal jet and eddies (Veth *et al.*, 1997), a source term within the oceans is suggested. Recent contact of these waters with continental margin sediments of the Argentine basin is conceivable (de Baar *et al.*, 1995; Löscher *et al.*, 1997).

This upstream margin source is also supported by similar, but weaker, particulate Al maxima in the boundary region (at *c.* 56°S) between the ACC and the Weddell gyre. At this ACC – Weddell Gyre Boundary (AWB) at 6°W the particulate Al reaches values as high as *c.* 12 nM (Fig. 4.9). The stronger expression of particles in this AWB is more westward at the Weddell–Scotia Confluence studied previously (Fig. 4.5). This increasing intensity towards the

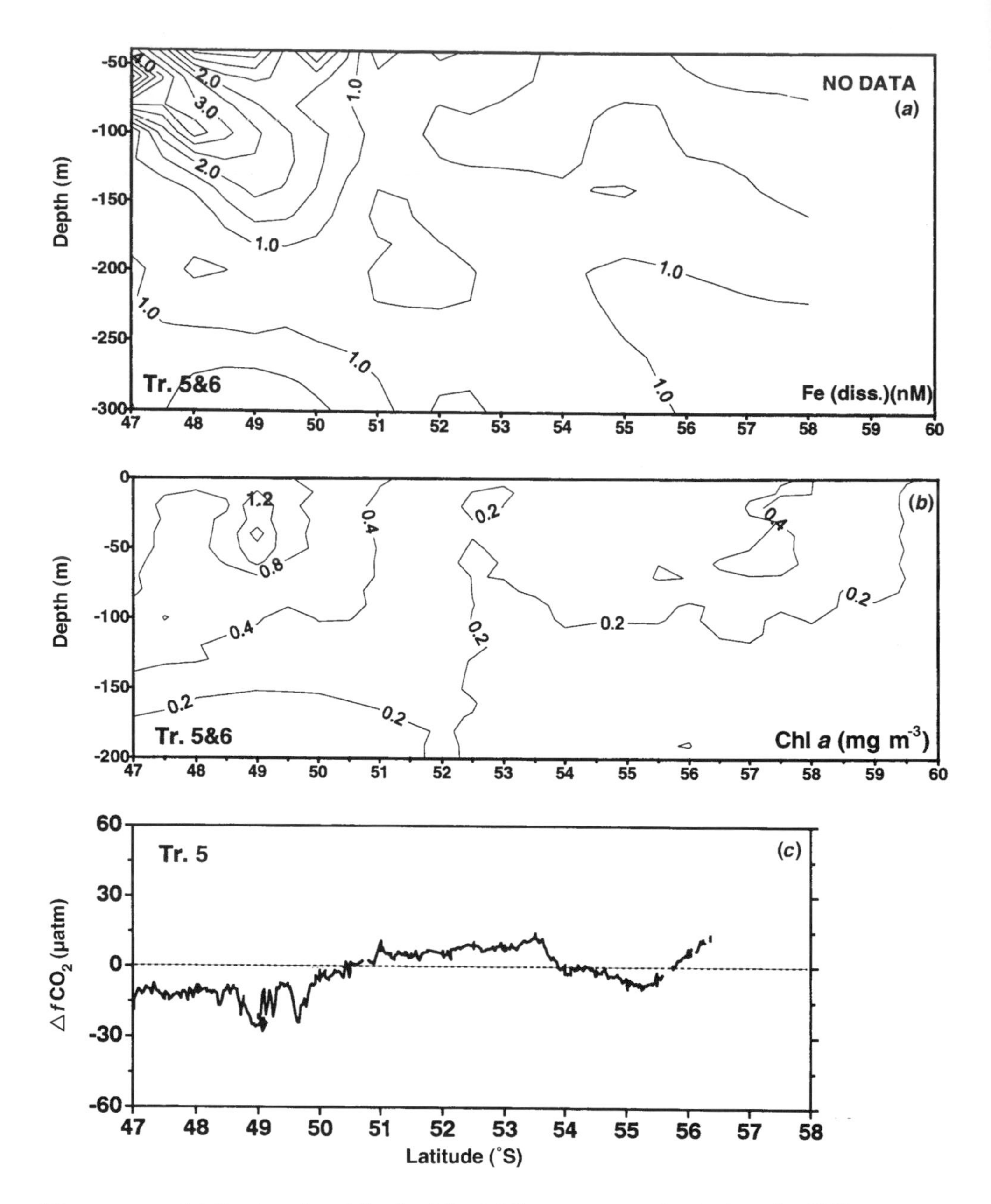

Figure 4.7 (a) *Section plot of dissolved Fe (nM) at 40–300 m depth along the 6°W meridian from 47 to 58°S, observed in the period 24 October–6 November 1992, JGOFS ANT X/6 transects 5/6 combined.* (b) *Section plot of chlorophyll* a *(μg kg^{-1} seawater) at depth 0–200 m and* (c) *difference in CO_2 partial pressure between surface waters (depth 10 m) and marine air at the same section. Taken from de Baar* et al. *(1995) with permission of* Nature, *MacMillan Press.*

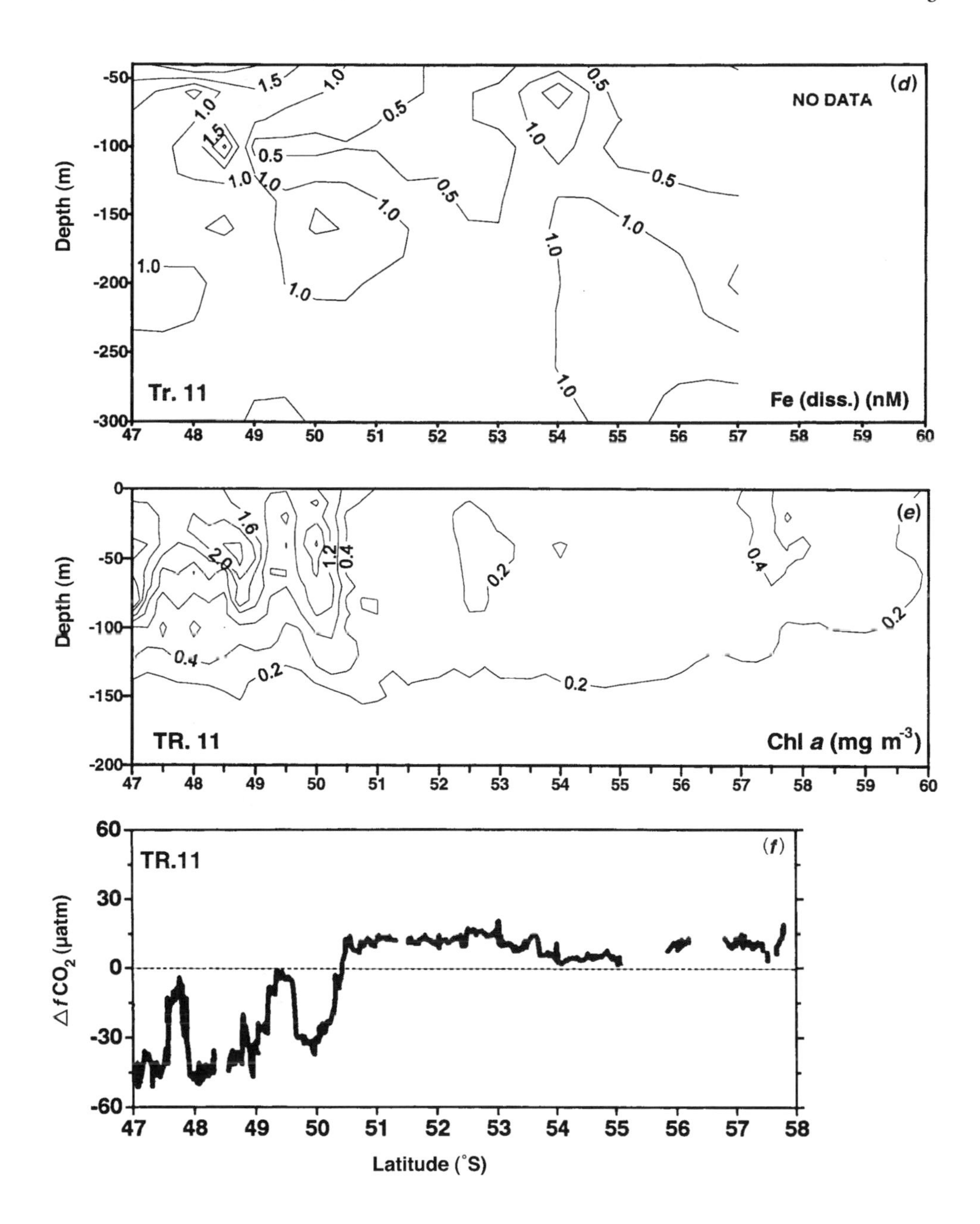

(d–f) As (a–c) but now for the period 10–21 November 1992, final transect 11. (d) Dissolved Fe (nM), (e) Chlorophyll a *(μg kg*$^{-1}$ *seawater), and (f) difference in* CO_2 *partial pressure. See Fig. 4.6 for chronology of transects, numbered 2–11, all at 6°W latitude. Taken from de Baar* et al. *(1995) with permission of* Nature, *MacMillan Press.*

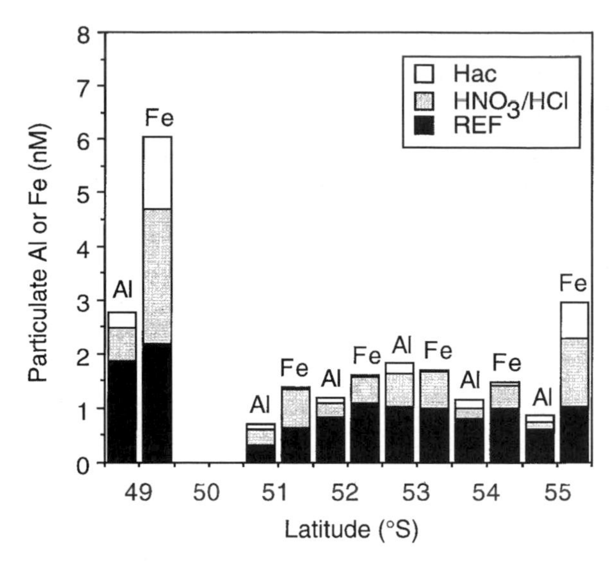

Figure 4.8 *Particulate Fe and Al in easily leachable (Hac), dissolvable (HNO_3/HCl) and refractory (REF) portions of total particulate matter collected at depth 40 m, section 5 of JGOFS 1992 ANT X/6. Taken from Löscher* et al. *(1997).*

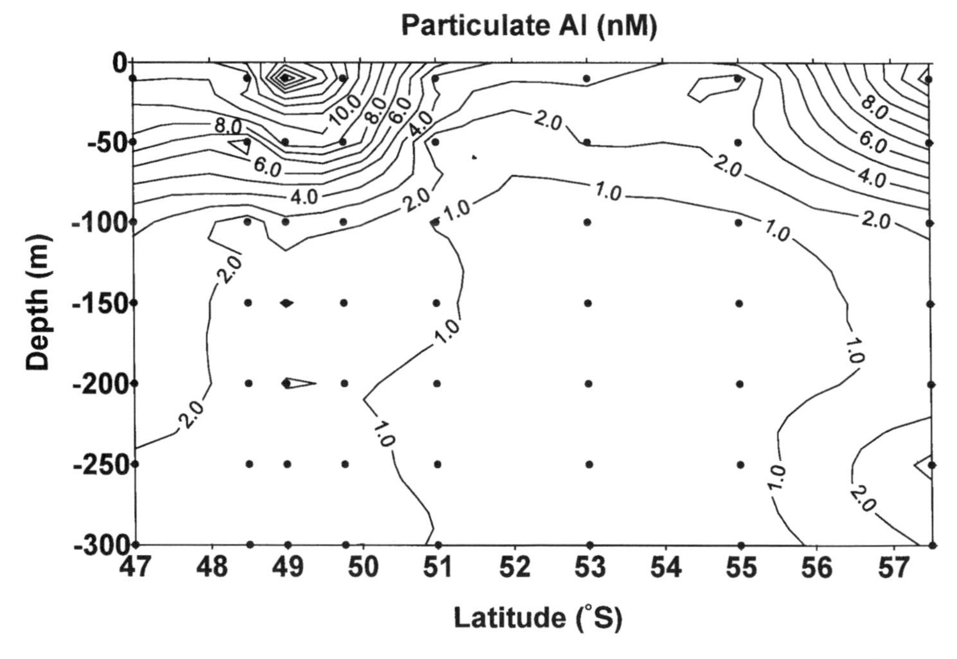

Figure 4.9 *Total particulate Al distribution from samples collected at depth 50 m at the same stations of section 5 of JGOFS ANT X/6 with different samplers at different hydrocasts and analysed with independent methods by Dr F. Dehairs (Brussels). The trend towards higher values at the Polar Front, consistent with that of particulate Al and Fe in Fig. 4.8, confirms the overall reliability of various independent methods. From Löscher* et al. *(1997).*

west is consistent with the shelves around the Antarctic Peninsula acting as the source for both dissolved Fe and particulate Al and Fe.

For the Polar Front, alternatively, an aeolian input may be invoked. However, from the then existing published estimate of aeolian input (Duce & Tindale, 1991) an upper limit of the Fe input of 3 mg m^{-2} yr^{-1} would upon 10% dissolution provide a five-fold lower input than the supply from below by upwelling (de Baar *et al.*, 1995). Recently, from careful dating of sediment cores, a modern, Holocene sedimentation rate of Fe was calculated (Kumar *et al.*, 1995). In the Polar Front this was as high as *c.* 50 mg m^{-2} yr^{-1}; in more southerly regions, rates of *c.* 30 mg m^{-2} yr^{-1} were determined. Firstly, these deposition rates for the Holocene are an order of magnitude higher than the present-day estimates of Duce & Tindale (1991). This apparent discrepancy probably reflects the uncertainties arising from inevitable assumptions inherent in the calculations of both estimates. Secondly, the higher inputs into the Polar Front itself, and the more northerly Polar Frontal Zone, were ascribed to aeolian input from the Patagonian Desert. Indeed, independent surface-water sections of total Al in the western South Atlantic basin tend to show concentrations higher (*c.* 8 nM) than the *c.* 2 nM found in the eastern part of the South Atlantic basin (Helmers & Rutgers van der Loeff, 1993). This also hints at a South American source term, although a higher removal rate of Al in the east part may also be responsible for the east–west difference. The Polar Frontal Zone ranges between, and includes, the Subantarctic Front and the Polar Front (i.e. from *c.* 40° to *c.* 50°S in Fig. 4.2) in the Atlantic sector. It may well receive significant aeolian Fe input. Hence this PFZ may be richer in productivity, and serve as a strong CO_2 sink in the Southern Ocean. Obviously, neither the suggested aeolian input nor the suggested diagenetic margin sediment source term of the high Fe in the Polar Front is well quantified, and further investigations are needed.

At the end of the 1992 spring survey, major blooms of large diatoms *Corethron criophilum* and *Fragilariopsis kerguelensis* had developed at the Polar Front. The total CO_2 had decreased by about 10–15 μM and silicate by about 10 μM compared with section 5/6 some 22 d before (not shown). An increase of particulate biogenic silica (opal) was also observed (Quéguiner *et al.*, 1997). The decrease in partial pressure of CO_2 was very distinct, with values as low as *c.*40–60 μatm undersaturation (Fig. 4.7). This demonstrates that iron-supported photosynthesis is more than counteracting the expected increase in p_{CO_2} due to seasonal warming of surface waters. Bakker *et al.* (1997) observed an average daily decrease of p_{CO_2} at 0.8 μatm d^{-1} over the 20.8 d period between the earlier transect 5 and the final transect 11. This change would have been almost twice as large, if a rise in water temperature of 0.7 °C had not counteracted the decrease with an increase of 0.5 μatm d^{-1}. In addition, CO_2 was supplied from the atmosphere at an average rate of 0.09 μatm d^{-1}. Hence, the daily residual change due to net biological uptake and removal was calculated to be about 1.40 μatm d^{-1}. Using the equation of the regional Revelle factor

$$\beta = \delta \ln(p_{CO_2})/\delta \ln(T_{CO_2}), \qquad (4.8)$$

after Bakker *et al.* (1997), the corresponding decrease in total dissolved inorganic carbon (T_{CO_2} or DIC) is 0.62 μM d^{-1} or 12.9 μM over the 80 m depth of the mixed layer in the 20.8 d period. This is in fair agreement with the above-mentioned observed decrease in T_{CO_2} between sections 5 and 11. Integrated over an upper mixed layer of depth 80 m this amounts to net T_{CO_2} uptake of 1.06 mol C m^{-2} in 20.8 d. This trend is consistent with the high and seasonally increasing primary productivity, at values as high as *c.* 1200–3000 mg C m^{-2} d^{-1}, at the Polar Front (Jochem *et al.*, 1995). The latter measured values are from bottle incubations and are deemed to be somewhat lower than the true primary production, owing to some respiration in the bottles. Integrated over the 20.8 d between sections 5 and 11, the measured primary productivity in the region amounted to 2.02 mol C m^{-2}. This is about twice the net T_{CO_2} removal, i.e. about half of the measured primary productivity is remineralised again. The intense photosynthesis in these waters is also apparent from the significant oversaturation of dissolved oxygen in the surface waters at the Polar Front (Bakker *et al.*, 1997).

From ^{234}Th deficiencies in the blooms, Rutgers van der Loeff *et al.* (1997*a*) estimated an export removal of 0.43–0.86 mol C m^{-2} from the surface waters over the 20.8 d from section 5 to section 11. From measured POC at consecutive transects (Bathmann *et al.*, 1997) the net increase of biomass in the upper 80 m of the Polar Front was estimated at 0.66 mol C m^{-2} (Bakker *et al.*, 1997). Combining the 20.8 d biomass budget

$$\text{biomass} = (\text{photosynthesis} - \text{respiration})_{\text{algae}} - \text{regeneration} - \text{sedimentation} \qquad (4.9)$$

with the net T_{CO_2} decrease, one finds fair agreement of the changes (expressed as mol C m^{-2}):

$$\text{net } T_{CO_2} = \text{biomass increase} + \text{sedimentation}$$
$$= (\text{P} - \text{R})_{\text{algae}} - \text{regeneration}; \qquad (4.10)$$

$$1.06 = 0.66 + (0.43 \text{ to } 0.86) = (2.02)_{\text{algae}} - \text{regeneration} \qquad (4.11)$$

This agreement is encouraging, but the inevitable intrinsic sources of error must be realised. Firstly, the heterogeneity due to dynamic meandering of the Polar Frontal jet (Veth *et al.*, 1997) complicates the simple comparison of water masses between consecutive sections. Secondly, there is some uncertainty in the underlying estimate of the C: ^{234}Th ratio in the exported material compared with that in the standing stock (see Rutgers van der Loeff *et al.*, 1997*a*).

Nevertheless, the net T_{CO_2} loss of 1.06 mol C m^{-2} over 20.8 d, by and large equating with the net biogenic uptake of 1.09 – 1.52 mol C m^{-2}, can be compared with the net loss of dissolved Fe between sections (Fig. 4.7). Brand (1991) found a cellular iron requirement ranging from 10^{-4} to $10^{-5.1}$ for Fe : C of neritic eukaryotic algae. Muggli *et al.* (1996) reported a ratio from *c.* $10^{-3.5}$ to $10^{-4.3}$ for a large oceanic diatom in Fe-replete and Fe-stressed conditions, respectively. Assuming a ratio of 5×10^{-5} and the 80 m upper water column, the calculated net iron loss of 0.66 nM or net biogenic uptake of 0.68–0.95 nM over 20.8 d is in the order of magnitude of the observed decrease of 0.73 nM over 22 d between the average measured concentrations of 1.87 and 1.14 nM of sections 5/6 and 11, respectively (Fig. 4.7). Again, this simple calculation may not be quite correct. For example, the Fe : C ratio of exported matter is probably different from that of the suspended matter. Nevertheless these straightforward budget estimates are in keeping with the notion of either neritic diatoms or large oceanic diatoms (both with high cellular iron requirements) thriving because of the iron supply and more suitable light climate (Jochem *et al.*, 1995; Lancelot *et al.*, 1996; Quéguiner *et al.*, 1997) of the Polar Front. Here the blooms appear to be the major driving force for transporting CO_2 from the atmosphere, via the surface waters, into the underlying intermediate and deep waters. The Polar Front also being the site of Antarctic Intermediate Water (AAIW) formation, it may be assumed that a combined biological–physical pump for driving C into the ocean interior is operating here.

At the time of writing it is not clear, however, whether this one-off observation of spring 1992 is typical for the Atlantic Polar Front. Firstly, the trend towards decreasing dissolved Fe with ongoing spring season would suggest lower Fe concentrations in summer. Lower dissolved Fe was indeed recently observed later in the summer (January 1996) and more towards the east (region 10 in Fig. 4.2) at the Polar Front (de Jong *et al.*, 1998). This would be analogous to the classical north Atlantic spring bloom of diatoms. In both cases, the better light climate in spring would trigger the bloom, which eventually succumbs owing to the lack of silicate or iron in the NE Atlantic and Antarctic Polar Front, respectively. Secondly, in 1992 the abundance of icebergs was anomalously high in the Polar Front. This may or may not relate to the interannual Antarctic oscillations (White & Peterson, 1996) of surface temperature, wind velocity and sea-ice cover, all factors of crucial importance for phytoplankton blooms and CO_2 exchange. Additional observations in other seasons and years, as well as other Antarctic regions, are needed before concluding that the Atlantic Polar Front is one of the major CO_2 sinks of the Southern Ocean.

Nevertheless these 1992 observations have shown that a more or less continuous supply of Fe is necessary for plankton blooms and CO_2 drawdown in

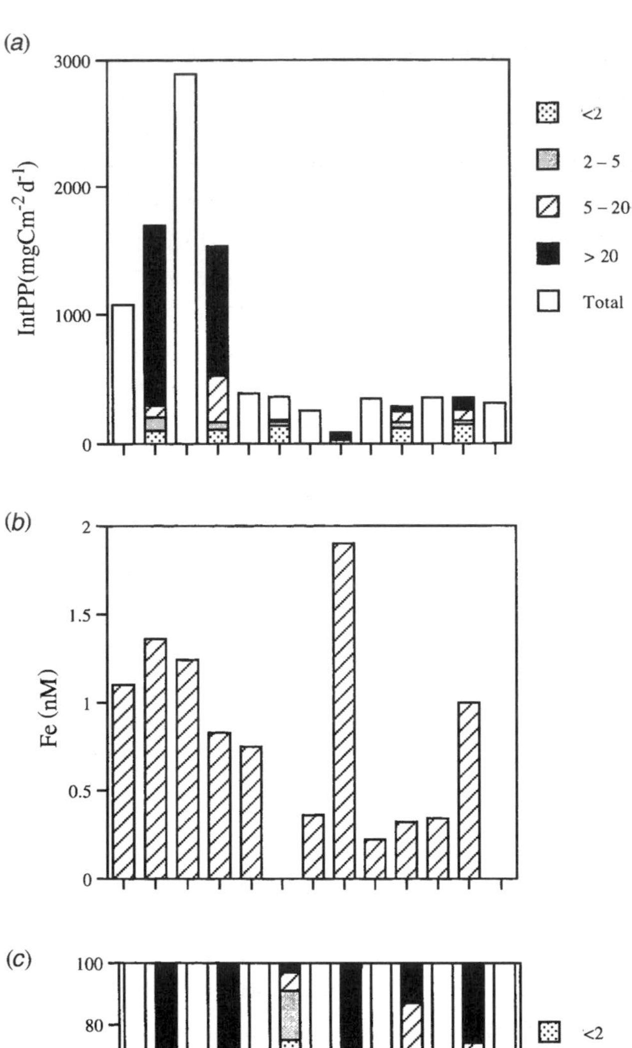

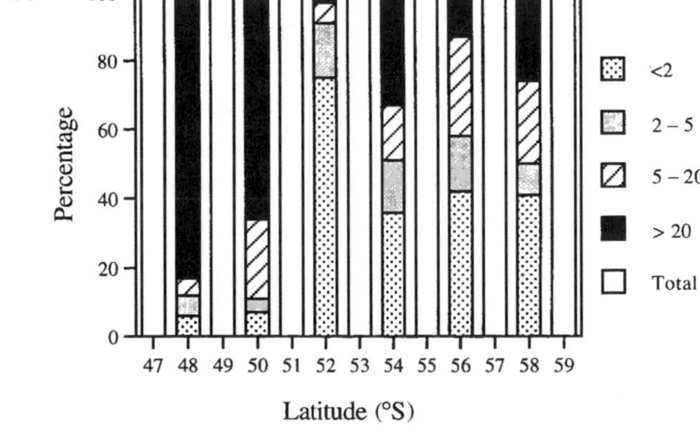

Figure 4.10 (a) *The size-fractionated primary production (IntPP) integrated over the water column at stations of even latitude along the final section 11 of the 6°W research transect. Also shown is the total integrated primary production at the odd-latitude stations, where no size fractionations were done. Redrawn after Jochem* et al. *(1995).* (b) *The mean value of dissolved Fe in the upper 120 m of the water column of the same section 11. Note the higher concentrations at (48–50°S) and north (47°S) of the Polar Front, coinciding with the*

the Southern Ocean (de Baar *et al.*, 1995). The evidence presented above thus far is strictly based on natural rate and state variables, without the need to extrapolate from artificial perturbation experiments, whether *in situ* or in bottles. Some 60 years after the first statement (Gran, 1931; Hart, 1934) the Antarctic iron hypothesis for low productivity in offshore waters and enhanced productivity in neritic waters (Buma *et al.*, 1991) was proven to be correct (de Baar *et al.*, 1995). Some four years after its first statement (Martin & Gordon, 1988), the iron hypothesis for concomitant CO_2 drawdown of the Southern Ocean was also found to be largely correct. The additional insights about the distinction between large diatoms and small phytoplankton are discussed below.

Synergies of light and iron and the size of organisms in the food web

The iron hypothesis had been suggested for several HNLC regions: the subarctic Pacific, the equatorial Pacific and the Antarctic Ocean. In all regions loss due to grazing also plays a major role (Miller *et al.*, 1991; Frost, 1991; Buma *et al.*, 1991; Dugdale & Wilkerson, 1990; Frost & Franzen, 1992; Banse, 1992, 1995). However, the Antarctic Ocean is special in that light limitation is probably much more pronounced owing to very deep wind mixing (60–140 m is common even in spring and summer), in combination with dense cloud cover and extended seasonal ice cover (Tranter, 1982; Mitchell *et al.*, 1991; Nelson & Smith, 1991; Lancelot *et al.*, 1993, 1996; de Baar, 1994). Obviously, light is a major limiting factor on its own for the Southern Ocean. Moreover, it has additional synergistic implications as under light stress the cellular iron requirement is predicted (Raven, 1990) and reported to be higher (Muggli & Harrison, 1996; Kudo & Harrison, 1997; van Leeuwe & Stefels, 1998; Stefels & van Leeuwe, 1998; van Leeuwe, 1997).

The observed seasonal increase of size-fractionated productivity at the PF was almost exclusively due to the largest size classes (diameter over 20 μm; Jochem *et al.*, 1995) of phytoplankton, i.e. the microplankton range of the large diatoms presumably thriving at adequate iron supply (Fig. 4.10). Very clearly,

Caption for Figure 4.10 (*cont*).

above dominance of the largest size fractions in productivity. No dissolved (filtered) data were available at 52°S and 59°S. Sea-ice cover was present at 58°S and 59°S. High dissolved Fe at 54°S is ascribed to recent sea-ice melting, but not accompanied by high integrated production owing to most unfavourable wind conditions. Data from Löscher et al. *(1997). (c) Percentage distribution of size-fractionated primary productivity. At the Polar Front (48–50°S), note the dominance of the largest size fraction. Such a tendency also occurs at 54°S and 58°S, where dissolved Fe is also higher but there is no bloom, owing to an unfavourable light climate as the result of wind mixing (54°S) and ice cover (58°S). Redrawn after Jochem* et al. *(1995). Nominal size cutoffs range from 2 to 20 μm.*

the largest size class of greater than 20 μm is dominant only in those waters where dissolved Fe exceeds an apparent threshold level of about 0.6 ± 0.2 nM. However, iron is not the only factor. At this frontal system the admixture of warmer subpolar waters from the north causes the upper water column to be more stable (Fig. 4.11), thus retaining the phytoplankton within the euphotic zone (Jochem *et al.*, 1995; Lancelot *et al.*, 1996). This will favour all phytoplankton size classes, but especially the larger microplankton, which reportedly may have a lesser quantum absorption efficiency (Morel & Bricaud, 1981; Jochem *et al.*, 1995). The stability (Jochem *et al.*, 1995) of the upper water column provides the more favourable light regime in the Polar Front (48–50°S). Figs. 4.10 and 4.11 jointly illustrate the natural *in situ* expression of the three terms in the earlier Equation 4.5 where the light climate is critical and limiting for all size classes. The ambient Fe appears adequate for the smallest size class (less than 2 μm), but apparently becomes severely limiting when below about 0.6 ± 0.2 nM for the largest size class, over 20 μm. Indeed when Equation 4.4 is combined with vertical profiles of light intensity derived from the wind-mixed layer (WML) depth in a simulation model at each degree latitude, the spring evolution at the Polar Front of observed large diatom blooms was supported (Lancelot *et al.*, 1996). Otherwise the strong predominance of large diatoms over picoplankton in the blooms is consistent with the rapid response of microzooplankton keeping the picoplankton grazed down (Becquevort, 1997). On the other hand, the large mesozooplankton grazers had virtually no impact on the bloom (Dubischar & Bathmann, 1997). It is generally understood (see, for example, Banse, 1995) that large grazers apparently take more time to accumulate in response to a diatom bloom (Jochem *et al.*, 1995).

For overall productivity versus iron (Fig. 4.10), there is one notable exception of the high dissolved Fe at the 54°S ice-edge-melting event of section 11. Here the strong wind mixing had caused the WML depth and light regime to deteriorate, hence reducing productivity. This was discussed before and is further illustrated by the very low Stability Index at this 54°S site (Fig. 4.11).

Figure 4.10 would suggest that the threshold for phytoplankton blooming is at about 0.6 ± 0.2 nM Fe, above which blooms may or may not occur when light is or is not favourable. However, at all sites the local light regime is less than the optimal *c.* 100 μmol photons $m^{-2} s^{-1}$ for Antarctic plankton communities, as determined on shipboard from P–I curves. In other words, at optimum light (i.e. not encountered here for the *in situ* wind-mixed layers ranging from 60 to 140 m in depth), the true threshold for diatom blooms would conceivably be somewhat lower than *c.* 0.6 nM, say Fe *c.* 0.4 nM. Low wind velocities of 4–6 m s^{-1} would favour such an optimum depth of the mixed layer. For mixed layers of *c.* 20 m only, in relatively nearshore waters with [Fe] > 1 nM, blooms are then certainly to be expected. In contrast, in offshore waters, and with strong

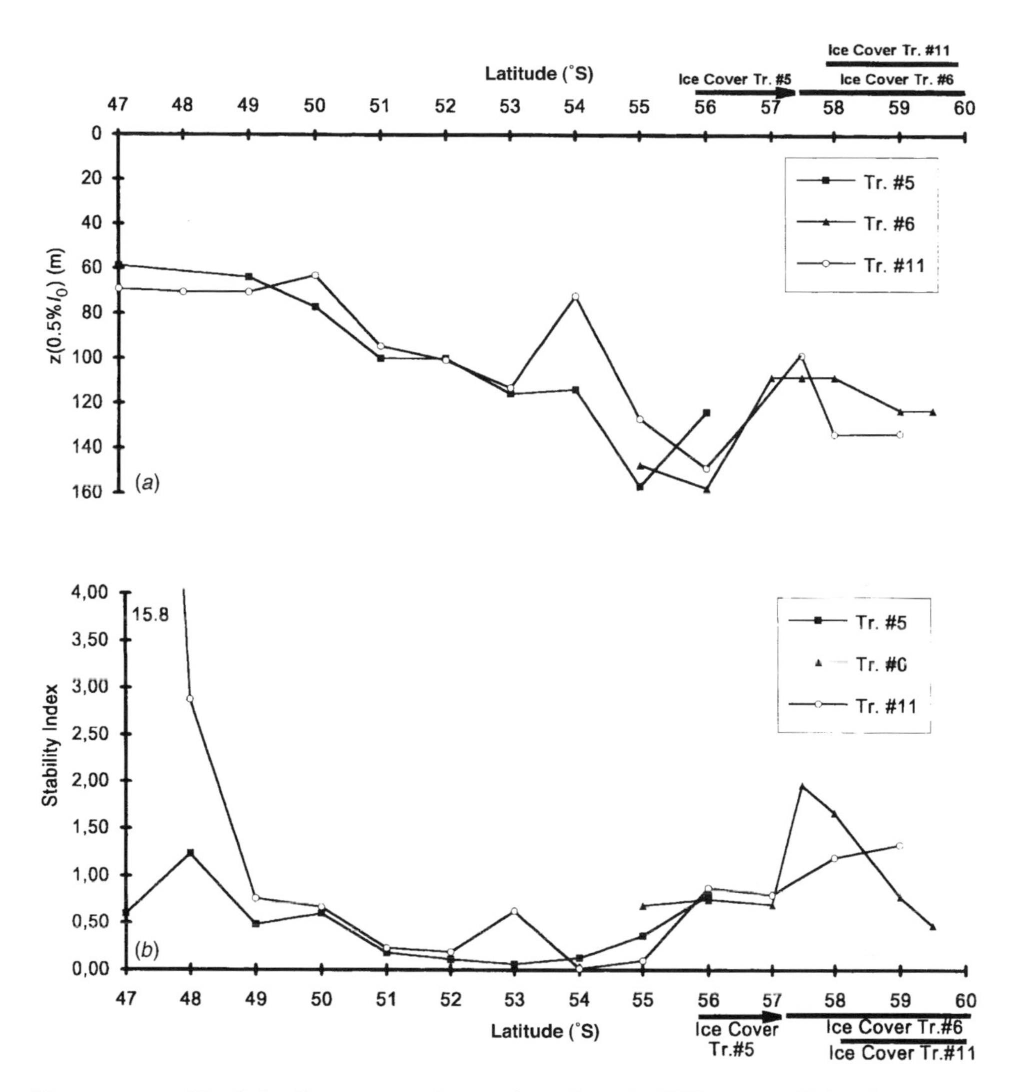

Figure 4.11 *The light climates at various stations along the 6°W transect. Taken from Jochem* et al. *(1995). (*a*) Depth of the 0.5% irradiance level (I_o) at transects 5, 6 and 11. The depth of the euphotic zone increases from 60–70 m at the Polar Front to 140–160 m in southern ACC waters. (*b*) The Stability Index at consecutive sections 5, 6 and 11. The Stability Index was derived from the observed density structure at the depth of the 0.5% I_o according to the equation: $SI = (\Delta G_{(\theta)})/\Delta z) \times 1000$.*

summer stratification (e.g. off the Ross Sea with mixed layers also of *c.* 20 m; see $\sigma\theta$ in Fig. 5a of Martin *et al.*, 1991*a*), one might occasionally observe the demise or lack of blooms solely owing to ambient iron being less than the threshold value of, say, *c.* 0.4 nM. However, in such hydrographic conditions, the upward supply of other nutrients (nitrate and silicate) is also suppressed, and when these are also rapidly removed, these major nutrients may become limiting as well (Nelson & Tréguer, 1992). Intense stratification and silicate depletion have most often been reported for the Ross Sea and may provide an ecological niche for other species, notably *Phaeocystis* sp., which blooms in the Ross Sea.

To disentangle the effects of light and iron on phytoplankton growth requires controlled laboratory experiments of essential diatom species. Unfortunately, the large diatoms *Corethron criophilum* and *Fragilariopsis kerguelensis* have proven to be most difficult to maintain in culture. Monod growth curves as function of iron at various light levels have yet to be achieved. During the expedition, several efforts to obtain pure strains of either species failed (R. Scharek, M. A. van Leeuwe & R. M. Crawford, unpublished results). However, a southern coastal diatom, *Thalassiosira antarctica*, and the oceanic diatom *Proboscia inermis* (both taken from the collection of the Alfred Wegener Institut) in shipboard incubations (100 μmol $m^{-2} s^{-1}$ in 16–8 h light–dark cycle) in local filtered seawater showed a very different response to addition of 2 nM iron (R. Scharek, M. A. van Leeuwe & H. J. W. de Baar, unpublished results). The coastal species responded very strongly to Fe addition, whereas in the controls it had a very slow division rate of once per 8 d. In contrast, the oceanic *Proboscia inermis* grew very rapidly with equal rates in the control and the Fe 2 nM enrichment. These findings for polar diatoms are consistent with those reported by Sunda *et al.* (1991) for diatoms of temperate waters. Previously, Buma *et al.* (1991) had found the largest diatoms to thrive upon significant Fe enrichment (see caption to Fig. 4.4). Recently, Muggli *et al.* (1996) found a high cellular Fe requirement of a subarctic diatom, *Actinocyclus*, which is oceanic in character, as further discussed below. In addition, in incubations at the Polar Front, the microscopic observations showed enhancement of *Fragilariopsis kerguelensis* upon Fe enrichments (described below).

MESOCOSM INCUBATIONS During 1992 JGOFS the response of the plankton community to iron enrichment was studied in a suite of shipboard incubations in 20 l mesocosms (van Leeuwe *et al.*, 1997; Scharek *et al.*, 1997). Near the Polar Front triplicate enrichments with 2 nM Fe showed a significant, reproducible, response compared with triplicate untreated controls (Fig. 4.12). The excellent reproducibility of enrichments and controls confirmed the overall reliability of clean procedures. The initial ambient dissolved Fe concentration was relatively high at 0.9 nM, but an additional 2 nM was still observed to stimulate the growth of the larger diatoms over the first 9 d of the experiment (Fig. 4.12). From simultaneous uptake experiments with radiotracer ^{55}Fe, the rate of Fe assimilation was estimated to be about 0.16 ± 0.06 nM (μg Chl *a*)$^{-1}$ d^{-1} (van Leeuwe *et al.*, 1997). Combining the *in situ* seasonal trend of increasing Chl *a* and decreasing Fe (Figs. 4.7 and 4.10), one would expect that in the control the initial 0.9 nM would have strongly decreased within days, leading to suboptimal growth of the diatom-dominated community. In the 2 nM Fe

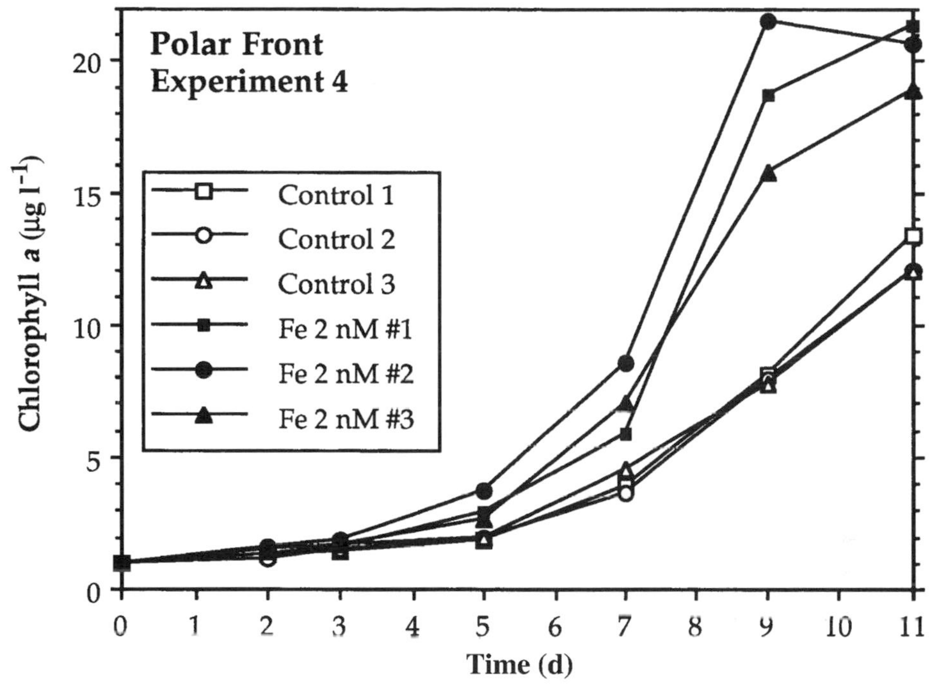

Figure 4.12 *The effect of 2 nM Fe addition (at* t = 0*) on the growth of plankton (as Chl* a*) in three replicate enrichments and three untreated controls. Mesocosms (20 l) under optimal light (*c. *100 µmol* m^{-2} s^{-1}*) in 16–8 h light–dark cycle. Taken from van Leeuwe* et al. *(1997). Microscopic counts at t = 8 showed significant Fe enhancements of* Fragilariopsis kerguelensis *and* Pseudonitzschia heimii/lineola *but for* Corethron *spp. and* Thalassionema *spp. the cell numbers were about the same in controls and enrichments (R. Scharek, M. van Leeuwe and H. de Baar, unpublished results). At t = 0 all these diatom species had been present as the dominant* in situ *plankton forms of the 'mixed-species' plankton bloom. (N.B. By the final day, t = 11, the silicate had largely disappeared from the Fe enrichments such that Si rather than Fe had eventually become the major limitation in those bottles.)*

enrichments, the growth should remain optimal for several more days (Fig. 4.12). This would explain the lower net increase (0.35 d^{-1}) for Chl *a* in the controls compared with 0.45 d^{-1} in the 2 nM Fe enrichments, as calculated from Fig. 4.12 (van Leeuwe *et al.*, 1997). Apparently these diatom species do require Fe of the order of 0.4 nM or more for optimal growth, as also discussed above based on observations *in situ* (Figs. 4.10 and 4.11).

At day $t = 8$ of this experiment, sub-samples were used to assess the rates of 'true' gross production versus microzooplankton grazing with a dilution experiment after Landry & Hassett (1982). In analogy to Equation 4.1 these rates are expressed as:

net growth = gross production − microzooplankton grazing; (4.12)

$$\mu = k - g. \quad (4.13)$$

In the control, the gross production rate (k) was 0.71 d^{-1} and the grazing rate, g, was 0.34 d^{-1}. In the 2 nM Fe enrichment, $k = 0.89$ d^{-1} and $g = 0.30$ d^{-1} (Scharek *et al.*, 1997). Thus, gross production rate appeared to be significantly higher in the Fe enrichment, whereas grazing rates were about the same. This is consistent with the observation of iron mostly stimulating the larger size classes (diatoms) and not the smaller algae, which remain to be kept in check by the microzooplankton. The resulting net growth rates, μ, at 0.37 d^{-1} in the control and 0.59 d^{-1} in the enrichment matched quite well the above mentioned net rates of Chl *a* synthesis of 0.35 d^{-1} and 0.45 d^{-1}, respectively, over the previous days.

In retrospect the optimal k term at 0.89 d^{-1} under optimal light and iron conditions is in keeping with the expectation from the experimental literature (Eppley, 1972) of *c.* 0.9 d^{-1} for optimal growth at 1 °C (Banse, 1991*c*, his Fig. 3). The net growth rates of 0.37 d^{-1} and 0.59 d^{-1} in the controls and enrichments, respectively, are similar to the values previously reported (de Baar *et al.*, 1990) in the neritic waters of the Weddell and Scotia Seas (see Banse, 1991*c*, his Table 1). In latter experiments the division rates of the diatoms are again somewhat higher (Buma *et al.*, 1991, their Table 2), most notably for the largest diatoms approaching the value above *c.* 0.89 d^{-1}. Thus, it appears there was virtually no grazing loss for the largest diatoms, in both the 1988 and 1992 studies. For the 1992 shipboard incubations, no systematic assessment of the size classes was done, but the complementary field observations (Fig. 4.10) of the size class over 20 μm are consistent with the notion of Fe mostly affecting the largest size classes.

In this, and in four other incubation series, the addition of iron always led to enhanced synthesis of Chl *a* and more rapid nutrient uptake (van Leeuwe *et al.*, 1997). The response was generally least in experiments starting in the low-iron environment of the southern ACC branch, with lowest initial biomass of the plankton community (Scharek *et al.*, 1997). This is consistent with the notion of these communities consisting largely of smaller size classes of algae, hardly being iron-stressed, and responsible for most of the primary productivity *in situ* (Jochem *et al.*, 1995; see also Fig. 4.10). In contrast, the experiments starting with intermediate or high initial biomass (Fig. 4.12) constitute a significant portion of large diatoms able to respond quickly to iron enrichment. Similarly, the short-term ^{55}Fe radiotracer uptake in general confirmed a high cellular iron quota for the iron-replete natural communities with high initial biomass at the Polar Front, and a low cellular iron quota for Fe-depleted communities with low initial biomass in the southern ACC region (van Leeuwe *et al.*, 1997).

Parallel experiments with enrichments of other essential trace metals, Mn, Co and Zn, also in various combinations with iron to probe conceivable synergies, in general showed less stimulatory effect. This confirms that iron is the most crucial trace element for phytoplankton growth in the Antarctic Ocean (Scharek *et al.*, 1997). Similarly, de Baar *et al.* (1990) and Buma *et al.* (1991) reported a lesser effect for Mn, Cu and Zn for the Weddell and Scotia Seas.

Upon iron enrichment, the growth of Chl *a* and POC was enhanced, and the major nutrients nitrate, phosphate and silicate were taken up more rapidly. This was also shown by enhancement of ^{15}N-labelled nitrate assimilation (van Leeuwe *et al.*, 1997). However, the distribution of ^{14}C-labelled bicarbonate into different biochemical pools (lipids, polysaccharides, proteins, low-molecular-weight molecules (LMWM)) remained in the same relative proportions (van Leeuwe *et al.*, 1997). In other words, under iron-stressed or iron-replete conditions the algae responded by slowing or accelerating, respectively, the overall rate of growth, while maintaining uniform biochemical composition (i.e. C: N ratio, fatty-acid composition, proteins and lipids). There are only two exceptions to this uniform composition. Firstly, there is an enhancement of Chl *a* synthesis by iron. Secondly, the assimilation of phosphate appears to be unaffected by iron (de Baar *et al.*, 1997). The ensuing non-uniform trends of Chl: C, C: N and N: P ratios had been observed before, albeit with different absolute values (de Baar *et al.*, 1990; Martin *et al.*, 1991*a*). In other words, upon Fe stimulation the Chl *a*: POC ratio of the cell increases. In addition, the cells are allowed to assimilate nitrate at a virtually ideal Redfield N: P ratio as opposed to a lower ratio in the Fe-depleted incubations (de Baar *et al.*, 1997). The latter effect is probably due to impairment of the Fe-containing enzymes of nitrate and nitrite reductases (Timmermans *et al.*, 1994). In general, during evolution of phytoplankton blooms *in situ*, the ambient Fe concentration is expected to decrease. This leads to less efficient nitrate assimilation, and hence to increasing ratios of ambient nitrate: phosphate, as indeed observed at bloom stations (not shown). This, in combination with AAIW formation and water mass transport, may be one of the mechanisms for maintaining an overall lower Redfield ratio of *c.* 14 in the Antarctic Ocean (de Baar *et al.*, 1997).

Indian Antarctic and Australian Subantarctic regions

INDIAN OCEAN SECTOR During the February 1994 French JGOFS *Antares II* expedition the total (unfiltered) Fe was determined in the upper 250 m along the 62°E meridian (Fig. 4.2). The total dissolvable Fe varied between 0.4 and 6.2 nM, with a tendency to higher values with increasing depth from the surface waters into the Circumpolar Deep Water (Sarthou *et al.*, 1997).

AUSTRALIAN SUBANTARCTIC WATERS Sedwick *et al.* (1997) report values for Fe and Mn at 4 stations along the 140°W meridian, from a Subantarctic station at 53°S to a subtropical station at 40°S, as well as at an additional station over the continental slope off northwestern Tasmania (Fig.

4.2). After filtration (0.2 μm) the dissolved Fe concentration was 0.1–0.2 nM in Subantarctic surface waters and increased to 0.8 nM in subtropical waters and over the continental slope. The unfiltered total dissolvable Fe was higher in the range 0.17–1.3 nM, and sometimes more than twice the dissolved Fe in the same samples. For total Fe, a similar increase was found in subtropical waters and towards the continental slope. This was also observed for dissolved Mn, which increased from 0.5 nM in the Subantarctic to more than 2 nM in shallow waters over the continental slope. The coinciding trends of dissolved Fe and Mn, as well as total Fe, all increasing towards the continental slope, suggest that the sediments at the latter slope are the major source of Fe and Mn in these waters.

Pacific Sector of Antarctic Circumpolar Current and Fronts (1995 JGOFS)
Recently, a north–south transect at 89°W was sampled (Fig. 4.2), crossing the Subantarctic Front and Polar Front areas, the Antarctic Circumpolar Current (ACC) and closer to the continent, entering into the Bellingshausen Sea (RV *Polarstern* ANT XII/4 from 21.03.95 to 14.05.95). The 89°W transect is deemed to be very far away from any continental sources (reducing shelf sediments or aeolian dust input) of Fe, and was studied at the end of the austral summer season.

IRON AVAILABILITY It was possible to measure iron on shipboard, by using two different techniques newly implemented in the Southern Ocean. Flow injection analysis using in-line preconcentration on a chelating resin column followed by chemiluminescence detection (FIA–CL; Obata *et al.*, 1993) was employed to measure total dissolved iron in acidified samples. The very low concentrations ranged from 0.05 nM near the surface to 0.5 nM in deeper waters (Fig. 4.13), as expected for very remote waters far away from upstream and upwind sources on land (de Baar *et al.*, 1999). Along the transect (52°S–69°S), the dissolved iron showed enhanced concentrations in the Polar Front, as well as near the Antarctic continental margin (0.6–1.0 nM). In between, the southern ACC branch was depleted in iron; here the concentrations in surface waters were uniform at 0.1–0.2 nM. The organic speciation of FE(III) was measured with competitive ligand equilibration adsorptive stripping voltametry (Nolting *et al.*, 1998). The amount of iron-binding ligands was at least six times, and generally more than ten times, that of the total dissolvable Fe concentration. The conditional stability constant K'_{FeL} for the organic ligands complexing iron was calcualted to be between 1021 and 1022.4 at the 89°W sites, similar to those found recently in surface waters of other ocens. In the Southern Ocean this was sufficient to complex *c.* 82–98% of the available iron.

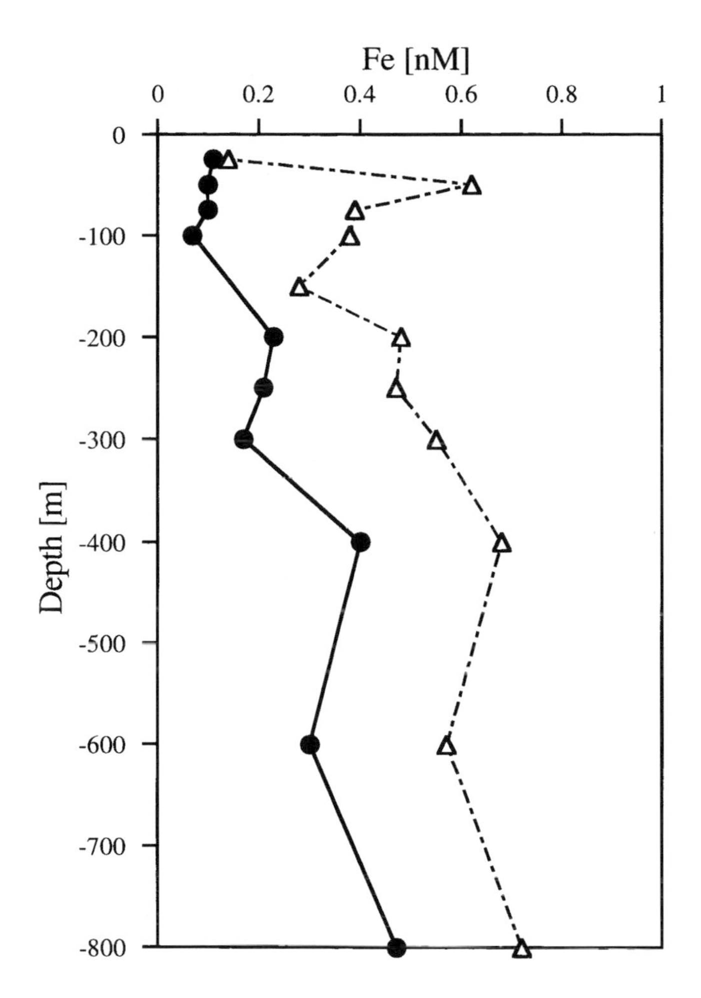

Figure 4.13 *Vertical distribution of dissolved (0.4 μm filtered) (circles) and total dissolvable iron (triangles) (nM) at 57°58'S, 91°50'W in the Pacific sector of the Antarctic Ocean (JGOFS ANT XII/4 at Polarstern, 20.03–15.05.1995; de Baar* et al., *1998). Part of a sampling programme at 10 stations at standard depths of 25, 50, 75, 100, 150, 200, 300, 400, 600 and 800 m.*

Obviously these conditions of low Fe availability in the Pacific Antarctic were close to those desired for studying HNLC mechanisms. The very deep wind-mixed layers, of the order of *c.* 100 m, were detrimental to photosynthesis (de Baar *et al.*, 1998). Timmermans *et al.* (1998), using bioassays under more optimal lighting, of highly Fe-stress-sensitive physiological indicators nitrate reductase,[15] N-nitrate assimilation and flavodoxin, have shown that the whole indigenous phytoplankton community is Fe-stressed. This was further confirmed by van Leeuwe *et al.* (1998*a*), who observed a decrease in fluorescence upon experimental Fe addition. Thus not only does Fe-stress severely restrict the larger microphytoplankton, but in addition the nanophytoplankton and picophytoplankton are Fe-stressed. Cell numbers as determined by flow cytometry were very low (van Leeuwe *et al.*, 1998*a*). At three stations around 57°S, well north of the Subantarctic Front (SAF), the average cell numbers were between 5000 and 7000 cells ml^{-1}, with a notable absence of diatoms (van Leeuwe *et al.* 1998*b*) owing to additional silicate limitation (de Baar *et al.*, 1999). At truly Antarctic stations south of the Polar Front the numbers of cells were lower at about 3000 cells ml^{-1}. However, for the southernmost stations (*c.* 68°S) near the southern Polar Front (sPF) there existed an additional dominant population of large diatoms as identified by a combination of HPLC photopigment analyses and micropscopy (van Leeuwe *et al.*, 1998*b*). This higher diatom abundance nicely coincided with higher levels of Chl *a* and the larger deficiencies in major nutrients and ^{234}Th (de Baar *et al.*, 1999), all indicative of recent diatom blooms. In addition, the somewhat shallower upper mixed layer at about 60 m would have been more favourable for plant growth. Previously plankton blooms had also been observed in this region during the austral spring period of 1992 (Turner *et al.*, 1995).

At the late summer time of the 1995 study, the deficiency of both light energy and Fe would explain the very low abundance of phytoplankton (Chl *a*) in the region and the conspicuous absence of plankton blooms. In the subantarctic waters north of the Polar Front the diatoms were furthermore prevented from blooming owing to the absence of dissolved silicate.

The northeast subarctic Pacific Ocean

Studies before JGOFS

Before the onset of the JGOFS programme, the pelagic biology of the NE subarctic Pacific region was comparatively well studied relative to other oceanic regions (Parslow, 1981; Frost, 1987; Banse, 1991*a*); a time series over more than 20 years exists for physical, chemical and biological measurements in this region. The time series of chlorophyll *a* and major nutrients in the upper water column at Ocean Station Papa (OSP, 50°N, 145°W in Fig. 4.1) demonstrate unequivocally that nitrate remains high and phytoplankton biomass remains low throughout the annual cycle (Frost, 1991; Wong *et al.*, 1994). This question had been perplexing scientists for many years. Indeed, McAllister *et al.* (1960) observed that the removal of larger zooplankton (mesozooplankton) from water samples before bottle incubations allowed the growth of phytoplankton until all the major nutrients were used up. This was probably the first iron enrichment (albeit inadvertent) to be carried out in this HNLC region. In the 1980s, the Subarctic Pacific Ecosystem Research (SUPER) programme made considerable advances in understanding pelagic ecosystem structure and function. Concurrently, work done by Martin's group as part of the Vertical Exchange (VERTEX) study in this region showed for the first time that iron limitation was probably responsible for the HNLC conditions consistently observed in this region (Martin *et al.*, 1989).

THE FINDINGS OF THE SUPER PROGRAMME SUPER was designed to investigate the major grazer hypothesis based on the observations of Heinrich (1957, 1962), which suggested that the life-history patterns of the large resident filter-feeding copepods were responsible for the low year-round phytoplankton biomass in this oceanic province. A total of six cruises, three in each of spring and summer, were conducted in the vicinity of the former site of ocean weather station Papa (OSP), located at 50°N 145°W (Fig. 4.1), during the mid to late 1980s (Miller, 1993). SUPER contributed to our understanding of the function of the pelagic ecosystem in several ways. On the basis of their observations, the major grazer hypothesis was rejected; microscopic analyses and other rate measurements indicated that the dominant grazers were the microzooplankton (Landry *et al.*, 1993) and the dominant phytoplankton were the autotrophic flagellates smaller than 5 μm (Booth *et al.*, 1993). In addition, SUPER provided evidence that the dominant phytoplankton were growing at close to maximum rates (Booth *et al.*, 1993). The latter part of the SUPER programme coincided with the seminal work by Martin *et al.* (1989) on the relation between iron supply and the magnitude of phytoplankton stocks in this region. There were attempts to carry out Fe-clean experiments by the SUPER participants, but these were ruled to be unsuccessful owing to inadvertent Fe contamination (Miller *et al.*, 1991). Thus, in experiments to assess grazing pressure by zooplankton on phytoplankton (Landry & Lehner-Fournier, 1988; Welschmeyer *et al.*, 1991; Landry *et al.*, 1993), no measurement of iron concentrations in the experimental vessel and hence no assessment of inadvertent iron contamination was available. Thus, SUPER made a major contribution to understanding the relationship between phytoplankton stocks and grazing pressure by both micro- and mesozooplankton, but were unable to also include Fe enrichment successfully in their experimental strategies.

IRON ENRICHMENTS AT 45–56°N, 143–148°W Two iron-enrichment studies have previously been carried out in the NE subarctic Pacific (Martin & Fitzwater, 1988; Martin *et al.*, 1989; Coale, 1991). Martin *et al.* (1989) carried out bottle experiments at three sites in this region (T-6, T-7 and T-8; see Fig. 4.1) and demonstrated marked utilisation of major nutrients (NO_3, SiO_4 and PO_4) corresponding to increases in chlorophyll in the experiments at each locale. These observations provided tantalising evidence of iron deficiency limiting phytoplankton growth in this HNLC region. The results of the simultaneous study by Coale (1991) supports that of Martin *et al.* (1989). In addition, Coale investigated the effects on phytoplankton stocks and

Table 4.3. *Summary of the major findings of experimental manipulations at OSP (50°N 145°W)*
p/c, Polycarbonate; p/e, polyethylene.

Study	Martin[1]	Coale[2]	Welsch[3]	Landry[4]	Boyd[5]
date	Aug '87	Aug '87	Sept '87	May/Sept '84	May '93
exp. duration	6 d	7 d	5 d	7 d	6–8 d
type	iron	iron	grazing	grazing/NH_4	grazing
vessel	2 l p/c	20 l p/c	2.7 l p/c	60 l p/e	25 l p/c
Fe addition	1, 5, 10 nM	0.89 nM	none	none	2 nM
chlorophyll ($\mu g\, l^{-1}$)					
initial	0.69	0.23	> 0.5	< 0.5	< 0.4
final (iron)	> 4.5	4	—	—	> 3.0
final (control)	1.07	0.47	> 2.0	> 2.0	< 1.5
nitrate (μM)					
initial	> 8.0	—	—	—	> 9.0
final (iron)	< 1.5	—	—	—	< 1.0
final (control)	< 6.5	—	—	—	> 4.5
main spp.	*Nitzschia*	diatoms	diatoms	diatoms	*Nitzschia*
^{14}C ($\mu g\, C\, l^{-1}\, d^{-1}$)					
initial	21.9	18.4	—	—	< 25
final (iron)	—	100	—	—	> 250
final (control)	—	29	—	—	< 50
C: Chl *a*					
initial	—	75	—	—	40
final	—	25	—	—	60

Sources: [1]Martin *et al.* (1989); [2]Coale (1991); [3]Welschmeyer *et al.* (1991); [4]Landry & Lehner-Fournier (1988); [5]P. Boyd *et al.* (personal communication).

size-fractionated production of other trace metals (Fe, Mn, Cu and Zn) and of chelators (EDTA). Coale (1991) demonstrated that Fe supply clearly had the most pronounced effect on the magnitude of both primary production and algal standing stocks. These incubation studies (Martin *et al.*, 1989; Coale, 1991) both suggested that iron enrichment is taxon-specific, preferentially enhancing the growth rates of large diatoms in the NE subarctic Pacific. Taxon-specific enhancements of growth rate may cause changes in the partitioning of biomass or production between phytoplankton size classes (Coale, 1991). Thus, these studies demonstrated that iron was likely to limit phytoplankton stocks, but did not investigate the crucial issue of grazer representation in the bottle enclosures (Table 4.3).

DISSOLVED IRON DISTRIBUTIONS Martin & Gordon (1988) demonstrated that dissolved iron was present in low concentrations (sub-nanomolar) in surface waters at OSP and increased with depth, albeit relatively slowly. They were able to put together a comprehensive iron budget demonstrating that atmospheric, rather than marine, supply of Fe was the most probable source near OSP. Before the JGOFS Canada programme, the eastward extent of this low-iron region was not known. However, work done on trace metal distributions in surface waters from coastal oceanic waters westwards to OSP indicate an order of magnitude decrease in zirconium (Zr) and hafnium (Hf) seaward from the coast (McKelvey & Orians, 1993; McKelvey, 1994). By analogy, a similar decrease may exist for Fe.

THE IRON VERSUS GRAZING DEBATE In the late 1980s and early 1990s there was considerable debate about the interpretation of results of the iron enrichments, with Banse (1990, 1991*a,b,c*) and Martin (Martin *et al.*, 1990*c*, 1991*b*; Martin, 1992) being two of the main protagonists. After re-analysis of both the data and experimental protocols several points were raised.

(1) There was the possibility of bottle effects causing a shift in species towards more hardy phytoplankters, for example pennate diatoms, as reflected by changes (decreases) in the assimilation index during Coale's 6 d bottle incubation (Banse, 1990).
(2) Upon iron enrichment, why were there no iron-mediated increases in rates of primary production in ^{14}C uptake experiments of 24 h duration (Frost, 1991)?
(3) Why were phytoplankton growth rates iron-limited when Booth *et al.* (1993) had shown cells to be growing at near-maximal rates?
(4) Why did algal biomass increase over time in control bottles? Was this due to the exclusion of grazers and/or trace metal contamination?
(5) There was no examination of (under-) representation of grazers in the bottle experiments.

Thus, the interpretation of such iron-enrichment experiments to define the role of iron supply in controlling phytoplankton stocks was viewed as contentious (Cullen, 1991; Martin *et al.*, 1994). The design of future iron-enrichment experiments would have to take into account these criticisms.

Canada JGOFS: iron–grazing interactions

The first phase of the JGOFS Canada programme commenced in 1992 and included a NE subarctic Pacific component. In this region, a series of cruises was carried out along a transect from the coast to OSP and included detailed process studies at 50°N, 145°W. The development of this programme benefited from the iron versus grazing hypothesis debate, and bottle enrichment experiments were performed in May 1993 and May 1994 at OSP. The assessment of the representation of grazers in the carboys was addressed in these experiments. In addition, although Martin *et al.* (1989) had measured the disappearance of nitrate from the carboys during their iron enrichments, no one had examined the effect of iron supply on the preferential utilisation of inorganic nitrogen species. Martin *et al.* (1989) and Coale (1991) had observed iron-mediated shifts in the species composition of the phytoplankton based on microscopical counts and on differences between size fractions over 25 μm and under 25 μm. However, neither study had investigated the partitioning of elements within the phytoplankton in sufficient detail to assess the quantitative implications of these shifts. As small cells (under 5 μm) are observed to dominate algal biomass and production in the NE subarctic Pacific (Booth *et al.*, 1993), any shift in the phytoplankton size spectra towards a dominance of large cells may alter the structure of the pelagic ecosystem, and the flux of carbon to depth (Boyd & Newton, 1995). These aspects were incorporated into the JGOFS Canada programme. In the NE subarctic Pacific, the mesozooplankton exhibit a strong seasonal migration, such that they are most abundant in the upper water column in late May – early June (Fulton, 1978). The iron-enrichment studies conducted by Martin *et al.* (1989) and Coale (1991) were performed in August, when the observed pelagic abundance of large copepods is *c.* 50% of the annual maximum (Fulton, 1978). The period of late May to early June was therefore selected as the best time to assess the relation between iron supply and grazing pressure on phytoplankton stocks (Boyd *et al.*, 1996). Water samples were obtained in 30 l samplers at dusk, transferred to carboys enriched with iron and then incubated for up to 8 days (Boyd *et al.*, 1996). *Neocalanus* spp. were observed to be actively moving in all carboys and a zooplankton net haul taken at dusk indicated that pelagic abundances of mesozooplankton in the water column were indeed of the same order as those observed by others in previous years (Fulton, 1978; Dagg, 1993).

THE ROLE OF THE MICROZOOPLANKTON Although microzooplankton and mesozooplankton grazing rates have been estimated in HNLC regions, including the NE subarctic Pacific (Miller *et al.*, 1991; Strom & Welschmeyer, 1991) it was not known how changes in phytoplankton biomass or size structure observed in iron-enrichment experiments would influence micro- or mesozooplankton grazing rates. During the study by Boyd *et al.* (1996), samples were withdrawn every two days and used to perform microzooplankton-grazing experiments. The results from the May 1993 microzooplankton grazing experiments indicated that phytoplankton growth rates at $t = 0$ (when the autotrophic flagellates < 5 μm accounted for over 70% of the biomass) were close to the theoretical maximum (Boyd *et al.*, 1996). Measured grazing rates were around 70% of growth ($g/\mu = 0.71 \pm 0.25$) throughout this study in the control, indicating a close, if not a tight coupling, comparable to data obtained by Landry *et al.* (1993) during more than 20 dilution experiments during the SUPER programme. In the iron-enriched samples, on day 2 the value of g/μ was 0.7, as in the control, but by day 4 g/μ had decreased to less than 0.6, suggesting a less close coupling (Boyd *et al.*, 1996). Concurrently, the phytoplankton were increasing in biomass and there were floristic shifts, resulting in the large diatoms becoming dominant. Boyd *et al.* (1996) suggested that this increase in the abundance of the larger cells was probably responsible for the weakening of the coupling between growth and grazing. They concluded that the microzooplankton, although probably responsible for keeping the small cells in check, were unlikely to graze many of the large cells that dominate the population after the introduction of iron.

THE ROLE OF THE MESOZOOPLANKTON In the May 1993 experiment, it was assumed that mesozooplankton abundance in the carboy was similar to that in the water column at this time, and grazing estimates were made indirectly (Boyd *et al.*, 1996). Although *Neocalanus* spp. were observed to be active in all carboys, they were not counted at the end of the experiment and thus, although it appeared that phytoplankton could escape grazer control, even by the large mesozooplankton, there was no direct proof. In May 1994, aspects of this experiment were repeated, with mesozooplankton abundances and health in each carboy being assessed at the end of the 6 d experiment (Boyd *et al.*, 1996). These results indicated that phytoplankton stocks were able to increase (Fig. 4.14) despite the presence of *Neocalanus* at abundances (and presumably grazing pressure) comparable to those observed in the water column in net hauls, at a time of year when the pelagic abundance of large grazers was highest.

THE ROLE OF IRON IN MICROALGAL C AND N UPTAKE Boyd *et al.* (1996), in a second enrichment experiment conducted in May 1993, studied the partitioning of C, N and Fe between different size classes of phytoplankton. The findings of this experiment supported the conclusions drawn by Price *et al.* (1991, 1994) in the equatorial Pacific, and suggested that they might be applicable to HNLC regions in general. Thus, nitrate uptake by small algal cells, such as nanoflagellates, was unaffected by iron enrichment, with small cells preferentially utilising ammonium. In contrast the larger phytoplankton were physiologically stimulated by iron supply and took up nitrate at elevated rates, relative to ambient conditions, during bottle iron enrichments (Fig. 4.15). In the May 1993 study, dissolved and particulate iron concentrations were measured in the carboys. Although dissolved iron was only measured at $t = 0$ and $t = 2$ d in the control carboy, the results indicated that there was slight inadvertent contamination of the control (0.2 nM higher than ambient concentrations at $t = 0$), suggesting that this was the explanation for the

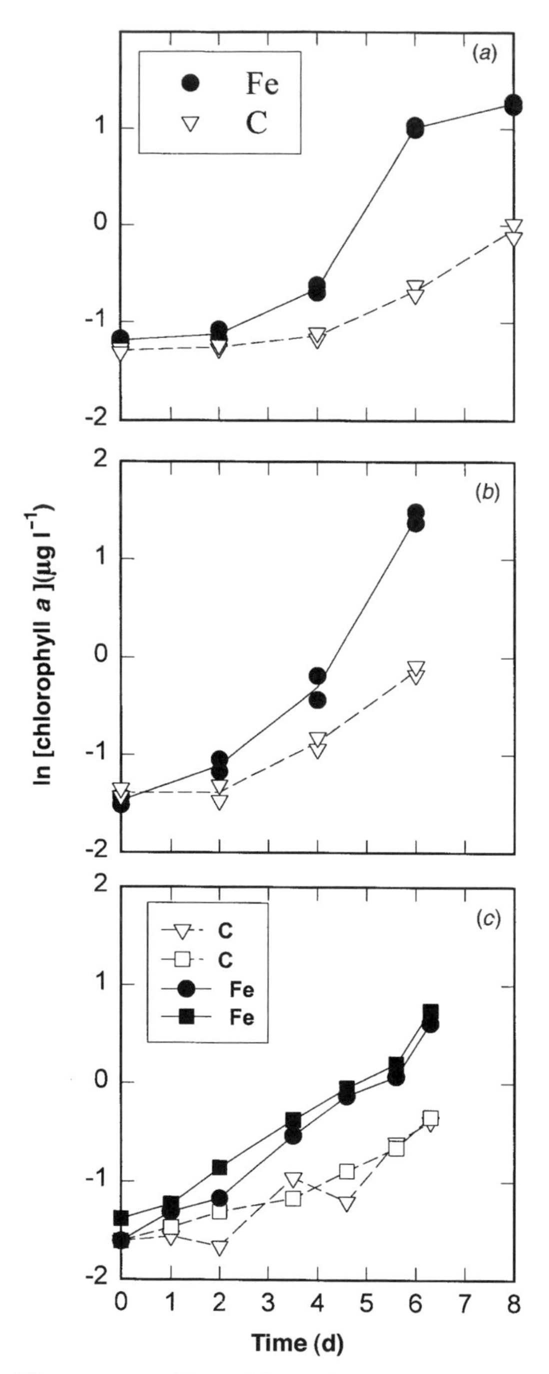

Figure 4.14 *Natural logarithm of chlorophyll* a *concentration during the course of 6–8 d for controls (open triangles) and Fe-enrichment carboys (filled dots).* (a) *May 1993, experiment 1.* (b) *May 1993, experiment 2.* (c) *May 1994, duplicate experiments with controls (open symbols) and Fe enrichment (filled symbols), each bottle also containing representative numbers of mesozooplankton. All experiments were conducted at station P. (Figure reproduced from Boyd* et al. *1996.)*

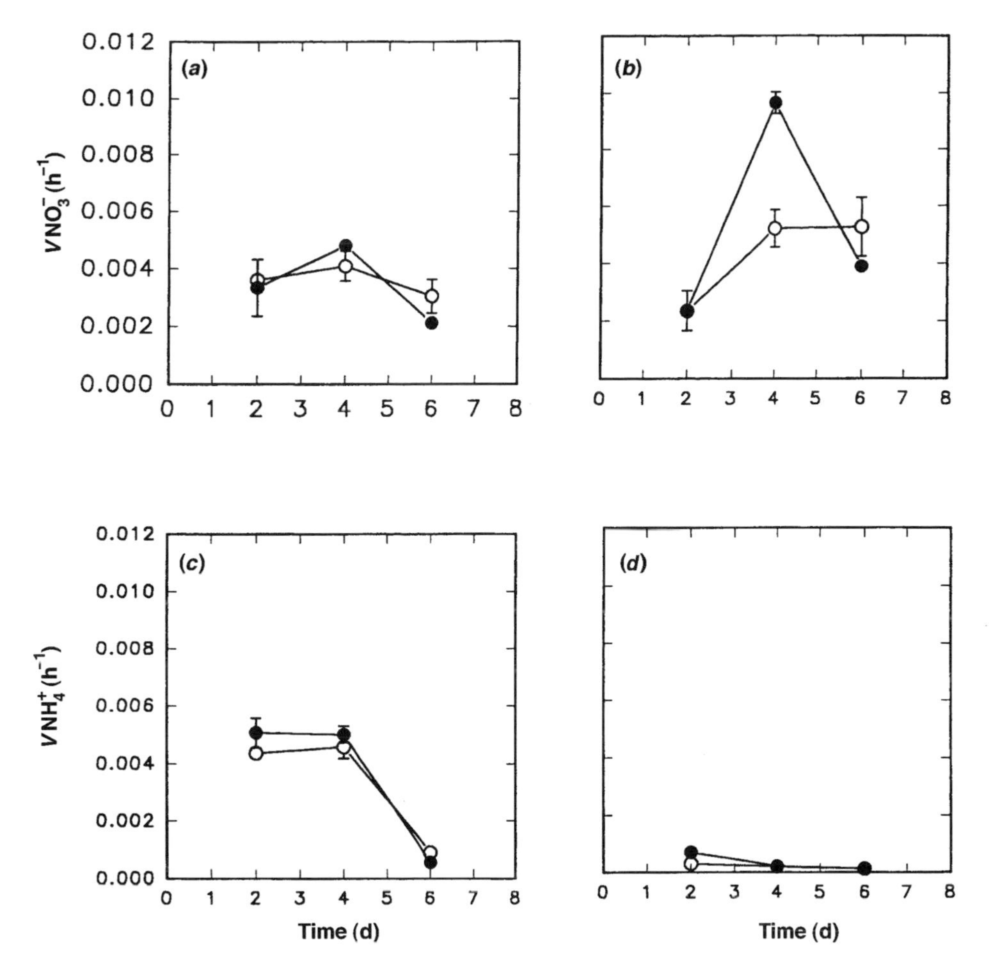

Figure 4.15 *Nitrogen-specific uptake rates in controls (open circles) and Fe enrichments (filled circles) during the course of 6 d of experiment 2 in May 1993.* (a) *Nitrate* (VNO_3^-) for the size class 0.7–5 *μm.* (b) *Nitrate for the size class* > *5 μm.* (c) *Ammonium* (VNH_4^+) for the size class 0.7–5 *μm.* (d) *Ammonium for the size class* > *5 μm.*

increases in phytoplankton biomass in the control carboy, rather than reduced grazing pressure. In addition, the partitioning of biomass and C uptake in the control indicated that there was a shift towards a dominance of the large cells (similar to that observed in the Fe carboys but less pronounced), which also suggested that Fe contamination had occurred.

PHYTOPLANKTON GROWTH RATES AND IRON QUOTAS

Phytoplankton taxonomic and size-fractionated chlorophyll data from six days during the May 1993 experiment of Boyd *et al.* (1996) were used as a proxy to calculate the specific growth rates of the phytoplankton in the controls and iron additions (Fig. 4.16). The large diatoms appeared to be growing at sub-optimal

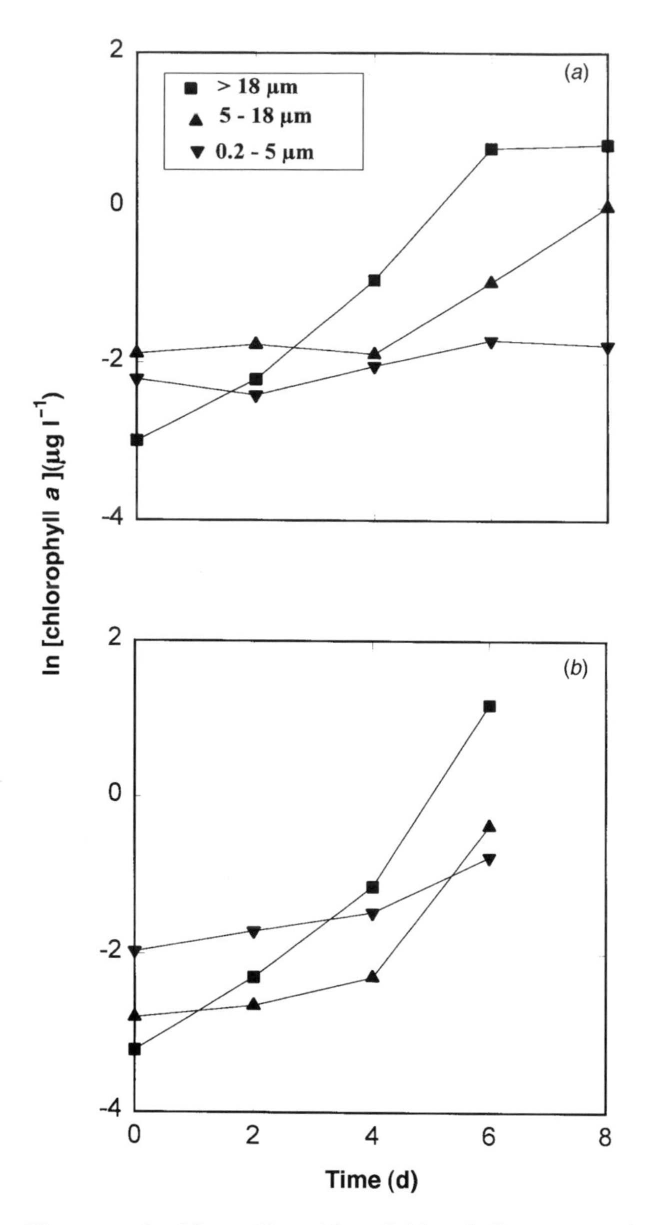

Figure 4.16 *Natural logarithm of chlorophyll* a *concentration for three size fractions from the Fe-enriched carboy during the course of 6 d of experiment 1* (a) *and experiment 2* (b) *in May 1993 at station P. No replicates. (Figure reproduced from Boyd* et al., *1996.)*

rates in the control, but at rates close to their theoretical maximum in the Fe-enriched carboys (Boyd *et al.* 1996). Recent work by Muggli and co-workers, in one of the first studies using phytoplankton species isolated from an HNLC region, indicated that the Fe quotas for a coccolithophorid (*Emiliania huxleyi*) and a large, oceanic, centric diatom (*Actinocyclus* sp.) were different (Muggli & Harrison, 1996; Muggli *et al.*, 1996). Fe quotas were low (90–130 μmol Fe: mol

C) for the coccolithophore (within the size range of the community of cells $< 5\,\mu m$, which dominate algal biomass under ambient conditions at OSP) and growth rates were relatively high ($0.02\,h^{-1}$) and unaffected by Fe supply. At low ambient Fe, the coccolithophorid is able to adapt by decreasing its cell size. In contrast, the Fe quota for the large oceanic diatom was high (330–370 μmol Fe: mol C), but the growth rate of these cells did increase markedly upon Fe supply. Apparently the structural constraints of the opaline skeleton prevent the diatom from adjusting its cell size; thus it cannot respond effectively to low ambient Fe and thrives only at high Fe supply. The growth rates of *Actinocyclus* under iron-depleted and replete conditions in the laboratory (Muggli *et al.*, 1996) were similar to those observed in corresponding size fractions by Boyd *et al.* (1996) for large diatoms at the OSP site.

These observations provide a refinement of concepts for iron regulation of diatom blooms. Thus far the group of coastal neritic diatoms were deemed to have a high cellular Fe requirement as opposed to the low Fe requirement of the group of oceanic species (Sunda *et al.*, 1991). Now it is apparent that a third group of truly oceanic, large, species with nevertheless high cellular Fe requirements do exist. It is likely that, for the two oceanic classes, cell size also plays a role in tolerance to low Fe supply.

THE BIOGEOCHEMICAL IMPLICATIONS OF FE SUPPLY Boyd *et al.* (1996) reported that the differential response of phytoplankton, within different size classes, to iron enrichment was responsible in this instance for a shift in the partitioning of production and biomass between size fractions (Fig. 4.17). This has important implications for grazer dynamics and the proportion of daily production that may potentially be exported to the deep ocean (Boyd & Newton, 1995). Estimates of column-integrated production, based on rates measured in the Fe carboy and scaled exponentially (using irradiance and attenuation coefficient data from OSP) for the water column, suggest that, at the peak of the 'bloom', rates of primary production of *c.* $3\,g\,C\,m^{-2}\,d^{-1}$ would be observed. As this rate is around three times higher than the highest measured at OSP (Welschmeyer *et al.*, 1993) and as cells larger than 18 μm were responsible for the majority of production (as opposed to cells smaller than 5 μm; see Welschmeyer *et al.*, 1993), it is likely that, if these large cells can escape mesozooplankton grazing pressure and form aggregates, the potential export of carbon to depth would be substantial. This in turn, has implications for the fate of the iron (supplied from the atmosphere) taken up by the phytoplankton. The majority of this may be incorporated into large cells and thus eventually sink out of the upper water column. Sinking of diatoms could be enhanced by Fe limitation, when the pulse of Fe supplied from the atmosphere becomes depleted by the phytoplankton. Muggli *et al.* (1996) found the large diatom

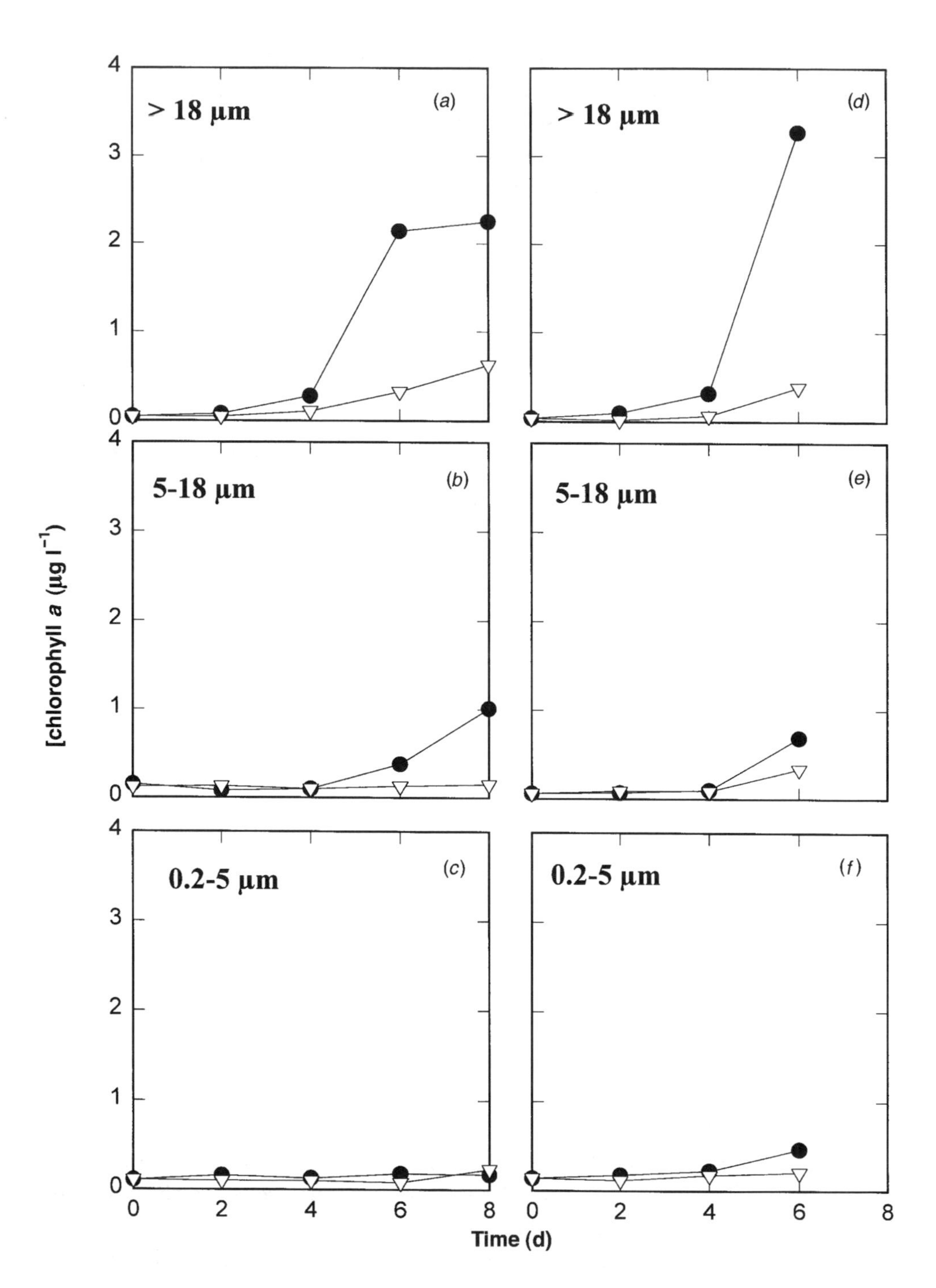

Figure 4.17 *Size-fractionated chlorophyll* a *concentration during 6–8 d from experiments 1 (*a–c*) and 2 (*d–f*) for three size fractions for Fe-enriched (filled dots) and control (open triangles) carboys in May 1993 at station P. No replicates. (Figure reproduced from Boyd* et al., *1996.)*

Actinocyclus sp. isolated from OSP, increased its sinking rate five times when it became Fe-limited. In contrast, the small coccolithophore *Emiliania huxleyi* became smaller when it was iron-limited and hence its sinking rate did not increase under Fe limitation. This suggests that large diatoms may be the major component of the biological pump at Ocean Station Papa.

RECENT DEVELOPMENTS The second phase of the Canada JGOFS programme commenced in September 1995 and has already resulted in some interesting findings. Tortell *et al.* (1996) found high amounts of iron, and high iron uptake rates, relative to phytoplankton, in heterotrophic bacteria at OSP. These results suggest that, although heterotrophic bacteria probably play a major role in the biogeochemical cycling of iron, their metabolism (specifically their growth efficiency) may be limited by iron supply. Most of the published work on iron limitation has focused on one site, that of OSP. However, the objectives of the JGOFS Science Plan are to make measurements that may be related to basin-scale events. La Roche *et al.* (1996), using an *in situ* marker, flavodoxin, for iron stress in phytoplankton, have provided an estimate of the areal extent of HNLC waters in the NE subarctic Pacific. Their findings indicate that iron stress in diatoms increases markedly in the region west of the 135°W meridian (about halfway between the coast and the OSP site) a region where dissolved iron concentrations in surface waters are less than 1 nM.

CONCLUSIONS The work conducted during the Canada JGOFS programme sheds light on some of the questions raised during the iron versus grazing debate. These include the conclusion that Fe contamination of some of the Canada JGOFS controls, rather than grazing pressure, is probably responsible for increased phytoplankton stocks in those controls. Primary production rates in bottles to which Fe has been added do not increase after 24 h because the phytoplankton that are Fe-limited compose less than 10% of the population under ambient conditions and 2–3 d are required before a significant increase in either production rate or stocks is observed. Similarly, the phytoplankton population under ambient conditions is growing at rates close to maximum because it is dominated (over 70%) by small cells, which are not (or are much less) iron-limited. More importantly, the results of the Canada JGOFS programme, tend to confirm the ecumenical iron hypothesis (Morel *et al.*, 1991*b*).

Equatorial Pacific Ocean

The equatorial Pacific Ocean is a site of intense upwelling of waters that, owing to deep respiration and mineralisation, are rich in nutrients and dissolved carbon dioxide (Murray *et al.*, 1995*a*). This region is the largest natural source of CO_2 to the atmosphere, supplying of the order of about 1 Gt C per year. At the same time it is a region where new production is fixing a similar amount of CO_2 per year, but this is less than the potential new production, as much of the upwelling nutrients are not utilised (Murray *et al.*, 1995*a*). There is a climate connection, as the productivity and fluxes of CO_2 and nutrients vary interannually with the El Niño – Southern Oscillation (Murray *et al.*, 1995*a*; Barber & Chavez, 1983).

The east equatorial Pacific has long been recognised (Walsh, 1976; Thomas, 1979) as an anomalous region that is now called a HNLC region. Thomas (1969) had suggested the Fe hypothesis for this HNLC region but the testing was not successful, probably because of inadvertent contamination. Barber & Ryther (1969) had suggested inhibition of growth by trace metals, notably Cu, in newly upwelled waters, to be overcome by formation of organic chelators once waters aged for a few days at the surface. The experimental testing included enrichments with a variety of Fe, trace metal and chelator solutions but, in retrospect, were deemed not successful either (Barber & Chavez, 1991). In the light of the new interest in the Fe hypothesis, Barber & Chavez (1991) analysed a large number of existing observations of nutrients, productivity and chlorophyll *a* and arrived at the conclusion that productivity is a linear function of nitrate supply in the West Equatorial Pacific, but severely limited by another factor, i.e. Fe, in the East Equatorial Pacific. However, productivity downstream from the Galapagos Islands was not limited, and a sedimentary Fe source at the Galapagos Plateau was suggested. The high abundance of Chl *a* west off the Galapagos Islands had previously been measured in October 1983 by the Coastal Zone Color Scanner (Feldman, 1991).

IRON ENRICHMENTS At the time it had been strongly advocated (Martin, 1990, 1991*a*) that most of the Fe input was from episodic atmospheric inputs (Donaghay *et al.*, 1991; their Fig. 2). If true, then instantaneous Fe enrichments, whether in bottles (Chavez *et al.*, 1991; Martin *et al.*, 1991*a*; Price *et al.*, 1991) or *in situ* (Watson *et al.*, 1991), could test this idea. In June–July 1990 experiments were done on the equator at 140°W where 1 nM of Fe, with and without sorbitol chelator, as well as 1–2 nM atmospheric-dust-leachate, were added to bottles and placed in deck incubators at simulated light depths corresponding to 20–80 m sampling depths for periods of 5 and 7 days (Martin *et al.*, 1991*a*). After this time, the phytoplankton biomass in all untreated controls was greater than the initial biomass. Apparently, in the field, other factors, such as mesozooplankton grazing, were again at play here. The Fe enrichments in the bottles showed strong stimulation of final C yield, notably for the aerosol-leachate with somewhat higher Fe concentration. In comparison with the natural dominance of small pico- and nanophytoplankton in the equatorial Pacific, the Fe enrichments showed a floristic shift towards larger, but still relatively small, pennate diatoms (*Nitzschia* sp.), as well as some large pennate and centric diatoms (Chavez *et al.*, 1991). The final C: N and C: P ratios were higher in the enrichments than in the controls, confirming the experimental trends observed in the Southern Ocean (de Baar *et al.*, 1990). In retrospect, the very low Redfield uptake ratios, NO_3: PO_4, of about 9–10 in the controls compared with the normal ratio of *c.* 16 in the Fe enrichments (Martin *et al.*, 1991*a*, their Table 2) is in keeping with nutrient fractionations observed in the Antarctic Polar Front (de Baar *et al.*, 1997). Thus, Chavez *et al.* (1991) considered that the tight balancing of high photosynthetic growth rates and loss terms, such as grazing and sinking, would be an explanation for the low and constant phytoplankton stocks in the equatorial Pacific. Apparently only a little added Fe enhances growth rates and may upset the balance and lead to biomass build-up. Finally, a small-scale Fe fertilisation experiment west of the Galapagos was proposed (Chavez *et al.*, 1991).

In parallel experiments during the same equatorial Pacific cruise it was found that Fe additions specifically stimulated uptake of nitrate and nitrite, whereas uptake of ammonium was unaffected (Price *et al.*, 1991). In a subsequent cruise, one year later in summer 1991, the nitrate uptake rate of the small phytoplankton in the region was inhibited by the high abundance of ammonium, and because the biomass was kept low by intense grazing, which contributed to the ammonium pool (Price *et al.*, 1994). The addition of Fe increased the short-term nitrate uptake rates by 2–9 times. This stimulation was confined for the larger (over 3 μm) size class, which thus must have been Fe-limited *in situ*. Hence the small phytoplankton appear not to be Fe-limited, but are grazed efficiently and primarily utilise ammonium, whereas the larger algae are Fe-limited and this prevents the uptake of nitrate. In other words, this HNLC region would be a grazer-controlled phytoplankton population in an iron-limited ecosystem (Price *et al.*, 1994). This may largely be correct, but some Fe stress also in the smallest size classes of phytoplankton cannot be ruled out, and was in fact suggested by Greene *et al.* (1994) on the basis of variations in quantum yield of fluorescence of the indigenous community.

The ambient dissolved Fe in the equatorial Pacific is very low; concentrations of 0.02 – 0.08 nM were found at 140°W (Martin, 1992). Johnson *et al.* (1994) studied the photochemistry of various chemical forms of added Fe, and its implications for biological productivity. Diurnal patterns in dissolved Fe had clear maxima at mid-day, probably owing to photo-reductive dissolution of colloidal Fe, which affected the availability of Fe for uptake into the cells. Takeda & Obata (1995) performed a suite of three bottle incubations at 160°W near the equator. In all cases the dissolved Fe in the controls was below the detection limit (less than 50 pM). The Fe enrichments were subtle at only 100–800 pM and did yield significant enhancement of the Chl *a* biomass, except for one experiment where the biomass in both the control and the Fe enrichment actually decreased over the 5 d incubation. The other three controls did show a modest or large increase of biomass, compared with the values for Chl *a in situ*. The Fe enrichments caused a large increase in Chl *a* values of the largest (over 10 μm) and medium-sized (3–10 μm) phytoplankton. This effect was increasing linearly with increasing amounts of Fe added. Apparently, these picomolar amounts are still below the amount required to fully overcome Fe deficiency.

JGOFS PROCESS STUDY The US JGOFS Process Study in the Equatorial Pacific Ocean has provided many insights and much information about the CO_2 budget, and primary and new production in the region (Murray, 1995*a* and articles therein). In addition, the effect of the 1991–93 El Niño event on these processes was well established (Kessler & McPhaden, 1995). During this El Niño, the outgassing flux of CO_2 was depressed about three-fold (Feely *et al.*, 1995; Wanninkhof *et al.*, 1995). Fitzwater *et al.* (1996) and Zettler *et al.* (1996) studied the role of Fe directly in bottle experiments. The latter found that most phytoplankton cells were physiologically affected by the low Fe concentrations, but only pennate diatoms showed significant increases in cell numbers due to the Fe enrichment. The non-diatom species did respond to the Fe enrichment by increasing their cell size, consistent with observations by others (Muggli & Harrison, 1996; van Leeuwe & Stefels, 1998). Over a period of five days the Fe enrichments showed an increasing response by the largest pennate diatoms (Zettler *et al.*, 1996, their Figs. 4 and 5), the smaller pennates also responding quite well but presumably being grazed more effectively in some of their experiments (Chavez *et al.*, 1991). For experiment T8F3, the cell numbers of small and large pennates increased significantly in the *c.* 2–2.5 nM Fe enrichment relative to the 0.03 nM Fe control. In the same experiment, the cell numbers of all other phytoplankton groups did not respond to Fe enrichment. In experiment T12F2 the small and large pennates were dividing at about 1 d^{-1} in the 2–2.5 nM enrichment compared with 0.45 d^{-1} (small) and 0.6 d^{-1} (large), respectively, in the control at 0.25 nM. Yet on average for 4 experiments, the growth in Fe enrichments was 0.69 d^{-1}. Thus in some of the experiments, even the 2–2.5 nM Fe was apparently not enough to sustain such high growth rates over the 5 d incubations. These findings for large pennates are consistent with the observation that large diatoms in the Southern Ocean apparently need an ambient Fe supply exceeding about 0.4–0.6 nM for optimal growth (Fig. 4.10).

Fitzwater *et al.* (1996) also found that the dominant group responding to Fe enrichments in the equatorial Pacific was diatoms. Otherwise the Fe additions were stimulating the overall community growth rates, with resultant increases of chlorophyll *a* and decreases of major nutrients. For the overall community the maximum rates of cell doubling in individual experiments were varying from 0.7 to 1.2 d^{-1} at total dissolved Fe concentrations of 2 nM or more in the enrichments. The Michaelis–Menten fit (Equation 4.3) to all data yields an apparent maximum rate of 0.97 d^{-1} with a half-saturation value K_{Fe} of 0.12 nM for the community. In comparison, Zettler *et al.* (1996) found on average 0.69 d^{-1} for all pennates in all four Fe experiments with final concentrations of 2–2.5 nM, and on average 0.44 d^{-1} in all controls (at 0.03–0.25 nM) or 0.23 d^{-1} for their lowest control at 0.03 nM. Thus for the Fe enrichments, it appears the community growing at 0.97 d^{-1} may have been composed of diatoms with, on average, slower growth rates of 0.69 d^{-1} and likely higher $K_{Fe} > 0.12$ nM, and pico–nanoplankton with rates exceeding 0.97 d^{-1} and lower $K_{Fe} < 0.12$ nM. However, in both experiments, simulated *in situ* light was applied in deck incubators and the derived parameter values in fact comprise the combined effects of light and iron. In addition, the Michaelis–Menten fit of Fitzwater *et al.* (1996) extends over an Fe range as high as 10 nM, and even 1000 nM, and may not necessarily be comparable with the Fe values of less than 2.5 nM applied by Zettler *et al.* (1996). More importantly, both studies (Fitzwater *et al.*, 1996; Zettler *et al.*, 1996) fully confirm the universal trend of Fe enrichments in open ocean waters most strongly stimulating the large diatoms.

In 1995, Lindley *et al.* provided indirect evidence about the *in situ* photosynthetic performance. This led to the conclusion that Fe limitation is more or less severe in three zones of the equatorial Pacific, where in two zones (5–10°N and 5–10°S) co-limitation by nitrate and Fe was also suggested (Lindley *et al.*, 1995, their Fig. 11). In the equatorial zone itself (5°S to 5°N) nitrate is in ample supply and only Fe would be limiting.

IRONEX I AND II EXPERIMENTS In October 1993, a study of the distributions of dissolved Fe and Chl *a* near and downstream of the Galapagos Islands was combined with an *in situ* perturbation experiment (initiated by the late John Martin) at a site some 500 km south of these islands (Martin *et al.*, 1994). Downstream of the islands the concentration of dissolved Fe was found to be as high as 1.3 nM, compared with only 0.06 nM upstream. This was accompanied by a shift in fluorescence from downstream of the Galapagos Islands to adjacent low-iron waters (Kolber *et al.*, 1994). In the low-iron waters, the efficiency of photosystem II of the natural phytoplankton assemblage was significantly (over 50%) inactivated (Kolber *et al.*, 1994). The highest Fe concentration of 3 nM in between the islands was accompanied by Chl *a* concentrations of over 13 $\mu g\,l^{-1}$ and nearly complete depletion of nitrate. In combination with the higher abundances of Chl *a* and higher primary productivity, this confirmed the suggestion that sedimentary Fe was a key factor in productivity (Barber & Chavez, 1991). Yet in the downstream plume, the nitrate at *c.* 10 µM and the CO_2 signal were not very different from adjacent HNLC waters to the north and south. Thus, compared with HNLC waters, the sedimentary Fe helped relieve the low chlorophyll and low productivity, but hardly affected the high nutrients and the CO_2 budget (see also Liss & Turner, Chapter 5, this volume).

At the experimental site (IronEx I), the labelled patch of surface water was enriched once, yielding initial Fe concentrations of about 4–6 nM. This had an immediate effect on the measured changes in fluorescence, again confirming suboptimal efficiency of PS II owing to iron stress (Kolber *et al.*, 1994). In addition, a doubling in plant biomass; a three-fold increase in Chl *a*; and a four-fold increase in primary productivity were found within the patch (Martin *et al.*, 1994). This was accompanied by various shifts in the plankton community, which were different from expectations based on earlier studies with bottle incubations (Martin *et al.*, 1994). Because only modest geochemical signals were observed and the community shifts led to many unresolved questions (Banse, 1995; Hutchins, 1995), a second experiment was deemed useful. The concentrations of nutrients *in situ* were hardly affected within the patch. Similarly, decreases in the total dissolved CO_2 and fugacity of CO_2 were barely significant and well below the apparent expectations based on Redfield stoichiometry (Watson *et al.*, 1994). Unfortunately, within about three days, the enriched patch was subducted by a cold water mass and this ended the experiment prematurely. Dissolved Fe decreased initially in accordance with mixing expectations and some days after the subduction it had dropped below the detection limit of *c.* 0.3 nM. Thus it was hypothesised that the single pulse of Fe in the IronEx I experiment may not have been sufficiently long-lasting, whereas a more continuous source of Fe would have had a larger effect on CO_2 (Watson *et al.*, 1994).

During the second *in situ* perturbation experiment, IronEx II, in May 1995, the fertilised patch remained at the surface, and the Fe was now added in three sequential infusions over a week to produce a more sustained increase in Fe within the patch. This led to a massive bloom dominated by diatoms, which apparently were not controlled by grazing (Coale *et al.*, 1996*b*). This confirmed the previous observations in the natural ecosystem of the unperturbed Antarctic Polar Front (de Baar *et al.*, 1995; Jochem *et al.*, 1995; Crawford, 1995). As in the previous experiment, the iron enrichment triggered an immediate biophysical response, by improved performance of photosystem II of the algae *in situ* (Behrenfeld *et al.*, 1996). This immediate response occurred before any changes in species composition and demonstrated that the local picoplankton community, primarily *Prochlorococcus* sp. and *Synechococcus* sp., was physiologically stimulated by Fe enrichment. Furthermore, a decrease of fugacity of CO_2 in surface water from a background value of about 510 μatm to approximately 420 μatm within the patch was found (Cooper *et al.*, 1996). Finally, during IronEx II a strong increase of dimethylsulfide was observed (Turner *et al.*, 1996; see also Liss & Turner, Chapter 5, this volume).

THE SUPPLY OF IRON The suggestion (de Baar, 1995) of a more distant sedimentary source near New Guinea in the western Pacific was consistent with the lower residual dissolved Fe (*c.* 0.35 nM) and particulate Al (*c.* 0.6 nM) at 140°W, as well as the lesser energetic intensity of the equatorial Pacific as compared to the Polar Front (see Löscher *et al.*, 1997). This, and similar upwelling calculations, led to the conclusion that upwelled Fe originating from the equatorial undercurrent was the major supply of Fe to equatorial Pacific surface waters (Coale *et al.*, 1996*a*).

Fe-biota studies in low-nutrient regions

Most Fe studies have been focused on the HNLC regions, but in low-nutrient waters co-limitation by trace metals may be expected. The central gyres in temperate zones of all major oceans would be suitable candidates for such co-limitation.

Pacific Ocean

Dissolved Fe in the central North Pacific gyre exhibits a concentration of about 0.37 nM in surface waters, then decreases rapidly to values of 0.02 nM at about 70–100 m, and increases again to about 0.45 nM at 1000 m (Bruland *et al.*, 1994). In the same surface waters, nitrate and silicate are depleted to values well below 1 μM (near the detection limit). For this region, Rue & Bruland (1995) found that 99.97% of the dissolved Fe was organically complexed, based on CSV titrations and subsequent calculations.

In the North Pacific central gyre observations have been reported of episodic dust inputs followed by increases of primary productivity by as much as 60% (Young *et al.*, 1991; DiTullio & Laws, 1991). Atmospheric input of both Fe and nitrate was suggested as the stimulus of productivity (DiTullio & Laws, 1991). DiTullio *et al.* (1993) subjected the plankton community to various treatments with extra Fe, a mixture of nutrients (NO_3, PO_4, and SiO_4) and the combination

(Fe, NO_3, PO_4 and SiO_4). Partly based on determination of pigments specific for prochlorophytes and diatoms respectively, the response to the nutrient mixture was ascribed to the small prochlorophytes. Apparently, low ambient Fe did not limit these responses. Following the Fe + nutrients treatment, and a 3 d lag, a large increase in large diatoms was observed. Thus, the diatoms *in situ* are severely limited by Fe as well as by the major nutrients. Calculations suggested that the community of small prochlorophytes were strongly controlled by intense grazing.

As mentioned above, Lindley *et al.* (1995) provided indirect evidence for co-limitation by Fe and nitrate in the zones at 5–10°N and 5–10°S.

North Atlantic Ocean

In the northwest Atlantic Ocean, Wu & Luther (1994) found dissolved Fe concentrations of less than 1 nM in surface waters. In autumn, the surface waters were depleted of Fe, leading to a nutrient-type profile over the complete water column; however, in mid-summer, the surface water was enriched with Fe and there was a subsurface minimum.

In surface waters of the northeast Atlantic Ocean, the winter concentrations of nutrients are significant and decrease rapidly when the spring bloom passes, except in waters north of a frontal system at about 50°N. In those northerly surface waters there is a spring bloom as well, but elevated nutrient concentrations persist in late spring and summer. In May 1989, during the JGOFS North Atlantic Bloom Experiment, the nitrate and phosphate were near depletion at the 47°N site but were still present at the 59°N site. The silicate was less than 1 μM at both sites, suggesting that the spring diatom bloom had already declined. The ambient dissolved Fe in the euphotic zone was about 0.07–0.23 nM at 47°N and 0.08–0.12 nM at 59°N (Martin *et al.*, 1993). At both sites, the dissolved Fe steadily increased with depth. Notably, at the 47°N site, the Fe correlated well with the nitrate. Upon enrichment of incubation bottles with 2 nM Fe, the final yield of plankton biomass (as C) was somewhat higher in the enrichments at both sites (Martin *et al.*, 1993). Thus Fe had stimulated growth but not as much as had been observed in some of the various HNLC regions.

In the HNLC regions, the Fe enrichments as a rule stimulated large diatoms the most. In the North Atlantic Ocean the spring diatom bloom had already passed. In retrospect, it appears conceivable that there was not enough silicate remaining for diatoms to thrive once more upon Fe addition. The half-saturation value for Si limitation varies considerably for various species reported in the literature, but Brzezinski & Nelson (1996) reckon that 2.3 μM is the representative K_s for temperate oceanic species.

Indian Ocean

The surface circulation of the Northwest Indian Ocean is driven by the northeast monsoon in the December–February period and the southwest monsoon prevailing during June–September. During the southwest monsoon, intense upwelling, notably off Oman, brings ample nutrients to the surface and productivity is very high.

During the northeast monsoon surface water concentrations of nitrate were found to be very low, between 0.05 and 0.78 μM, at a suite of stations at 65°E in the northwestern Indian Ocean (Takeda *et al.*, 1995). Dissolved Fe was 0.15 – 0.47 nM, and rapidly increased with depth to 5.16 nM at 200 m. These values are comparable to those reported by Saager *et al.* (1989), apart from improved reproducibility of the recent data. The one station of Saager *et al.* (1989) was occupied during the upwelling season and therefore may have had some elevated Fe supply from below.

The enrichment experiments of Takeda *et al.* (1995) were done with additions of ammonium, or nitrate, or Fe, or combined Fe and nitrate. For Fe alone, no stimulation was found, as apparently a source of N was lacking. The additions of NH_4 and NO_3 alone did yield various responses; the Fe + NO_3 addition provided the strongest response, also for diatoms. However, the dominance of coccolithophorids and *Phaeocystis* sp. may be interpreted as silicate deficiency of the diatoms. High N:P ratios of phytoplankton in N-enriched cultures indicated some P-limited growth. Overall, the various results of various treatments are consistent with co-limitation by Si, N, P and Fe in the surface waters during the northeast monsoon.

Co-limitation

The overall findings of experiments in temperate low-nutrient waters are consistent with various degrees of co-limitation by several nutrients including Fe. An adequate supply of all nutrients, N, P, Si and Fe, yields diatom blooms, as also found in HNLC waters; an obvious exception to the rule occurs where the Si supply is inadequate.

Validation of the iron hypothesis

Discussion

From the findings, it appears that the three major HNLC regions have several trends in common, but also major differences do exist.

The regular supply of iron and the microbial food web

In all regions the supply of dissolved iron from below (e.g. Antarctic Circumpolar Current) is adequate for sustaining the recycling community of mostly small sized algae being efficiently grazed upon by microzooplankton. In general, the small algae appear to be growing well, i.e. they are not iron-limited *per se* and grazing largely controls their biomass. Nevertheless, in all three HNLC regions, there are indications from various physiological parameters that even these small algae experience at least some Fe stress.

The earlier suggestion of dominant aeolian Fe input (Martin, 1990, 1991*a*) into both the equatorial Pacific and Southern Ocean has now given way to the idea of coastal margin sediments and upwelling, or the combination thereof, being the major sources of Fe supply (de Baar *et al.*, 1995; Coale *et al.*, 1996*a*). Thus only for the subarctic Pacific Ocean does aeolian input still remain to be considered the major local source term (de Baar *et al.*, 1995). One caveat here is the aforementioned discrepancy between low aeolian input rates (Duce & Tindale, 1991) and high sedimentation rates in the Southern Ocean (Kumar *et al.*, 1995) and equatorial Pacific Ocean (Murray *et al.*, 1995*b*).

Otherwise, for the Atlantic–Antarctic Ocean south of the Polar Front the new production sustained by Fe upwelling was found to be 14 mmol $C\,m^{-2}\,d^{-1}$ (for Fe : C = 1 : 100 000), about half of the potential new production as assessed on the basis of simultaneous upwelling of nitrate (de Baar *et al.*, 1995). By analogy in the equatorial Pacific the upwelled Fe can support new production of about 20 mmol $C\,m^{-2}\,d^{-1}$ (for Fe : C = 1 : 167 000), which would only be about 20% of the potential of simultaneous nitrate supply from below (Coale *et al.*, 1996*a*).

The enhanced supply of iron and diatom blooms

Upon addition of iron, either naturally in the Polar Front and neritic nearshore Antarctic waters, or in shipboard incubations, or by *in situ* perturbations, the larger size class of mostly diatoms is able to outgrow the background population of small algae. Apparently the larger mesozooplankton are not capable of keeping the larger diatoms fully in check, in keeping with the notion of longer generation times of mesozooplankton compared with microzooplankton. In the bottle incubations, this trend may be further enhanced by the exclusion of some of the mesozooplankton grazers, although their intentional inclusion in NE Pacific experiments did not influence the increase in diatoms after Fe enrichment (Boyd *et al.*, 1996). In addition, some bottle incubations are being done in deck incubators at simulated 'natural' light depths using screening. Alternatively, several other bottle incubations, to distinguish the effect of iron stress from that of light stress, were designed to avoid light limitation by setting the light regime at optimal conditions for growth (typically *c.* 100 $\mu mol\,m^{-1}\,s^{-1}$ in a 16–8 h light–dark cycle). In all bottles (controls and enrichments), this would favour algal growth compared with the conditions *in situ* most notably in the Antarctic Ocean, where light limitation is deemed to be crucial. Thus, it is not uncommon to observe the biomass increasing in the untreated controls as well. Nevertheless, in virtually all iron-enrichment incubations reported thus far, there is always an Fe stimulation of algal growth compared to the controls, with the exception of the experiments done at very nearshore, truly neritic sites (near Seal Island; Helbling *et al.*, 1991).

The almost selective stimulation of larger diatoms is a strikingly consistent trend in all Fe studies done thus far. It appears in keeping with the concept of the lower surface: volume ratio of the large diatom cells. This assumes spherical shapes; the adjustments that can be made for non-spherical strings and pennate diatoms do not modify the surface: volume ratio very much (Chisholm, 1992). Obviously the specific surface area does play an important role: where Muggli *et al.* (1996) have shown that the morphology of the opaline skeleton largely prevents a large oceanic diatom from decreasing its cell size in response to Fe stress. In contrast, other algae respond strongly to Fe stress by reducing their size (Muggli & Harrison, 1996; van Leeuwe & Stefels, 1998). Apart from cell size, it is also important to realise that coastal neritic diatoms probably have higher cellular Fe requirements per unit biomass than some of their oceanic counterparts (Sunda *et al.*, 1991). On the other hand, the largest oceanic diatoms, whether pennates in the equatorial Pacific (Zettler *et al.*, 1996), centric *Actinocyclus* sp. in the subarctic Pacific (Muggli *et al.*, 1996) or chain forming *Fragilariopsis kerguelensis* in the Polar Front, also appear to have a high Fe requirement for optimal growth, this corresponding to ambient concentrations of dissolved Fe apparently equal to or exceeding the 0.25–0.6 nM range.

The supply of iron to the subarctic Pacific

The upwelling supply of Fe in the NE Pacific Ocean is relatively small compared with the two other regions. However, episodic Fe input from aerosols might lead to significant blooms with uptake of CO_2 and major nutrients, but this has never been directly observed in the region. Such aeolian Fe input is conceivable, though, judging from direct observations and subsequent transient stimulation of growth in the central North Pacific Ocean (Young *et al.*, 1991; DiTullio & Laws, 1991). Indirectly, the records of phytoplankton abundance may provide a clue. On the one hand, Miller *et al.* (1991) mentioned that in 30 years of observations the Chl *a* at Ocean Station P was commonly in the range 0.3–0.4 mg m^{-3}, and had never exceeded 2 mg m^{-3}. On the other hand, occasional blooms of large diatoms have been reported (Clemons & Miller, 1984) but their significance over years and the whole region is not well known. Booth *et al.* (1993) also recognise the occurrences of higher than average numbers of diatoms in surface waters, but reckon that these represent only low levels of POC and Chl *a* such that the term 'bloom' would be somewhat of a misnomer. However, the downward flux of frustules of large diatom species, as collected in sediment traps, would suggest that diatom blooms are regular events in the subarctic Pacific region (Takahashi *et al.*, 1990). In summary, the suggested intermittent aeolian supply of Fe to the region may well give rise to diatom blooms and sedimentation events, but the existing evidence is not yet conclusive.

Iron and the geochemical budgets of carbon dioxide and nutrients

In the Antarctic Polar Front, the natural effect of Fe on the biological uptake of CO_2 and nutrients in major diatom blooms has been observed (de Baar *et al.*, 1995). The corresponding CO_2 budget is consistent with the other JGOFS observations at this site (Jochem *et al.*, 1995; Bakker *et al.*, 1997; Rutgers van der Loeff *et al.*, 1997*a*). Moreover, the Polar Front is a site where formation of deep water (AAIW) is taking place such that a combined biological–physical pump for drawdown of CO_2 may be at work.

In the equatorial Pacific, an enhanced phytoplankton standing stock downstream of the Galapagos Islands has been reported (Martin *et al.*, 1994), but this is not accompanied by anomalies of CO_2 and major nutrients. However an intentional *in situ* perturbation experiment with repeated iron additions did yield significant anomalies of CO_2 and nutrients in the equatorial Pacific region (Coale *et al.*, 1996*b*). Within this Fe-enriched patch of IronEx II, the fugacity of CO_2 in surface water was observed to decrease from a background value of about 510 μatm to approximately 420 μatm (Cooper *et al.*, 1996). This evidence, albeit more circumstantial, is in keeping with the original direct observations of a natural decrease from about 356 μatm to values as low as approximately 310 μatm, exactly coinciding with the Fe-rich Polar Front blooms (de Baar *et al.*, 1995; Bakker *et al.*, 1997). An important difference is that the equatorial surface waters shifted from strongly over-saturated (relative to the atmosphere) to less over-saturated, whereas the Polar Front had shifted from approximate equilibrium to distinct under-saturation. Hence in the IronEx II patch the *in situ* perturbation led to a transient suppression of natural outgassing at the equator, whereas the natural *in situ* bloom at the Polar Front led to an influx of CO_2 into the Southern Ocean.

The evolution of diatom blooms appears to be important for understanding the export into the deep sea. Apparently, at the end of a bloom period, when Fe may have become depleted, the Antarctic diatom *Corethron criophilum* reaches a stage of sexual reproduction accompanied by shedding and settling of diatom frustules (Crawford, 1995). This would strongly affect the Si budget. In the laboratory, Muggli *et al.* (1996) observed a five-fold increase of settling velocity of a subarctic diatom when grown under Fe-stress. In the situation *in situ* this would affect the budgets of not only Si, but also C, N and P.

Common trends and differences between HNLC regions

In general in the three major HNLC regions the same mechanisms are at play, only their prevalence varies (Table 4.4). This also applies to the low-nutrient waters of 'traditional' oligotrophic waters (mostly the temperate gyres), where the suite of mechanisms is further complicated by additional limitations of major nutrient elements N, P and Si. Overall, the various ecosystems can be described reasonably well by the combined ecological concepts of bottom-up and top-down control for different size classes of algae and zooplankton. The wax and wane of the diatom bloom is furthermore influenced by the intrinsic strategy of life cycle and reproduction, with significant effects on losses due to sedimentation.

In HNLC waters the large oceanic diatoms increase upon Fe enrichment; blooms develop and suffer little from grazing owing to the different generation times of diatoms (days) and mesozooplankton (months). In all three HNLC regions, the microzooplankton grazers (e.g. protozoa) have generation times similar to those of the algae and appear well capable of keeping the pico- and nanoplankton in check (see, for example, Banse, 1995). In the Antarctic Ocean, the supply of iron from below is significant. This is true also in the equatorial Pacific Ocean. The upward supply is relatively small in the strongly stratified NE subarctic Pacific. In all three HNLC regions, this upward supply from below is deemed adequate for sustaining the efficiently recycling microbial foodweb. On the other hand, in the Antarctic Ocean the deep wind-mixed layer and seasonal sea-ice cover make the Antarctic severely light-limited compared with the two other regions. Additional iron supply in coastal waters, and in those frontal systems downstream from land sources, is supporting most Antarctic blooming, whereas intermittent aeolian supply is mostly responsible for blooming in the open equatorial Pacific and Subarctic Pacific. For the Southern Ocean at large, an aeolian supply may also contribute to the Atlantic Subantarctic region and even further south into the Polar Frontal Zone, but this conceivable source term needs further quantification.

Iron hypotheses for HNLC ecosystems: true or false?

Returning now to the above set of hypotheses (Table 4.1, hypotheses 1–8) to be validated, we may be able to confirm some and falsify others.

HYPOTHESIS 1. MOSTLY FALSIFIED In both regions, there is always an adequate supply of iron to support at least the growth of some small algae, supporting the always-existing recycling ecosystem. These small algae are not Fe-limited *per se*, although physiological evidence hints at some degree of Fe stress. Grazing by microzooplankton was found to consume up to 70% of primary production.

HYPOTHESIS 2. CONFIRMED In all three HNLC regions, the additional supply of dissolved Fe, whether natural or experimental, stimulates phytoplankton, notably the larger diatoms. In the Antarctic region some of the

Table 4.4. *A qualitative assessment of the relative importance of various mechanisms in the three major HNLC regions, as well as some other HNLC regions, and the low nutrient waters of the oligotrophic central gyres of temperate zones*

One to five pluses for more Fe input, more severe stress or more common occurrence of blooms and export system superimposed on the always-operating recycling system of the microbial food web. Minus sign, trivial impact; 'nil' for no impact. The shelf source for the equatorial Pacific upwelling region is given in brackets as an indirect source off NE Guinea fuelling the direct supply by upwelling (see text). Grazing losses in the microbial food web are not listed, as these are important in all regions and seasons. The loss due to sedimentation is implicit in the bloom and export condition. For underlying reasoning, see text.

	Iron supply to surface waters			Other stress than iron			System
	From below	Shelf source	Aeolian	Light	Silicate	NO_3 and PO_4	Bloom and export
Southern Ocean							
Polar Frontal Zone	++	++	++	+++	+++	++	+
Polar Front	+++	++++	+	+++	+	–	+++
Antarctic region proper:							
open southern ACC	+++	–	–	++++	nil	nil	–
Weddell–ACC Front	+++	++++	+	+++++	nil	nil	+
coastal shelves	–	+++++	+	+++++	nil	nil	++++
Equatorial systems							
Pacific upwelling	++	(+)	++	nil	nil	nil	+
Galapagos Plateau	–	+++++	++	?	?	?	+++
Atlantic upwelling	+++	(+)	+++++	nil	++++	++++	–

NW Indian upwelling							
Subarctic regions							
subarctic NE Pacific	+	−	++	++	−	nil	+
northern N Atlantic (60°N)	−	−	+	++	+++	nil	+
Central gyres							
N Pacific	−	nil	++	+	+++++	+++++	+
S Pacific	−	nil	+	+	+++++	+++++	+
N Atlantic	−	nil	++	+	+++++	+++++	+
S Atlantic	−	nil	++	+	+++++	+++++	+

surface waters have higher ambient iron owing to a fairly steady supply derived from sediment sources; these are neritic coastal surface waters and those fronts that are downstream of continental margins. Here diatom blooms were found and these have significant effects on budgets of nitrate, silicate and CO_2 in surface waters, where both export production and CO_2 drawdown from the atmosphere have been documented. Light is also very important, most notably in the Southern Ocean. At high wind velocities in the Antarctic Ocean, blooms cannot exist in deeply mixed layers owing to light stress, even when the Fe supply is adequate. However, the highly variable Antarctic wind conditions do allow brief interludes of stratification in spring and summer, allowing blooms and some removal of nutrients and CO_2. Life cycle and sexual reproduction of the diatoms also play a role here. The seasonal diatom bloom and export system at the Atlantic Polar Front coincides with the region of formation of Antarctic Intermediate Water and this may well be a major sink region for atmospheric CO_2 into the Southern Ocean.

Upon intentional and repeated *in situ* Fe enrichment, a similar diatom bloom with significant changes of CO_2 and nutrients was also found in the equatorial Pacific Ocean. The bottle incubations in all three HNLC regions had consistently shown the same trend of large diatoms being stimulated by Fe enrichment. This is also true in nutrient-depleted temperate waters, provided some silicate is present. Over the time course of several days the change within the *in situ* patch of IronEx II largely confirmed the findings in previous bottle incubations. In other words, this one-off *in situ* experiment had demonstrated the findings of bottle incubations over several day periods to be correct, i.e. not suffering from conceivable artifacts due to so-called bottle effects.

Although hypothesis 2 has been confirmed, it must be realised that hypothesis 1 has not been completely falsified. This is due to the consistent observations of several physiological indicators of Fe stress *in situ* (^{15}N-nitrate assimilation, fluorescence response, enzyme activity and flavodoxin) in all three HNLC regions. The latter observations had been ascribed to the indigenous small (pico)plankton community at large and would suggest that this whole community suffers somewhat from Fe deficiency, although it is mostly affected by intense grazing. Whether or not the physiological stress is indeed a community-wide phenomenon would however require further unravelling by taxa and/or size classification. In many reports of physiological stress indicators, part of the community consisted of some diatoms, which upon Fe enrichment gave rise to massive blooms. In Fe-depleted waters, this sub-population of mostly large diatoms may consist of species with a high cellular iron quota (as in neritic-type diatoms), in which diffusion limitation also plays a role. Apparently these diatoms experience Fe stress, and may partly or wholly account for the physiological stress observed for the whole community.

HYPOTHESIS 3. FALSIFIED Despite adequate supply of iron to the Polar Front, the major nutrients, although they are being utilised significantly, are not being exhausted. Similarly, the fixation of dissolved CO_2 is less than some had expected, and the calculated CO_2 drawdown from the atmosphere of initial global models (Joos *et al.*, 1991*a,b*; Sarmiento & Orr, 1991; Peng & Broecker, 1991*a,b*; Kurz & Maier-Reimer, 1993) are at best well beyond the upper limits, if not irrelevant (Mitchell *et al.*, 1991). Apparently other factors, most notably light stress at high wind mixing rates, prevent the full utilisation of major nutrients and concomitant CO_2 drawdown, no matter how much Fe is made available to Antarctic waters, either naturally or by perturbation experiments. Matters are further complicated by the fact that elemental ratios (C:N:P:Fe) in solution, in standing plankton stock and in settling export of particles are much more likely to vary greatly, rather than obeying Redfield stoichiometry throughout. The different roles of Fe in assimilation and enzymatic conversion of C or N versus P by the cell also makes this obvious, such that Fe stress will cause shifts in major element ratios (C:N:P) in the various pools.

On the necessary supply of iron, we have learned some more.

HYPOTHESIS 4. REJECTED Aeolian input of continental aerosols is not the major supply of Fe to the Antarctic and equatorial Pacific. However, intermittent dust input may well be the major supply into the subarctic Pacific and central north Pacific, superimposed on a smaller annual supply from below by mixing and upwelling. This aeolian supply would give rise to occasional blooms, notably of diatoms. The eastward transport of dust from the Sahara provides a more steady Fe supply to the equatorial Atlantic Ocean (Helmers & Rutgers van der Loeff, 1993; Powell *et al.*, 1995; Helmers, 1996; Rutgers van der Loeff *et al.*, 1997*b*), which is deemed to be not Fe limited and not a HNLC region either. Accumulated aeolian supply on winter sea-ice in the Antarctic Ocean may, upon sudden spring melting, supply dissolved Fe to the Marginal Ice Zone, allowing blooms when suitable vertical stratification permits. From sedimentation rates, it appears that the aeolian supply into the Atlantic Polar Frontal Zone is much larger than hitherto assumed.

HYPOTHESIS 5. CONFIRMED Input of dissolved iron from below is the major supply to the Antarctic Circumpolar Current. Similarly upwelling is the major supply of Fe into surface waters of the equatorial Pacific Ocean, with the underlying equatorial undercurrent in turn being supplied from coastal margin sedimentary sources.

HYPOTHESIS 6. CONFIRMED Lateral transport of dissolved iron from ocean margin sediments may contribute extra iron to the Atlantic Polar Front, although aeolian supply from South America cannot be ruled out as an alternative major source. In the nearshore waters of the Weddell Sea, nearby islands of the Antarctic Peninsula, as well as the frontal system downstream of the Peninsula (Weddell–Scotia Convergence and Scotia Front, e.g. EPOS at 49°W, 49°S; Fig. 4.5; and more easterly Weddell–ACC front, e.g. at 6°W, see Fig. 4.9), there undoubtedly is a significant supply of dissolved iron (as well as mineral particles) from margin sediments, which under suitable light conditions may give rise to blooms. Frontal systems between the Weddell gyre and the ACC often coincide with the seasonal sea-ice edge and may thus affect the Marginal Ice Zone blooms sometimes, but not always, observed in spring and summer. In other words, the requirement of adequate Fe supply is being met at this extended Weddell gyre – ACC frontal system. During one expedition (EPOS 1988; Hempel, 1993) or the other (JGOFS 1992; Bathmann *et al.*, 1994*a*; Smetacek *et al.*, 1997) one may (Lancelot *et al.*, 1993) or may not (Fig. 4.7) encounter MIZ blooms, depending largely on wind-mixed layer depths. Yet over decadal time scales blooms have been observed from satellite to occur more often in the Fe-replete frontal region downstream of the Peninsula (Plates 1 and 7 of Comiso *et al.*, 1993).

In general the aeolian supply, as well as the margin sediments source, needs further investigation at all sites. In comparison, in the equatorial Pacific Ocean, the general upwelling is stronger than in the Antarctic or subarctic (Gargett, 1991). Hence at the equatorial region the major supply of iron also appears to come from below by mixing and upwelling (Coale *et al.*, 1996*a*), in combination with an Asian margin sediment source.

HYPOTHESIS 7. UNDECIDED The seventh issue, the recycling efficiency of iron vis-à-vis that of other elements C, N and P, is well recognised but not resolved. These elements are all involved in the biological cycle, but the additional adsorption and oxidation of Fe at the surfaces of biogenic particles would lead one to expect that Fe is more rapidly removed from surface waters. Thus, Fe would be recycled less efficiently. On the other hand, the *c.* 99% organic complexation would tend to counteract adsorption on settling particles. Moreover, several studies following the fate of Fe-labelled phytoplankton prey have shown that owing to predator uptake particulate Fe can be recycled quite efficiently (Hutchins & Bruland, 1994, 1995; Hutchins *et al.*, 1995). Here the generally more acidic digestive tract of the zooplankton may bring the Fe back into solution, not only from biological prey, but also from Fe-colloids (Barbeau *et al.*, 1996). Thus grazing activity by protozoa would provide Fe that otherwise would have remained unavailable as colloids (Wells *et al.*, 1991, 1995). In fact, the digestion may also give rise to the existence of organic complexes of Fe in seawater. Thus the fate and recycling of Fe in the food web has been shown in laboratory experiments, and is now to be confirmed in the open ocean. Photoreduction of colloidal and particulate Fe also brings Fe back into truly dissolved state available for phytoplankton uptake, the daily maximum of dissolved Fe conveniently coinciding with the mid-day maximum of photosynthesis (Johnson *et al.*, 1994; Waite *et al.*, 1995).

HYPOTHESIS 8. REJECTED The organic complexation of Fe in the Pacific sector of the Antarctic Ocean suggests that about 95% of the dissolved iron is complexed by organics, thus probably affecting the rate of iron uptake by algae. This is consistent with but significantly lower than the *c.* 99% organic complexation reported for temperate oceans. Nevertheless, it now appears that the majority (90–99%) of dissolved iron is associated with natural dissolved organics. Whether or not this natural organic complexation affects the rate of Fe uptake by the cell is currently being investigated. On the one hand, the artificial chelator EDTA is well known to bind dissolved Fe very strongly, such that only the remaining, inorganic, dissolved Fe fraction is proportional to rate of uptake and growth of phytoplankton. On the other hand the natural organic ligands do not appear to interfere as severely with Fe-limited rates of growth, i.e. either the natural Fe–ligand complexes are dissociating with very rapid kinetics, or the cell is able to assimilate the undissociated natural Fe–ligand complex.

Conclusion

It took some 60 years before it was possible to validate the Fe hypothesis (Gran, 1931). From the study of Martin & Fitzwater (1988) onwards it took less than a decade to test, re-define and finally firmly demonstrate that Fe plays a key role in regulating the biological productivity and the carbon budget of the global ocean. At low dissolved iron concentrations in the open subarctic Pacific, the open ACC, and in the equatorial Pacific, the surface waters are characterised by a recycling ecosystem. This is dominated by small phytoplankton, which are closely coupled with microzooplankton. This ecosystem structure has little impact on the global biogeochemical budgets of N, Si, CO_2 and Fe itself. The small phytoplankton are growing at rates close to maximum and are under microzooplankton grazer control, although there are several physiological indications of algal Fe stress. The overall experimental and observational evidence consistently shows that adequate light and Fe allow large diatoms to alter the ecosystem structure in HNLC waters from a recycling system to one dominated by blooms and export production. This shift appears to determine the essential role of the major HNLC regions in global geochemical cycling and climate.

After a short period of turbulent scientific progress, Fe has now become as intrinsic to plankton ecology as the well-known factors light, major nutrients and grazing. Thus, Fe is now seen as one of several factors governing biological production in all regions of the world oceans, rather than as the single limiting factor just for the HNLC regions. The consistent response of large diatoms to Fe enrichment in all experiments in all oceans is most striking and has major implications for global budgets not only of CO_2 but also of silicate. The brief turmoil in the past decade has paved the way for the next period, during which the ongoing investigations will more gradually evolve into a thorough understanding of the function of Fe at various trophic levels, in various regions, seasons and years.

Acknowledgements

This chapter is merely a synthesis of the original research of several colleagues. The reviews of Geider & La Roche (1994), Hutchins (1995) and de Baar (1994) were often used, but the original articles were consulted and cited whenever possible. Several sources have been cited literally, but quote marks have been omitted for the sake of readability. The authors gratefully acknowledge the efforts of their contributors Dorothee C. E. Bakker, Uli V. Bathmann, R. H. Goldblatt, Paul J. Harrison, Frank J. Jochem, Jeroen T. F. de Jong, Christiane Lancelot, Maria A. van Leeuwe, Robert F. Nolting, Michiel M. Rutgers van der Loeff, Renate Scharek, Victor Smetacek, Klaas R. Timmermans and Cees Veth in providing reprints, pre-prints, and other written reports of their research, sometimes not yet published, as well as for giving constructive criticism of various versions of the manuscript. This gratitude extends to several other colleagues for the same reasons. Special thanks are due to Paul Harrison. The Netherlands Institute for Sea Research, the University of Groningen, the Netherlands Antarctic Research Committee and the Netherlands GeoSciences Foundation of the Netherlands Organisation for Scientific Research have provided support. P. Boyd acknowledges support from the Natural Sciences and Engineering Research Council of Canada via the Canadian JGOFS Special Collaborative Programme. This is NIOZ contribution number 3138.

References

Andersen, M. A. & Morel, F. M. M. (1982). The influence of aqueous iron chemistry on the uptake of iron by the coastal diatom *Thalassiosira weissflogii*. *Limnology*

and Oceanography, **27**, 789–813.
Baar, H. J. W. de (1994). von Liebig's Law of the Minimum and Plankton Ecology (1899–1991). *Progress in Oceanography*, **33**, 347–86.
Baar, H. J. W. de, Franz, H. G., Ganssen, G. M., Gieskes, W. W. C., Mook, W. G. & Stel, J. H. (1989*a*). Towards a Joint Global Ocean Flux Study: Rationale and Objectives. In *Oceanography 1988*, ed. A. Ayala-Castanares, W. S. Wooster & A. Yanez-Arancibia, pp. 11–31. Proceedings of the Joint Oceanographic Assembly, Acapulco, Mexico. Mexico DF: Universidad Nacional Autonoma de Mexico.
Baar, H. J. W. de, Buma, A. G. J., Jacques, G., Nolting, R. F. & Tréguer, P. J. (1989*b*). Trace Metals – Iron and Manganese effects on phytoplankton growth. *Berichte zur Polarforschung*, **65**, 34–44.
Baar, H. J. W. de, Buma, A. G. J., Nolting, R. F., Cadée, G. C., Jacques, G. & Tréguer, P. J. (1990). On iron limitation of the Southern Ocean: experimental observations in the Weddell and Scotia Seas. *Marine Ecology Progress Series*, **65**, 105–22.
Baar, H. J. W. de, de Jong, J. T. M, Bakker, D. C. E., Löscher, B. M., Veth, C., Bathmann, U. & Smetacek, V. (1995). Importance of Iron for Phytoplankton Spring Blooms and CO_2 Drawdown in the Southern Ocean. *Nature*, **373**, 412–15.
Baar, H. J. W. de, van Leeuwe, M. A., Scharek, R., Goeyens, L., Bakker, K. M. J. & Fritsche, P. (1997). Iron availability may affect the nitrate/phosphate ratio (A. C. Redfield) in the Antarctic Ocean. *Deep-Sea Research*, Part II, **44**, 229–60.
Baar, H. J. W. de, de Jong, J. T. M., Nolting, R. F., van Leeuwe, M. A., Timmermans, K. R., Bathmann, U. V., Rutgers van der Loeff, M. M. & Sildam, J. (1999). Low dissolved Fe and the absence of diatom blooms in remote Pacific waters of the Southern Ocean. *Marine Chemistry* (in press).
Bakker, D. C. E., Baar, H. J. W. de & Bathmann, U. V. (1997). Changes of carbon dioxide in surface waters during spring in the Southern Ocean. *Deep-Sea Research*, Part II, **44**, 91–128.
Banse, K. (1990). Does iron really limit phytoplankton production in the offshore subArctic Pacific? *Limnology and Oceanography*, **35**, 772–5.
Banse, K. (1991*a*). Iron availability, nitrate uptake and exportable new production in the subarctic Pacific. *Journal of Geophysical Research*, **96**, 741–8.
Banse, K. (1991*b*). Iron, nitrate uptake by phytoplankton, and mermaids. *Journal of Geophysical Research*, **96**, 20 701.
Banse, K. (1991*c*). Rates of phytoplankton cell division in the field and in iron enrichment experiments. *Limnology and Oceanography*, **36**, 1886–98.
Banse, K. (1992). Grazing, temporal changes of phytoplankton concentrations, and the microbial loop in the open sea. In *Primary Productivity and Biogeochemical Cycles in the Sea*, ed. P. G. Falkowski & A. D. Woodhead, pp. 409–40. (*Environmental Science Research*, **43**.) New York: Plenum Press.
Banse, K. (1995). Community response to IronEx. *Nature*, **375**, 112.
Barbeau, K., Moffett, J. W., Caron, D. A., Croot, P. L. & Erdne, D.L. (1996). Role of protozoan grazing in relieving iron limitation of phytoplankton. *Nature*, **380**, 61–4.
Barber, R. T. & Ryther, J. H. (1969). Organic chelators: factors affecting primary production in the Cromwell Current upwelling. *Journal of Experimental Marine Biology and Ecology*, **3**, 191–9.
Barber, R. T. & Chavez, F. P. (1983). Biological consequences of El Niño. *Science*, **222**, 1203–10.
Barber, R. T. & Chavez, F. P. (1991). Regulation of primary productivity rate in the equatorial Pacific. *Limnology and Oceanography*, **36**, 1803–15.
Bathmann, U. V., Scharek, R., Crawford, R., Peeken, I., Bakker, D., & van Franeker, J.A. (1994*a*). Phytoplankton spring blooms at the Polar Front and the Ice Edge north of the Weddell Sea. *Eos*, **75**, 158.

Bathmann, U. V., Smetacek, V., Baar, H. de, Fahrbach, E. & Krause, G. (eds) (1994*b*). *The expeditions ANTARKTIS X/6–8 of the Research Vessel 'POLARSTERN' in 1992/1993.* (*Berichte zur Polarforschung*, **135**.) 236 pp.

Bathmann, U., Scharek, R., Dubischar, C., Klaas, C., & Smetacek, V. (1997). Chlorophyll and phytoplankton species distribution in the Atlantic sector of the Southern Ocean in spring. *Deep-Sea Research*, Part II, **44**, 51–68.

Becquevort, S. (1997). Nanoproto-zooplankton in the Atlantic sector of the Southern Ocean during early spring: Biomass and feeding activities. *Deep-Sea Research*, Part II, **44**, 355–74.

Becquevort, S., Mathot, S. & Lancelot, C. (1993). Interactions in the microbial community of the marginal ice zone of the northwestern Weddell Sea through size distribution analysis. *Polar Biology*, **12**, 211–18.

Behrenfeld, M. J., Bale, A. J., Kolber, Z. S., Aiken, J. & Falkowski, P. (1996). Confirmation of iron limitation of phytoplankton photosynthesis in the equatorial Pacific Ocean. *Nature*, **383**, 508–11.

Berg, C. M. G. van den (1995). Evidence for organic complexation of iron in seawater. *Marine Chemistry*, **50**, 139–57.

Booth, B. C., Lewin, J. & Postel, J. R. (1993). Temporal variation in the structure of autotrophic and heterotrophic communities in the subArctic Pacific. *Progress in Oceanography*, **32**, 57–99.

Boyd P. & Newton, P. (1995). Evidence of the potential influence of planktonic community structure on the interannual variability of particulate carbon flux. *Deep-Sea Research*, Part I, **42**, 619–39.

Boyd, P. W., Muggli, D. L., Varela, D. E., Goldblatt, R. H., Chretien, R., Orians, K. J. & Harrison, P. J. (1996). In vitro iron enrichment experiments in the NE Subarctic Pacific. *Marine Ecology Progress Series*, **136**, 179–93.

Brand, L. E. (1991). Minimum iron requirements of marine phytoplankton and the implications for the biogeochemical control of new production. *Limnology and Oceanography*, **36**, 1756–71.

Bruland, K. W., Orians, K. J. & Cowen, J. P. (1994). Reactive trace metals in the stratified central North Pacific. *Geochimica et Cosmochimica Acta*, **58**, 3171–82.

Bruland, K. W. & Wells, M. L. (1995). The Chemistry of Iron in Seawater and its Interaction with Phytoplankton. *Marine Chemistry*, **50**, 241 pp.

Brzezinski, M. A. & Nelson, D. M. (1996). Chronic substrate limitation of silicic acid uptake rates in the western Sargasso Sea. *Deep-Sea Research*, Part II, **43**, 437–53.

Buma, A. G. J. (1992). Factors controlling phytoplankton growth and species composition in the Antarctic Ocean. Ph.D. Thesis, University of Groningen.

Buma, A. G. J., Baar, H. J. W. de, Nolting, R. F. & van Bennekom, A. J. (1991). Metal enrichment experiments in the Weddell-Scotia Seas: effects of iron and manganese on various plankton communities. *Limnology and Oceanography*, **36**, 1865–78.

Chamberlin, T. C. (1890). The method of multiple working hypotheses. *Science*, **15**, 92–8.

Chavez, F.P. *et al.* (1991). Growth rates, grazing, sinking and iron limitation of equatorial Pacific phytoplankton. *Limnology and Oceanography*, **36**, 1865–34.

Chisholm, S. W. (1992). Phytoplankton Size. In *Primary Productivity and Biogeochemical Cycles in the Sea*, ed. P. G. Falkowski & A. D. Woodhead, pp. 213–37. (*Environmental Science Research*, **43**.) New York: Plenum Press.

Chisholm, S. W. & Morel, F. M. M. (1991). What controls phytoplankton production in nutrient-rich areas of the open sea? *Limnology and Oceanography*, **36**, 1507–70.

Clemons, M. J. & Miller, C. B. (1984). Blooms of large diatoms in the oceanic, subarctic Pacific. *Deep-Sea Research*, **31**, 85–95.

Coale, K. H. (1991). Effects of iron, manganese, copper and zinc enrichments on productivity and biomass in the subarctic Pacific. *Limnology and Oceanography*, **36**, 1851–64.

Coale, K. H., Fitzwater, S. E., Gordon, R. M., Johnson, K. S. & Barber, R. T. (1996*a*). Control of community growth and export production by upwelled iron in the equatorial Pacific Ocean. *Nature*, **379**, 621–4.

Coale, K. H. *et al.* (1996*b*). A massive phytoplankton bloom induced by an ecosystem-scale iron fertilisation experiment in the equatorial Pacific Ocean. *Nature*, **383**, 495–501.

Comiso, J. C., McClain, C. R., Sullivan, C. W., Ryan, J. P. & Leonard, C. L. (1993). Coastal zone color scanner pigment concentrations in the Southern Ocean and relationships to geophysical features. *Journal of Geophysical Research*, **98**, 2419–51.

Cooper, L. H. N. (1937). On the ratio of nitrogen to phosphorus in the sea. *Journal of the Marine Biological Association of the United Kingdom*, **22**, 177–82.

Cooper, D. J., Watson, A. J. & Nightingale, P. D. (1996). Large decrease in ocean-surface CO_2 fugacity in response to *in situ* iron fertilisation. *Nature*, **383**, 511–13.

Crawford, R. M. (1995). The role of sex in the sedimentation of a marine diatom bloom. *Limnology and Oceanography*, **40**, 200–4.

Crawford, R. M., Hinz, F. & Rynearson, T. (1997). Spatial and temporal distribution of assemblages of the diatom *Corethron criophilum* in the Polar Frontal region of the South Atlantic. *Deep-Sea Research*, Part II, **44**, 479–98.

Cullen, J. J. (1991). Hypotheses to explain high-nutrient conditions in the open sea. *Limnology and Oceanography*, **36**, 1578–99.

Cullen, J. J. (1995). Status of the iron hypothesis after the Open-Ocean enrichment experiment. *Limnology and Oceanography*, **40**, 1336–43.

Dagg, M. (1993). Grazing by the copepod community does not control phytoplankton production in the subArctic Pacific Ocean. *Progress in Oceanography*, **32**, 163–83.

Deacon, G. E. R. (1984). *The Antarctic Circumpolar Ocean*. Cambridge University Press.

De Angelis, M., Barkov, N. I. & Petrov, V. N. (1987). Aerosol concentrations over the last climatic cycle (160kyr) from an Antarctic ice core. *Nature*, **325**, 318–21.

Dehairs, F., Goeyens, L., Stroobants, N. & Mathot, S. (1992). Elemental composition of suspended matter in the Scotia-Weddell Confluence area during spring and summer 1988 (EPOS Leg 2). *Polar Biology*, **12**, 25–33.

Denman, K., Hofmann, E. & Marchant, H. (1996). Marine biotic responses to environmental change and feedbacks to climate. In *Climate Change 1995, The Science of Climate Change*, ed. J. T. Houghton, L. G. Meira Filho, B. A. Callander, N. Harris, A. Kattenberg & K. Maskell, pp. 483–516. Cambridge University Press.

DiTullio, G. R. & Laws, E. A. (1991). Impact of an atmospheric-oceanic disturbance on phytoplankton community dynamics in the North Pacific central gyre. *Deep-Sea Research*, **38**, 1305–29.

DiTullio, G. R., Hutchins, D. A. & Bruland, K. W. (1993). Interaction of iron and major nutrients controls phytoplankton growth and species composition in the tropical North Pacific Ocean. *Limnology and Oceanography*, **38**, 495–506.

Donaghay, P. L., Liss, P. S., Duce, R. A., Kester, D. R., Hanson, A. K., Villareal, T., Tindale, N. W. & Gifford, D. J. (1991). The role of episodic atmospheric nutrient inputs in the chemical and biological dynamics of oceanic ecosystems. *Oceanography*, **4**(2), 62–70.

Droop, M. R. (1974). The nutrient status of algal cells in continuous cultures. *Journal of the Marine Biological Association of the United Kingdom*, **54**, 825–55.

Dubischar, C. D. & Bathmann, U. V. (1997). Grazing impact of copepods and salps on phytoplankton in the Atlantic sector of the Southern Ocean. *Deep-Sea Research*, Part II, **44**, 415–34.

Duce, R. A. & Tindale, N. W. (1991). Atmospheric transport of iron and its

deposition in the ocean. *Limnology and Oceanography*, **36**, 1715–26.
Dugdale, R. C. & Wilkerson, F. P. (1990). Iron addition experiments in the Antarctic: a re-analysis. *Global Biogeochemical Cycles*, **4**, 13–19.
Dugdale, R. C., Wilkerson, F. P. & Minas, H. J. (1995). The role of the silicate pump in driving new production. *Deep-Sea Research*, **42**, 697–719.
Eicken, H. (1992). The role of sea-ice in structuring Antarctic ecosystems. *Polar Biology*, **12**, 3–13.
Eppley, R. W. (1972). Temperature and phytoplankton growth in the sea. *Fishery Bulletin*, **70**, 1063–85.
Eppley, R. W., Renger, E. H., Venrick, E. L. & Mullin, M. (1988). A study of plankton dynamics and nutrient cycling in the central gyre of the North Pacific Ocean. *Limnology and Oceanography*, **18**, 534–51.
Fanning, K. A. (1992). Nutrient provinces in the sea: concentration ratios, reaction ratios, and ideal covariation. *Journal of Geophysical Research*, **97**, 5693–712.
Feldman, G. (1991). Color image of HNLC waters with bloom west off the Galapagos Islands. In *What Controls Phytoplankton Production in Nutrient-rich Areas of the Open Sea?*, ed. S. W. Chisholm & F. M. M. Morel, pp. 1507–970. (*Limnology and Oceanography*, **36**.)
Feely, R. A., Wanninkhof, R., Cosca, C. E., Murphy, P. P., Lamb, M. F. & Steckley, D. M. (1995). CO_2 distributions in the equatorial Pacific during the 1991–1992 ENSO event. *Deep-Sea Research*, Part II, **42**, 365–86.
Fischer, G., Fuetterer, D., Gersonde, R., Honjo, S., Osterman, D. & Wefer, G. (1988). Seasonal variability of particle flux in the Weddell Sea and its relation to ice cover. *Nature*, **335**, 426–8.
Fitzwater, S. E., Coale, K. H., Gordon, R. M., Johnson, K. S. & Ondrusek, M. E. (1996). Iron deficiency and phytoplankton growth in the equatorial Pacific. *Deep-Sea Research*, Part II, **43**, 995–1015.
Frost, B. W. (1987). Grazing control of phytoplankton stock in the open subarctic Pacific Ocean: A model assessing the role of mesozooplankton, particularly the large calanoid copepods, *Neocalanus* spp. *Marine Ecology Progress Series*, **39**, 49–68.
Frost, B. W. (1991). The role of grazing in nutrient-rich areas of the open sea. *Limnology and Oceanography*, **36**, 1616–30.
Frost, B. W. & Franzen, N. C. (1992). Grazing and iron limitation in the control of phytoplankton stock and nutrient concentration: a chemostat analogue of the Pacific equatorial upwelling zone. *Marine Ecology Progress Series*, **83**, 291–303.
Fulton, J. D. (1978). Seasonal and annual variations of net zooplankton at Ocean Station P, 1965–1976. Canada Fisheries Marine Service Data Report, **49**.
Gargett, A. E. (1991). Physical processes and the maintenance of nutrient-rich euphotic zones. *Limnology and Oceanography*, **36**, 1527–45.
Geider, R. J. & La Roche, J. (1994). The role of iron in phytoplankton photosynthesis, and the potential for iron-limitation of primary productivity in the sea. *Photosynthesis Research*, **39**, 275–301.
Gledhill, M. & van den Berg, C. M. G. (1994). Determination of complexation of iron (III) with natural organic ligands in seawater using cathodic stripping voltametry. *Marine Chemistry*, **47**, 41–54.
Gledhill, M. & van den Berg, C. M. G. (1995). Determination of the redox speciation of iron in seawater by catalytic cathodic stripping voltametry. *Marine Chemistry*, **50**, 51–61.
Gordon, A. L., Molinelli, J. & Baker, T. N. (1986). *Southern Ocean Atlas*. Columbia University Press, New York (1982); Rotterdam: A. A. Balkema Publishers (1986).
Goeyens, L., Tréguer, P., Baumann, M. E. M., Baeyens, W. & Dehairs, F. (1995). The leading role of ammonium in the nitrogen uptake regime of Southern Ocean marginal ice zones. *Journal of Marine Systems*, **6**, 345–61.
Gran, H. H. (1931). On the conditions for the production of plankton in the sea. *Rapports et Procès verbaux des Réunions, Conseil International pour l'exploration de la Mer*, **75**, 37–46.

Greene, R. M., Kolber, Z. S., Swift, D. G., Tindale, N. W. & Falkowski, P. (1994). Physiological limitation of phytoplankton photosynthesis in the eastern equatorial Pacific determined from variability in the quantum yield of fluorescence. *Limnology and Oceanography*, **39**, 1061–74.

Hart, T. J. (1934). On the phytoplankton of the Southwest Atlantic and the Bellingshausen Sea, 1929–1931. *Discovery Reports*, 8, 1–268.

Hart, T. J. (1942). Phytoplankton periodicity in Antarctic surface waters. *Discovery Reports*, **21**, 261–365.

Helbling, E. W., Villafane, V. & Holm-Hansen, O. (1991). Effect of iron on productivity and size distribution of Antarctic phytoplankton. *Limnology and Oceanography*, **36**, 1879–85.

Heinrich, A. K. (1957). The propagation and the development of the common copepods in the Bering Sea. *Trudy Vsesoyuznogo Gidrobiologicheskogo Obshchestva*, **8**, 143–62.

Heinrich, A. K. (1962). The life history of planktonic animals and seasonal cycles of plankton communities in the oceans. *Journal du Conseil International pour l'exploration de la Mer*, **27**, 15–24.

Helmers, E. (1996). Trace metals in suspended particulate matter of Atlantic Ocean surface water. *Marine Chemistry*, **53**, 51–67.

Helmers, E. & Rutgers van der Loeff, M. M. (1993). Lead and aluminium in Atlantic surface waters (50°N to 50°S) reflecting anthropogenic and natural sources in the eolian transport. *Journal of Geophysical Research*, **89**, 20261–73.

Hempel, G. (ed.) (1993). Weddell Sea Ecology. Results of EPOS European Polarstern Study. *Polar Biology*, **12**, 333 pp.

Houghton, J. T., Meira Filho, L. G., Callander, B. A., Harris, N., Kattenberg, A. & Maskell, K. (eds) (1996). *Climate Change 1995, The Science of Climate Change*. Cambridge University Press.

Hudson, R. J. M. & Morel, F. M. M. (1990). Iron transport in marine phytoplankton: kinetics of cellular and medium coordination reactions. *Limnology and Oceanography*, **35**, 1002–20.

Hudson, R. J. M. & Morel, F. M. M. (1993). Trace metal transport by marine microorganisms: implications of metal coordination reactions. *Deep-Sea Research*, **40**, 129–50.

Hutchins, D. A. (1995). Iron and the marine phytoplankton community. In *Progress in Phycological Research*, vol. **11**, ed. F. E. Round & D. J. Chapman, pp. 1–48. Bristol, UK: Biopress Ltd.

Hutchins, D. A., DiTullio, G. R. & Bruland, K. W. (1993). Iron and regenerated production: evidence for biological iron recycling in two marine environments. *Limnology and Oceanography*, **38**, 1242–55.

Hutchins, D. A. & Bruland, K. W. (1994). Grazer-mediated regeneration and assimilation of Fe, Zn and Mn from planktonic prey. *Marine Ecology Progress Series*, **110**, 259–69.

Hutchins, D. A. & Bruland, K. W. (1995). Fe, Zn, Mn, and N transfer between size classes in a coastal phytoplankton community: trace metal and major nutrient recycling compared. *Journal of Marine Research*, **53**, 297–313.

Hutchins, D. A., Wang, W. & Fisher, N. S. (1995). Copepod grazing and the biogeochemical fate of diatom iron. *Limnology and Oceanography*, **40**, 989–94.

Jochem, F. J., Mathot, S. & Quéguiner, B. (1995). Size-fractionated primary production in the open Southern Ocean in austral spring. *Polar Biology*, **15**, 381–92.

Johnson, K. S., Coale, K. H., Elrod, V. A. & Tindale, W (1994). Iron photochemistry in seawater from the equatorial Pacific. *Marine Chemistry*, **46**, 319–35.

Jong, J. T. M. de, den Das, J., Bathmann, U. V., Stoll, M. H. C., Kattner, G., Nolting, R. E. & de Baar, H. J. W. (1998). Dissolved iron at subnanomolar levels in the Southern Ocean as determined by shipboard analysis. *Analytica et Chimica acta*, **377**, 113–24.

Joos, F., Sarmiento, J. L. & Siegenthaler, U. (1991*a*). Estimates of the effect of Southern Ocean iron fertilisation on atmospheric CO_2 concentrations. *Nature*,

349, 772–5.

Joos, F., Siegenthaler, U. & Sarmiento, J. L. (1991*b*). Possible effects of iron fertilisation in the Southern Ocean on atmospheric CO_2 concentration. *Global Biogeochemical Cycles*, **5**, 135–50.

Kamykowski, D. & Zentara, S. J. (1989). Circumpolar plant nutrient covariation in the Southern Ocean: patterns and processes. *Marine Ecology Progress Series*, **58**, 101–11.

Kessler, W. S. & McPhaden, M. J. (1995). The 1991–1993 El Niño in the central Pacific. *Deep-Sea Research*, Part II, **42**, 295–333.

Kolber, Z. S., Barber, R. T., Coale, K. H., Fitzwater, S. E., Greene, R. M., Johnson, K. S., Kudo, S. & Falkowski, P. (1994). Iron limitation of phytoplankton photosynthesis in the equatorial Pacific Ocean. *Nature*, **371**, 145–8.

Kudo, I. & Harrison, P. J. (1997). Effect of iron nutrition on the marine cyanobacterium *Synechococcus* grown on different N sources and irradiances. *Journal of Phycology*, **33**, 232–40.

Kumar, N., Anderson, R. F., Mortlock, R. A., Froelick, P. N., Kubik, P., Dittrich-Hannen, B. & Suter, M. (1995). Increased biological productivity and export production in the glacial Southern Ocean. *Nature*, **378**, 675–80.

Kurz, K. D. & Maier-Reimer, E. (1993). Iron fertilisation of the austral ocean – the Hamburg model assessment. *Global Biogeochemical Cycles*, **7**, 229–44.

Lalli, C. M. (ed.) (1991). *Enclosed Experimental Ecosystems; A Review and Recommendations*. New York: Springer-Verlag.

Lalli, C. M. & Parsons, T. R. (1993). *Biological Oceanography, an Introduction*. Oxford: Pergamon Press.

Lancelot, C. (1995). The mucilage phenomenon in the continental coastal waters of the North Sea. *Science of the Total Environment*, **165**, 83–102.

Lancelot, C., Mathot, S., Veth, C., & de Baar, H. J. W. (1993). Factors controlling phytoplankton ice-edge blooms in the marginal ice-zone of the northwestern Weddell Sea during sea ice retreat 1988: field observations and mathematical modeling. *Polar Biology*, **13**, 377–87.

Lancelot, C., Becquevort, S., Menon, P., Mathot, S. & Dandois, J. M. (1996). *Ecological modeling of the planktonic microbial food-web*. Report of contract A3/11/001 of the Third Antarctic Research Programme of Belgium.

Lancelot, C., Rousseau, V., Keller, M., Smith, W. & Mathot, S. (1998). Auto-ecology of *Phaeocystis* sp. blooms. In *Physiological Ecology of Harmful Algal Blooms*, ed. D. M. Anderson, A. D. Cembella & G. M. Hallegraef (NATO ASI Series, vol. G41). Berlin: Springer-Verlag.

Landry, M. R. & Hassett, R. P. (1982). Estimating the grazing impact of marine micro-zooplankton. *Marine Biology*, **67**, 283–8.

Landry, M. R. & Lehner-Fournier, J. M. (1988). Grazing rates and behaviors of *Neocalanus plumchrus*: Implications for phytoplankton control in the subArctic. *Pacific Hydrobiology*, **167/168**, 9–19.

Landry, M. R., Monger, B. C. & Selph, K. E. (1993). Time dependency of microzooplankton grazing and phytoplankton growth in the subArctic Pacific. *Progress in Oceanography*, **32**, 205–22.

La Roche, J., Boyd, P. W., McKay, R. M. L. & Geider, R. J. (1996). Flavodoxin as an in situ marker for iron stress in phytoplankton. *Nature*, **382**, 802–5.

Leeuwe, M.A. van (1997). A barren ocean: iron and light interactions with phytoplankton growth in the Southern Ocean. Ph.D. Thesis, University of Groningen.

Leeuwe, M. A. van, Scharek, R., de Baar, H. J. W., de Jong, J. T. M. & Goeyens, L. (1997). Iron enrichment experiments in the Southern Ocean: physiological responses of a plankton community. *Deep-Sea Research*, Part II, **44**, 189–208.

Leeuwe, M. A. van & Stefels, J. (1998). Effects of iron and light stress on the biochemical composition of Antarctic *Phaeocystis* sp. (Prymnesiophyceae). II. Pigment composition and fucoxanthin markers. *Journal of Phycology*, **34**,

496–503.
Leeuwe, M. A. van, Timmermans, K. R., Witte, H. J., Kraay, G. W., Veldhuis, M. J. W. & de Baar, H. J. W. (1998*a*). Effects of iron stress on chromatic adaptation by natural phytoplankton communities in the Pacific region of the Southern Ocean. *Marine Ecology Progress Series*, **166**, 43–52.
Leeuwe, M. A. van, de Baar, H. J. W. & Veldhuis, M. J. W. (1998*b*). Pigment distribution in the Pacific region of the Southern Ocean (autumn 1995). *Polar Biology*, **19**, 348–53.
Levitus, S., Conkright, M. E., Reid, J. L., Najjar, R. G. & Mantyla, A. (1993). Distribution of nitrate, phosphate and silicate in the world oceans. *Progress in Oceanography*, **31**, 245–73.
Lindley, S. T., Conkright, R. R. & Barber, R. T. (1995). Phytoplankton photosynthesis parameters along 140°W in the equatorial Pacific. *Deep-Sea Research*, Part II, **42**, 441–63.
Löscher, B. M., de Jong, J. T. M., de Baar, H. J. W., Veth, C. & Dehairs, F. (1997). The distribution of Iron in the Antarctic Circumpolar Current. *Deep-Sea Research*, Part II, **44**, 143–88.
Martin J. H. (1990). Glacial-Interglacial CO_2 change: the iron hypothesis. *Paleoceanography*, **5**, 1–13.
Martin, J. H. (1991*a*). Iron still comes from above. *Nature*, **353**, 123.
Martin, J. H. (1991*b*). Iron, Liebig and the greenhouse. *Oceanography*, **4**, 52–5.
Martin, J. H. (1992). Iron as a limiting factor in oceanic productivity. In *Primary Productivity and Biogeochemical Cycles in the Sea*, ed. P. G. Falkowski & A. D. Woodhead, pp. 123–37. (*Environmental Science Research*, **43**.) New York: Plenum Press.
Martin, J. H. & Fitzwater, S. E. (1988). Iron deficiency limits phytoplankton growth in the northeast Pacific subarctic. *Nature*, **331**, 341–3.
Martin, J. H. & Gordon, R. M. (1988). Northeast Pacific iron distributions in relation to phytoplankton productivity. *Deep-Sea Research*, **35**, 177–96.
Martin, J. H., Gordon, R. M., Fitzwater, S. E. & Broenkow, W. W. (1989). VERTEX: phytoplankton/iron studies in the Gulf of Alaska. *Deep-Sea Research*, **35**, 649–80.
Martin, J. H., Gordon, R. M. & Fitzwater, S. E. (1990*a*). Iron in Antarctic waters. *Nature*, **345**, 156–8.
Martin, J. H., Fitzwater, S. E. & Gordon, R. M. (1990*b*). Iron deficiency limits phytoplankton growth in Antarctic waters. *Global Biogeochemical Cycles*, **4**, 5–12.
Martin, J. H., Broenkow, W. W. Fitzwater, S. E. & Gordon, R. M. (1990*c*). Yes it does: a reply to the comment by Banse. *Limnology and Oceanography*, **35**, 775 7.
Martin, J. H., Gordon, R. M. & Fitzwater, S. E. (1991*a*). The case for iron. *Limnology and Oceanography*, **36**, 1793–802.
Martin, J. H., Fitzwater, S. E. & Gordon, R. M. (1991*b*). We still say iron deficiency limits phytoplankton growth in the subarctic Pacific. *Journal of Geophysical Research*, **96**, 699–700.
Martin, J. H., Fitzwater, S. E., Gordon, R. M., Hunter, C. N. & Tanner, S. J. (1993). Iron, primary production, and carbon-nitrogen flux studies during the JGOFS North Atlantic Bloom Experiment. *Deep-Sea Research*, **40**, 115–34.
Martin, J. H. *et al.* (1994). Testing the iron hypothesis in ecosystems of the equatorial Pacific Ocean. *Nature*, **371**, 123–9.
McAllister, C. D., Parsons, T. R. & Strickland, J. D. H. (1960). Primary productivity and fertility at station P in the north-east Pacific Ocean. *Journal du Conseil International pour l'exploration de la Mer*, **35**, 240–59.
McKelvey, B. A. (1994). *The marine geochemistry of zirconium and hafnium.* University of British Columbia, BC, Canada. 125 pp.
McKelvey, B. A. & Orians, K. J. (1993). Dissolved zirconium in the North Pacific Ocean. *Geochimica et Cosmochimica Acta*, **57**, 3801–5.
Michaelis, M. & Menten, M. L. (1913). Kinetics of invertase action. *Zeitschrift für Biochemie*, **49**, 333.

Michaels, A. F. & Knap, A. (1995). Overview of the U.S. JGOFS Bermuda Atlantic Time-series Study and the Hydrostation S program. *Deep-Sea Research*, Part II, **43**, 183.

Miller, C. B. (1993). Pelagic production processes in the subArctic Pacific. *Progress in Oceanography*, **32**, 1–15.

Miller, C. B. *et al.* (1991). Ecological dynamics in the subArctic Pacific, a possibly iron limited system. *Limnology and Oceanography*, **36**, 1662–77.

Mills, E. (1989). *Biological Oceanography. An early history, 1870–1960*. Ithaca: Cornell University Press.

Minas, H. J., Minas, M. & Packard, T. T. (1986). Productivity in upwelling areas deduced from hydrographic and chemical fields. *Limnology and Oceanography*, **31**, 1182–206.

Mitchell, B. G., Brody, E. A., Holm-Hansen, O., McClain, C. & Bishop, J. (1991). Light limitation of phytoplankton biomass and macronutrient utilisation in the Southern Ocean. *Limnology and Oceanography*, **36**, 1662–77.

Monod, J. (1950). La technique de la culture continue: Theorie et applications. *Annales de l'Institut Pasteur, Lille*, **79**, 390–410.

Morel, A. & Bricaud, A. (1981). Theoretical results concerning light absorption in a discrete medium and application to specific absorption of phytoplankton. *Deep-Sea Research*, **28**, 1375–93.

Morel, F. M. M., Hudson, R. J. & Price, N. M. (1991*a*). Limitation of productivity by trace metals in the sea. *Limnology and Oceanography*, **36**, 1742–55.

Morel, F. M. M., Reuter, J. G. & Price, N. M. (1991*b*). Iron nutrition of phytoplankton and its possible importance in the ecology of ocean regions with high nutrients and low biomass. *Oceanography*, **4**, 56–61.

Muggli, D. M. & Harrison, P. J. (1996). Effects of nitrogen source on the physiology and metal nutrition of *Emiliania huxleyi* grown under different iron and light conditions. *Marine Ecology Progress Series*, **130**, 255–67.

Muggli, D. M., LeCourt, M. & Harrison, P. J. (1996). The effects of iron and nitrogen source on the sinking rate, physiology and metal composition of an oceanic diatom from the subArctic Pacific. *Marine Ecology Progress Series*, **132**, 215–27.

Murray, J. W., Johnson, E. & Garside, C. (1995*a*). A U.S. JGOFS Process Study in the equatorial Pacific (EqPac): Introduction. *Deep-Sea Research*, **42**, 275–93.

Murray, R. W., Leinen, M., Murray, D. W., Mix, A. C. & Knowlton, C. W. (1995*b*). Terrigenous Fe input and biogenic sedimentation in the glacial and interglacial equatorial Pacific Ocean. *Global Biogeochemistry Cycles*, **9**, 667–84.

Nelson, D. M. & Smith, W. O. (1991). Sverdrup revisited: Critical depths, maximum chlorophyll levels, and the control of the Southern Ocean productivity by the irradiance-mixing regime. *Limnology and Oceanography*, **36**, 1650–61.

Nelson, D. M. & Tréguer, P. (1992). Role of silicon as a limiting nutrient to Antarctic diatoms: Evidence for kinetic studies in the Ross Sea ice edge zone. *Marine Ecology Progress Series*, **62**, 283–92.

Nolting, R. F., de Baar, H. J. W., van Bennekom, A. J. & Masson, A. (1991). Cadmium, copper and iron in the Scotia Sea, Weddell Sea and Weddell/Scotia Confluence (Antarctica). *Marine Chemistry*, **35**, 219–43.

Nolting, R. F., Gerringa, L. J. A., Swagerman, M. J. W., Timmermans, K. R. & de Baar, H. J. W. (1998). Fe(III) speciation in the High Nutrient Low Chlorophyll Pacific region of the Southern Ocean. *Marine Chemistry*, **62**, 335–52.

Obata, H., Karatani, H. & Nakayama, E. (1993). Automated determination of iron in seawater by chelating resin concentration and chemiluminescence detection. *Analytical Chemistry*, **65**, 1524–8.

O'Neill, R. V., DeAngelis, D. L., Pastor, J. J., Jackson, B. J. & Post, W. M. (1989). Multiple nutrient limitations in ecological models. *Ecological Modelling*, **46**, 147–63.

Parslow, J. S. (1981). Phytoplankton-zooplankton interactions: data analysis and modelling (with particular reference to Ocean Station P (50°N, 145°W) and controlled ecosystem experiments. Ph.D. Univ. British Columbia.

Peng, T. H. & Broecker, W. S. (1991*a*). Dynamic limitations on the Antarctic iron fertilisation strategy. *Nature*, **349**, 227–9.

Peng, T. H. & Broecker, W. S. (1991*b*). Factors limiting the reduction of atmospheric CO_2 by iron fertilisation. *Limnology and Oceanography*, **36**, 1919–27.

Peterson, R. G. & Stramma, L. (1991). Upper-level circulation in the South Atlantic Ocean. *Progress in Oceanography*, **26**, 1–73.

Platt, T., Gallegos, C. L. & Harrison, W. G. (1980). Photoinhibition of photosynthesis in natural assemblages of marine phytoplankton. *Journal of Marine Research*, **38**, 687–701.

Powell, R. T., Whitney King, D. & Landing, W. M. (1995). Iron distributions in surface waters of the south Atlantic. *Marine Chemistry*, **50**, 13–20.

Price, N. M., Andersen, L. F. & Morel, F. M. M. (1991). Iron and nitrogen nutrition of equatorial Pacific plankton. *Deep-Sea Research*, **38**, 1361–78.

Price, N. M., Ahner, B. A. & Morel, F. M. M. (1994). The equatorial Pacific Ocean: grazer-controlled phytoplankton populations in an iron-limited ecosystem. *Limnology and Oceanography*, **39**, 520–34.

Quéguiner, B., Tréguer, P., Peeken, I. & Scharek, R. (1997). Biogeochemical dynamics and the silicon cycle in the Atlantic sector of the Southern Ocean during austral spring 1992. *Deep-Sea Research*, Part II, **44**, 69–91.

Raven, J. A. (1988). The iron and molybdenum use efficiencies of plant growth with different energy, carbon, and nitrogen sources. *New Phytologist*, **109**, 279–87.

Raven, J. A. (1990). Predictions of Mn and Fe use efficiencies of phototrophic growth as a function of light availability for growth. *New Phytologist*, **116**, 1–18.

Reid, J. L. (1962). On the circulation, the phosphate-phosphorus content, and the zooplankton volumes in the upper part of the Pacific Ocean. *Limnology and Oceanography*, **7**, 287–306.

Reid, R. T. & Butler, A. (1991). Investigation of the mechanism of iron acquisition by the marine bacterium *Alteromonas lutioviolaceu*: characterization of siderophore production. *Limnology and Oceanography*, **36**, 1783–92.

Riley, G. A. (1946). Factors controlling phytoplankton populations on Georges Bank. *Journal of Marine Research*, **6**, 54–73.

Rue, E. L. & Bruland, K. W. (1995). Complexation of iron (III) by natural organic ligands in the Central North Pacific as determined by a new competitive ligand equilibration/adsorptive cathodic stripping voltametric method. *Marine Chemistry*, **50**, 117–38.

Rueter, J. G. & Ades, D. R. (1987). The role of iron nutrition in photosynthesis and nitrogen assimilation in *Scenedesmus quadricauda* (Chlorophyceae). *Journal of Phycology*, **32**, 452–7.

Rutgers van der Loeff, M., Friedrich, J. & Bathmann, U. V. (1997*a*). Carbon export during the spring bloom at the Antarctic Polar Front determined with the natural tracer ^{234}Th. *Deep-Sea Research*, Part II, **44**, 457–78.

Rutgers van der Loeff, M., Helmers, E. & Kattner, G. (1997*b*). Continuous transects of cadmium, copper, and aluminium in surface waters of the Atlantic Ocean, 50°N to 50°S. *Geochimica et Cosmochimica Acta*, **61**, 47–62.

Ruud, J. T. (1930). Nitrates and phosphates in the Southern Seas. *Rapports et Proces verbaux des Réunions, Conseil International pour l'exploration de la Mer*, **5**, 347–60.

Saager, P. M., de Baar, H. J. W. & Burkill, P. H. (1989). Manganese and iron in Indian Ocean waters. *Geochimica et Cosmochimica Acta*, **53**, 2259–67.

Sakshaug, E., Slagstad, D. & Holm-Hansen, O. (1991). Factors controlling the development of phytoplankton blooms in the Antarctic Ocean – a mathematical model. *Marine Chemistry*, **35**, 259–71.

Sambrotto, R. N., Savidge, G., Robinson, C., Boyd, P., Takahashi, T., Karl, D. M., Langdon, C., Chipman, D., Marra, J. & Codispoti, J. (1993). Elevated consumption of carbon relative to nitrogen in the surface ocean. *Nature*, **363**, 248–50.

Sarmiento, J. L. & Orr, J. C. (1991). Three-dimensional simulations of the impact of Southern Ocean nutrient depletion on atmospheric CO_2 and ocean chemistry. *Limnology and Oceanography*, **36**, 1928–50.

Sarthou, G., Jeandel, C., Brisset, L., Amouroux, D., Besson, T. & Donard, O. F. X. (1997). Fe and H_2O_2 distributions in the upper water column in the Indian sector of the Southern Ocean. *Earth and Planetary Science Letters*, **147**, 83–92.

Scharek, R., van Leeuwe, M. A. & de Baar, H. J. W. (1997). Responses of Antarctic phytoplankton to the addition of trace metals. *Deep-Sea Research*, Part II, **44**, 209–28.

Sedwick, P. N., Edwards, P. R., Mackey, D. J., Griffiths, F. B. & Parslow, J. S. (1999). Iron and manganese in surface waters of the Australian Subantarctic region. *Deep-Sea Research*, **44**(7), 1239–53.

Smetacek, V. (1985). The role of sinking in diatom life-history cycles: ecological, evolutionary and geological significance. *Marine Biology*, **84**, 239–51.

Smetacek, V., Scharek, R. & Nöthig, E.-M. (1990). Seasonal and regional variation in the pelagial and its relationship to the life history cycle of krill. In *Antarctic Ecosystems. Ecological Change and Conservation*, ed. K. R. Kerry & G. Hempel, pp. 103–14. Heidelberg: Springer-Verlag.

Smetacek, V., de Baar, H. J. W., Bathmann, U. V., Lochte, K. & Rutgers van der Loeff, M. M. (1997). Ecology and Biogeochemistry of the Antarctic Circumpolar Current during Austral Spring: a summary of Southern Ocean JGOFS Cruise ANT X/6 of RV Polarstern. *Deep-Sea Research*, Part II, **44**, 1–22.

Smith, W. O. (1991). Nutrient distributions and new production in polar regions: parallels and contrasts between Arctic and Antarctic. *Marine Chemistry*, **35**, 245–57.

Soria-Deng, S. & Horstmann, U. (1995). Ferrioxamines B and E as iron sources for the marine diatom *Phaeodactylum tricornutum. Marine Ecology Progress Series*, **127**, 269–77.

Stefels, J. & Leeuwe, M. A. van. (1998). Effects of iron and light stress on the biochemical composition of Antarctic *Phaeocystis* sp. (Prymnesiophyceae). I. Intracellular DMSP concentrations. *Journal of Phycology*, **34**, 486–95.

Strom, S. & Welschmeyer, N. A. (1991). Pigment-specific rates of phytoplankton growth and microzooplankton grazing in the open subArctic Pacific Ocean. *Limnology and Oceanography*, **36**, 50–63.

Sullivan, C. W., Arrigo, K. R., McClain, C. R., Comiso, J. C. & Firestone, J. (1993). Distribution of Phytoplankton Blooms in the Southern Ocean. *Science*, **262**, 1832–7.

Sunda, W. G. & Huntsman, S. A. (1995). Iron uptake and growth limitation in oceanic and coastal phytoplankton. *Marine Chemistry*, **50**, 189–206.

Sunda, W. G., Swift, D. & Huntsman, S. A. (1991). Iron growth requirements in oceanic and coastal phytoplankton. *Nature*, **351**, 55–7.

Takahashi, K., Billings, J. D. & Morgan, J. K. (1990). Oceanic province: Assessment from the time-series diatom fluxes in the northeastern Pacific. *Limnology and Oceanography*, **35**, 154–65.

Takeda, S. & Obata, H. (1995). Response of equatorial Pacific phytoplankton to subnanomolar Fe enrichment. *Marine Chemistry*, **50**, 219–27.

Takeda, S., Kamatani, A. & Kawanobe, K. (1995). Effects of nitrogen and iron enrichments on phytoplankton communities in the northwestern Indian Ocean. *Marine Chemistry*, **50**, 229–41.

Taylor, A. H., Harbour, D. S., Harris, R. P., Burkill, P. H. & Edwards, E. S. (1993). Seasonal succession in the pelagic ecosystem of the North Atlantic and the utilisation of nitrogen. *Journal of Plankton*

Research, **15**, 875–91.

Thomas, W. H. (1969). Phytoplankton nutrient enrichment experiments off Baja California and in the Eastern Equatorial Pacific Ocean. *Journal of the Fisheries Research Board, Canada*, **26**, 1133–45.

Thomas, W. H. (1979). Anomalous nutrient chlorophyll interrelationships in the off-shore eastern tropical Pacific. *Journal of Marine Research*, **37**, 327–35.

Timmermans, K. R., Stolte, W. & de Baar, H. J. W. (1994). Iron-mediated effects on nitrate reductase in marine phytoplankton. *Marine Biology*, **121**, 389–96.

Timmermans, K. R., Leeuwe, M.A. van, de Jong, J. T. M., McKay, R., Nolting, R. F., Witte, H., Ooyen, J. van, Swagerman, M., Kloosterhuis, H. & de Baar, H. J. W. (1998). Iron limitation in the Pacific region of the Southern Ocean. *Marine Ecology Progress Series*, **166**, 27–41.

Tortell, P. D., Maldonado, M. T. & Price, N. M. (1996). The role of heterotrophic bacteria in iron-limited ocean ecosystems. *Nature*, **383**, 330–2.

Townsend, D. W., Cammen, L. M., Holligan, P. M., Campbell, D. A. & Pettigrew, N. R. (1994). Causes and consequences of variability in the timing of spring phytoplankton blooms. *Deep-Sea Research*, **41**, 747–65.

Tranter, D. J. (1982). Interlinking of physical and biological processes in the Antarctic Ocean. *Oceanography and Marine Biology, Annual Review*, **20**, 11–35.

Trick, C. G., Anderson, R. J., Price, N. M., Gillam, A. & Harrison, P. J. (1983). Prorocentrin: an extracellular siderophore produced by the marine dinoflagellate *Prorocentrum minimum*. *Science*, **219**, 306–8.

Trick, C. G. & Wilhelm, S. W. (1995). Physiological changes in the coastal marine cyanobacterium *Synechococcus* sp. PCC 7002 exposed to low ferric iron levels. *Marine Chemistry*, **50**, 207–18.

Turner, D., Owens, N. & Priddle, J. (eds.) (1995). Southern Ocean JGOFS: The U.K. 'STERNA' Study in the Bellingshausen Sea. *Deep-Sea Research*, Part II, **42**, 905–1335.

Turner, S. M., Nightingale, P. D., Spoeks, L. J., Liddicoat, M. I. & Liss, P. (1996). Increased dimethyl sulphide concentrations in sea water from *in situ* iron enrichment. *Nature*, **383**, 513–17.

Veth, C., Peeken, I. & Scharek, R. (1997). Physical anatomy of fronts and surface waters in the ACC near the 6 degree West meridian during austral spring 1992. *Deep-Sea Research*, Part II, **44**, 23–50.

Waite, A. & Harrison, P. J. (1992). Role of sinking and ascent during sexual reproduction in the marine diatom *Ditylum brightwellii*. *Marine Ecology Progress Series*, **87**, 113–22.

Waite, T. D., Szymczak, R., Esprey, Q. I. & Furnas, J. M. (1995). Diel variations in iron speciation in northern Australian shelf waters. *Marine Chemistry*, **50**, 79–92.

Walsh, J. J. (1976). Herbivory as a factor in patterns of nutrient utilisation in the sea. *Limnology and Oceanography*, **21**, 1–13.

Wanninkhof, R., Feely, R. A., Atwood, D. K., Berberian, G., Wilson, D., Murphy, P. P. & Lamb, M. F. (1995). Seasonal and lateral variations in carbon chemistry of surface water in the eastern equatorial Pacific during 1992. *Deep-Sea Research*, Part II, **42**, 387–409.

Watson, A. J., Liss, P. & Duce, R. (1991). Design of a small-scale in situ iron fertilisation experiment. *Limnology and Oceanography*, **36**, 1960–5.

Watson, A. J. *et al.* (1994). Minimal effect of iron fertilisation on sea-surface carbon dioxide concentrations. *Nature*, **371**, 143–5.

Wefer, G. & Fischer, G. (1991). Annual primary production and export flux in the Southern Ocean from sediment trap data. *Marine Chemistry*, **35**, 597–614.

Wefer, G., Fischer, G., Fütterer, D. & Gersonde, R. (1988). Seasonal particle flux in the Bransfield Strait, Antarctica. *Deep-Sea Research*, **35**, 891–8.

Weiss, R. F. (1974). Carbon dioxide in water and seawater: the solubility of a non-ideal gas. *Marine Chemistry*, **2**, 203–5.

Wells, M. L., Mayer, L. M. & Guillard, R. R. L. (1991). Evaluation of iron as a triggering factor for red tide blooms. *Marine Ecology Progress Series*, **69**, 93–102.

Wells, M. L., Price, N. M. & Bruland, K. W. (1995). Iron chemistry in seawater and its relationship to phytoplankton: a workshop report. *Marine Chemistry*, **48**, 157–82.

Welschmeyer, N. A., Goericke, R., Strom, S. & Peterson, W. (1991). Phytoplankton growth and herbivory in the subArctic Pacific: a chemotaxonomic analysis. *Limnology and Oceanography*, **36**, 1631–49.

Welschmeyer, N. A., Strom, S., Goericke, R., DiTullio, G., Belvin, M. & Peterson, W. (1993). Primary production in the subarctic Pacific Ocean: Project SUPER. *Progress in Oceanography*, **32**, 101–36

Westerlund, S. & Öhman, P. (1991). Iron in the water column of the Weddell Sea. *Marine Chemistry*, **35**, 199–217.

Wheeler P. A. & Kokkinakis, S. A. (1990). Ammonium recycling limits nitrate use in oceanic subarctic Pacific. *Limnology and Oceanography*, **35**, 1267–78.

White, W. B. & Peterson, R. G. (1996). An Antarctic circumpolar wave in surface pressure, wind, temperature and sea-ice extent. *Nature*, **380**, 699–701.

Whitworth, T. III (1988). The Antarctic Circumpolar Current. *Oceanus*, **31**, 53–8.

Wong, C. S., Whitney, F. A., Iseki, K., Page, J. S. & Zeng, J. (1994). Analysis of trends in primary productivity and chlorophyll-a over two decades at Ocean Station P (50°N, 145°W) in the Subarctic Northeast Pacific Ocean. *Canadian Journal of Fisheries and Aquatic Science*, **121**, 107–17.

Wu, J. & Luther, G. W. III (1994). Size-fractionated iron concentrations in the water column of the Northwest Atlantic Ocean. *Limnology and Oceanography*, **39**, 1119–29.

Young, R. W. *et al.* (1991). Atmospheric iron inputs and primary productivity: phytoplankton responses in the north Pacific. *Global Biogeochemical Cycles*, **5**, 119–34.

Zettler, G. R., Olson, R. J., Binder, B. J., Chisholm, S. W., Fitzwater, S. E. & Gordon, R. M. (1996). Iron-enrichment bottle experiments in the equatorial Pacific: response of individual phytoplankton cells. *Deep-Sea Research*, Part II, **43**, 1017–29.

5 The influence of iron on ocean biology and climate: insights from the IronEx studies in the equatorial Pacific

P. S. Liss and S. M. Turner

Keywords: iron, carbon dioxide, dimethyl sulphide, phytoplankton, sulphur hexafluoride, IronEx, high nutrient – low chlorophyll

Introduction

Over much of the ocean, primary production is limited by the availability of light and nutrients (nitrate, phosphate and silicate). However, for about 20% of the area of the ocean, light and nutrients are abundant in surface waters, although chlorophyll concentrations are lower than expected. It has been hypothesised that in these regions marine plants cannot fully exploit the major nutrients because of limited availability of the trace element iron.

The IronEx I and II field experiments in the equatorial Pacific in October 1993 and June 1995 gave an excellent opportunity to test the iron limitation hypothesis *in situ*. A patch of ocean water approximately 100 km^2 was dosed with ferrous sulphate to raise its iron concentration by about a nanomole. As well as iron, sulphur hexafluoride was also added as an inert tracer to allow the iron-enriched patch of water to be tracked in near-real time from aboard ship.

Addition of even this minute amount of iron produced very significant changes in primary production in the water, which were particularly marked in IronEx II. Here chlorophyll increased by up to 30-fold, accompanied by large decreases in p_{CO_2} (by up to 90 μatm) and nitrate, and corresponding increases in water dimethyl sulphide (DMS) concentration by up to 3.5-fold.

It is clear from the IronEx results that, at least in the two areas of the equatorial Pacific studied, availability of iron was indeed limiting primary production. However, the implications of these results for both ocean biogeochemistry and climate, now, in the past and in the future are far from clear and substantial further investigation is required.

The findings summarised above are discussed further in this chapter, but first we make some brief general comments on the role of nutrient availability and limitation in plant productivity.

It is well known that, on land, one of the most important factors limiting the growth of plants is the availability of nutrient elements, in particular nitrogen (N) and phosphorus (P). This knowledge is exploited by farmers, who increase crop yields by addition of fertilisers containing these and other elements. In some special cases one of these other elements is iron, which is often deficient in lime-rich soils. For example, under such conditions rhododendrons and fruit trees need supplements of the element for proper growth.

In an analogous way the growth of plants (phytoplankton) in the sea is often limited by the availability of the same nutrient elements. From the early years of this century it was known that upwelling of water rich in N and P was a vital factor controlling plankton growth. In addition, in the late 1920s and early 1930s, marine scientists realised that some chemical factor besides N and P must be limiting plankton growth over large areas of the remote oceans. Thus, in the Southern Oceans it was often found that, in austral spring and summer, when light conditions were ideal for plankton to multiply, the biomass was low despite there being plenty of N and P in the water, a phenomenon called the 'Antarctic Paradox'. These regions have lately been called high nutrient – low chlorophyll (HNLC) areas. It is estimated that they cover about 20% of the total surface oceans, mainly in the equatorial and north Pacific and the Southern Oceans (de Baar & Boyd, Chapter 4, this volume).

Soon after the identification of such HNLC areas it was proposed that their existence was due to a shortage of iron in the water (see, for example, Gran, 1931; Harvey, 1933; Hart, 1934). However, unambiguous experimental verification of the hypothesis was difficult because of the problems at that time of measuring the very low (nanomolar) concentrations of iron that exist in seawater. Even now, when the analytical techniques are orders of magnitude more sensitive, considerable precautions must be taken in analysing for iron and in shipboard iron-enrichment experiments to avoid contamination from the research vessel, sampling apparatus and chemicals and glassware used in the experiments and subsequent chemical analyses (Fitzwater *et al.*, 1996).

These difficulties have meant that little progress was made on this topic until the advent of clean techniques for trace-metal sampling and analysis of seawater in the last twenty years. A pioneer worker in these more recent studies was the late John Martin. He made some of the very first reliable measurements of iron in the oceans and conducted shipboard incubation studies in which flasks of seawater had soluble iron added to them. The results were promising (for example, Martin & Fitzwater, 1988) and clearly indicated that additions of iron could lead to substantial increases in plankton growth, as judged by elevated chlorophyll concentrations in the flasks. However, there was uncertainty as to how well the shipborne incubations in small flasks mirrored conditions in the oceans, for example whether zooplankton grazing can be properly mimicked. Further, and rather disturbingly, in some cases the blank (no added iron) flasks also showed enhanced plankton growth, probably owing to contamination with iron when the water used in the flask experiments was collected (Fitzwater *et al.*, 1996).

In the following sections, we discuss the results of a novel approach to testing the iron hypothesis, concentrating on the implications of the findings for ocean biology and climate.

The IronEx experiments

Further progress in iron studies awaited a real ocean experiment, which would allow a small patch of surface ocean water to be dosed with iron, and then followed for several days. The approach used in the experiments called IronEx was to mark a small area of water (of the order of 100 square kilometres) with the gas sulphur hexafluoride (SF_6), as a tracer of the movement of the patch. Previously, SF_6 has been used as a purposeful tracer in a variety of studies, including determination of eddy diffusivity (Ledwell & Watson, 1991; Ledwell *et al.*, 1993) and air–sea gas exchange (Watson *et al.*, 1991*a*; Wanninkhof *et al.*, 1993).

Initially, the concentration of SF_6 in the IronEx patches was about 200 fmol (1 fmol = 10^{-15} mol), which was easily detectable given the very high sensitivity of the electron capture detector used in the analysis. Iron was added simultaneously with the tracer. Because sulphur hexafluoride can be rapidly measured aboard a research vessel, the output of the gas chromatograph was linked to the ship's navigational computer and the movement of the iron-enriched patch visualised in near-real time (Upstill-Goddard *et al.*, 1991). One of the major benefits of the approach is that by sailing the vessel into and out of the enriched patch, measurements of the active (within patch) and blank (outside patch) water can be made. In this way the water outside the patch acts as a true 'control' against which any changes observed in the iron-enriched patch can be judged. It should be pointed out that iron itself could not be used to follow the patch because it was not possible to detect it sensitively or rapidly enough.

The principle of the 'Lagrangian' approach described above was first outlined by Watson *et al.* (1991*b*). It is illustrated in Fig. 5.1, where the concentration of a biological parameter (in this case p_{CO_2}) in the fertilised water is plotted against SF_6 concentration at three times (0, 3 and 5 d) during the lifetime of the patch. As the patch ages, the concentration of SF_6 decreases owing to both lateral

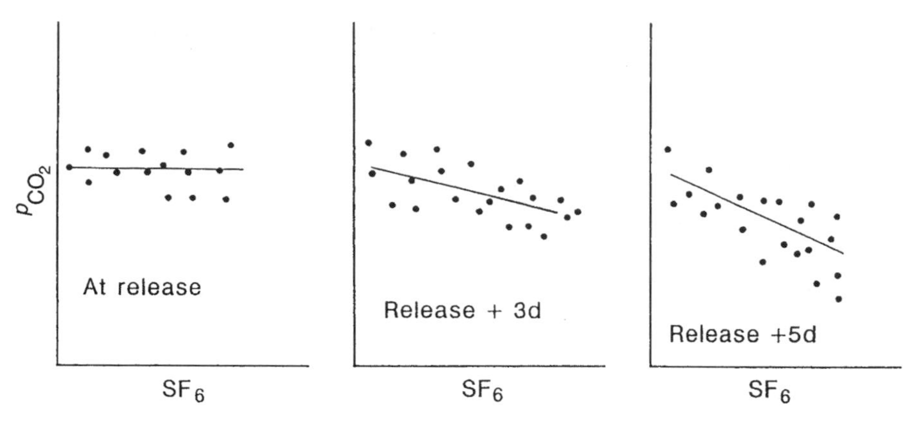

Figure 5.1 *Results from a hypothetical experiment to test whether iron fertilisation affects marine biological activity. Biologically affected chemical constituents such as CO_2 (illustrated here) and DMS should develop significant covariation with the SF_6 concentration in the fertilised patch in the days following addition, if the release has an effect on the biota* in situ *(from Watson* et al., *1991*b*).*

mixing and loss to the atmosphere. In this hypothetical example, enhanced primary production due to the added iron leads to p_{CO_2} decreases with time inside the patch. For a parameter such as dissolved dimethyl sulphide (DMS), concentrations might be expected to increase as a result of enhanced phytoplankton activity and the plots of DMS against SF_6 would show positive slopes, in contrast to the negative ones shown in Fig. 5.1 for CO_2. Real results from the use of this approach in the field are shown later (Fig. 5.3).

The IronEx experiments took place in the equatorial Pacific, a known HNLC area. They were performed twice, once in October 1993 (IronEx I) and again in June 1995 (IronEx II). In IronEx I, the enriched patch was subducted after about 4 d owing to a lens of fresher water submersing the labelled water to a depth of about 25 metres. This made interpretation of the results difficult, because the environmental conditions at this depth were clearly considerably different from those near the surface. In IronEx II the first of two marked patches was followed for up to 17 d, during which time it remained at the surface but deepened with time, and moved laterally by about 1500 km. In both experiments, the iron was added as an acidic ferrous sulphate solution to achieve an increase in iron concentration of the order of 1 nmol in the dosed patch of seawater. Further details of the experiments are given in Martin *et al.* (1994) for IronEx I and Coale *et al.* (1996) for IronEx II. Detailed results of the two experiments are also given in these and the accompanying papers (i.e. Kolber *et al.*, 1994 and Watson *et al.*, 1994 for IronEx I; Behrenfeld *et al.*, 1996, Cooper *et al.*, 1996 and Turner *et al.*, 1996 for IronEx II).

In the enriched IronEx I patch, biological activity showed a significant increase compared with outside; for example, chlorophyll concentrations were some three times higher. Nutrients, such as nitrate, phosphate and silicate, exhibited no significant difference between the inside and outside of the patch, although ammonium was consistently lower in the patch compared with the control. Of the climatically active trace gases, carbon dioxide was only slightly decreased in the patch (Watson *et al.*, 1994) and DMS concentrations did not show any significant changes, although its algal precursor (dimethylsulphonio-propionate, DMSPp) did increase 3–fold (Turner *et al.*, 1996). The small drawdown in carbon dioxide (about 5 μatm) is consistent with the increase in algal biomass, and represents only 10% of the potential drawdown if all the available nitrate and phosphate had been utilised. Because the rise in algal biomass was accompanied by increased grazing (50% increase in the microzooplankton biomass and rises in mesozooplankton numbers) it is likely that the system quickly reached steady state with carbon dioxide being recycled back into the water, i.e. carbon fixation and respiration rates quickly became equivalent. It is surprising that DMS concentrations did not increase, considering the 3-fold rise in its precursor. It is possible that bacterial utilisation of DMS was stimulated to the extent that no net change in DMS concentrations could be observed. The main conclusion of the IronEx I study was that iron did appear to lead to enhanced biological activity, but that the biological and geochemical responses were much smaller than might have been expected.

The IronEx II study was designed to test the hypotheses that had been advanced to explain the limited response in the first experiment (Coale *et al.*, 1996). Most of the added iron had very rapidly formed a suspension of colloidal oxyhydroxides, which are thought not to be available to the plankton. Further, when the patch was subducted to lower light intensities on the fourth day, photo-dissolution of the colloids would have been minimised. Thus, in IronEx II further increments of iron were injected into the patch 4 and 7 d after the initial addition. The results were dramatic, with chlorophyll concentrations rising by up to thirty-fold, with corresponding changes in related parameters such as nitrate. Indeed, it was possible by visual observation from the ship to know whether or not it was within the fertilised patch, because the colour of the water changed from blue outside to a green-brown colour inside, owing to the enhanced phytoplankton, particularly diatom, growth. That diatoms were the major group of organisms to benefit from the supplementary iron is shown clearly in Fig. 5.2. This indicates that, although most groups increased in biomass following fertilisation with iron, the pennate diatoms showed by far the biggest effect.

In IronEx II there was a large decrease in the partial pressure of carbon dioxide of up to 90 μatm, corresponding to a transient 60% reduction in the normal sea-to-air flux of the gas found in this outgassing region (Cooper *et al.*,

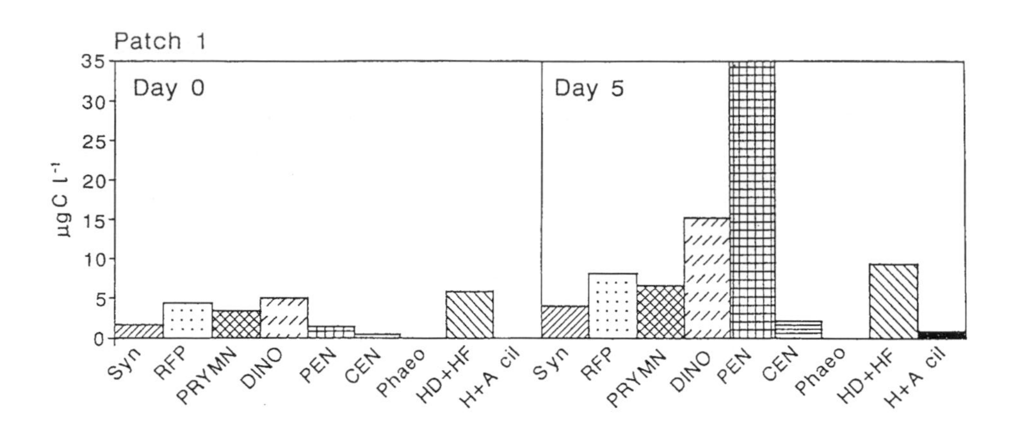

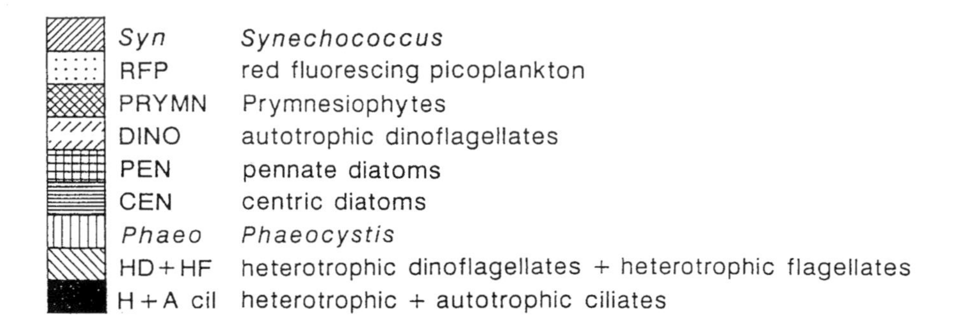

Figure 5.2 *Plankton community composition (expressed in µg c l^{-1}) within the main IronEx II patch before addition of iron (Day 0) and 5 d after fertilisation. The composition shown for Day 0 is similar to that observed in the 'outside patch' stations over time (after Coale* et al., *1996).*

1996). This effect on p_{CO_2} is clearly illustrated in Fig. 5.3*b*, where the results are plotted in a similar way to those shown in Fig. 5.1. The authors argue that, in the longer term and considering the experimental area in the wider context of the Pacific, the overall effects on atmospheric levels of carbon dioxide are likely to be very small, owing to the way in which the circulation system supplies nutrients to the surface waters. If nutrients are used more efficiently in the tropical Pacific then there are lesser amounts to drive production in sub-tropical areas. However, Cooper *et al.* (1996) recommend that an iron fertilisation experiment be performed in the Southern Oceans, because the circulation regime there is more likely to lead to a climatically significant drawdown of atmospheric carbon dioxide.

The results of IronEx I and II show a similar increase in DMSPp (3- to 4-fold) in the iron-enriched patches, even though there was a much greater change in biomass in IronEx II. The reason why DMSPp increased by a factor

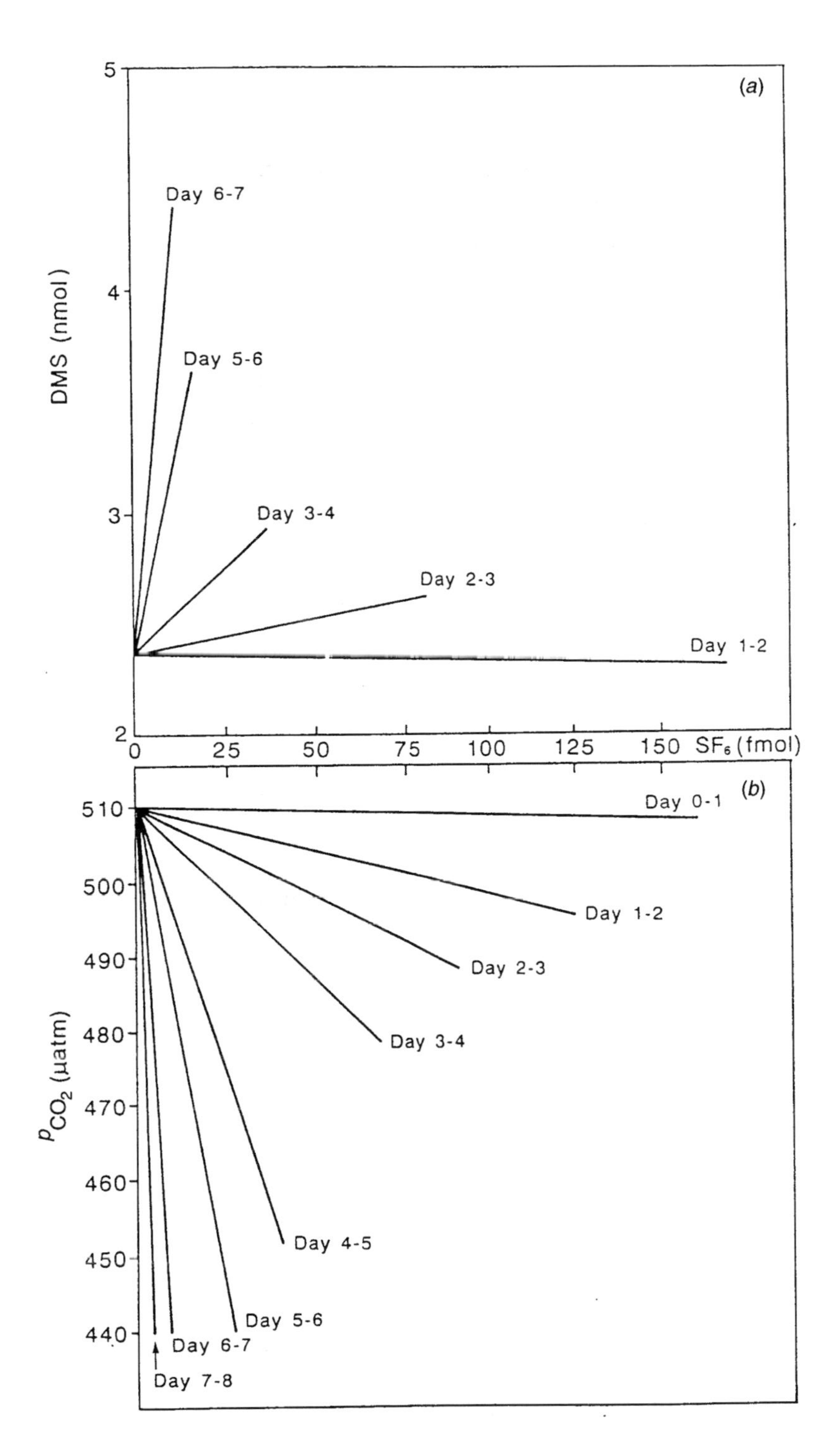

Figure 5.3 *Evolution of the DMS* (a) *and* p_{CO_2} (b) *increase and drawdown, respectively, in surface waters measured during IronEx II, plotted as a function of* SF_6 *concentration (after Turner* et al., *1996 and Cooper* et al., *1996). This diagram is a real example of the hypothetical situation illustrated for* CO_2 *in Fig. 5.1.*

of 10 less than that for chlorophyll is almost certainly because so much of the biomass enhancement was due to growth of diatoms (Fig. 5.2), which are known to be poor producers of DMSP and DMS (Liss *et al.*, 1994). The groups of phytoplankton that are known to include species with high DMSPp: cell carbon ratios are the prymnesiophytes and the dinoflagellates, and Fig. 5.2 shows that there were up to 3-fold increases in their biomass. DMS concentrations increased in both IronEx II patches and the increase was delayed until a couple of days after the rise in chlorophyll and DMSPp (Fig. 5.3*a*). This occurs because DMS is a degradation product of DMSP and does not appear in the water until the plankton cells have lysed or been consumed by zooplankton (Belviso *et al.*, 1990). The sea-to-air flux of DMS is mainly controlled by its concentration at the sea surface and wind speed. Thus, for a 3–fold increase in DMS with no change in wind speed, there would be a 3–fold increase in the amount of DMS entering the atmosphere. As mentioned earlier, iron enrichment of the equatorial and sub-tropical Pacific would have little overall effect on atmospheric CO_2, but it is not clear how the predicted change in nutrient supply would affect DMS fluxes in the longer term. Firstly, the processes of net CO_2 drawdown and DMS production are very different because the former requires increased export of surface ocean biomass into the deeper ocean, whereas the latter arises from the continuous cycle of the surface ocean microbial loop and does not require 'new production' *per se*. Secondly, there is some evidence to suggest that DMSP production is favoured in major nutrient limited ecosystems, both at the cellular and community structure levels (Turner *et al.*, 1988).

The climatic implications of enhanced DMS fluxes, due to iron fertilisation, arise from the fact that after degassing from seawater to the atmosphere, DMS is oxidised to form fine sulphate particles. These are a major source of cloud condensation nuclei (CCN) in remote oceanic areas (Charlson *et al.*, 1987). In regions distant from terrestrial sources, cloud albedo is sensitive to changes in the numbers of CCN, as calculated by Twomey (1991) and shown in Fig. 5.4. It is clear from the figure that the effect of CCN number on albedo is a nonlinear process, with by far the largest effect occurring where the existing number density is low. This applies to much of the southern hemisphere and reinforces the view expressed earlier, in the context of carbon dioxide, that it is vital to conduct IronEx-type experiments in the Southern Oceans.

Climatic implications of iron fertilisation

Apart from the IronEx results summarised here, there is growing evidence for the importance of iron as a control on marine primary production, as well as

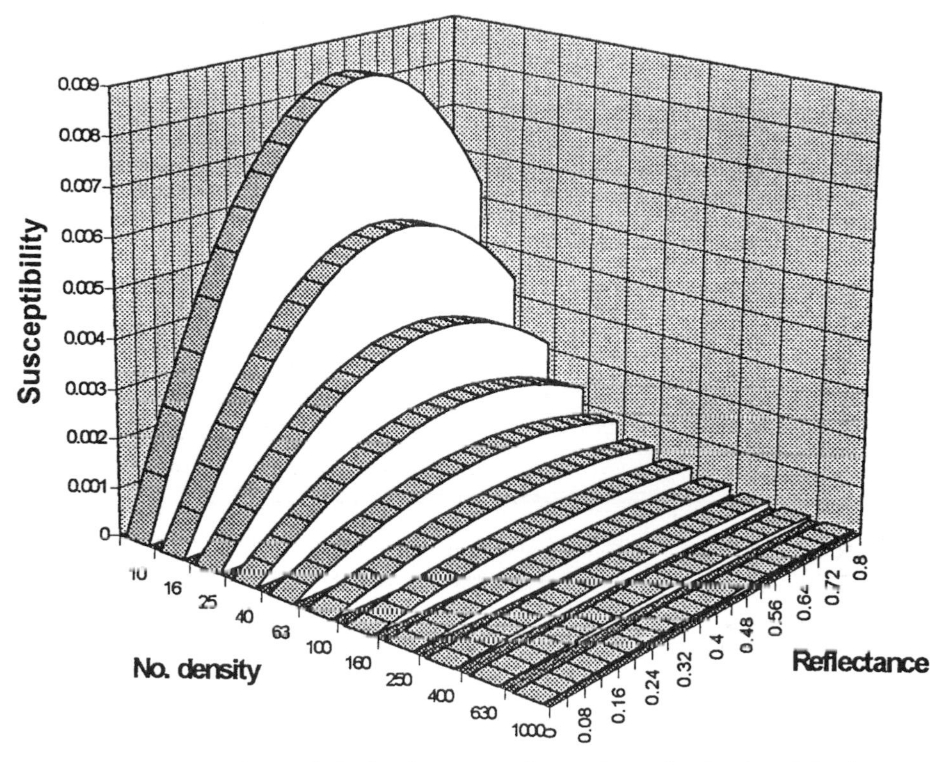

Figure 5.4 *Susceptibility of cloud reflectance to changes in number density of cloud condensation nuclei, for different total cloud reflectances. Susceptibility is the increase in reflectance per additional droplet* cm^{-3} *(redrawn from the equations given by Twomey, 1991).*

species distribution. Much of this information is given by de Baar & Boyd (Chapter 4, this volume). We first note that the major source of *new* iron to surface seawater far from river inputs (iron in which is rapidly removed to the sediments in estuaries and coastal zones) is terrestrial dust carried to the ocean via the atmosphere (Jickells, 1995; Donaghay *et al.*, 1991). Where iron is limiting production, any change in this aeolian flux clearly has the potential to significantly affect atmospheric concentrations of CO_2 and DMS. Whether this potential will be realised is not self-evident (see earlier discussion on CO_2 drawdown in the equatorial Pacific) and is a topic which requires considerable further study before definitive conclusions can be drawn.

However, the record of past environmental conditions preserved in polar ice cores and other data sources reveals some tantalising covariations of parameters, some of which are shown in Fig. 5.5. These are plotted against time from the present back to 30 000 years, i.e. through the Holocene and the end of the last ice age. The upper panel shows $\delta^{18}O$ as a proxy for temperature in an Antarctic ice core. What is striking is the close relation between the lower temperature

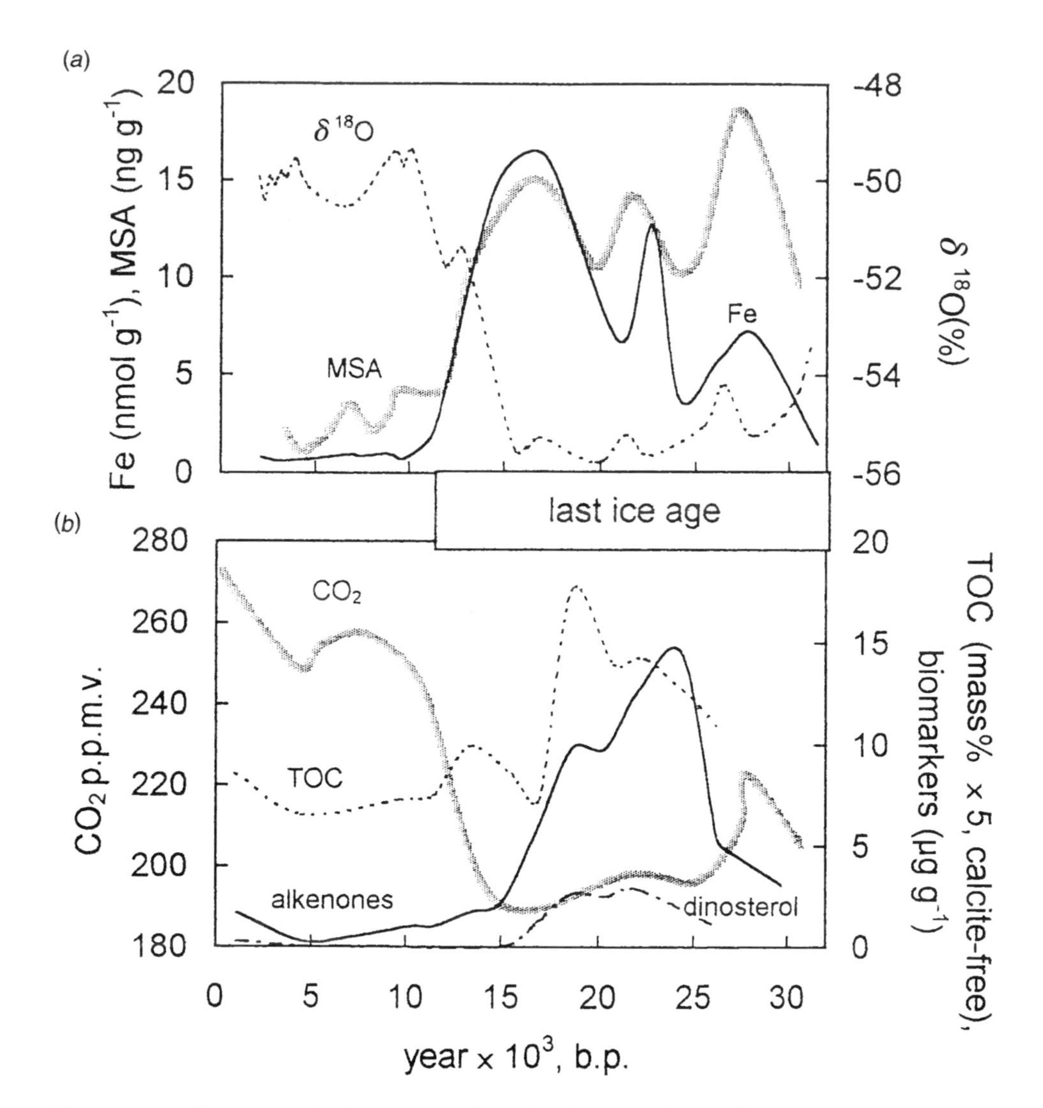

Figure 5.5 *Ice-core and sediment data from the literature for the Holocene and end of the last ice age. (*a*) Methanesulphonate (MSA), a proxy for atmospheric DMS (grey line), and $\delta^{18}O$, a proxy for temperature (dashed line) in an ice core from Dome C, east Antarctica (Saigne & LeGrand, 1987). Fe concentration (solid black line) estimated from Al levels in an ice core from Vostok, Antarctica (DeAngelis* et al., *1987). (*b*) CO_2 concentration (grey line) in an ice core from Vostok, Antarctica (LeGrand* et al., *1991). Total organic carbon (TOC, a proxy for surface ocean productivity; dashed line), alkenones (solid black line) and dinosterol (dot-dashed black line) in a sediment core from the eastern tropical Pacific (Prahl, 1992).*

during the ice age, decreased CO_2 in air bubbles trapped in the ice (Fig. 5.5*b*), together with higher iron and methanesulphonate (MSA, an atmospheric oxidation product of DMS, which has no other known source) in the ice. Non-sea-salt sulphate (much of which in this region will have been derived from marine DMS) also exhibits elevated levels in an Antarctic ice core during the last ice age. These results are not shown in Fig. 5.5 and can be found in LeGrand *et al.* (1991). Such covariations, however clear, cannot prove cause and effect, but they do support the hypothesis that elevated aeolian iron concentrations during the ice age led to enhanced oceanic uptake of atmospheric CO_2 and increased production and release of DMS from the oceans. Other data, e.g. those for total organic carbon (TOC) in a sediment core from the tropical Pacific (Fig. 5.5*b*), also support enhanced biological oceanic production during the last ice age. Similarly, marine sediment concentrations of biomarkers, such as dinosterol and alkenones, derived from planktonic species that are currently known to be strong producers of DMS (Liss *et al.*, 1994), show elevated concentrations 15–25 $\times$ 10^3 years ago.

Although such relations are thought-provoking, they must be treated with caution: the current interpretations of ice-core records from different sites are not all in agreement. This may suggest localised events rather than a global phenomenon. If so, then direct comparison of records from geographically diverse marine sediments and ice cores is problematical. None the less, diagrams such as Fig. 5.5 imply that such interpretation is possible and is the best that can be done with the amounts of data currently available. The drawing of globally applicable conclusions awaits information from future studies.

Conclusions

From the IronEx experiments and the results obtained the following conclusions and deductions can be drawn.

The hypothesis that iron limits marine productivity in HNLC areas can reasonably be said to have been proved in the two IronEx experiments.

Extrapolation of the IronEx results to other regions and times is problematical. It should be noted that the effects of similar iron additions in two equatorial Pacific IronEx studies produced quantitatively very different but significant biological and trace gas results. On the bases of marine productivity, nutrient supply and climatic importance, the Southern Oceans must be the next area to be studied, using the approaches developed and shown to work in the IronEx experiments. This presents a very considerable challenge: increased storminess and the strong seasonality in biological activity, in addition to obvious logistical difficulties, will make a Southern Ocean study significantly more difficult than those conducted in the equatorial Pacific. Such a study has recently (February 1999) been successfully performed in the Southern Ocean (in the vicinity of (141°E 61°S).

Further studies with greater spatial and temporal coverage are required before marine biogeochemical and climatic deductions from the IronEx results (i.e. two small areas of the equatorial Pacific) can be applied regionally or globally. In addition, it is necessary to know the longer-term effects of sustained iron fertilisation (noting that IronEx II lasted for only 17 d) and how other biogenic trace gases might be affected.

In some quarters, there is considerable interest in the idea of purposeful, large-scale iron fertilisation of 'susceptible' oceanic areas to engineer decreased atmospheric CO_2 concentrations. We do not discuss the idea here, because we believe that such an action would be folly, given our present poor understanding of how the system operates.

We also note that human activities are increasing dust and therefore iron inputs into the remote oceans as a result of land-use changes brought about by agriculture, industry and urbanisation. Andreae & Crutzen (1997) cite evidence that suggests that 30–50% of the current atmospheric mineral-dust loading is the result of human disturbance, especially in semi-arid regions. This implies that we may already be engaged in an unplanned fertilisation of the oceans; the consequences of this are essentially unknown, for the reasons given earlier.

The IronEx experiments have proved certain technical issues: (1) the ability to conduct a perturbation experiment in the open ocean, with a proper control provided by measurements outside the fertilised marked patch; (2) the possibility of testing other hypotheses in similar experiments, e.g. a 'ZincEx' (what is the role of Zn in marine productivity?); (3) the ability to use a SF_6-marked patch of ocean water to study natural temporal variations in, for example, biogeochemistry. This has recently been done as part of the NERC PRIME study in the North Atlantic (A. P. Martin *et al.*, 1998).

Acknowledgements

We thank Andrew Watson for re-plotting Fig. 5.4. The work reported here was supported by NERC grants GR3/10085 and GST/02/1059 (PRIME). The US ONR, US NSF and UK NERC funded the IronEx experiments.

References

Andreae, M. O. & Crutzen, P. J. (1997). Atmospheric aerosols: Biogeochemical sources and role in atmospheric chemistry. *Science*, **276**, 1052–8.

Behrenfeld, M. J., Bale, A. J., Kolber, Z. S., Aiken, J. & Falkowski, P.G. (1996). Confirmation of iron limitation of phytoplankton photosynthesis in the

equatorial Pacific Ocean. *Nature*, **383**, 508–11.
Belviso, S., Kim, S-K., Rassoulzadegan, F., Krajka, B., Nguyen, B. C., Mihalopoulos, N. & Buat-Menard, P. (1990). Production of dimethylsulfonium propionate (DMSP) and dimethylsulfide (DMS) by a microbial food web. *Limnology and Oceanography*, **35**, 1810–21.
Charlson, R. J., Lovelock, J. E., Andreae, M. O. & Warren, S. G. (1987). Oceanic phytoplankton, atmospheric sulphur, cloud albedo and climate. *Nature*, **326**, 655–61.
Coale, K. H., Johnson, K. S., Fitzwater, S. E., Gordon, R. M., Tanner, S., Chavez, F. P., Ferioli, L., Sakamoto, C., Rogers, P., Millero, F., Steinberg, P., Nightingale, P., Cooper, D., Cochlan, W. P., Landry, M. R., Constantinou, J., Rollwagen, G., Trasvina, A. & Kudela, R. (1996). A massive phytoplankton bloom induced by an ecosystem-scale iron fertilization experiment in the equatorial Pacific Ocean. *Nature*, **383**, 495–501.
Cooper, D. J., Watson, A. J. & Nightingale, P. D. (1996). Large decrease in ocean-surface CO_2 fugacity in response to *in situ* iron fertilization. *Nature*, **383**, 511–13.
DeAngelis, M., Barkov, N. I. & Petrov, V. N. (1987). Aerosol concentrations over the last climatic cycle (160 kyr) from an Antarctic ice core. *Nature*, **325**, 318–21.
Donaghay, P. L., Liss, P. S., Duce, R. A., Kester, D. R., Hanson, A. K., Villareal, T., Tindale, N. W. & Gifford, D. J. (1991). The role of episodic atmospheric nutrient inputs in the chemical and biological dynamics of oceanic ecosystems. *Oceanography*, **4**, 62–70.
Fitzwater, S. E., Coale, K. H., Gordon, R. M., Johnson, K. S. & Ondrusek, M. E. (1996). Iron deficiency and phytoplankton growth in the equatorial Pacific. *Deep-Sea Research*, Part II, **43**, 995–1015.
Gran, H. H. (1931). On the conditions for the production of plankton in the sea. *Rapports et Proces verbaux des Réunions, Conseil International pour l'exploration de la Mer*, **75**, 37–46.
Hart, T. J. (1934). On the phytoplankton of the Southwest Atlantic and the Bellingshausen Sea 1929–1931. *Discovery Reports*, **8**, 1–268.
Harvey, H. W. (1933). On the rate of diatom growth. *Journal of the Marine Biological Association of the United Kingdom*, **19**, 253–76.
Jickells, T. D. (1995). Atmospheric inputs of metals and nutrients to the oceans: their magnitude and effects. *Marine Chemistry*, **48**, 199–214.
Kolber, Z. S., Barber, R. T., Coale, K. H., Fitzwater, S. E., Greene, R. M., Johnson, K. S., Lindley, S. & Falkowski, P. G. (1994). Iron limitation of phytoplankton photosynthesis in the equatorial Pacific Ocean. *Nature*, **371**, 145–8.
Ledwell, J. R. & Watson, A. J. (1991). The Santa Monica Basin tracer experiment: a study of diapycnal and isopycnal mixing. *Journal of Geophysical Research*, **96**, 8695–718.
Ledwell, J. R., Watson, A. J. & Law, C. S. (1993). Evidence of slow mixing across the pycnocline from an open ocean tracer release experiment. *Nature*, **363**, 701–3.
LeGrand, M., Feniet-Saigne, C., Saltzman, E. S., Germain, C., Barkov, N. I. & Petrov, V. N. (1991). Ice-core record of oceanic emissions of dimethylsulphide during the last climate cycle. *Nature*, **350**, 144–6.
Liss, P. S., Malin, G., Turner, S. M. & Holligan, P. M. (1994). Dimethyl sulphide and *Phaeocystis*: a review. *Journal of Marine Systems*, **5**, 41–53.
Martin, A. P., Wade, I. P., Richards, K. J. & Heywood, K. J. (1998). The PRIME eddy. *Journal of Marine Research*, **56**, 439–62.
Martin, J. H., Coale, K. H., Johnson, K. S., Fitzwater, S. E., Gordon, R. M., Tanner, S. J., Hunter, C. N., Elrod, V. A., Nowicki, J. L., Coley, T. L., Barber, R. T., Lindley, S., Watson, A. J., Van Scoy, K., Law, C. S., Liddicoat, M. I., Ling, R., Stanton, T., Stockel, J., Collins, C., Anderson, A., Bidigare, R., Ondrusek, M., Latasa, M., Millero, F. J., Lee, K., Yao, W., Zhang, J. Z., Friederich, G., Sakamoto, C., Chavez, F., Buck, K., Kolber, Z., Greene, R.,

Falkowski, P., Chisholm, S. W., Hoge, F., Swift, R., Yungel, J., Turner, S., Nightingale, P., Hatton, A., Liss, P. & Tindale, P. W. (1994). Testing the iron hypothesis in ecosystems of the equatorial Pacific Ocean. *Nature*, **371**, 123–9.

Martin, J. H. & Fitzwater, S. E. (1988). Iron deficiency limits phytoplankton growth in the north-east Pacific subarctic. *Nature*, **331**, 341–3.

Prahl, F. G. (1992). Prospective use of molecular paleontology to test for iron limitation on marine primary production. *Marine Chemistry*, **39**, 167–85.

Saigne, C. & LeGrand, M. (1987). Measurements of methanesulphonic acid in Antarctic ice. *Nature*, **330**, 240–2.

Turner, S. M., Malin, G., Liss, P. S., Harbour, D. S . & Holligan, P. M. (1988). The seasonal variation of dimethyl sulfide and dimethylsulfoniopropionate in Antarctic waters and sea ice. *Deep-Sea Research*, **42**, 1059–80.

Turner, S. M., Nightingale, P. D., Spokes, L. J., Liddicoat, M. I. & Liss, P. S. (1996). Increased dimethyl sulphide concentrations in sea water from *in situ* iron enrichment. *Nature*, **383**, 513–16.

Twomey, S. (1991). Aerosols, clouds and radiation. *Atmospheric Environment*, **25A**, 2435–42.

Upstill-Goddard, R. C., Watson, A. J., Wood, J. & Liddicoat, M. I. (1991). Sulphur hexafluoride and helium-3 as seawater tracers: Deployment techniques and a method for the continuous underway analysis of SF_6. *Analytica Chimica Acta*, **249**, 555–62.

Wanninkhof, R., Asher, W., Weppernig, R., Chen, H., Schlosser, P., Langdon, C. & Sambrotto, R. (1993). Gas transfer experiment on Georges Bank using two volatile deliberate tracers. *Journal of Geophysical Research*, **98**, 20237–48.

Watson, A. J., Upstill-Goddard, R. J. & Liss, P. S. (1991*a*). Air-sea exchange in rough and stormy seas measured by a dual-tracer technique. *Nature*, **349**, 145–7.

Watson, A. J., Liss, P. S. & Duce, R. A. (1991*b*). Design of a small-scale *in situ* iron fertilization experiment. *Limnology and Oceanography*, **36**, 1960–5.

Watson, A. J., Law, C. S., Van Scoy, K. A., Millero, F. J., Yao, W., Friederich, G. E., Liddicoat, M. I., Wanninkhof, R. H., Barber, R. T. & Coale, K. H. (1994). Minimal effect of iron fertilization on sea-surface carbon dioxide concentrations. *Nature*, **371**, 143–5.

6 Testing the importance of iron and grazing in the maintenance of the high nitrate condition in the equatorial Pacific Ocean: a physical–biological model study

F. Chai, S. T. Lindley, J. R. Toggweiler and R. R. Barber

Keywords: physical model, biological model, model integration, zooplankton grazing, phytoplankton production, iron limitation, nitrate conditions

Introduction

Satellite maps of sea-surface temperature (SST) and ocean colour show that the waters of the central and eastern equatorial Pacific Ocean (10°S to 5°N) are unusually cool and slightly higher in biomass than the tropical waters north and south of the equatorial zone. That these two meridional anomalies are the result of equatorial upwelling was recognised by the earliest oceanographers studying the region (Graham, 1941; Cromwell, 1953; Montgomery, 1954; Sette, 1955). The westward trade winds blowing along the equator create a divergence in the surface flow field that causes the upwelling of sub-surface waters. The obvious manifestations of equatorial upwelling are lower SST, increased concentrations of surface macronutrients, and higher CO2 partial pressure (p_{CO_2}) near the equator. Equatorial upwelling supplies macronutrients, such as nitrate, phosphate and silicate, to the euphotic zone; these nutrients are assumed to enhance primary production in the region, accounting for the increased phytoplankton and zooplankton biomass (Graham, 1941).

Moreover, satellite maps of the entire Pacific basin reveal an east–west (zonal) asymmetry in thermal structure as well as in biological activity (Colin *et al.*, 1971; Vinogradov, 1981). The temperature of the surface water in the western Pacific warm pool is above 28 °C, and on the same latitude in the eastern equatorial Pacific it is between 22 and 24 °C most of the time. The surface nitrate concentration in the western Pacific warm pool rarely exceeds 0.2 mmol m^{-3}. On the other hand, surface nitrate concentrations in the central and eastern Pacific are moderately high, falling in the range 4–10 mmol m^{-3}. These persistent surface nitrate concentrations in the central and eastern Pacific support enhanced primary production levels in the region (Barber & Chavez, 1991). Just as the equatorial upwelling causes meridional anomalies in temperature, nitrate and biological productivity, so the zonal and vertical structure of the equatorial system, generated by the trade-wind stress and the basin-wide pressure field, are responsible for creating the zonal asymmetries in thermal structure and nutrient concentration.

The equatorial Pacific Ocean plays an important role in determining the global carbon cycle. The central and eastern Pacific Ocean supplies about 1 Gt of carbon per year as CO_2 to the atmosphere (Tans *et al.*, 1990) owing to the equatorial upwelling, and this sea-to-air flux of CO_2 varies interannually in connection with the El Niño – Southern Oscillation (ENSO) (Feely *et al.*, 1995). The flux of CO_2 at the air–sea interface is also influenced by the biological processes in the region, which is considered as a high nutrient – low chlorophyll (HNLC) regime.

Surface nitrate concentration in the central and eastern equatorial Pacific is persistently well above the half-saturation concentration for uptake by phytoplankton (Dugdale *et al.*, 1992; McCarthy *et al.*, 1996). The central and eastern equatorial Pacific Ocean is also characterised by a phytoplankton biomass that is lower than expected based on the nutrient concentration (Barber, 1992). However, the factor responsible for the relatively high nutrient and low biomass conditions in the equatorial Pacific was controversial when this modelling study was begun. It has been proposed that zooplankton grazing keeps the phytoplankton biomass so low that the absolute uptake rate cannot deplete nitrate at the characteristically high equatorial rates of physical supply (Walsh, 1976; Frost, 1991; Cullen *et al.*, 1992; Frost & Franzen, 1992; Price *et al.*, 1994). The alternative interpretation is that specific uptake rates are low (Dugdale *et al.*, 1992) and specific photosynthetic rates are also reduced in equatorial Pacific waters (Barber & Ryther, 1969; Barber & Chavez, 1991). Another interpretation was that the availability of a trace metal micronutrient limits specific rates so that phytoplankton do not deplete nitrate for physiological reasons (Barber & Ryther, 1969; Martin, 1990; Martin *et al.*, 1991; Duce, 1986).

Recent Iron Enrichment Experiment (IronEx I and IronEx II) results have shown enhanced specific photosynthetic and growth rates in iron-enriched patches in the equatorial Pacific (Kolber *et al.*, 1994; Lindley & Barber, 1999; Behrenfeld *et al.*, 1996). Therefore, it is clear that iron limitation in the central and eastern equatorial Pacific Ocean does regulate the specific uptake and growth rates of ambient phytoplankton. Results from the US JGOFS Process Study in the Equatorial Pacific (EqPac) done in 1992 at 140°W showed that both primary production and phytoplankton biomass increased along the equator when enhanced upwelling of iron occurred from the equatorial undercurrent (Barber *et al.*, 1996; Landry *et al.*, 1997). Results from the EqPac study in 1992 and IronEx experiments in 1993 and 1995 do much to answer the question of why the equatorial Pacific is not greener and the answers provided are compelling; but, when scaled up to the entire equatorial region, do they account for the persistent large-scale feature of anomalous nutrient concentrations? IronEx II results showed that an intensive addition of iron to a small area for a short time can trigger a phytoplankton bloom, and is capable of substantially reducing surface nitrate concentration. The ecosystem structure was disrupted because diatoms bloomed and there were not enough zooplankton to eat them. These results may not be easy to extrapolate to larger scales where coupling between phytoplankton and zooplankton might be maintained, and they might not represent the oceans' response to natural levels of variation in iron availability. Furthermore, the IronEx patch moved rapidly away from the equator, and by the time nutrients were consumed, it was near the South Equatorial Current front (Lindley & Barber, 1999). Therefore, the IronEx results do not tell us much about how large-scale stimulation of phytoplankton growth would affect the distribution of nutrients in the equatorial Pacific. Fundamental questions remain unanswered: is phytoplankton specific rate limitation by iron sufficient, by itself, to account for the persistently high surface nitrate concentrations? Or must rate limitation be combined with grazing to account for the nutrient anomaly?

This paper presents results from a series of experiments with a biological–physical model of the equatorial Pacific Ocean (Chai, 1995). The objective of these experiments was to understand how the nitrate distribution in the equatorial Pacific might be determined by the interplay between physical supply of nitrate, phytoplankton uptake, and zooplankton grazing. To begin, the results of two hypothetical conditions are shown: one is an abiotic simulation (without biological processes); the other is a simulation without equatorial circulation and mixing (without physical processes). To demonstrate the role of iron limitation of phytoplankton rates on the nitrate field, the potential maximum specific growth rate was increased as observed in the IronEx I experiment (Lindley & Barber, 1999). Asymmetric north–south growth rates were incorporated into the model to reflect the asymmetric pattern in equatorial iron flux (Duce & Tindale, 1991), and the impact of zooplankton grazing on the nitrate field was investigated by varying the maximum specific grazing rate.

A physical–biological model in the Equatorial Pacific

Physical model

This analysis used the Modular Ocean Model (MOM) (Pacanowski *et al.*, 1991), which is based on the Geophysical Fluid Dynamics Laboratory (GFDL) Ocean General Circulation Model (OGCM) developed by Bryan (1969) and Cox (1984). The MOM was modified by focusing on the tropical Pacific Ocean in a manner similar to that implemented by Philander *et al.* (1987). The spatial domain was from 50°N to 40°S and from 80°W to 130°E. Meridional resolution was 1/3° between 10°N and 10°S, and increased gradually poleward out of this region; longitudinal resolution was 1° over the entire domain. There were 27 levels in the vertical with 17 levels in the upper 300 m, providing a resolution of 10 m within the upper 120 m. Realistic topography was incorporated to the nearest model level. The explicit horizontal eddy diffusivity coefficient was 10^7 cm^2 s^{-1}. The vertical eddy diffusivity coefficient was Richardson-number-dependent according to the format described by Pacanowski & Philander (1981).

The model was initialised with January climatological temperature and salinity (Levitus, 1982) and with current velocities of zero. To study seasonal variation the model was forced with the Comprehensive Ocean Atmosphere Data Set (COADS) monthly mean wind and air temperature (Slutz *et al.*, 1985; Oort *et al.*, 1987). After several years of integration, or 'spin-up', the model achieved a quasi-equilibrium state in the upper ocean. The boundaries at 40°S and 50°N were closed walls and water was forced to flow along, not through them. A decay term $\kappa(T^* - T)$ was added to the temperature equation [$\kappa(S^* - S)$ for salinity equation], which restored it to the observed temperature T^* (salinity S^*) field at the two closed walls. The five ecosystem balance equations were solved simultaneously with the physical model.

Biological model

The biological model consists of five compartments describing phytoplankton (P), zooplankton (Z), non-living particulate organic nitrogen (D), and two forms of dissolved inorganic nitrogen: nitrate (NO_3) and ammonium (NH_4). The currency of the model is nitrogen. Nitrate and ammonium were treated as separate nutrients, thus enabling primary production to be divided into new production fuelled by nitrate, and regenerated production fuelled by ammonium (Dugdale & Goering, 1967). The ratio of new to total primary production in the model is referred to as the *f*-ratio, following the conventional usage of Eppley & Peterson (1979). The biological model is conceptually similar to the one developed by Fasham *et al.* (1990), but the oceanographic, phytoplankton and zooplankton parameters were adjusted from recent equatorial observations (Barber & Chavez, 1991; Lindley, 1994; Murray *et al.*, 1994, 1995; Landry *et al.*, 1995) and especially from recent IronEx results (Lindley & Barber, 1999). The equations describing inter-compartment flows of the modelled ecosystem and values of parameters are given in the Appendix.

Below the euphotic zone, sinking particulate organic nitrogen is converted to inorganic ammonia and eventually to nitrate. The parameterisation of the regeneration processes was similar to those introduced by Sarmiento *et al.* (1993), in which all particulate matter decays to ammonium and thence to nitrate. The flux of particulate material, $F(z)$, was specified by using an empirical function determined from Pacific Ocean sediment-trap observations by Martin *et al.* (1987).

The five ecosystem balance equations were solved simultaneously with the physical model. There were 12 layers in the euphotic zone, the upper 120 m, with 10 m vertical resolution. Below the euphotic zone, there were 15 layers through the water column. There were no mass fluxes across the air–sea boundary and for most of the model domain the water depth is several thousand metres, so fluxes across the ocean floor also were not evaluated in this analysis. No mass fluxes across the lateral boundaries at 50°N and 40°S were permitted. The initial condition for the nitrate field was the climatological nitrate field of Levitus *et al.* (1993). The initial values for P, Z, D, and NH_4 were 0.25 mmol m^{-3} at the surface, decreasing exponentially with a scale length of 120 m. Chai (1995) has shown that the initial conditions in the euphotic zone for the ecosystem do not alter the final results because the ecosystem reaches equilibrium faster than the equatorial physics does.

The model integration

The physical–biological model was integrated simultaneously from the initial conditions. With seasonally varying surface forcing (air temperature and wind stress), the model achieved a quasi-equilibrium state of its annual cycle within a few years, which is the time scale for Rossby waves to cross the Pacific basin (Gill, 1982). During the first few years of integration there is a trend of decreased annual averaged nitrate and a trend of increasing annual averaged temperature, but after five years the annual rates of change are less than 0.02 mmol m^{-3} yr^{-1} for nitrate and 0.02 °C yr^{-1} for temperature. The rates of change in nitrate and temperature are for the upper 120 m; when averaged over the upper 500 m the rates of change are even smaller because nitrate and temperature at depth do not change as rapidly as at the surface (Chai *et al.*, 1996).

The time course of achieving quasi-equilibrium in integrated (0–120 m) primary productivity in terms of total, new, and regenerated production at 140°W on the equator is shown in Fig. 6.1. The modelled primary production

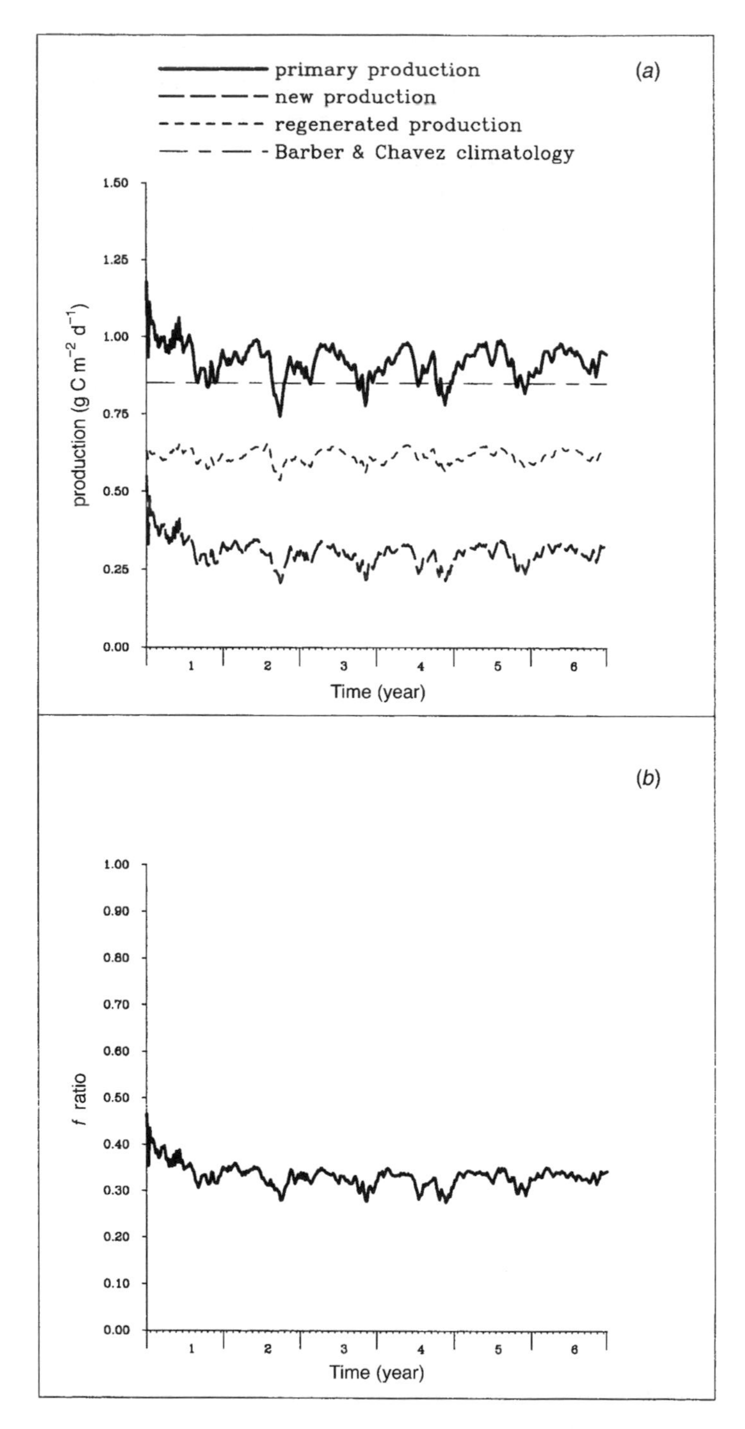

Figure 6.1 *Weekly mean vertical integrated (0–120 m) productions at 140°W on the equator over the first six years' simulation:* (a) *new, regenerated and primary production along with the Barber & Chavez (1991) climatological value of primary production (0.88 g C m^{-2} d^{-1});* (b) *vertical averaged f ratio (the ratio of new to primary production), the mean value of which is 0.33.*

agrees well with the climatological value of Barber & Chavez (1991). By the sixth year of the simulation, the annual rates of change for temperature, nitrate and productivity are at least two orders smaller than their seasonal variations; therefore, the sixth year was analysed for this study. More detailed comparisons between the modelled results and observations, such as temperature, nitrate concentration, nitrate budget and new production, can be found in Chai *et al.* (1996).

Fig. 6.2 (colour plate) shows a comparison of modelled annual mean (year six) surface chlorophyll (top panel) with chlorophyll derived from the Coastal Zone Color Scanner (CZCS) data (bottom panel). The modelled top layer (5 m) phytoplankton concentration, in nitrogen units, was converted to chlorophyll by using a nominal gram chlorophyll to mole nitrogen ratio of 1.59, which corresponds to a chlorophyll to carbon mass ratio of 1 : 50 and a C : N mole ratio of 6.625. The satellite image is a composite of all CZCS data collected during the 7.5 years of operation from November 1978 to June 1986 (Feldman *et al.*, 1989).

The overall pattern of the model-predicted annual mean surface chlorophyll agrees with the CZCS data in two fundamental aspects, the positions of the maximum gradients, both north–south and east–west. There are also two differences. First, the modelled chlorophyll was higher than the CZCS estimates. For example, along the equator in the central Pacific, chlorophyll is 0.3 mg m^{-3} in the model and 0.15 mg m^{-3} in the CZCS data. Chlorophyll measurements *in situ* showed the mean surface chlorophyll was between 0.2 and 0.3 mg m^{-3} in general (Barber & Chavez, 1991), suggesting that the CZCS under-estimated chlorophyll concentration in the equatorial Pacific. Second, the modelled surface chlorophyll maxima are at 3°S and 3°N, whereas the CZCS estimated maximum is right on the equator. The observations *in situ* (Bidigare & Ondrusek, 1996) showed that the surface chlorophyll values at two to three degrees away from the equator were higher than those right on the equator. The absence of off-equator chlorophyll maxima in the CZCS estimate may be a result of temporal and spatial averaging of the CZCS data.

Results and discussion

Two case studies

Biological uptake and export of nitrate act as a sink that prevents the accumulation of nitrate in the euphotic zone. To quantify the magnitude and importance of this biological sink term, a case study was carried out with biological uptake and export set to zero. Comparison of surface nitrate concentration between a biotic simulation (standard experiment with biological

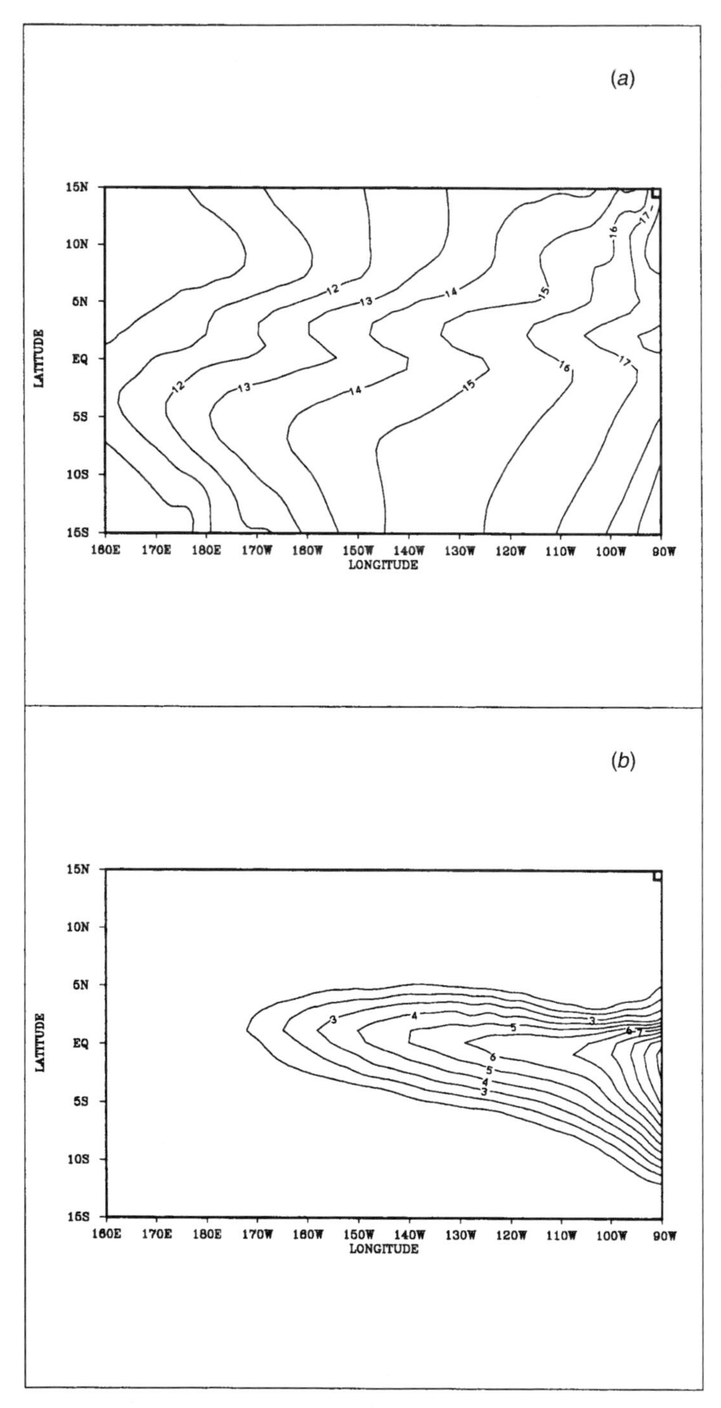

Figure 6.3 *The annual mean (year six averaged) surface nitrate concentration comparison between (a) an abiotic simulation (without biological uptake of nitrate); (b) the standard experiment (with normal biological uptake). Contour interval is 1.0 mmol m^{-3}. The difference between (a) and (b) is due to the biological uptake of nitrate during a period of six years.*

uptake) in the 15°S to 15°N, 90°W to 160°E region and an abiotic simulation (without biological uptake), both initialised with the same observed nitrate field, demonstrates the magnitude of the biological sink (see Fig. 6.3).

The annual nitrate concentration during year six of the abiotic simulation exceeds 11 mmol m^{-3} everywhere in the 15°S to 15°N domain across the Pacific basin to 160°E (see Fig. 6.3*a*). In contrast, the annual mean surface nitrate of the simulation with the biotic sinks has a classic 'cold tongue' pattern with nitrate concentrations of less than 0.5 mmol m^{-3} poleward of 10°S and 5°N and west of 170°W (see Fig. 6.3*b*). The large increase in nitrate concentration in Fig. 6.3*a* relative to Fig. 6.3*b* is due to the accumulation of nitrate in the euphotic layer in the abiotic simulation; this "excess" nitrate was stripped from the euphotic zone in the biotic simulation by the combined action of phytoplankton uptake and vertical export over the six-year period. The difference between the nitrate fields in Fig. 6.3*a* and Fig. 6.3*b* demonstrates the importance of the 'biological pump' (Volk & Hoffert, 1985) for creating the classical 'cold tongue' pattern of the equatorial nitrate field.

Obviously, the physical processes of equatorial upwelling and mixing are as necessary as biological sinks for maintaining the equatorial pattern. The interesting question to investigate via modelling is how quickly surface nitrate is reduced to zero when the physical source is eliminated. How quickly will the biological pump export a certain amount of nitrate, for example 6 mmol m^{-3}, from the surface layer? In other words, what is the biologically determined residence time of surface nitrate in the equatorial Pacific? Fig. 6.4 shows results from a case study without the equatorial circulation and mixing in which the initial nitrate condition was the same as in the standard experiment. After three months' model integration, the surface nitrate concentration was reduced by at least 50% from the initial value in the central and eastern Pacific (see Fig. 6.4*a*,*b*). After six months, surface nitrate was depleted (see Fig. 6.4*c*).

The model experiment without the equatorial circulation and mixing simulates, to certain degree, the physical effects on the nitrate field during ENSO events. A good example of quick reduction of nitrate is the 1982–83 El Niño, during which the surface nitrate concentration near the Galapagos Islands reduced to zero from 7 mmol m^{-3} within four months (Kogelschatz *et al.*, 1985). During El Niño the migration of the warm pool and upwelling of warm, nutrient-poor water weakens the normal zonal gradient of surface nitrate concentration across the basin. Reduced nutrient supply in the central and eastern Pacific Ocean during El Niño reduced biological productivity dramatically (Barber & Chavez, 1986). Because of the depression of thermocline and reduction of the upwelling along the equator, the CO_2 partial pressure (p_{CO_2}) during El Niño is lower than normal, and the net flux of CO_2 from the ocean to the atmosphere is considerably lower during ENSO period (Feely *et al.*, 1995).

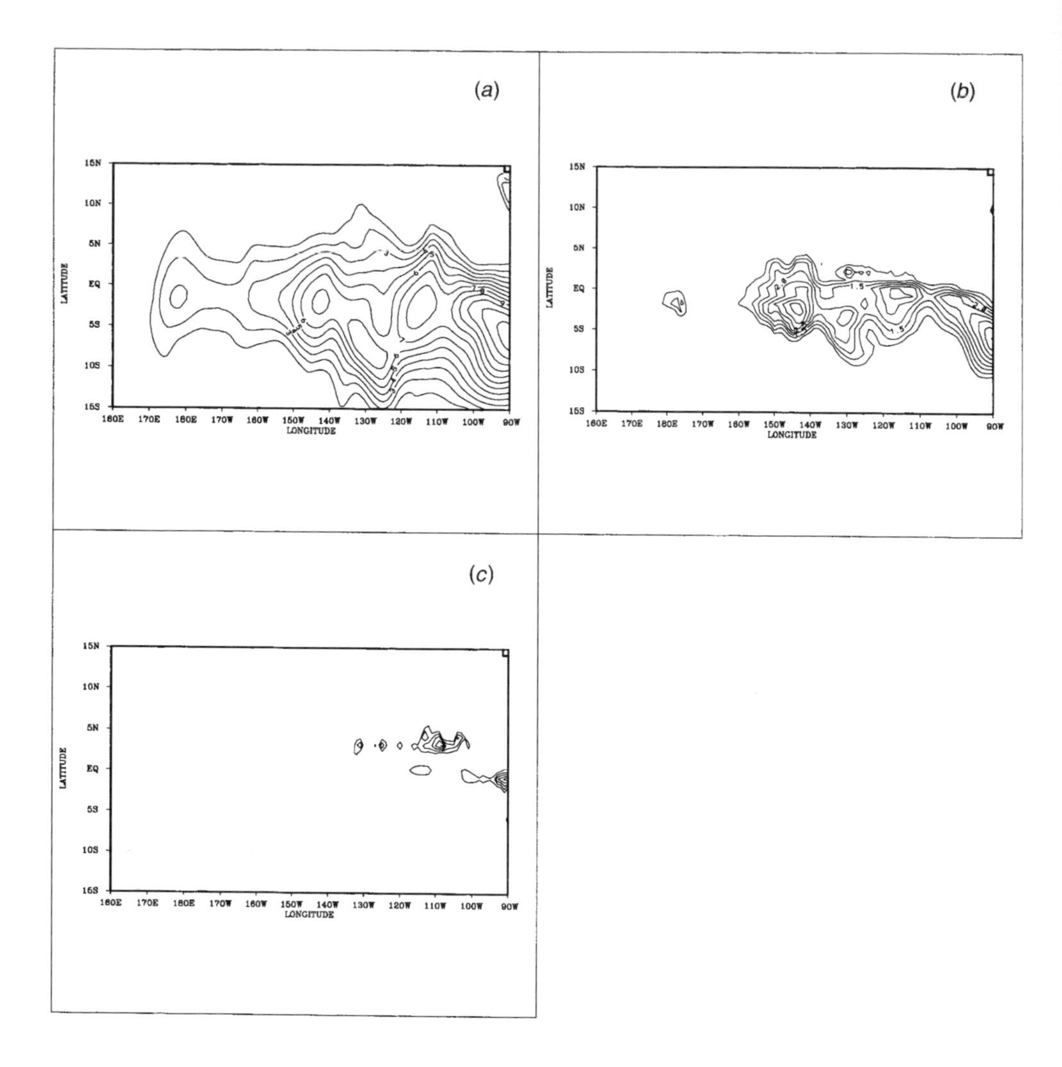

Figure 6.4 *The surface nitrate concentration change during the first six months' model integration in a simulation without physics:* (a) *the initial condition, contour interval* 1.0 mmol m^{-3}; (b) *at the end of three months, contour interval* 0.5 mmol m^{-3}; (c) *at the end of six months, contour interval* 0.5 mmol m^{-3}. *The model indicates that the biological uptake in the equatorial Pacific can consume the entire surface nitrate in less than six months.*

Fig. 6.5 shows time series of the surface nitrate concentration at 140°W on the equator for the three different model conditions (standard, without biological sinks, without physical sources). Without equatorial circulation and mixing, the nitrate decreased from 6 mmol m^{-3} to zero during the first five months, and remained depleted afterwards. Five months is therefore an estimate of the residence time of surface nitrate in the central equatorial Pacific in this model. Dugdale *et al.* (1992) estimated that nitrate turnover time in the entire euphotic zone at 150°W on the equator was 197 d, an estimate that this

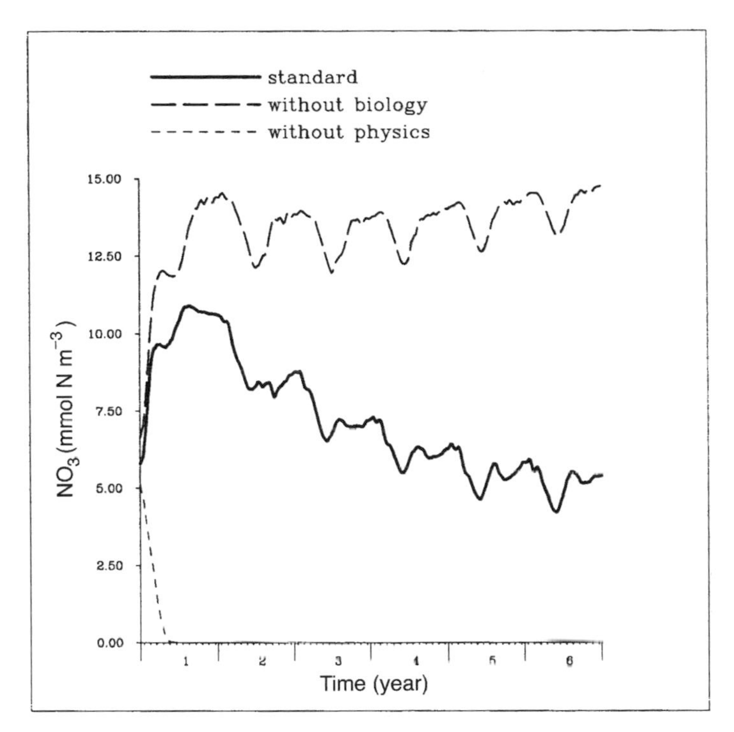

Figure 6.5 *Time series of weekly mean surface nitrate concentration at 140°W on the equator from three different experiments. Without the physics, the surface nitrate drops from 6 to 0 mmol m^{-3} within the first 5 months; this result suggests that the surface nitrate residence time in the central Pacific is 5 months. Without the biological uptake, the surface nitrate increases quickly in the first couple of months, and levels at 15 mmol m^{-3}.*

model result supports remarkably well. McCarthy *et al.* (1996) estimated that the euphotic-layer nitrate depletion times in the meridional region of elevated nitrate were 190 d during the EqPac fall cruise (non-El Niño conditions of 1992) and 305 d during the spring 1992 El Niño conditions.

Another interesting question concerns what surface nitrate values would be without a biological sink. After six years' integration, the surface nitrate in the abiotic simulation was almost tripled compared to the value of the standard simulation (see Fig. 6.5). The most dramatic change in surface nitrate took place within the first six months of the model integration; after that it levelled off at between 13 and 15 mmol m^{-3} depending upon season. This result indicates that 15 mmol m^{-3} is an upper bound of nitrate concentration set by the equatorial upper ocean circulation and mixing.

Iron limitation of phytoplankton growth

Having demonstrated the results that both the physical sources and biological sinks are necessary to simulate the equatorial nitrate field, and having shown that this model reproduces the observed equatorial nitrate field when observed equatorial rate parameters are used (Chai *et al.*, 1996), we now investigate how changes in biological rates affect the strength of the biotic sink and alter the pattern and concentration of the surface nitrate field. A rate change of particular interest that can be investigated is the relation between iron deficiency and anomalously high equatorial nitrate concentrations.

The photosynthetic performance of phytoplankton is described with two parameters: α, the slope of specific photosynthetic rate over irradiance at low irradiance; and I_k, the irradiance at light saturation. In the context of the model, α is the product of the maximum quantum yield of photosynthesis and the nitrogen-normalised phytoplankton light absorption coefficient. In the IronEx I experiment, Lindley & Barber (1999) observed that quantum yield approximately doubled inside the iron-enriched patch compared with that outside the patch, and chlorophyll concentrations per cell also increased, whereas I_k was about the same inside and outside the patch. To represent this observed response of ambient equatorial phytoplankton to an iron addition, the photosynthetic rate parameters in our 'iron addition' model experiment were changed to the values observed in IronEx I; that is, α was increased from 0.1 to 0.2 d^{-1} $(W\,m^{-2})^{-1}$ and I_k was kept the same. $P_{max} = \alpha/I_k$; therefore the observed IronEx I result of α doubling while I_k remained the same produces a doubling of P_{max}. In the convention used in photosynthetic physiology, P_{max} is the maximal specific photosynthetic rate measured by short exposures to saturating, but not inhibiting, light intensities (Jassby & Platt, 1976). In this model nitrogen is the currency of the ecosystem model. Specific maximal photosynthetic rate in nitrogen currency is N_{uptake} $N^{-1}{}_{phyto}$ $time^{-1}$, which reduces to reciprocal time (d^{-1}). The term μ_{max} in our nitrogen currency is the mathematical equivalent of maximum specific growth rate (P_{max}) and has the same units (d^{-1}), so the consequence of doubling α while keeping I_k the same is that μ_{max} is doubled. Only α and I_k were taken from our IronEx I observations because determination of α and I_k is relatively straightforward with conventional P–E methodology (Lewis & Smith, 1983) and a modified photosynthetron (Lindley, 1994). By keeping other model parameters the same, including especially the grazing formulation, the effects of iron limitation or iron-replete conditions on the surface nitrate field can be modelled.

With the increased α and the resultant increased μ_{max}, the model again reached a steady state after five years' integration; the annual mean during year six is presented for comparison with the results from the standard experiment. Comparison of surface nitrate concentrations between the standard (Fig. 6.3*b*) and the doubled α simulation (Fig. 6.6) shows that, with increased α, the area extent of the high-nitrate tongue was reduced. The 1 mmol m^{-3} contours are closer to the equator and on the equator the 1 mmol m^{-3} contour moved east from 170°W to 150°W. However, a large equatorial region of anomalously high

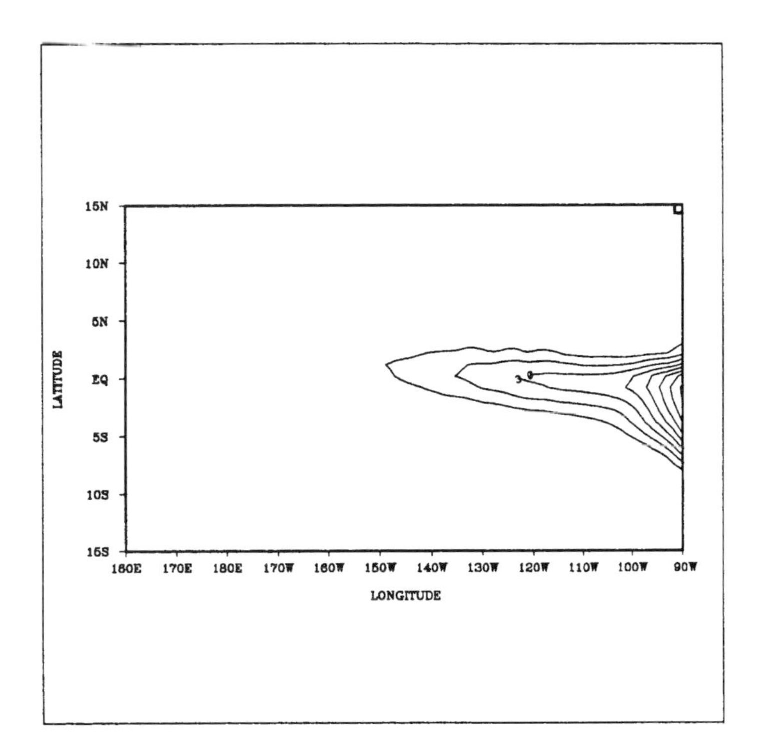

Figure 6.6 *The surface nitrate concentration from the experiment with doubled rates simulation ($\alpha = 0.2\ d^{-1}\ (W\ m^{-2})^{-1}$, and $\mu_{max} = 4.0\ d^{-1}$). Contour interval is* 1 mmol m^{-3}. *With a higher physiological performance of phytoplankton, the pool of the unused surface nitrate reduced as well as the concentration. A comparison can be made with the normal simulation ($\alpha = 0.1\ d^{-1}\ (W\ m^{-2})^{-1}$, and $\mu_{max} = 2.0\ d^{-1}$); see Fig. 6.3*b.

nitrate persisted east of 150°W between 5°S and 5°N when α was increased.

We believe that α (and therefore μ_{max}) were at saturated or iron-replete values in our 'iron-enriched' model experiment because the iron concentrations *in situ* in the patch during IronEx I exceeded the half-saturation (K_s) value determined by Coale *et al.* (1996*a*). This line of reasoning suggests that the iron-enriched photosynthetic performance values observed by Lindley & Barber (1999) were iron-saturated or iron-replete rates. The fast repetition-rate fluorometer results of Kolber *et al.* (1994) from IronEx I and of Behrenfeld *et al.* (1996) from IronEx II support this assumption. Persistence of a high-nitrate tongue in the 'iron-enriched' model experiment indicates that equatorial upwelling (Chai *et al.*, 1996) in the eastern equatorial Pacific would maintain unused nitrate in the surface waters, even if the dominant small phytoplankton were to grow twice as fast as they normally grow, as long as the zooplankton grazing rate was coupled to the phytoplankton growth rate. This last condition is an important caveat. In this model there is only one class of phytoplankton and they are vulnerable to zooplankton grazing. This model simulates only a 'balanced' system as

Table 6.1. *Comparison between the standard and iron-enriched simulation at 0°N, 140°W*

	Surface phytoplankton (mmol m^{-3})	Surface zooplankton (mmol m^{-3})	Integrated primary production (g C m^{-2} d^{-1})	Integrated zooplankton grazing (g C m^{-2} d^{-1})
standard experiment (normal α and μ_{max})	0.21	0.5	0.85	1.0
iron-enriched exp. (doubled α and μ_{max})	0.25	1.0	1.80	2.0

described by Frost & Franzen (1992) or Landry *et al.* (1997). Some recent model studies suggest that the iron has to be explicitly added into the model to simulate the variability of primary production and biomass in the HNLC regions (Lindley *et al.*, 1994; Loukos *et al.*, 1997).

Two factors determine the absolute rate of nitrate uptake by phytoplankton: the maximum specific uptake rate (V_{max}) and phytoplankton biomass (Dugdale, 1967). In the nitrogen currency of this model and with the assumption of balanced phytoplankton growth over the model's time steps, V_{max} as defined by Dugdale (1967) is the mathematical equivalent of μ_{max}. With μ_{max} saturated with respect to iron, the surface nitrate concentration at 140°W on the equator was reduced to 2 mmol m^{-3} but not depleted (Chai *et al.*, 1996). This result indicates the important contribution played by the other factor, phytoplankton biomass, in nitrate uptake and export.

Table 6.1 summarises the modelled phytoplankton and zooplankton biomass, primary production, and zooplankton grazing for both the standard experiment (with 'normal' α and μ_{max}) and the iron-enriched simulation (with doubled α and μ_{max}). The phytoplankton biomass increased only slightly (from 0.21 to 0.25 mmol m^{-3}) after α and μ_{max} doubled, explaining why nitrate remains at 2 mmol m^{-3} instead of going to depletion. Not enough phytoplankton biomass accumulates to deplete nitrate even though phytoplankton are taking up nitrate and growing at a maximal specific rate.

In contrast, zooplankton biomass doubled (from 0.5 to 1.0 mmol m^{-3}) after α and μ_{max} doubled. The increased abundance of grazers keeps phytoplankton biomass at low levels even though phytoplankton-specific uptake and growth rates increase. This result indicates that the faster phytoplankton grow, the more zooplankton consume. This can be seen clearly from integrated primary production and secondary production in the standard experiment and the iron-enriched simulation. The primary productivity rate doubled (from 1.0 to 2.0 g C m^{-2} d^{-1}) after α and μ_{max} doubled, but secondary productivity rate also doubled (from 0.85 to 1.8 $g\,C\,m^{-2}\,d^{-1}$). Using insight from this model experiment we suggest that the IronEx I observations of increased phytoplankton-specific rates (Lindley & Barber, 1999) but no significant drawdown of nitrate (Watson *et al.*, 1994) resulted because the zooplankton grazing rate stayed functionally coupled to the increased phytoplankton growth rates. That is, the realised ecosystem transfers during IronEx I were essentially identical to our modelled processes. Frost (1991) and Landry *et al.* (1997) have described this coupled phytoplankton growth and zooplankton grazing condition as 'balance'. It is clear that this 'balance' was maintained in IronEx I (Martin *et al.*, 1994) and in our iron-enriched simulation, but in IronEx II (Coale *et al.*, 1996*b*) 'balance' was not maintained and as a result there was a dramatic increase in diatom biomass and significant drawdown of nitrate (Cooper *et al.*, 1996). It is not possible to simulate the diatom bloom during the IronEx II experiment with this one class of phytoplankton (i.e. small phytoplankton) in the model. A more sophisticated model with size-dependent growth and grazing rates is needed to simulate diatom bloom formation and episodic productivity.

In the present model, the grazers suffer a mortality rate (m_2) that does not vary with their concentration. In a more realistically structured model, all levels of predators would respond to changes in the concentration of their prey, hence the 1 : 1 relation between changes in primary production and variations in zooplankton biomass would be unlikely. Therefore, predator suppression of grazers might allow phytoplankton to utilise all nitrate under iron-replete conditions. The above discussion needs to be tested with a more advanced model to address interactions among higher-level consumers.

In summary of this simplified model experiment, the modelled iron experiment indicates that, with iron-replete conditions and increased physiological performance of phytoplankton (doubled α and μ_{max}), primary production in the cold tongue of the equatorial Pacific would double; however, increased zooplankton grazing would prevent the development of a phytoplankton bloom with enough biomass to deplete surface nitrate if the grazing–growth balance were maintained. Additional iron supply would not necessarily eliminate the HNLC conditions unless the iron addition changed the taxonomic composition of the phytoplankton and a diatom bloom developed (Barber *et al.*, 1995).

The asymmetric surface nitrate pattern in the equatorial Pacific

Recently, Lindley et al. (1995) hypothesised that the central and eastern equatorial Pacific comprises three distinct regimes (see Fig. 11 in Lindley *et al.*, 1995) in terms of nitrate and iron limitation of physiological rates. In the oligotrophic regime north of 5°N, specific rates are limited by insufficiency of nitrate and other macronutrients. In the equatorial regime bounded roughly by 5°N and 5°S, macronutrients supplied by upwelling saturate phytoplankton uptake; yet the iron supply from upwelling and dust deposition, although somewhat increased, is insufficient to support maximal physiological rates. To the south of the enriched equatorial water, because of decreasing eolian iron flux (Duce & Tindale, 1991; Rea *et al.*, 1994) from north of the equator to the south, the lack of surface iron input may limit photosynthetic performance and primary production.

Based on the hypothesis of Lindley *et al.* (1995) and the data from Duce & Tindale (1991), if $2.0\,d^{-1}$ is the value of μ_{max} on the equator, the rate should be decreased in the Southern Hemisphere because aeolian and upwelling iron flux there is less by an order of magnitude. In the standard experiment discussed above, the same μ_{max} was used both north and south of the equator, based primarily upon the data of Barber & Chavez (1991) from stations located near the equator. The assumption of equal μ_{max} north and south of the equator generated surface nitrate concentrations that were lower than observed values south of the equator (Chai *et al.*, 1996). This model result and ocean observations suggest that μ_{max} is not constant, but that it varies as a function of aeolian iron flux, with higher rates north and lower rates south of the equator. To model this characteristic, μ_{max} was modified as a function of latitude (see Fig. 6.7) to take into account the meridional iron supply gradient, as Lindley *et al.* (1995) suggested. Unlike the iron addition experiment discussed above, α was kept constant in the meridional gradient experiment.

With meridionally varying μ_{max}, the annual mean surface nitrate concentration during year six is shown in Fig. 6.8. As expected, surface nitrate concentrations south of the equator were higher in response to decreased μ_{max}. In addition, the asymmetry of nitrate about the equator enhanced slightly with a latitude-dependent μ_{max}. The major difference was south of the equator, especially in the eastern part of the basin (east of 120°W). In the central Pacific, along 140°W, the similarity of modelled and observed nitrate concentrations was improved slightly with a modified μ_{max} (see Fig. 6.9) south of the equator, but observations south of the equator are still higher than those modelled by the modified μ_{max} experiment. Apparently a lower α value in the south of the equator in this modified μ_{max} experiment (for example, α at 15°S of $0.05\,d^{-1}\,(W\,m^{-2})^{-1}$ instead of $0.1\,d^{-1}\,(W\,m^{-2})^{-1}$) would enhance the nitrate asymmetry more.

These model results suggest that the meridional gradient in iron flux, with the resultant gradient in μ_{max}, may not be a major factor for higher surface nutrient concentrations south of the equator. The physical processes are more likely to generate the nitrate asymmetry across the equator. Eastward flow of low-nitrate water in the North Equatorial Countercurrent (NECC) reduces

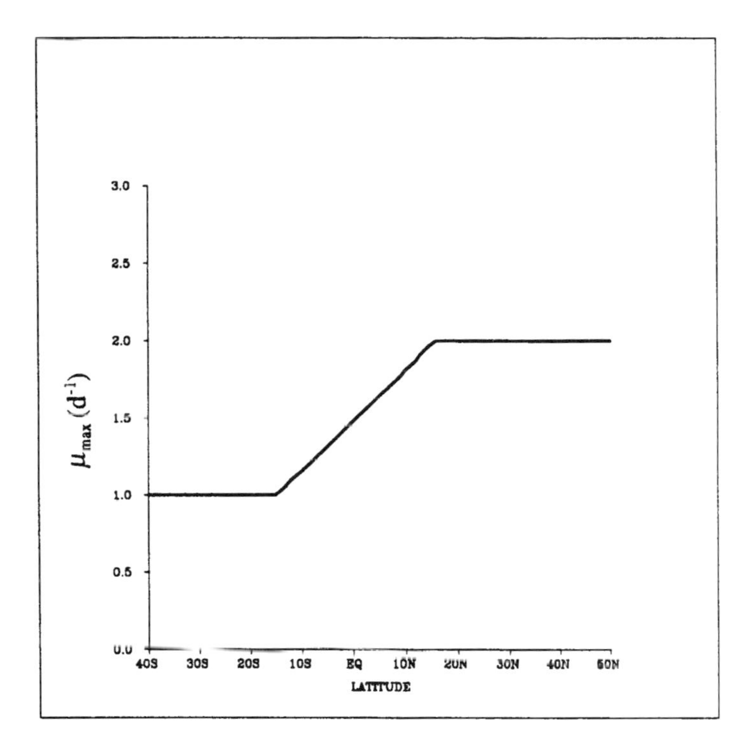

Figure 6.7 *Latitudinally modified maximum potential growth rate of phytoplankton (μ_{max}) according to Lindley's hypothesis (Lindley* et al., *1995) and the atmospheric iron flux to the ocean.*

nitrate concentrations north of the equator, and westward flow of high-nitrate water associated with the South Equatorial Current (SEC) increases nitrate concentrations south of the equator. Toggweiler *et al.* (1991) discussed a relation between the asymmetry pattern in ^{14}C measurements and the equatorial flow field, and their discussion could also be applied to the surface nitrate field. The asymmetry of the nitrate field is generated and maintained by the equatorial asymmetry of the flow field (NECC or SEC) and origins of water masses (the western equatorial Pacific water or water from the coast of Peru).

Testing the role of zooplankton grazing

The idea that zooplankton grazing contributes to the control of phytoplankton biomass has existed for over half a century. Riley (1946, 1947) formulated the first mathematical model describing the dynamics of phytoplankton and grazer assemblages and quantitatively demonstrated the potential capability of grazers in maintaining low stocks of phytoplankton. Walsh (1976) was the first to argue that grazing (or over-grazing), not vigorous circulation or micronutrient deficiency, is the process that keeps equatorial phytoplankton biomass at anomalously low levels. Frost (1987, 1991) analysed both model results and field

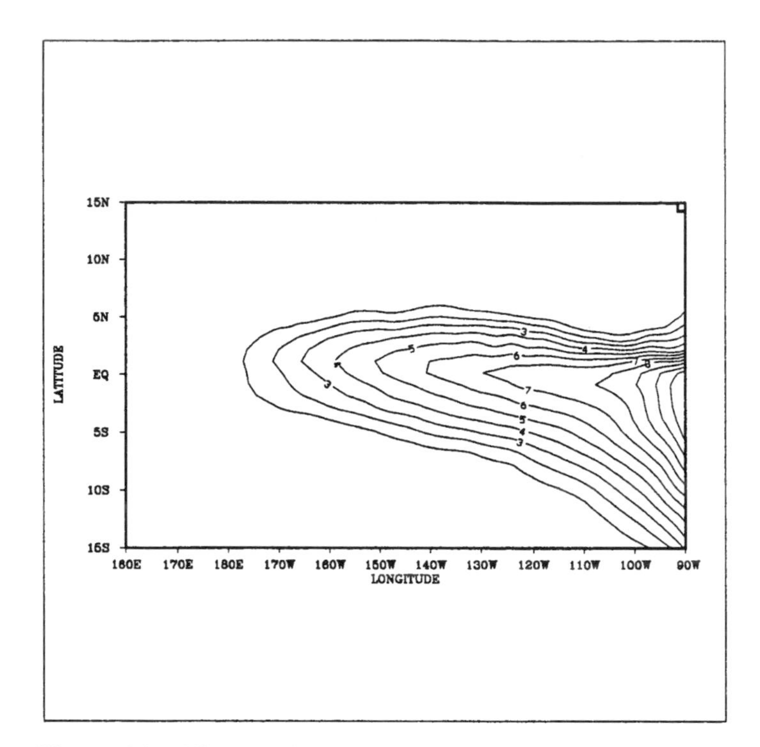

Figure 6.8 *The annual mean (year six) surface nitrate from the experiment with the latitudinal modified maximum growth rate. Contour interval is 1.0 mmol m^{-3}. The surface nitrate concentration was increased owing to less biological uptake. The asymmetrical feature of surface nitrate was enhanced in the central and eastern Pacific. A comparison can be made with the normal simulation (constant maximum growth rate); see Fig. 6.3*b.

data at Ocean Station P (50°N, 145°W) and summarised the role of grazing in the nutrient-rich areas of the open ocean with his grazing hypothesis (Frost, 1991): 'In the nutrient-rich areas of the open sea, phytoplankton net specific growth rate and specific grazing mortality tend towards approximate balance that may be perturbed, but not fully disrupted, by episodic forcing events external or internal to the pelagic food web.' Two questions arise that can be investigated with our equatorial model. What is the perturbation range within which the phytoplankton growth and zooplankton grazing balance can be maintained? How does the nutrient concentration change when balance is lost? In this section, model results from two experiments with different maximum grazing rates (G_{max}) are presented to address these questions.

In the standard experiment, μ_{max} is 2.0 d^{-1} and G_{max} is 1.0 d^{-1}. The most direct way to examine the role of zooplankton grazing is to alter G_{max} in the model while keeping μ_{max} unchanged. In the first experiment, G_{max} was increased from 1.0 d^{-1} to 1.5 d^{-1}; in the second experiment, G_{max} was decreased from 1.0 d^{-1} to 0.5 d^{-1}.

Fig. 6.10 shows the annual mean (year six) surface nitrate concentration from

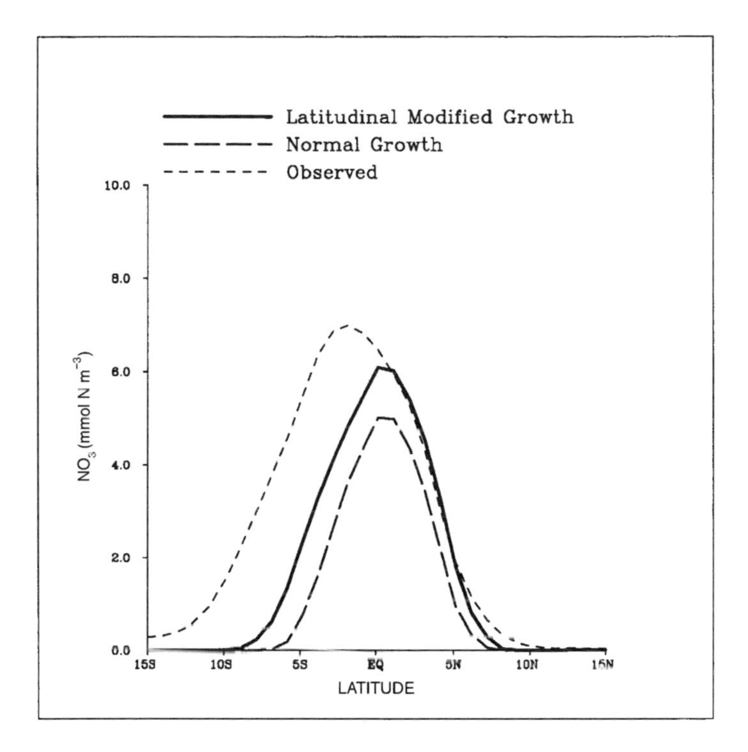

Figure 6.9 *The annual mean (year six) surface nitrate concentration along 140°W. In the simulation with latitudinally modified maximum growth rate, the surface nitrate concentration comparison with the observed values was improved slightly.*

the experiment with enhanced maximum grazing rate. Nitrate increased by more than 50% with increased G_{max} and was depleted in the simulation with reduced maximum grazing (thus not shown because of the absence of contour lines). The balance between μ_{max} and G_{max} was disrupted in the second experiment; therefore, the model produced unrealistic surface nitrate concentrations, namely zero surface nitrate everywhere. On the other hand, the first grazing experiment with increased G_{max} suggested that while the balance still existed, the higher nitrate concentrations indicated that a new (lower) steady state of nitrogen export was reached. Equations 4–8 of Lindley *et al.* (1995) describe mathematically how this new steady state is achieved.

Fig. 6.11 shows time series of nitrate concentration at 140°W on the equator during the six years of model integration in three different grazing experiments. By the third year, the surface nitrate was depleted and remained depleted (less than 0.2 mmol m^{-3}) with reduced G_{max}. Conversely, a higher nitrate concentration was maintained with increased G_{max}. With lower G_{max}, phytoplankton biomass increased until nitrate was depleted, whereas with higher G_{max} phytoplankton biomass decreased (see Fig. 6.12). Interestingly, zooplankton biomass increased with reduced G_{max} and decreased with enhanced

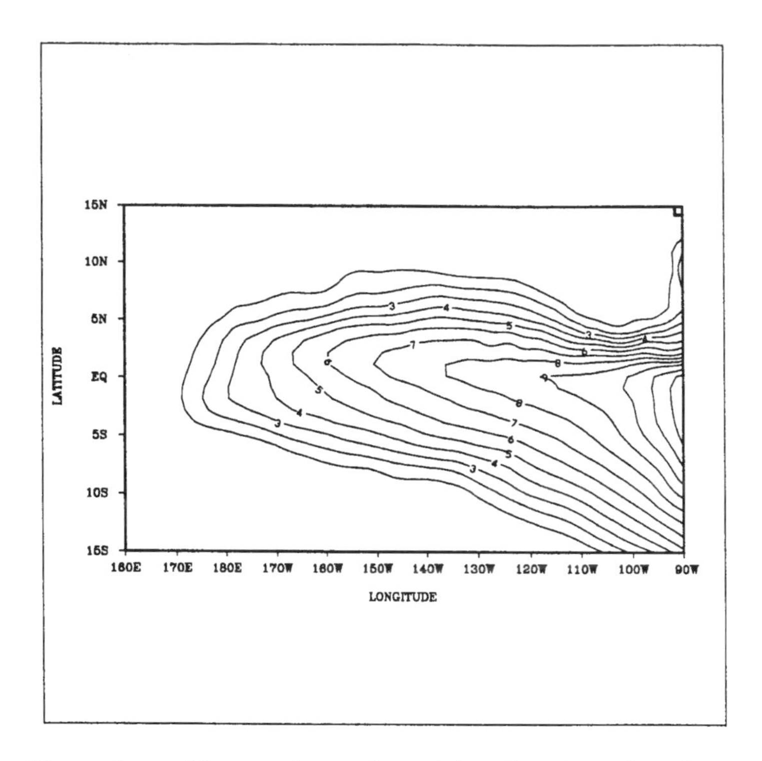

Figure 6.10 *The annual mean (year six) surface nitrate from the experiment with the enhanced maximum grazing* (G_{max}). *Contour interval is 1.0 mmol* m^{-3}. *The surface nitrate concentration was increased owing to the increase in the maximum grazing pressure; see Fig. 6.3b for comparison with the normal simulation. Higher grazing pressure keeps phytoplankton concentration at a lower level and there is therefore a higher surface nitrate concentration.*

G_{max}. Why do G_{max} and zooplankton biomass respond in opposite directions?

Parallel to the discussion of μ_{max} and phytoplankton biomass in the section on iron limitation, there are two factors that determine the zooplankton biomass: G_{max} and food availability (phytoplankton biomass). If G_{max} is high (as in the enhanced grazing experiment), new phytoplankton are grazed faster than they grow, so phytoplankton biomass is held at a low level. This results in food limitation for zooplankton growth, low phytoplankton biomass and high nitrate. On the other hand, if the zooplankton grazing is lower, phytoplankton accumulate enough biomass to deplete nitrate and provide abundant food for zooplankton. In this case, phytoplankton biomass is high, zooplankton biomass is high, and nitrate is depleted. This explanation is supported by the primary productivity and zooplankton grazing (secondary production) rates from all three experiments (see Fig. 6.13). Primary productivity rate doubled with the reduced G_{max} experiment and zooplankton grazing also doubled. The primary productivity rate and the grazing rate decreased by half in the increased G_{max}

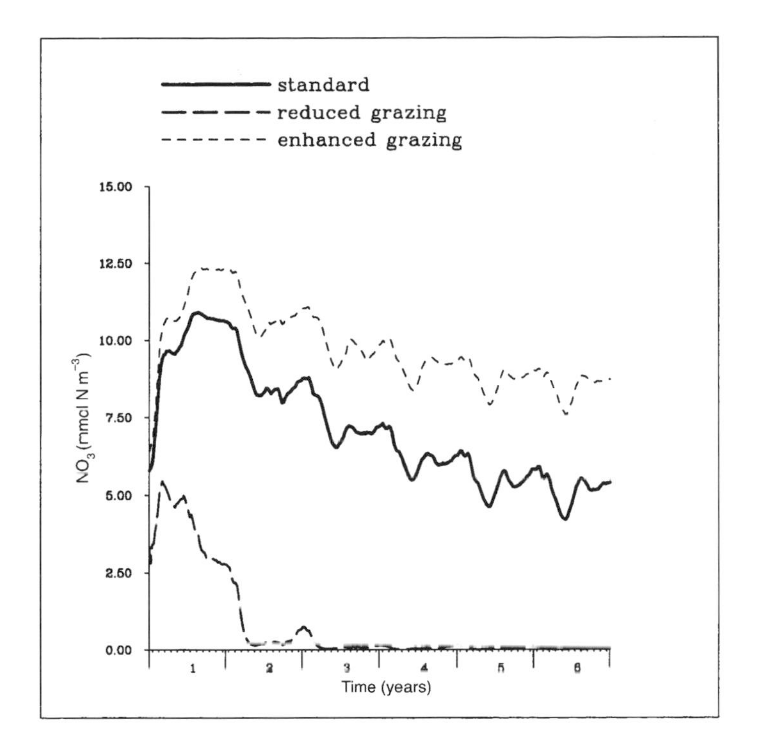

Figure 6.11 *Time series of weekly mean surface nitrate concentration at 140°W on the equator from three different experiments. The surface nitrate drops to zero within two years and stays at depletion afterwards in the reduced maximum grazing simulation. The surface nitrate concentrations increased by 50% in the enhanced maximum grazing simulation.*

experiment. Frost & Franzen (1992) presented a similar discussion about growth and grazing, but they used a simplified model.

To reproduce realistic nitrate fields with the five-component ecosystem model, the quotient G_{max}/μ_{max} must be between 0.5 and 0.75 in our equatorial Pacific model. The model produced unrealistically low nitrate values if G_{max}/μ_{max} was below 0.5 and unrealistically high nitrate ones if G_{max}/μ_{max} was above 0.75. Becuase we did not test many possible combinations of G_{max}/μ_{max} systematically, our conclusion about the dependency of realistic results on the range for the G_{max}/μ_{max} may hold up in a simple model but may not be the way it works in nature.

Combining the results shown in Fig. 6.10 through Fig. 6.13 with those in Table 6.1 (the iron-enriched experiment), the similarity of our model results with observations suggests that phytoplankton growth and zooplankton grazing are tightly coupled in the equatorial Pacific ecosystem, and that this coupling or 'balance' is a fundamental characteristic of the HNLC condition.

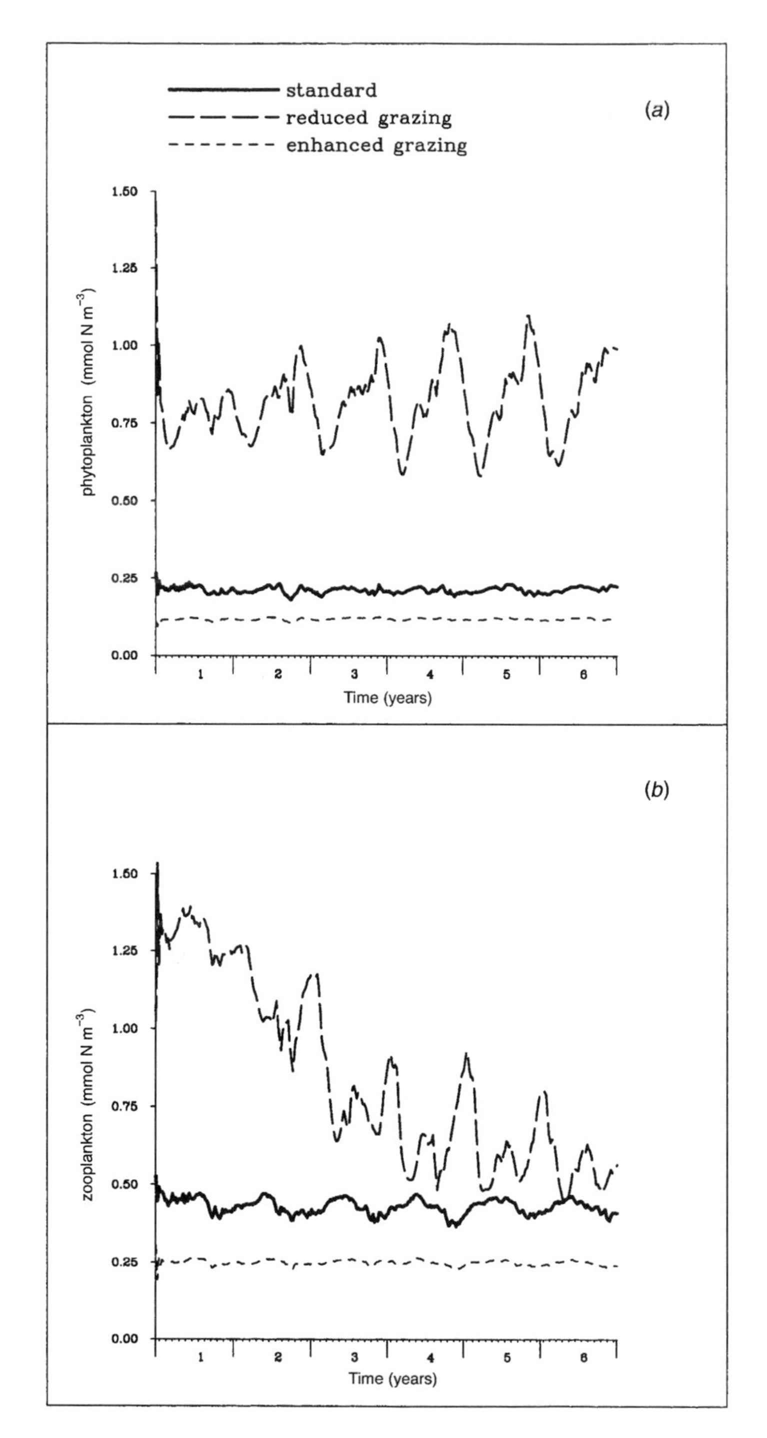

Figure 6.12 *Time series of weekly mean surface (*a*) phytoplankton (*b*) zooplankton at 140°W on the equator from three different experiments to test the grazing hypothesis. The surface phytoplankton concentration was increased by a factor of 3 in the reduced maximum grazing simulation; this, in turn, causes the surface nitrate depletion (Fig. 6.11). The enhanced maximum grazing results in a decrease in both phytoplankton and zooplankton.*

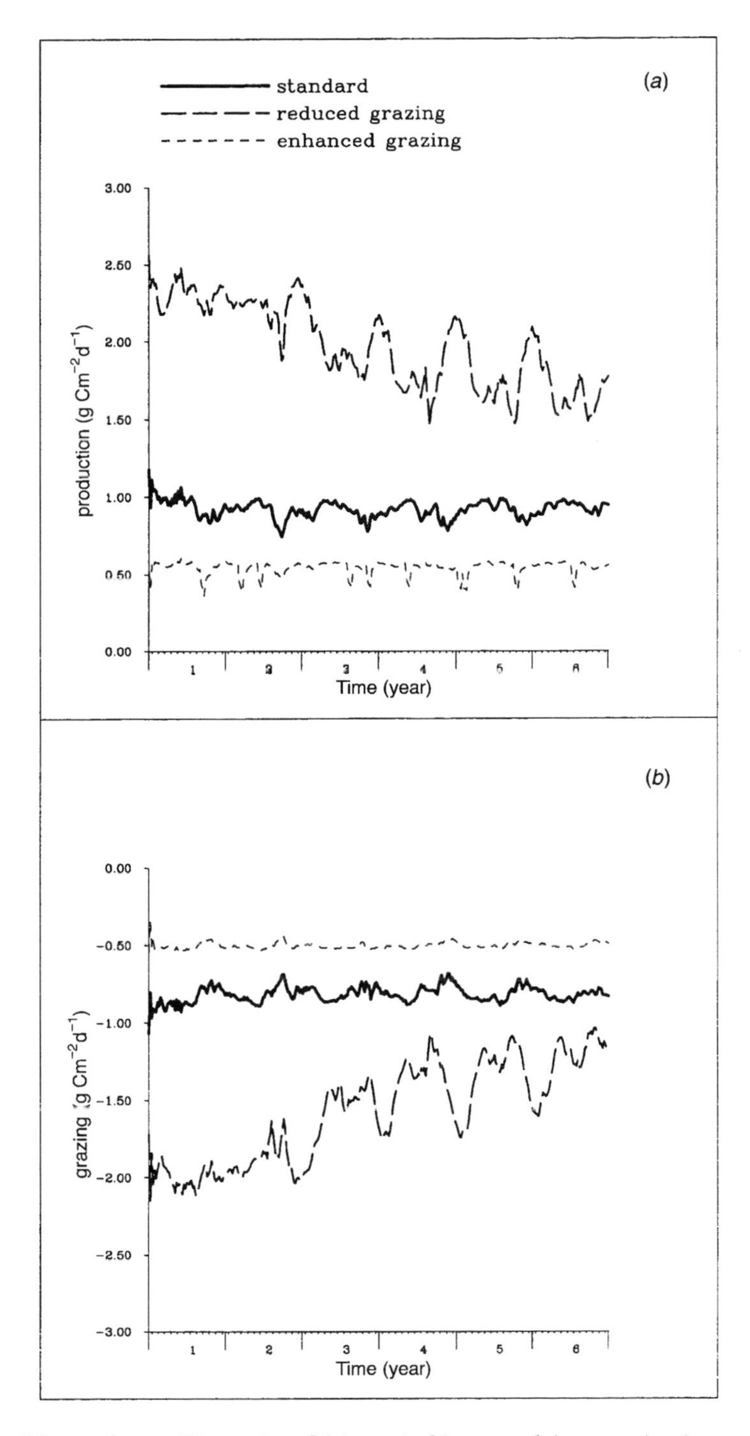

Figure 6.13 *Time series of* (a) *vertical integrated (0–120 m) primary production* (b) *vertical integrated zooplankton grazing on phytoplankton at 140°W on the equator for three different experiments. With the enhanced maximum grazing, the primary production was decreased, as was the grazing. Both primary production and grazing were doubled in the reduced maximum grazing experiment. The phytoplankton growth and zooplankton are hand-checked closely in the equatorial Pacific.*

Appendix

The biological model consisted of five compartments describing phytoplankton (P), zooplankton (Z), non-living particulate organic nitrogen (D), and two forms of dissolved inorganic nitrogen: nitrate (NO_3) and ammonium (NH_4). The five equations describing inter-compartment flows all take the form:

$$\frac{\partial C_i}{\partial t} = \text{PHYSICAL}\,(C_i) + \text{BIOLOGY}\,(C_i) \qquad i = 1,\ldots,5 \tag{A.1}$$

(For example, $C_1 = NO_3$, $C_2 = NH_4$, and so on.) The term PHYSICAL (C_i) represents the concentration change due to physical processes, including advection and diffusion.

$$\text{PHYSICAL}\,(C_i) \equiv \underbrace{-u\frac{\partial C_i}{\partial x} - v\frac{\partial C_i}{\partial y}}_{\text{horizontal advection}} \underbrace{- w\frac{\partial C_i}{\partial z}}_{\text{vertical advection}}$$

$$\underbrace{+\frac{\partial}{\partial x}\left(K_h\frac{\partial C_i}{\partial x}\right) + \frac{\partial}{\partial y}\left(K_h\frac{\partial C_i}{\partial y}\right)}_{\text{horizontal eddy diffusion}} \underbrace{+ \frac{\partial}{\partial z}\left(\frac{K_v}{\delta p}\cdot\frac{\partial C_i}{\partial z}\right)}_{\text{vertical eddy diffusion}} \tag{A.2}$$

The u, v, and w here are the velocity field. K_h and K_v are the horizontal and vertical diffusion coefficients. The value of $\delta p = 1$ except when the convection adjustment occurs, in which case $\delta p = 0$.

The term BIOLOGY(C_i) represents biological sources and sinks for the concentration field. In the euphotic zone (the upper 120 m), the biological terms, BIOLOGY(C_i), are:

$$\text{BIOLOGY}\,(NO_3) = \underbrace{-\,\text{NP}}_{\text{new production}} \tag{A.3}$$

$$\text{BIOLOGY}\,(NH_4) = \underbrace{-\,\text{RP}}_{\text{regenerated production}} \quad \underbrace{+\,\mu Z}_{NH_4 \text{ regeneration}} \tag{A.4}$$

$$\text{BIOLOGY}\,(\text{P}) = \underbrace{+\,(\text{NP} + \text{RP})}_{\text{total production}} \underbrace{-\,G_1}_{\text{grazing}} \underbrace{-\,\frac{\partial}{\partial z}(w_1\text{P})}_{\text{sinking}} \underbrace{-\,m_1\text{P}}_{\text{mortality}} \tag{A.5}$$

$$\text{BIOLOGY}\,(\text{Z}) = \underbrace{+\,\gamma(G_1 + G_2)}_{\text{net grazing}} \underbrace{-\,\mu Z}_{NH_4 \text{ regeneration}} \underbrace{-\,\frac{\partial}{\partial z}(w_2 Z)}_{\text{sinking}} \underbrace{-\,m_2 Z}_{\text{mortality}} \tag{A.6}$$

$$\text{BIOLOGY}\,(\text{D}) = \underbrace{+\,(1-\gamma)(G_1 + G_2)}_{\text{unassimilated grazing}} \underbrace{-\,G_2}_{\text{grazing}} \underbrace{-\,\frac{\partial}{\partial z}(w_3\text{D})}_{\text{sinking}} \underbrace{-\,m_1\text{P}}_{\text{mortality}} \tag{A.7}$$

The dimensionless parameter γ is assimilation efficiency; zooplankton excretion rate of ammonium μ (d^{-1}), sinking velocities w_1 (m d^{-1}), w_2 (m d^{-1}), w_3 (m d^{-1}), and mortality rate m_1 (d^{-1}), m_2 (d^{-1}) are specified in Table A.

Growth functions were written as follows.

$$\text{New production: NP} = \mu_{max}\underbrace{\frac{NO_3}{NO_3 + K_{NO_3}}}_{NO_3\text{ regulation}}\underbrace{\exp(-\psi NH_4)}_{NH_4\text{ inhibition}}\underbrace{\left(1 - \exp\left(-\frac{\alpha}{\mu_{max}}I\right)\right)}_{\text{light regulation}}P \tag{A.8}$$

$$\text{Regenerated production: RP} = \mu_{max}\underbrace{\frac{NH_4}{NH_4 + K_{NH_4}}}_{NH_4\text{ regulation}}\underbrace{\left(1 - \exp\left(-\frac{\alpha}{\mu_{max}}I\right)\right)}_{\text{light regulation}}P \tag{A.9}$$

The parameter μ_{max} represents the phytoplankton maximum specific growth rate when nitrogen and light are saturating for growth. Conceptually, μ_{max} may be regulated by other factors not included explicitly in the model, such as iron. We can explore the potential effects of these implicit limiting agents by changing μ_{max} directly.

Grazing functions were written as follows:

$$\begin{aligned} G_1 &= G_{max}\frac{\zeta_1 P}{K_{gr} + \zeta_1 P + \zeta_2 D}\frac{P}{P_{ave}}Z \\ G_2 &= G_{max}\underbrace{\frac{\zeta_2 D}{K_{gr} + \zeta_1 P + \zeta_2 D}}_{\text{food limitation}}\underbrace{\frac{D}{D_{ave}}}_{\text{depth modification}}Z \end{aligned} \tag{A.10}$$

where ζ_1 and ζ_2 are preferences for a given food type, and defined as follows:

$$\begin{aligned} \zeta_1 &= \frac{\rho_1 P}{\rho_1 P + \rho_2 D} \\ \zeta_2 &= \frac{\rho_2 D}{\rho_1 P + \rho_2 D} \end{aligned} \tag{A.11}$$

The values for ρ_1 and ρ_2 are given in Table A. P_{ave} and D_{ave} are depth-averaged (within the euphotic zone) phytoplankton and detritus concentration, which are defined as:

$$P_{ave} = \frac{1}{z'}\int_{-z'}^{0} P\,dz \tag{A.12}$$

Table A. *Biological model parameters*
Note: μ_{max} is used in this paper for the maximum specific growth rate, where P_{max} was used in Chai *et al.* (1996) for the maximum specific growth rate. The value and unit are the same for μ_{max} and P_{max}.

Parameter	Symbol	Value	Unit	Source
Maximum specific growth rate	μ_{max}	2.0	d^{-1}	(1, 2)
Initial slope of P–I curve	α	0.1	$d^{-1} (W m^{-2})^{-1}$	(2)
Ammonium inhibition parameter	ψ	6.0	$(mmol m^{-3})^{-1}$	(4)
Light attenuation due to water	k_1	0.046	m^{-1}	(3, 5)
Light attenuation by phytoplankton	k_2	0.03	$m^{-1} (mmol m^{-3})^{-1}$	(3, 5)
Phytoplankton specific mortality rate	m_1	0.05	d^{-1}	(6)
Phytoplankton sinking speed	w_1	1.0	$m d^{-1}$	(7)
Half-saturation for nitrate uptake	K_{NO_3}	0.25	$mmol m^{-3}$	(8)
Half-saturation for ammonium uptake	K_{NH_4}	0.05	$mmol m^{-3}$	(6)
Averaged surface noontime irradiance	I_0^{Noon}	350	$W m^{-2}$	(9)
Zooplankton maximum grazing rate	G_{max}	1.0	d^{-1}	(10)
Assimilation efficiency	γ	0.75	—	(6)
Zooplankton excretion rate to ammonium	μ	0.3	d^{-1}	(6)
Zooplankton sinking speed	w_2	1.0	$m d^{-1}$	(7)
Zooplankton specific mortality rate	m_2	0.05	d^{-1}	(6)
Half-saturation for zooplankton ingestion	K_{gr}	0.25	$mmol m^{-3}$	(11)
Grazing preference for phytoplankton	ρ_1	0.75	—	(5)
Grazing preference for detritus	ρ_2	0.25	—	(5)
Detritus sinking speed	w_3	10.0	$m d^{-1}$	(11)
Aphotic zone ammonium regeneration rate	β	0.25	d^{-1}	(11)
Aphotic zone detritus flux decay exponent	δ	0.858	(12)	

Sources: (1) Barber & Chavez, 1991; (2) Lindley, 1994; (3) Evans & Parslow, 1985; (4) Hofmann & Ambler, 1988; (5) Fasham *et al.*, 1990; (6) Fasham, 1995; (7) Jamart *et al.*, 1977, 1979; (8) Eppley *et al.*, 1969; (9) Chavez *et al.*, 1996; (10) Landry *et al.*, 1995; (11) Sarmiento *et al.*, 1993; (12) Martin *et al.*, 1987.

$$D_{ave} = \frac{1}{z'} \int_{-z'}^{0} D\,dz$$

By modifying the grazing functions with terms P/P_{ave} and D/D_{ave}, we found that the predator-prey oscillation could be reduced significantly in the model, but reasons are not clear. The variable z' is the depth of the euphotic zone, and is set to 120 m in the model. $I(z, t)$ is the irradiance, and is given by:

$$I(z,t) = I_0(t)\exp(-k_1 z - k_2 \int_{-z}^{0} P\,dz) \qquad (A.13)$$

Here k_1 is the attenuation coefficient due to water, and k_2 is the self-shading parameter due to the phytoplankton. Values for k_1 and k_2 are given in Table A. $I_0(t)$ is the surface irradiance, and has the form during a 24 h period:

$$I_0(t) = I_0^{Noon} \sin\left(\frac{t-6}{12}\pi\right) \text{ during day (6 am to 6 pm)}$$

$$= 0 \qquad \text{during night (6 pm to 6 am)} \qquad (A.14)$$

I_0^{Noon} is the annual mean noontime averaged surface irradiance for the equatorial Pacific, and the value is given in Table A.

Below the euphotic zone, the biological terms are:

$$\text{BIOLOGY }(NO_3) = \underbrace{+\beta NH_4}_{NO_3 \text{ regeneration}} \qquad (A.15)$$

$$\text{BIOLOGY }(NH_4) = \underbrace{+\beta(P+Z) - \frac{\partial F(z)}{\partial z}}_{NH_4 \text{ regeneration}} \underbrace{-\beta NH_4}_{NO_3 \text{ regeneration}} \qquad (A.16)$$

$$\text{BIOLOGY }(P) = \underbrace{-\beta P}_{NH_4 \text{ regeneration}} \qquad (A.17)$$

$$\text{BIOLOGY }(Z) = \underbrace{\beta Z}_{NH_4 \text{ regeneration}} \qquad (A.18)$$

$$\text{BIOLOGY }(D) = \underbrace{0}_{\text{No biological processes}} \qquad (A.19)$$

$F(z)$ is the downward flux of particulate nitrogen in the form of the sum of sinking phytoplankton (P), zooplankton (Z) and detritus (D). Below the euphotic zone, particulate nitrogen between two layers in the model was converted to increase ammonium concentration. $F(z)$ is an empirical function determined from Pacific Ocean sediment-trap observations by Martin *et al.* (1987):

$$F(z) = F(z')\left(\frac{z}{z'}\right)^{-\delta} \qquad \text{(A.20)}$$

$F(z')$ ($z' = 120$ m) is the downward flux of the particulate material at the base of the euphotic zone, and it includes sinking P, Z and D, as well as zooplankton mortality loss in the euphotic zone.

$$F(z') = w_1\mathrm{P}(z_t) + w_2\mathrm{Z}(z') + w_3\mathrm{D}(z') + \int_{-z'}^{0} m_2\mathrm{Z}\,\mathrm{d}z \qquad \text{(A.21)}$$

Parameter values used in the standard experiment are given in Table A.

Acknowledgements

This research was supported by the Harvey W. Smith Endowment at Duke University, by NASA Grant NAGW-3655 to R. T. Barber, and by a Cray Research Fellowship awarded to F. Chai at the North Carolina Supercomputing Center. NSF Grant OCE-9024373 to R. T. Barber supported the sea work. We thank Hugh Ducklow and an anonymous reviewer for providing comments that improved the final version of the manuscript. The authors thank Roger Hanson for his editorial assistance during the final stage of editing and formatting the article. This publication is US JGOFS contribution number 393.

References

Barber, R. T. (1992). Introduction to the WEC88 cruise: an investigation into why the equator is not greener. *Journal of Geophysical Research*, **97**, 609–10.

Barber, R. T. & Chavez, F. P. (1986). Ocean variability in relation to living resources during the 1982–83 El Niño. *Nature*, **319**, 279–85.

Barber, R. T. & Chavez, F. P. (1991). Regulation of primary productivity rate in the equatorial Pacific. *Limnology and Oceanography*, **36**, 1803–15.

Barber, R. T. & Ryther, J. H. (1969). Organic chelators: factors affecting primary production in the Cromwell Current upwelling. *Journal of Experimental Marine Biology and Ecology*, **3**, 191–9.

Barber, R. T., Chai, F., Lindley, S. T. & Bidigare, R. R. (1995). Regulation of equatorial primary production. In *Global Fluxes of Carbon and Related Substances in the Coastal Sea-Ocean-Atmosphere System*, ed. I. Koike, pp. 294–300. Tokyo: Science Council of Japan.

Barber, R. T., Sanderson, M. P., Lindley, S. T., Chai, F., Newton, J., Trees, C. C., Foley, D. G. & Chavez, F. P. (1996). Primary productivity and its regulation in the equatorial Pacific during and following the 1991–1992 El Niño. *Deep-Sea Research*, Part II, **43**, 933–69.

Behrenfeld, M. J., Bale, A. J., Kolber, Z. S., Aiken, J. & Falkowski, P. G. (1996). Confirmation of iron limitation of phytoplankton photosynthesis in the equatorial Pacific Ocean. *Nature*, **383**, 508–11.

Bidigare, R. R. & Ondrusek, M. E. (1996). Spatial and temporal variability of phytoplankton pigment distributions in the central and eastern equatorial Pacific Ocean. *Deep-Sea Research*, Part II, **43**, 809–34.

Bryan, K. (1969). A numerical method for the study of the world ocean. *Journal of Comparative Physics*, **4**, 347–76.

Chai, F. (1995). Origin and maintenance of high nitrate condition in the equatorial Pacific, a biological-physical model study. Ph.D. Dissertation, Duke University, Durham, NC, 170 pp.

Chai, F., Lindley, S. T., Barber, R. T. (1996). Origin and maintenance of high nitrate condition in the equatorial Pacific. *Deep-Sea Research*, Part II, **43**, 1031–64.

Chavez, F. P., Buck, K. R., Service, S. K., Newton, J. & Barber, R. T. (1996). Phytoplankton variability in the central and eastern tropical Pacific. *Deep-Sea Research*, Part II, **43**, 835–70.

Chisholm, S. W. & Morel, F. M. M. (1991). What controls phytoplankton production in nutrient-rich areas of the open sea? *Limnology and Oceanography*, **36**(8). (Preface.)

Coale, K. H., Fitzwater, S. E., Gordon, R. M., Johnson, K. S. & Barber, R. T. (1996*a*). Iron limits new production and community growth at picomolar levels in the equatorial Pacific Ocean. *Nature*, **379**, 621–4.

Coale, K. H. *et al.* (1996*b*). A massive phytoplankton bloom induced by an ecosystem-scale iron fertilization experiment in the equatorial Pacific Ocean. *Nature*, **383**, 495–501.

Colin, C., Hisard, C. & Oudot, P. (1971). Le Courant de Cromwell dans le Pacifique central en fevrier. *Cahiers ORSTOM, Series Oceanography*, **9**, 167–87.

Cooper, D. J., Watson, A. J. & Nightingale, P. D. (1996). Large decrease in ocean-surface CO_2 fugacity in response to *in situ* iron fertilization. *Nature*, **383**, 511–13.

Cox, M. D. (1984). A primitive equation, 3-dimensional model of the ocean. GFDL Ocean Group Technical Report, No. 1.

Cromwell, T. (1953). Circulation in meridional plane in the central equatorial Pacific. *Journal of Marine Research*, **12**, 196–213.

Cullen, J. J., Lewis, M. R., Davis, C. O. & Barber, R. T. (1992). Photosynthetic characteristics and estimated growth rates indicate grazing is the proximate control of primary production in the equatorial Pacific. *Journal of Geophysical Research*, **97**, 639–54.

Donaghay, P. L., Liss, P. S., Duce, R. A., Kester, D. R., Hanson, A. K., Villareal, T., Tindale, N. & Gifford, D. J. (1991). The role of episodic atmospheric nutrient inputs in the chemical and biological dynamics of oceanic ecosystems. *Oceanography*, **4**, 62–70.

Duce, R. A. (1986). The impact of atmospheric nitrogen, phosphorus, and iron species on marine biological productivity. In *The Role of Air-Sea Exchange in Geochemical Cycling*, ed. P. Buat-Menard, pp. 497–529. Dordrect: D. Reidel Publishing Company.

Duce, R. A. & Tindale, N. W. (1991). Atmospheric transport of iron and its deposition in the ocean. *Limnology and Oceanography*, **36**, 1715–26.

Dugdale, R. C. (1967). Nutrient limitation in the sea: dynamics, identification, and significance. *Limnology and Oceanography*, **12**, 685–95.

Dugdale, R. C. & Goering, J. J. (1967). Uptake of new and regenerated forms of nitrogen in primary productivity. *Limnology and Oceanography*, **12**, 196–206.

Dugdale, R. C., Wilkerson, F. P., Barber, R. T. & Chavez, F. P. (1992). Estimating new production in the equatorial Pacific Ocean at 150°W. *Journal of Geophysical Research*, **97**, 681–6.

Eppley, R. W. & Peterson, B. J. (1979). Particulate organic matter flux and planktonic new production in the deep ocean. *Nature*, **282**, 677–80.

Eppley, R. W., Rogers, J. N., & McCarthy, J. J. (1969). Half saturation constants for uptake of nitrogen and ammonium by marine phytoplankton. *Limnology and Oceanography*, **14**, 912–20.

Evans, G. T. & Parslow, J. S. (1985). A model of annual plankton cycles. *Biological Oceanography*, **24**, 483–94.

Fasham, M. J. R. (1995). Variations in the seasonal cycle of biological production in subarctic oceans: A model sensitivity analysis. *Deep-Sea Research*, **42**, 1111–49.

Fasham, M. J. R., Ducklow, H. W. & McKelvie, S. M. (1990). A nitrogen-based model of plankton dynamics in the oceanic mixed layer. *Journal of Marine Research*, **48**, 591–639.

Feely, R. A., Wanninkhof, R., Cosca, C. E., Murphy, P. P., Lamb, M. F. & Steckley, M. D. (1995). CO_2 distributions in the equatorial Pacific during the 1991–1992 ENSO event. *Deep-Sea Research*, Part II, **42**, 365–86.

Feldman, G. *et al.* (1989). Ocean color: availability of global data set. *Eos, Transactions of the American Geophysical Union*, **70**(23), 634.

Frost, B. W. (1987). Grazing control of phytoplankton stock in the open subarctic Pacific Ocean: a model assessing the role of mesozooplankton, particularly the large calanoid copepods, *Neocalanus* spp. *Marine Ecology Progress Series*, **39**, 49–68.

Frost, B. W. (1991). The role of grazing in nutrient-rich areas of the open sea. *Limnology and Oceanography*, **36**, 1616–30.

Frost, B. W. & Franzen, N. C. (1992). Grazing and iron limitation in the control of phytoplankton stock and nutrient concentrations; a chemostat analog of the Pacific equatorial upwelling zone. *Marine Ecology Progress Series*, **83**, 291–303.

Gill, A. E. (1982). *Atmosphere-Ocean Dynamics*. New York: Academic Press.

Graham, H. W. (1941). Plankton production in relation to character of water in the open Pacific. *Journal of Marine Research*, **4**, 189–97.

Hofmann, E. E. & Ambler, J. W. (1988). Plankton dynamics on the outer southeastern US continental shelf Part II: a time-dependent model. *Journal of Marine Research*, **9**, 235–48.

Jamart, B. M., Winter, D. F., Banse, K., Anderson, G. C. & Lam, R. K. (1977). A theoretical study of phytoplankton growth and nutrient distribution in the Pacific Ocean of the northwestern US coast. *Deep-Sea Research*, **24**, 753–73.

Jamart, B. M., Winter, D. F. & Banse, K. (1979). Sensitivity analysis of a mathematical model of phytoplankton growth and nutrient distribution in the Pacific Ocean of the northwestern US coast. *Journal of Plankton Research*, **1**, 267–90.

Jassby, A. D. & Platt, T. (1976). Mathematical formulation of the relationship between photosynthesis and light for phytoplankton. *Limnology and Oceanography*, **21**, 540–7.

Kogelschatz, J., Solorzano, L., Barber, R. T. & Mendoza, P. (1985). Oceanographic conditions in the Galapagos Islands during the 1982/1983 El Niño. In *El Niño in the Galapagos Islands: The 1982/1983 Event*, ed. G. Robinson and E. M. del Pino, pp. 91–123. Quito, Ecuador: Charles Darwin Foundation for the Galapagos Islands.

Kolber, Z. S., Barber, R. T., Coale, K. H., Fitzwater, S. E., Greene, R. M., Johnson, K. S., Lindley, S. & Falkowski, P. G. (1994). Iron limitation of phytoplankton photosynthesis in the equatorial Pacific Ocean. *Nature*, **371**, 145–9.

Landry, M. R., Constantinou, J. & Kirshtein, J. (1995). Microzooplankton grazing in the central equatorial Pacific during February and August 1992. *Deep-Sea Research*, Part II, **42**, 657–72.

Landry, M. R., Barber, R. T., Bidigare, R. R., Chai, F., Coale, K. H., Dam, H. G., Lewis, M. R., Lindley, S. T., McCarthy, J. J., Roman, M. R., Stoecker, D. K., Verity, P. G. & White, J. R. (1997). Iron and grazing constraints on primary production in the central equatorial pacific: an EqPac synthesis. *Limnology and Oceanography*, **42**, 405–18.Levitus, S. (1982). Climatological atlas of the world oceans. NOAA Prof.

Paper No. 13. Washington D.C.: US Government Printing Office. 137 pp.

Levitus, S., Reid, J. L., Conkright, M. E. & Najjar, R. (1993). Distribution of nitrate, phosphate and silicate in the world ocean. *Progress in Oceanography*, **31**, 245–73.

Lewis, M. R. & Smith, J. C. (1983). A small volume, short-incubation-time method for measurement of photosynthesis as a function of incident irradiance. *Marine Ecology Progress Series*, **13**, 99–102.

Lindley, S. T. (1994). Regulation of the maximum quantum yield of phytoplankton photosynthesis by iron, nitrogen, and light in the eastern equatorial Pacific. Ph.D. Dissertation, Duke University, Durham, NC. 136 pp.

Lindley, S. T. & Barber, R. T. (1999). Phytoplankton response to natural and experimental iron addition. *Deep-Sea Research*, Part II (in press).

Lindley, S. T., Chai, F. & Barber, R. T. (1994). John Martin's Iron Hypothesis: Modeling the Response of the Equatorial Pacific to Iron Addition. *Eos*, **75**(3), 135.

Lindley, S. T., Bidigare, R. R. & Barber, R. T. (1995). Phytoplankton photosynthesis parameters along 140W in the equatorial Pacific. *Deep-Sea Research*, Part II, **42**, 441–64.

Loukos, H., Frost, B., Harrison, E. & Murray, J. W. (1997). An ecosystem model with iron limitation of primary production in the equatorial Pacific at 140°W. *Deep-Sea Research*, Part II **44**, 2221–50.

Martin, J. H. (1990). Glacial–Interglacial CO_2 change: the iron hypothesis. *Paleoceanography*, **5**, 1–13.

Martin, J. H., Knauer, G. A., Karl, D. M. & Broenkow, W. W. (1987). VERTEX: Carbon cycling in the northeast Pacific. *Deep-Sea Research*, **34**, 267–85.

Martin, J. H., Gordon, R. M. & Fitzwater, S. F. (1991). The case of iron. *Limnology and Oceanography*, **36**, 1793–802.

Martin, J. H. *et al.* (1994). Testing the iron hypothesis in ecosystems of the equatorial Pacific Ocean. *Nature*, **371**, 123–9.

McCarthy, J. J., Garside, C., Nevins, J. L. & Barber, R. T. (1996). New production along 140°W in the equatorial Pacific during and following the 1992 El Niño event. *Deep-Sea Research*, Part II, **43**, 1065–93.

Montgomery, R. B. (1954). Analysis of a Hugh M. Smith oceanographic section from Honolulu southward across the equator. *Journal of Marine Research*, **13**, 67–75.

Murray, J. W., Barber, R. T., Bacon, M. P., Roman, M. R. & Feely, R. A. (1994). Physical and biological controls on carbon cycling in the equatorial Pacific. *Science*, **266**, 58–65.

Murray, J. W., Johnson, E. & Garside, C. (1995). A US JGOFS Process Study in the equatorial Pacific (EqPac): Introduction. *Deep-Sea Research*, Part II, **42**, 275–94.

Oort, A. H., Pan, Y. H., Reynolds, R. W. & Ropelewski, C. F. (1987). Historical trends in the surface temperature over the oceans based on the COADS. *Climate Dynamics*, **2**, 29–38.

Pacanowski, R. C. & Philander, S. G. H. (1981). Parameterization of vertical mixing in numerical models of tropical oceans. *Journal of Physical Oceanography*, **11**, 1443–51.

Pacanowski, R., Dixon, K. & Rosati, A. (1991). The GFDL modular ocean model users guide, version 1.0. Technical Report No. 2, Ocean Group, Geophysical Fluid Dynamic Laboratory, Princeton.

Philander, S. G. H., Hurlin, W. J. & Seigel, & A. D. (1987). A model of the seasonal cycle in the tropical Pacific ocean. *Journal of Physical Oceanography*, **17**, 1986–2002.

Price, N. M., Ahner, B. A. & Morel, F. M. M. (1994). The equatorial Pacific ocean: grazer-controlled phytoplankton populations in an iron-limited ecosystem. *Limnology and Oceanography*, **39**, 520–34.

Rea, D. K., Hovan, S. A. & Janecek, T. R. (1994). Late quaternary flux of eolian dust to the pelagic ocean. In *Studies in Geophysics, Material Fluxes on the Surface of the Earth*, pp. 116–24. National Research Council, Washington: National Academy Press.

Riley, G. A. (1946). Factors controlling phytoplankton populations on Georges Bank. *Journal of Marine Research*, **6**,

54–73.

Riley, G. A. (1947). A theoretical analysis of the phytoplankton populations on Georges Bank. *Journal of Marine Research*, **6**, 104–13.

Sarmiento, J. L., Slater, R. D., Fasham, M. J. R., Ducklow, H. W., Toggweiler, J. R. & Evans, G. T. (1993). A seasonal three-dimensional ecosystem model of nitrogen cycling in the North Atlantic euphotic zone. *Global Biogeochemical Cycles*, **7**, 417–50.

Sette, O. E. (1955). Consideration of midocean fish production as related to oceanic circulatory system. *Journal of Marine Research*, **14**, 398–414.

Slutz, R. T., Lubker, S. J., Hiscox, J. D., Steurer, P. M. & Elms, J. D. (1985). Comprehensive Ocean–Atmosphere Data Set: Release 1. NOAA Environmental Research Laboratory, Climate Research Program, Boulder, CO.

Tans, P.P., Fung, I.Y. & Takahashi, T. (1990). Observational constraints on the global atmospheric CO_2 budget. *Science*, **247**, 1431–8.

Toggweiler, J. R., Dixon, K., & Broecker, W. S. (1991). The Peru upwelling and the ventilation of the South Pacific thermocline. *Journal of Geophysical Research*, **96**, 20467–97.

Vinogradov, M. E. (1981). Equatorial upwelling ecosystems. In *Analysis of Marine Ecosystems*, ed. A. Longhurst, pp. 69–93. London: Academic Press.

Volk, T. & Hoffert, M. I. (1985). Ocean carbon pumps: analysis of relative strengths and efficiencies in ocean-driven atmospheric CO_2 changes. In *The Carbon Cycle and Atmospheric CO_2: Natural Variations Archean to Present*, ed. E. T. Sundquist & W. S. Broecker, pp. 99–110. Washington: American Geophysical Union.

Walsh, J. J. (1976). Herbivory as a factor in patterns of nutrient utilization in the sea. *Limnology and Oceanography*, **21**, 1–13.

Watson, A. J., Law, C. S., Van Scoy, K. A., Millero, F. J., Yao, W., Friederich, G. E., Liddicoat, M. I., Wanninkhof, R. H., Barber, R. T. & Coale, K. H. (1994). Minimal effect of iron fertilization on sea-surface carbon dioxide concentrations. *Nature*, **371**, 143–5.

7 Continental margin carbon fluxes

K.-K. Liu, K. Iseki and S.-Y. Chao

Keywords: particulate organic carbon, primary production, new production, carbon exchange, coastal upwelling, river runoff, carbon budgets and models

Introduction

The Joint Global Ocean Flux Study (JGOFS) was established in 1988 to better understand the marine biogeochemical cycles that influence global climate (SCOR, 1990). During the past decade, the role of continental margins within the marine biogeochemical cycles has been increasingly emphasised (Walsh, 1988; Wroblewski & Hofmann, 1989; Mantoura *et al.*, 1991). Relative to the open ocean, continental shelves usually have large standing stocks of particulate organic carbon and fast rates of primary production. It has been suggested that continental margins may be a significant source of organic carbon for the open ocean (Walsh, 1989). Therefore, the coastal oceans are believed to have a strong impact on open-ocean processes. In 1992, continental margin studies were adopted by the SCOR Scientific Steering Committee for JGOFS as one of the operational elements of JGOFS (SCOR, 1992). A joint task team was established (Hall & Smith, 1995) to advocate continental margin studies for JGOFS and Land–Ocean Interaction in the Coastal Zone (LOICZ), two core projects of the International Geosphere–Biosphere Programme (IGBP). The goals of the continental margin studies are as follows (SCOR, 1994):

1. to quantify the exchange of carbon between given marginal zones, shelves and the open ocean;
2. to understand the seasonal and inter-annual fluctuation of carbon fluxes and nutrient cycles due to physical and biogeochemical variation;
3. to evaluate the importance of carbon deposition and benthic processes on the continental slope;
4. to determine the air–sea CO_2 exchange rates in strong upwelling areas;
5. to determine the source function of dimethyl sulphide, methane and nitrous oxide on the shelf; and

6. to characterise those features of continental margins that enable extrapolation to the global scale.

A continental margin is defined in marine geology as the transitional zone between the continental crust and the oceanic crust, including the continental shelf, slope and rise and coastal plains (Kennett, 1982). In the past, most of the oceanographic studies on the continental margins have emphasised continental shelves (Walsh, 1988). Continental shelves are considered to extend from the coastline to a depth of 200 m (Fig. 7.1), occupying 7.6% of the seafloor (Sverdrup *et al.*, 1942). For the study of the global carbon cycle, the definition should be dictated by modelling. In this pragmatic context, continental margins are defined as parts of the ocean between the open ocean and the land that are neither represented in a global ocean biogeochemical model nor covered by the terrestrial ecosystem model. However, continental margins provide the most variable environment in the ocean, and the most active biogeochemical processes. The significance of continental margins in the global carbon cycle is difficult to quantify precisely in view of the complexity, variability and individuality of each coastal unit. Consequently, this chapter provides two views: one reflecting the reality of present knowledge towards the goals of JGOFS continental margin studies and the other pointing to a strategy for approaching these goals.

A historical perspective

Walsh *et al.* (1981) proposed that organic deposition on the continental slope could represent the 'missing billion metric tons of carbon' in global CO_2 budgets. The proposition was based on measurements of carbon production, metabolism and exchange along food webs. They suggested that coastal eutrophication coupled with a nutrient shuttle operating between the shelf and slope may account for such an efficient sink, and that about 50% of the shelf production may be exported to the slope and the regenerated anthropogenic nutrients may be used cyclically to sequester carbon.

Their hypothesis inspired the Shelf Edge Exchange Processes (SEEP) experiment, which took place at the continental margin south of New England. Results from SEEP (Rowe *et al.*, 1986), however, challenged the hypothesis by raising the question, 'Do continental shelves export organic matter?' The measured primary production was not enough to support the estimated consumption by zooplankton, pelagic microbes and benthic fauna, not to mention a large fraction of particulate organic carbon export. However, these estimates are subject to large uncertainties. If the upper limits of the primary production and the lower limits of the consumption are adopted, there could be some surplus of primary production available for export to the slope region.

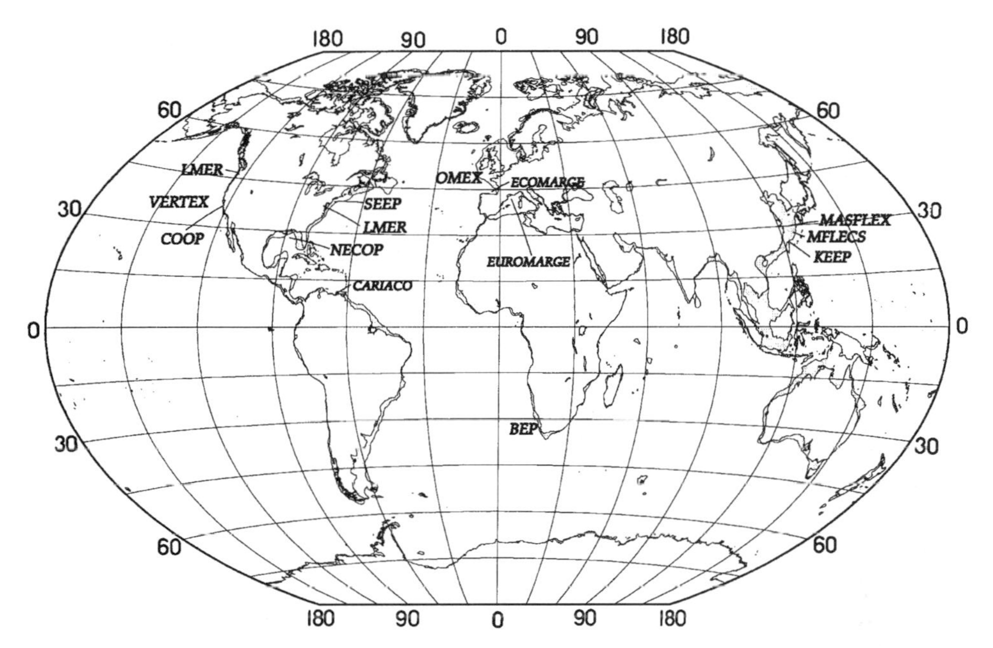

Figure 7.1 *Map of continental margins. Stippled areas represent the continental shelves with depths less than 200 m. Selected programmes for continental margin studies, both completed and on-going, are shown.*

Therefore, they concluded that only a small fraction of the phytodetritus might be exported from the shelf. Using a different approach, Berner (1992) demonstrated by simple calculation that continental margins cannot be the principal sinks for the missing carbon. If the missing carbon (150 Pg C; 1 Pg = 1 Gt = 10^{15} g) were deposited in the coastal sediments (1200–2500 Gt) in the past 100 years, the organic carbon percentage of the sediments would be as high as 6–12% by mass. This percentage is about ten times higher than the measured average organic carbon content (0.5–1.5%) of the continental margin sediments.

The role of continental margin carbon fluxes

Instead of being a major sink for the missing anthropogenic carbon, continental margins could serve as a link between the terrestrial and marine ecosystems. Sarmiento & Sundquist (1992) suspected that the riverine flux of carbon (0.8 Gt C yr^{-1}) at the continental margin was a major component making up the gap between two independent estimates of the oceanic uptake of anthropogenic CO_2 (Tans *et al.*, 1990; Houghton *et al.*, 1990). However, the geographic distribution of riverine carbon fluxes is not homogeneous, the degree of reactivity of the terrigenous carbon is variable (Spitzy & Ittekkot, 1991), and its dispersion in the ocean may be quite limited. Similarly, the net CO_2 fluxes across the air-sea interface are not uniform either (Tans *et al.*, 1990). Whether the spatial distribution of riverine carbon fluxes is compatible with the observed distribution of carbon transfer in the ocean is not clear. Smith & Hollibaugh (1993) suggested the observed invasion fluxes of anthropogenic CO_2 in the coastal and open oceans be increased by 0.08 and 0.2 Gt C yr^{-1}, respectively, to accurately represent the perturbed carbon fluxes, because, under natural conditions, they were net sources of CO_2. These additional fluxes arise in conjunction with the lateral transport of organic carbon from land to the shelf and from the shelf to the open ocean.

Smith & Hollibaugh's (1993) estimates of the laterally transported carbon fluxes were based on carbon budgets rather than on direct observation of cross-shelf transport of carbon. Large uncertainties along this line of estimates seem inevitable. Nevertheless, their approach to the issue demonstrated the usefulness of a compartmentalised ocean model. A carbon budget for each compartment may be constructed with exchange fluxes not only at the sea surface and between compartments (Broecker & Peng, 1992) but also at the continental margins. Better constraints should yield better estimates of the anthropogenic CO_2 invasion into the ocean.

Leaving the issue of anthropogenic CO_2 uptake slightly aside, the most important role of the continental margins may be their capacity as an auxiliary biological pump (Walsh, 1989) as illustrated in Fig. 7.2. Coastal upwelling in eastern boundary current systems and shelf break upwelling in western boundary current systems may bring nutrients from the subsurface water to the shelf sea (Csanady, 1990). Both should be balanced by a seaward transport of organic matter or, to a lesser extent, removed over the shelf by burial or denitrification (Walsh, 1991). The export of organic carbon fixed by shelf production to the subsurface water of the open ocean is essentially a biological pump, which is termed the 'coastal biological pump' to distinguish from the vertically operating biological pump of the open ocean (Longhurst & Harrison, 1989). The unused nutrients may be spread out to the ocean interior and enhance the biological pump in the open ocean.

According to Sarmiento & Siegenthaler (1992), the biological pump in the ocean only assists the drawdown of anthropogenic CO_2 indirectly by reducing the CO_2 fugacity in the surface water, which in turn favours dissolution of atmospheric CO_2. The biological pump may affect the drawdown of anthropogenic CO_2 only if the carbon-fixing capacity of the marine ecosystem changes in time and causes change in the mean CO_2 fugacity of surface water. The marine ecosystem may be changed directly by human activity, such as eutrophication and over-fishing, or altered indirectly owing to climate change. Continental margins are conceivably the most vulnerable ecosystems in the marine environment because they are most susceptible to human activity and arguably the most sensitive to climate change. Their potential responses to global change deserve attention.

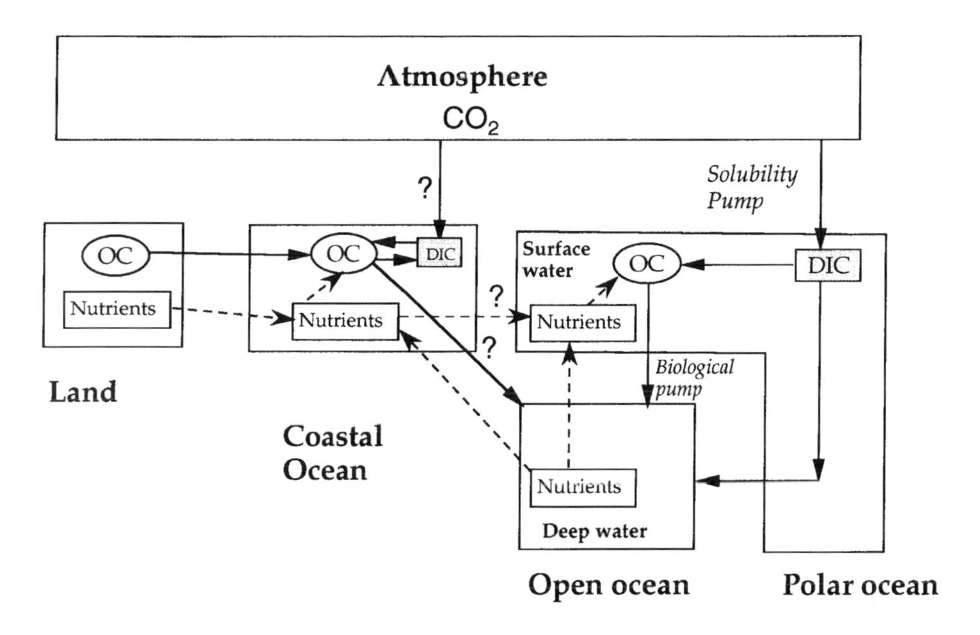

Figure 7.2 *The conceptual model of the transfer of atmospheric* CO_2 *into the ocean. The coastal ocean is shown as an auxiliary biological pump, which may account for about 20% of the global capacity.*

In summary, continental margins play three different roles in the carbon cycle: a sink for carbon, a link between the land and the ocean, and an auxiliary biological pump that is susceptible to human alteration. For convenience, the continental margins are collectively called the coastal ocean (Smith & Hollibaugh, 1993) in the ensuing discussion (Fig. 7.2). This chapter reviews the carbon fluxes and related physical and biogeochemical processes in the coastal ocean, with emphasis on the third role. Some regional studies are used to illustrate the variability of the continental margins. From these regional observations, fluxes on a global scale have been estimated. Although the estimates are qualitatively consistent, they do show large variations and, therefore, are subject to large uncertainties. Many of the discrepancies stem from the gap between the measurable fluxes and the conceptual fluxes. In order to understand continental margin fluxes better, regional models have to be invoked to assimilate the observations of the physical, chemical and biological processes at the continental margins.

Systematic of fluxes and processes

Continental margin carbon fluxes represent the exchange of carbon between a continental margin and its surroundings: the ocean interior, the atmosphere, the

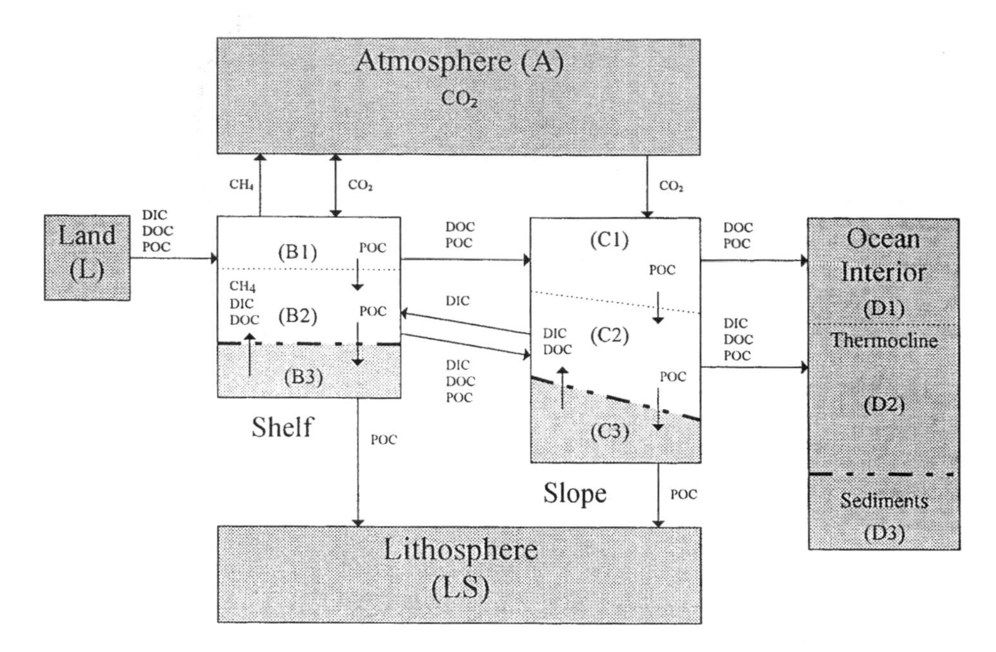

Figure 7.3 *The exchange of carbon between a continental margin, consisting of the shelf and the slope, and its surroundings. The fluxes are designated by the carbon species and compartments involved. For example, the POC flux from the land to the shelf sea is designated as POC(L–B1).*

land and the lithosphere (Fig. 7.3). A continental margin may be further divided into a shelf area and a slope region. The carbon species of interest include three categories: the gaseous compounds (CO_2 and CH_4), the dissolved species (dissolved inorganic and organic carbon, DIC and DOC) and particulate organic carbon (POC). The fluxes of these species between a continental margin and its surroundings are exemplified below by the processes involved and by comparisons between the conceptual and the measured fluxes. Each flux is abbreviated according to the carbon species and the compartments involved. Numerals 1, 2 and 3 refer to the upper layer interacting with the atmosphere, the main water column beneath it and the sediment layer in communication with the water column. For example, the POC flux from the land to the shelf sea is designated as *POC (L–B1)*. Also of interest are the nutrients closely associated with the carbon cycle. Nitrogen is representative of the nutrient elements and its biogeochemical fluxes are briefly discussed in relation to the carbon cycle.

Land–sea fluxes

The transfer of biophilic elements from land to sea has a long-term influence on the seawater composition and a short-term impact on the biogeochemistry of the coastal environment. The continental margin is often referred to as a filter, which retains and removes dissolved and suspended loads in the river runoff (Billen *et al.*, 1991). The complicated biogeochemical processes controlling the chemical species of carbon in the river system have been described by a chromatographic model in which materials are transformed and fractionated (Richey & Victoria, 1993). In the end, most of the terrigenous materials are retained and buried in the land-to-sea continuum while a small fraction are discharged to the sea through river runoff and ground water seepage. Terrigenous materials may also be transported to the ocean via glacial movement and aerosol transfer. Among all processes, the riverine discharge is by far the most important.

The present riverine fluxes of sediments and water are about 15–20 Gt yr^{-1} and 35000 $km^3\ yr^{-1}$, respectively (Milliman, 1991; Milliman & Syvitski, 1992). The river runoff carries bicarbonate ions, DOC and POC as well as nutrients to the ocean. The riverine bicarbonate ions and inorganic nutrients are the main sources of these species in the marine environment. The riverine fluxes may be the most easily measurable fluxes. Their transport is essentially one-way (from land to ocean) and most susceptible to anthropogenic influences (Meybeck & Jussieu, 1993).

The global riverine DIC flux has been estimated in the range 0.429–0.439 Gt C yr^{-1}; the removal of DIC by marine carbonate sedimentation has been estimated in the range 0.221–0.236 Gt C yr^{-1} (Berner *et al.*, 1983; Meybeck, 1987). The excess DIC plus the potential CO_2 flux associated with decarbonation is then in the range 0.19–0.28 Gt C yr^{-1} (Sarmiento & Sundquist, 1992). The global riverine flux of organic carbon has been estimated to be 0.38 Gt C yr^{-1} partitioned about 4 : 5 between POC and DOC (Ludwig *et al.*, 1996). The labile component makes up 30% of POC and 35% of DOC (Spitzy & Ittekkot, 1991). These authors suggest that refinements in these estimates are not necessary for global carbon-cycle modelling because riverine organic carbon flux is relatively insignificant compared with anthropogenic CO_2 emission. Whether the estimate of riverine flux of organic carbon is precise enough for global carbon-cycle modelling is debatable on two grounds. First, the uncertainties of riverine organic carbon flux may not be insignificant relative to the missing carbon of about 2 Gt C yr^{-1}. The other is that the geographic coverage of riverine organic carbon flux is not sufficient and its temporal variation is poorly known (Meybeck & Jussieu, 1993). Refinement is nevertheless forthcoming, as Meybeck & Jussieu (1993) suggested that as much as 0.1 Gt C yr^{-1} additional organic carbon may enter the ocean owing to human influence. Improved knowledge of the riverine fluxes of organic carbon may be important in understanding the anthropogenic perturbation of the terrestrial ecosystem and the regional coastal zone.

The estimates of riverine POC and DOC fluxes are based on the percentage of organic carbon content in sediments and river water discharged into the ocean. The geographic distribution of riverine fluxes varies greatly (Milliman, 1991). About 80% of the sediments and 65% of the fresh water are discharged from southeastern Asia, Oceania and northeastern South America, but most of the rivers in these areas, except the Chinese rivers and the Amazon, are not properly documented. An example of insufficient information is illustrated in the estimation of the DOC yield in Oceania. It has been estimated to be very small (less than 0.1 $g\,C\,m^{-2}\,yr^{-1}$) by Spitzy & Ittekkot (1991). However, a recent study (Kao & Liu, 1997) has shown the DOC yield in a typical Oceania watershed in Taiwan to be 4.1 $g\,C\,m^{-2}\,yr^{-1}$, which is as high as the Amazon (Degens *et al.*, 1990). The human-enhanced riverine fluxes are also evident in these areas of high sediment yield (Milliman *et al.*, 1984; Meybeck & Jussieu, 1993; Kao & Liu, 1996).

The natural flux of fixed nitrogen discharged from the river has been estimated in the range 36–48 $Tg\,N\,yr^{-1}$ of which 60% is particulate nitrogen (Meybeck & Jussieu, 1993). An additional 7 $Tg\,N\,yr^{-1}$ may be discharged to the sea from anthropogenic sources that are estimated to have a total capacity of 140 $Tg\,N\,yr^{-1}$ (Matthews, 1994). The natural flux of phosphorus from the river has been estimated to be 21 $Tg\,P\,yr^{-1}$, of which 95% is in particulate form; an additional 1 $Tg\,P\,yr^{-1}$ may be discharged from anthropogenic sources (Meybeck & Jussieu, 1993). The input of anthropogenic nutrients potentially has a strong impact on the coastal primary production (Ortner & Dagg, 1995) but its overall effect on the global carbon cycle is not clear (Smetacek *et al.*, 1991). Recently, more information has become available (see, for example, Galloway *et al.*, 1996; Howarth *et al.*, 1996; Zhang, 1996) and may shed new light on this issue.

Air–sea exchange

The shelf-edge export of organic carbon to the deep ocean may not contribute to sequestering of anthropogenic CO_2 unless the organic carbon export can be cashed in as CO_2 intake in the shelf sea (Sarmiento & Siegenthaler, 1992). The two basic parameters that control the air–sea exchange flux are p_{CO_2} and the gas transfer coefficient. The CO_2 partial pressure (p_{CO_2}) of surface water and the gas transfer coefficient may be affected by factors unique in coastal oceans, such as salinity and alkalinity gradients, terrigenous carbon sources, high turbidity, eutrophication, benthic demineralisation, suboxic respiration and pollution.

Primary production, respiration and vertical mixing control the CO_2 partial pressure in the surface water. The CO_2 uptake may be enhanced by eutrophication of the coastal water. CO_2 partial pressure as low as 88 μatm due to a phytoplankton bloom induced by anthropogenic nutrients has been observed in the German Bight (Kempe & Pegler, 1991). These authors reported a mean CO_2 partial pressure of 323 μatm in the surface water of the North Sea, which yields a CO_2 uptake flux of 16 g C m^{-2} yr^{-1} during the warm season. If the North Sea were devoid of life, it would have been a source of CO_2 with a mean outgassing flux of 38 g C m^{-2} yr^{-1} in the warm season. Thus, the net change in CO_2 flux induced by eutrophication is 54 g C m^{-2} yr^{-1}. However, the fate of the CO_2 taken up is not clear. About 40% of this influx enters the sediments as POC and PIC and the rest is stored in the water body (Kempe & Pegler, 1991). If the North Sea water is exchanged with the sub-surface Atlantic water, then the CO_2 taken up can be stored beneath the thermocline. Otherwise, the absorbed CO_2 may be released back to the atmosphere in cold seasons but these authors suspect that the North Sea may also be a CO_2 sink in the cold season as well. More study is needed to understand air–sea CO_2 exchange in continental margins.

Recent results elsewhere are also encouraging. Relatively low CO_2 partial pressure (300 μatm) has also been observed in the East China Sea (Chen & Wang, 1997). Therefore, some continental margins could be a net CO_2 sink.

Although coastal upwelling induces high productivity, it also brings CO_2 super-saturated water to the surface. However, not all upwelling conditions lead to net CO_2 outgassing because CO_2 outgassing is a rather slow process. In instances of moderate upwelling, intensive biological uptake during a relatively short exposure time of the upwelling water at the surface may result in rapid decrease of CO_2 partial pressure in the surface water. Consequently, a net drawdown of atmospheric carbon dioxide may occur during a coastal upwelling event (Friederich *et al.*, 1995). In addition to coastal upwelling, other vertical processes, such as buoyancy-induced circulation, storm-induced vertical mixing, and overturn of shelf water by cooling, also affect the CO_2 partial pressure in the surface water, but little is known about their effects.

Besides CO_2, other gases such as methane, nitrous oxide and dimethyl sulphide are also climate-related trace gases, for which the continental margins are probably important sources. Their biogeochemical fluxes are beyond the scope of this review but by no means insignificant.

Continental margin – ocean interior exchange

Although the coastal ocean is more productive than the open ocean in general, it is not necessarily a more efficient biological pump. The downward transfer of organic carbon over the shelf may be released back to the atmosphere when it is mineralised and mixes back with the surface water. This process neutralises the biological pump unless there is a net transfer from the shelf water to the sub-surface water of the ocean interior. In other words, exchange with the sub-surface ocean interior dictates the effectiveness of the biological pump over the shelf.

Such an exchange involves two-way traffic horizontally and one-way transfer vertically. The former can only be estimated from proxy measurements or from model calculations. Because the shelf sea has higher concentrations of POC and DOC, the net transfer of them must be seaward. The net transfer of DIC between the shelf sea and the open ocean can be either way. Exchange of nutrients is also important in affecting carbon biogeochemistry. The margin-to-interior transport of carbon may be depicted as a two-step process: from shelf to slope and from slope to the ocean interior, as shown in the schematic diagram (Fig. 7.3). Because the latter is a faster process, the former is the bottleneck. Ideally, a margin-to-interior flux may be described for each shelf unit as a depth-dependent source function from the sea surface at the shelf break down to the ocean bottom along the continental slope and rise, serving as a boundary condition for global ocean models of the carbon cycle. This remains as a distant goal. At present, even the depth-integrated cross-shelf carbon flux is difficult to quantify.

The cross-shelf transport of carbon has many different pathways for different carbon species (Fig. 7.3). Both POC and DOC may escape from the shelf to the slope water in the surface layer (*DOC (B1–C1)* and *POC (B1–C1)*). Only a fraction of the POC may sink below the thermocline (*POC (C1–C2)*). Little is known about the fate of the dispersed DOC. The POC on the shelf may sink to the bottom first. After resuspension, it may be transported off the shelf and deposited on the slope (*POC (B2–C2)*). DOC may be released from the pore water to the bottom water on the shelf and transported to the open ocean by water exchange (*DOC (B2–C2)*). All organic carbon is subject continuously to biological oxidation into DIC until it is buried deep. In view of the complexity of carbon transport, techniques available to measure the fluxes hardly suffice for the needs. The adequacy of each type of measurement is discussed below.

Particle flux in the water column

Sediment traps of different types are useful tools for collecting settling particles and measuring their fluxes in the ocean. Floating traps were designed to measure the flux of the sinking particles from the euphotic zone (Knauer *et al.*, 1979). The sequential traps deployed on moorings provide depth-dependent continuous records of particle fluxes (Zeitzschel *et al.*, 1978; Baker & Melburn, 1983; Honjo & Doherty, 1988; Heussner *et al.*, 1990). All traps are subject to technical difficulties: collecting efficiency, *in situ* decomposition and dissolution, biogenic contaminants (such as swimmers) and chemical interference from preservatives (Gardner, 1980, 1985; Iseki *et al.*, 1980; Gardner, Chapter 8, this volume). These technical problems aside, the particle fluxes measured in the coastal ocean are subject to uncertainties resulting from lateral particle movement in the continental margin.

Floating traps have been used in the Vertical Transport and Exchange (VERTEX) experiment to catch the sinking POC in the coastal upwelling region off California (Martin *et al.*, 1987). Under intensive upwelling conditions, POC flux as high as 85 g C m^{-2} yr^{-1} was measured at 100 m at a station off Point Sur, decreasing exponentially with depth. At 2000 m, the POC flux was only 7 g C m^{-2} yr^{-1}. Whether this 'coastal' station is representative of the coastal ocean is debatable because its bottom depth exceeds 2000 m. However, coastal upwelling in the California Current System is not properly represented in the coarse-grid GCM model often used for the global carbon cycle study (J. R. Toggweiler, personal communication). In the latter sense, the region can be viewed as a coastal zone by default. The VERTEX floating traps may catch particles directly sinking from the upper water column, but may miss those transported offshore near the bottom (*POC (B2–C2)* in Fig. 7.3). (See discussion in **Regional and global studies**, below.)

The near-bottom transport of POC may not be important when coastal upwelling is strong; the offshore advection may carry most of the POC away from the shelf and sink out of the euphotic zone in the slope water. During non-upwelling conditions, offshore transport near the bottom becomes important.

The bottom-intensified offshore transport of POC has been measured with moored sediment traps on the continental slopes in experiments such as SEEP (Biscaye *et al.*, 1988; Biscaye & Anderson, 1994) and ECOMARGE (Monaco *et al.*, 1990*a,b*). The measured fluxes usually increase downward and decrease offshore. The total transport has been estimated for the Mid-Atlantic Bight by integrating the fluxes over the slope. Over-trapping and under-trapping introduce uncertainties in the estimate. Schematic analysis shows that the height of a trap dictates the amount of material it collects under conditions dominated by lateral transport (Fig. 7.4). Higher traps cannot collect the particles that move close to the bottom. Lower traps may collect re-suspended particles from the upper portion of the slope. When integrated over the entire slope, over-estimates may result from repeated trapping of cascading particles near the seafloor. On the other hand, under-trapping may result regardless of the trap height for the fine particles not settled in the slope region. In fact, offshore transport of unsettled fine particles has been observed as mid-depth maximum of POC concentration over the continental slope in the southern East China Sea (Liu *et al.*, 1995). According to Biscaye *et al.* (1988), the sediment traps deployed on the slope 50 m above the seafloor on the slope may catch 30–45% of the exported particle flux measured at the shelf break. Under-trapping seems more serious than over-trapping.

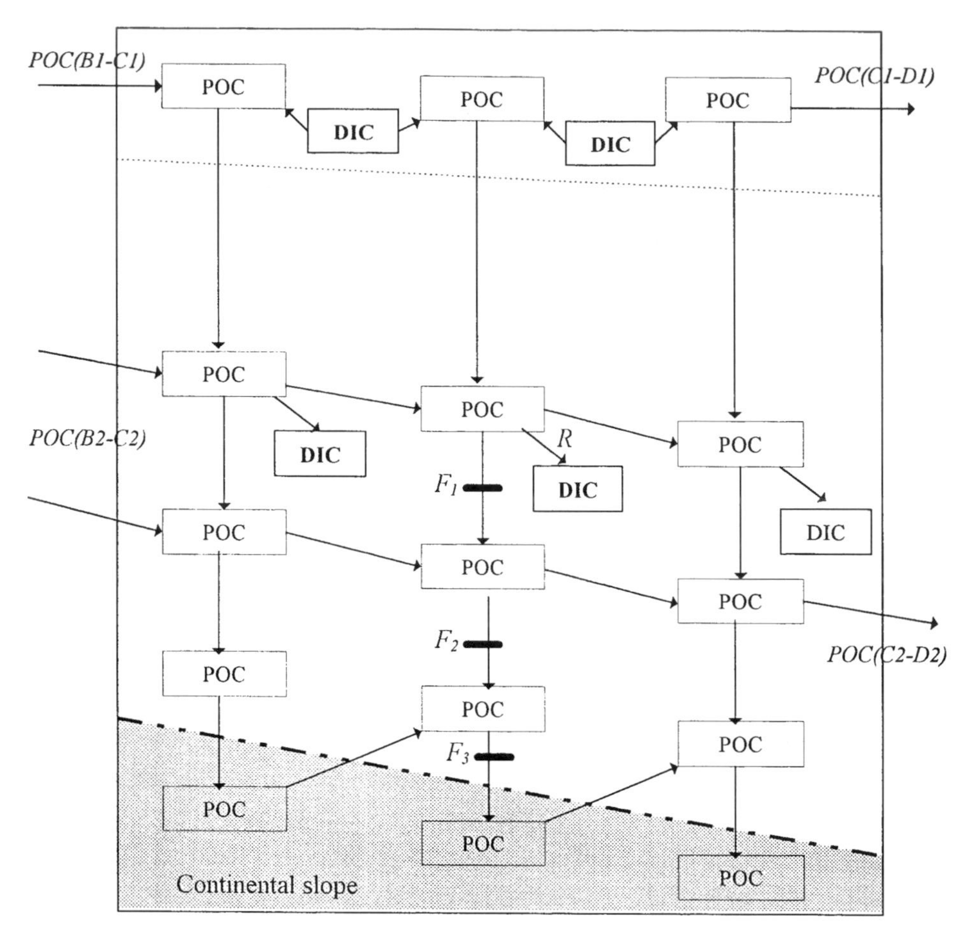

Figure 7.4 *Transfer of POC across the continental slope and collection by moored sediment traps (modified from Heussner, 1995). F_1, F_2 and F_3 represent the POC fluxes caught in moored sediment traps. The sources of POC are in the upper part (the euphotic zone over the slope) and on the upper left corner (shelf and upper slope).*

Benthic fluxes

Another way to estimate the cross-shelf transport is to determine the POC rain rate on the seafloor as the sum of POC burial and consumption during benthic respiration (Reimers *et al.*, 1992). The POC burial flux may be determined from the TOC content in the sediments and the sedimentation rate, and the organic carbon consumption flux may be calculated from benthic respiration rates. The latter may be in turn determined from the profiles of oxygen and other electron acceptors in the pore water or by using benthic chambers. The POC fluxes thus determined for the continental margin off California range from 5 to 12 g C m^{-2} yr^{-1} (Reimers *et al.*, 1992).

This approach avoids the possible error of over-trapping by sediment traps due to re-suspension of sediments (Jahnke, 1995). It nevertheless has drawbacks. Patchiness makes precise estimation of the total benthic flux difficult. One example is the 'hot spot' of POC deposition off Cape Hatteras (DeMaster *et al.*, 1994), which has very high benthic DIC flux (Blair *et al.*, 1994) but was missed by previous measurements. Another example is the large discrepancy between two sets of oxygen flux measurements in similar regions in the western North Atlantic by Smith (1978) and by Hales *et al.* (1994). Both sets of data show a trend of decreasing oxygen consumption flux as the seafloor depth increases from 2000 m to 5000 m at the continental margin. However, the former (0.018–0.22 mol $O_2\,m^{-2}\,yr^{-1}$, corresponding to organic carbon consumption flux of 0.2–2.6 $g\,C\,m^{-2}\,yr^{-1}$) is considerably lower than the latter set (0.1–0.46 mol $O_2\,m^{-2}\,yr^{-1}$, corresponding to organic carbon consumption flux of 1.2–5.4 $g\,C\,m^{-2}\,yr^{-1}$). It is not clear whether this discrepancy is caused by extreme patchiness or by technical problems.

A schematic analysis (see Fig. 7.3) reveals that the POC rain rate on the seafloor does not include all the organic carbon entering the deep ocean from the shelf sea, because some of the organic carbon flux is in dissolved form (*DOC (B2–C2)*), which is dispersed in the water column, and some particulate carbon is oxidised in the water column. As a result, neither carbon pools will reach the seafloor. It is also clear that the sum of the flux of the POC burial and the flux of organic carbon oxidation during respiration may be less than the POC deposited on the sea floor. It is believed that some of the deposited POC may be recycled as DOC (*DOC (C3–C2)*) and not oxidised (Fig. 7.3). The benthic DOC fluxes measured at the continental margin off California are as large as 13–17 $g\,C\,m^{-2}\,yr^{-1}$ (Bauer *et al.*, 1995), and those measured on the northwestern Atlantic continental slope account for over 50% of recycled organic carbon (Martin & McCorkle, 1993). Obviously, the DOC flux must be included in the future for a complete assessment of benthic carbon fluxes.

Carbon budgeting

Measurements of carbon fluxes are often difficult and uncertain as demonstrated above. Alternative methods to determine the export flux by mass balance have been employed. The exported organic carbon can be calculated from the organic carbon budget of the shelf sea. Further, the carbon budget may be inferred from the nitrogen budget, provided the C:N ratios of the organic matter involved are known (Walsh, 1991).

The organic carbon export from the shelf may be calculated as the surplus of production over consumption (Rowe *et al.*, 1986; Kemp *et al.*, 1994). In earlier approaches to the carbon budget (Walsh, 1981; Walsh *et al.*, 1981), the consumers include zooplankton, phytophagous fish and benthic fauna but not pelagic and benthic microbes. Consequently, the calculation resulted in a large surplus of primary production (30–50%), which was assumed to be exported

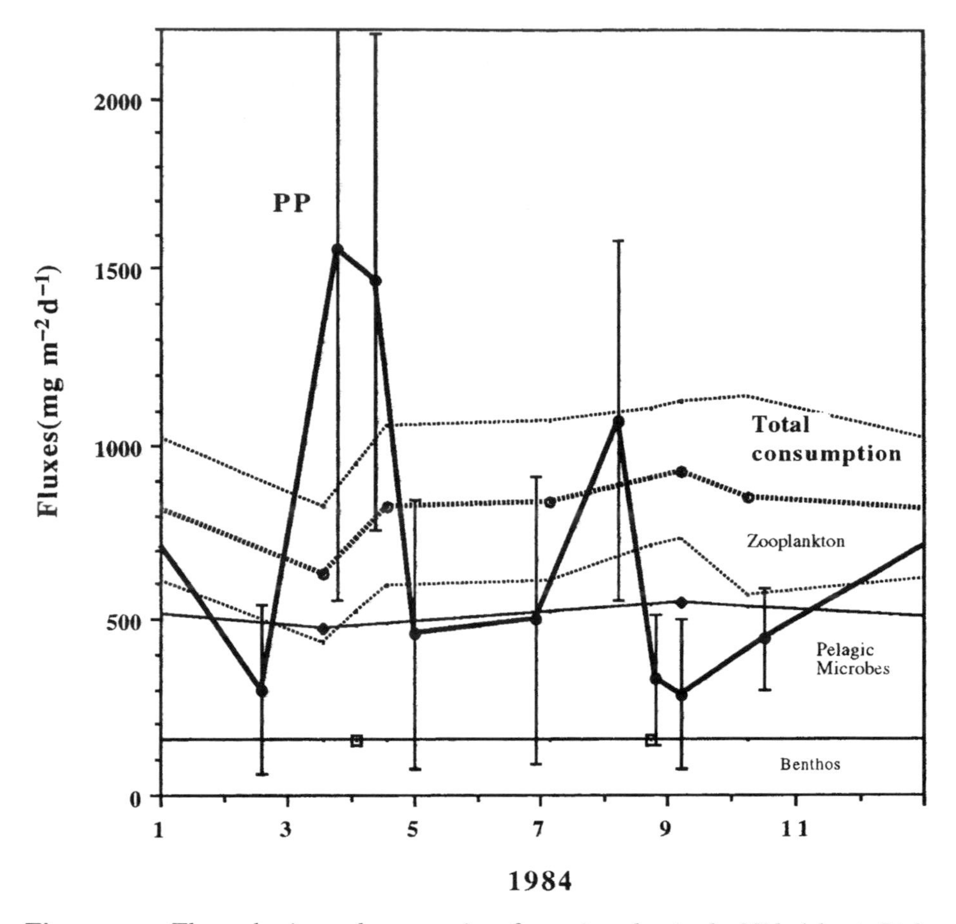

Figure 7.5 *The production and consumption of organic carbon in the Mid-Atlantic Bight observed by Rowe* et al. *(1986). The vertical bars represent the estimated uncertainties of primary production. The heavy dotted line represents the total consumption by zooplankton, pelagic microbes and benthos. The dotted lines indicate the interval of uncertainty of total consumption.*

from the shelf or buried in the shelf sediments. When bacterial consumption was included (Rowe *et al.*, 1986), the surplus was almost completely gone (Fig. 7.5). Recent carbon budget for shelf seas also include an additional type of consumer, the protists, which feed on both small phytoplankton and bacteria (Kemp *et al.*, 1994).

The close coupling between production and consumption often results in a relatively small fraction of primary production (less than 10% for the Mid-Atlantic Bight) (Falkowski *et al.*, 1994; Kemp *et al.*, 1994) being available for export. When the surplus is small, the accuracy of the estimate is in doubt, because considerable uncertainties exist in estimating both production and consumption (Fig. 7.5). Often neglected is the exudation of DOC during

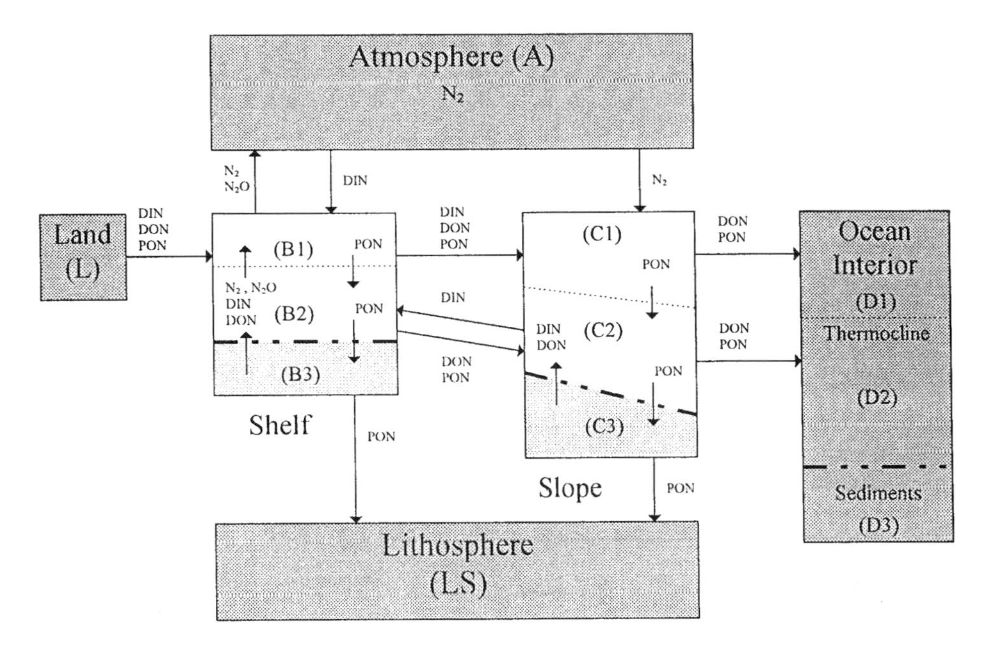

Figure 7.6 *As Fig. 7.4, but for nitrogen.*

primary production, accounting for up to 15% of primary production (Wood *et al.*, 1992). The temporal and spatial variabilities of primary production are as large as 40–60% of the mean primary production (Fig. 7.5). As for consumption, many conversion factors are involved in the estimation of the organic carbon consumption rates by zooplankton, pelagic microbes and benthos. For example, determination of the bacterial consumption rate of organic carbon often involves three conversion factors between thymidine incorporation rate and organic carbon consumption rate, each with a large range (Ducklow & Carlson, 1992). The large ranges of empirical conversion factors make extrapolation from rates of bacterial production to rates of carbon consumption highly uncertain (Jahnke & Craven, 1995).

An entirely different approach is to use the nitrogen budget as a proxy for the carbon budget (Walsh, 1991; Wollast, 1993). The input of fixed nitrogen, mostly in the form of DIN, to the shelf sea must be matched by an equal amount of output, mostly in the form of PON and DON (Fig. 7.6), which may be translated into POC and DOC. The input of nitrogenous nutrients comes from two major sources: nutrient-laden sub-surface offshore water, and riverine discharge. A small fraction of the nitrogen may be lost in the form of N_2 and N_2O, mostly during denitrification, and an even smaller fraction is buried on the shelf. Under favourable conditions, the rest of the nitrogen may all be converted to organic matter and returned to the ocean interior. However, in most cases, nitrate is not exhausted and some is returned back to the open ocean. An

Table 7.1. *Nitrogen budget for the Mid- and South Atlantic Bight*
The corresponding fluxes of organic carbon are calculated by using the Redfield ratio.

Type of flux	Nitrogen flux ($g\,N\,m^{-2}\,yr^{-1}$)	Carbon flux ($g\,C\,m^{-2}\,yr^{-1}$)
Input		
Nitrate	15.5	—
Output		
Nitrate export	7.3	—
DOM[a] export	1.3	20.7[b]
POM[a] export	5.2	29.7
Burial	0.2	0.9
Denitrification	1.4	—
Total output	15.5	—
Total OM[a] output	6.7	51.3
PP[a]$_{POM}$	—	274
PP_{DOM}	—	55
PP_{total}	—	329
f_x[c]	—	0.16

[a]DOM, dissolved organic matter; POM, particulate organic matter; OM, organic matter; PP, primary production.
[b]The C: N atomic ratio of the DOM is assumed to be 18 for the inert DOM.
[c]The f_x ratio designates the fraction of PP of the shelf buried on the shelf or exported to the sub-surface ocean interior.
Source: Walsh (1994).

example (Table 7.1) is provided for the Mid- and South Atlantic Bights (Walsh, 1994). The corresponding fluxes of organic carbon are related to the organic nitrogen by the Redfield ratio. For the dissolved organic matter, which is assumed to be of the inert type in this example, the C: N ratio is 18: 1 instead of the normal 6.6: 1. The fraction of primary production exported or buried, designated as f_x, is then computed to be 0.16. Because the exported organic nitrogen is a major fraction of the total flux of nitrogen across the shelf, its estimation may be more reliable than the direct estimation of the organic carbon export, which is the difference between two large numbers, the production and the consumption. A similar approach using phosphorus as the proxy has also been advocated (Hall *et al.*, 1996). However, all nutrient-budgeting approaches rely on sufficient understanding of exchange processes at the shelf edge.

Regional and global studies

Continental margins of North America

Typical eastern and western boundary currents border the Pacific and the Atlantic margins of the North America, respectively. An equatorward-flowing surface current (the California Current) characterises the California Current System at the Pacific Margin. Coastal upwelling, subduction and filaments make the coastal zone a highly variable and dynamic zone (Brink & Cowles, 1991). The rates of primary production and new production are closely related to the upwelling and other dynamic processes in the coastal transition zone (Chavez *et al.*, 1991). By contrast, the Gulf Stream flows poleward along the shelf break in the South Atlantic Bight (SAB) off North America and separates from the continental margin at Cape Hatteras. Although coastal upwelling along the SAB is not as widespread and persistent as that over the Pacific margin, a shoreward supply of nutrients is effected by frontal eddies and bottom intrusions (Csanady, 1990). Thus, the primary production is elevated at both margins and results in a clear imprint on the benthic biogeochemistry. The re-mineralisation rates on the seafloor (Fig. 7.7), estimated with benthic chamber or pore water chemistry, are high (10–20 $g\,C\,m^{-2}\,yr^{-1}$) on the upper slope near the shelf break and decrease rapidly seaward from the continental margins (Smith, 1978; Jahnke *et al.*, 1990; Martin & McCorkle, 1993; Anderson *et al.*, 1994; Hales *et al.*, 1994; Blair *et al.*, 1994; Jahnke, 1995). Despite similar seaward decreasing trends, the two margins show very different temporal and spatial variations in carbon fluxes.

The VERTEX experiment provided a transect of particle fluxes from Point Sur, California, to the central North Pacific north of Hawaii (Martin *et al.*, 1987). The observed POC flux decreased from top to bottom in the water column and from the continental margin to the open ocean (Fig. 7.8). The POC flux at a coastal station decreases as a power function of depth (Martin *et al.*, 1987):

$$POC = POC_{100}(Z/100)^{-0.85}, \qquad (7.1)$$

where Z is in units of metres and POC_{100} is the flux at 100 m. The observed POC_{100} values were 42–85 $g\,C\,m^{-2}\,yr^{-1}$ (Martin *et al.*, 1987). The high POC flux near the continental margin is obviously related to the high primary production (250–420 $g\,C\,m^{-2}\,yr^{-1}$) caused by coastal upwelling in the California Current System. The fractional primary production export from the euphotic zone is about 17–20%. During the Coastal Transition Zone Study, detailed spatial and temporal variations of primary productivity were examined (Chavez *et al.*, 1991). When the offshore advection is intensified within a filament, the primary production is intensified to 220–730 $g\,C\,m^{-2}\,yr^{-1}$ in the coastal transition zone.

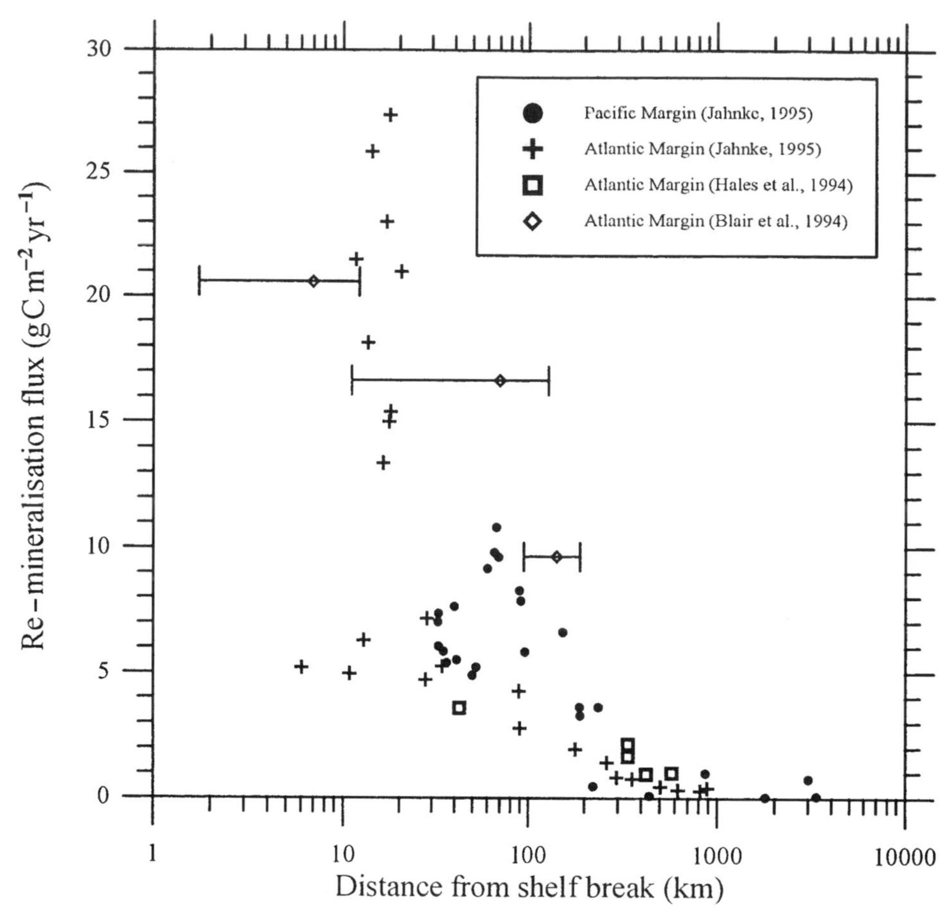

Figure 7.7 *Re-mineralisation fluxes of carbon observed at the Pacific and Atlantic margins of North America. Data are taken from Hales* et al. *(1994), Blair* et al. *(1994) and the compilation of Jahnke (1995) who reports data from Smith (1978), Jahnke* et al. *(1990), Martin & McCorkle (1993), and Anderson* et al. *(1994).*

When the longshore flow dominates, the primary production decreases to 150–370 g C m^{-2} yr^{-1}. Because of the coupling between offshore advection and high primary production, a large fraction of the phytodetritus may be transported to the deep ocean without being retained on the shelf. Vertical settling may be the most important process of exporting POC to the deep water in the California Current System.

The Shelf Edge Exchange Processes experiments, SEEP-I and -II, were originally designed to test the hypothesis of export of a large proportion of the shelf primary productivity in the Mid-Atlantic Bight (MAB) between New England and Cape Hatteras (Walsh *et al.*, 1988; Biscaye *et al.*, 1994). SEEP-I was conducted in the northern MAB off Long Island, and SEEP-II in the southern MAB (Fig. 7.9). The 95% confidence interval for mean primary

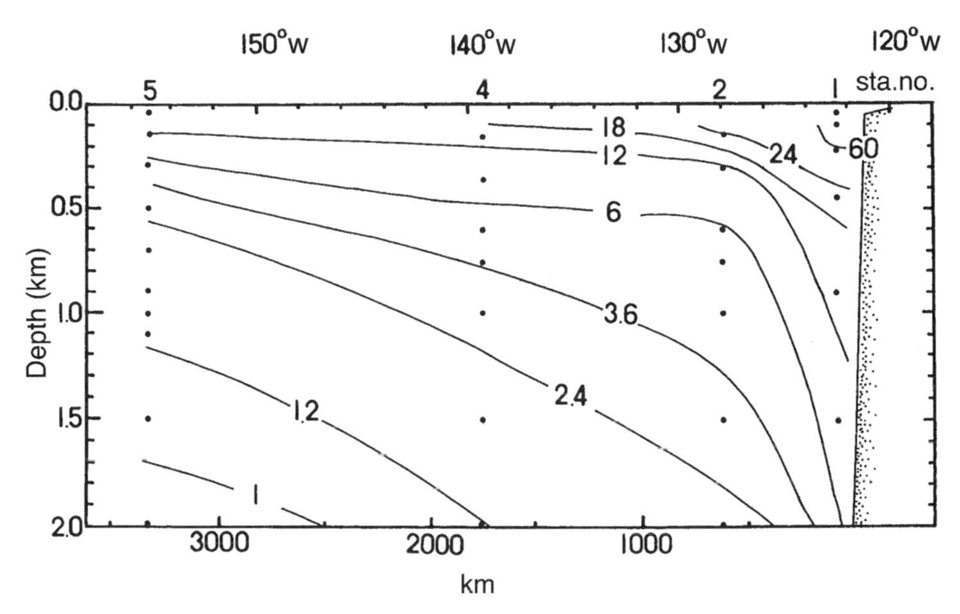

Figure 7.8 *POC fluxes observed between Point Sur, California, and Hawaii in the VERTEX programme (modified after Martin* et al., *1987).*

production in the MAB is 321–354 g C m^{-2} yr^{-1}, 15% of which is DOC production (Campbell & O'Reilly, 1989). Two modes of transport operate for the organic carbon export: diffusive and advective transports (Walsh, 1994). Arrays of sediment traps and moored sensors were deployed to measure the diffusive transport. The observations showed that the export particle fluxes were not very high and only a small fraction of carbon is exported across the shelf break – slope front to the slope depocentre. Most of shelf carbon is oxidised on the shelf. The POC fluxes measured with the moored sediment traps were mostly 4–12 g C m^{-2} yr^{-1} at the base of the euphotic zone and 5–24 g C m^{-2} yr^{-1} near the bottom (Fig. 7.10). The flux decreases seaward, but increases towards the seafloor, suggesting that the transport is mainly seaward and along the seafloor. The integration of POC flux across the continental slope yields a total POC export of 4.8 Tg C yr^{-1} for the MAB. This POC export represents 6.4% of the total shelf production, 75 Tg C yr^{-1} (Falkowski *et al.*, 1994). However, this estimate does not include the DOC export and some modes of particulate transport that may evade the sediment traps as discussed above. Therefore, this estimate of 6.4% primary production export is probably low. According to the observations of the SEEP-II mooring programme, most shelf waters leave the shelf before it reaches Cape Hatteras, and with it a fraction of the suspended and dissolved carbon (Biscaye & Anderson, 1994). Using a barotropic circulation model, Walsh (1994) estimated that 12–25% of primary production may be exported from the MAB and SAB.

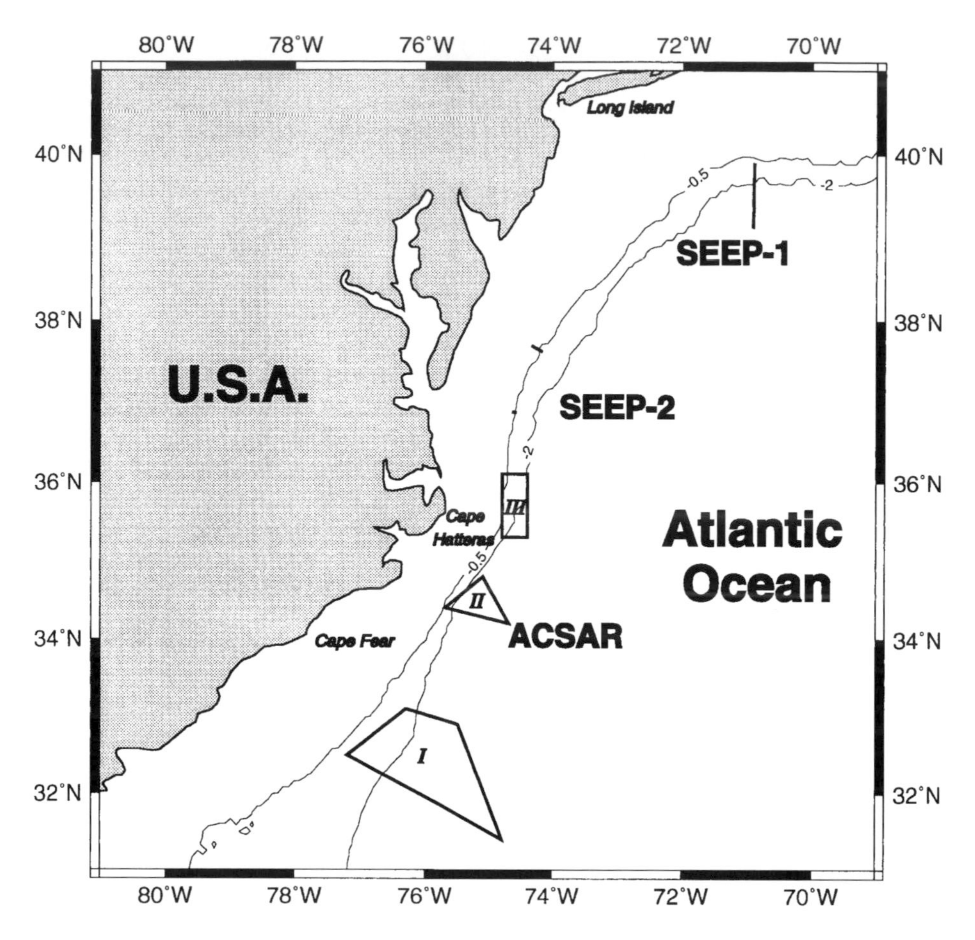

Figure 7.9 *The location of mooring lines for the SEEP-I and -II programmes (Biscaye* et al., *1988; Biscaye & Anderson, 1994) and sediment sampling areas for the ACSAR programme at the Atlantic Margin of North America (DeMaster* et al., *1994).*

The productivity of, and the fraction of primary production exported from, the Atlantic Margin seem lower than those of the Pacific Margin. This is conceivable in view of the lack of strong coastal upwelling and the associated offshore advection on the Atlantic Margin. The contrast is apparently reflected in the benthic respiration flux (see Fig. 7.7). The organic carbon re-mineralisation fluxes in the Pacific and the Atlantic Margins compiled by Jahnke (1995) show that the re-mineralisation flux is very high (up to $27\,g\,C\,m^{-2}\,yr^{-1}$) on the upper slope of the Atlantic Margin but decreases very rapidly offshore of the shelf break. The flux decreases to less than $5\,g\,C\,m^{-2}\,yr^{-1}$ when the distance is more than 80 km from the shelf break. A similar trend was also observed for the POC flux in the water column (Falkowski *et al.*, 1994). For the Pacific Margin, the maximum re-mineralisation flux is not as high as that of the Atlantic Margin but the decreasing trend is more gradual. Hinga (1985) has

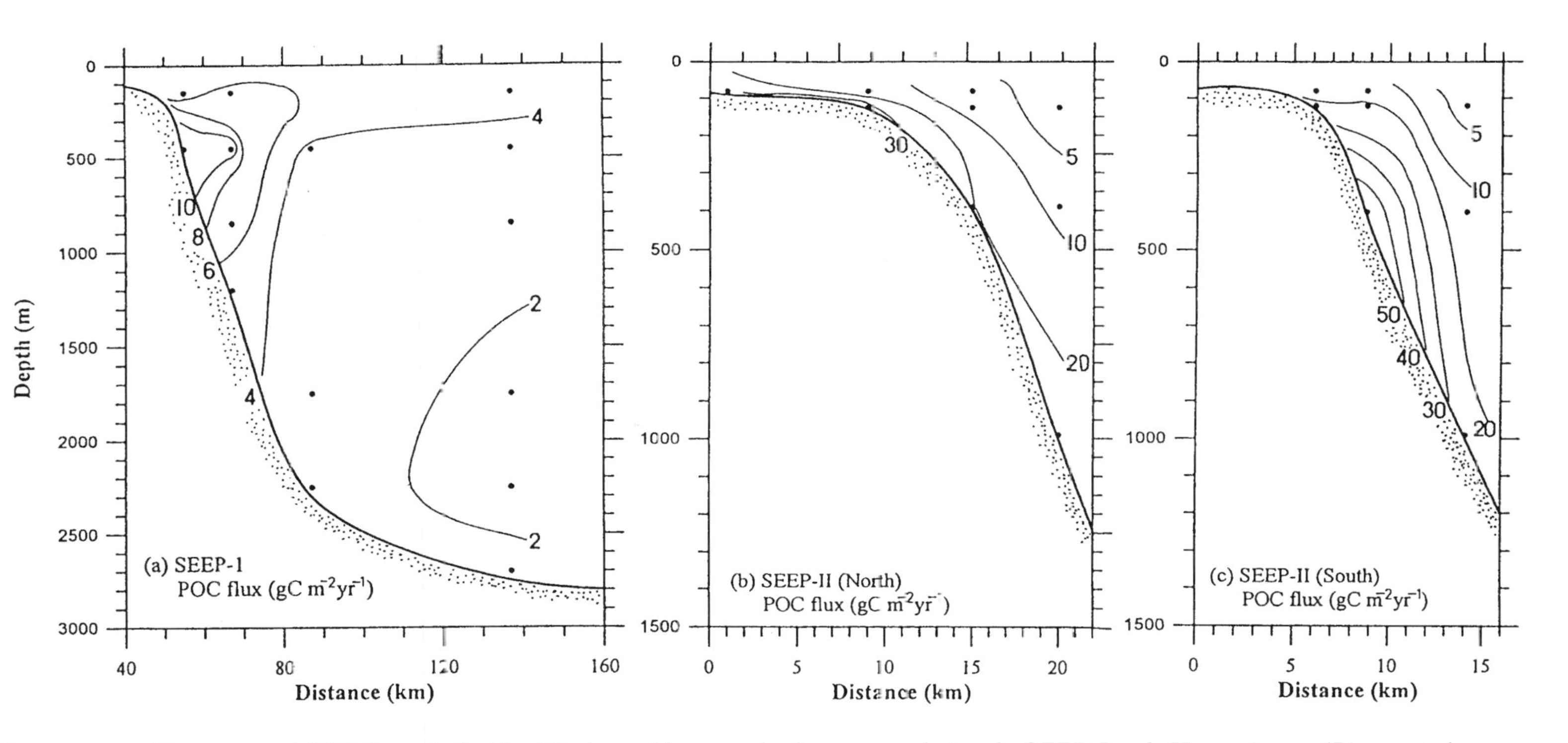

Figure 7.10 *Time-averaged POC fluxes* ($g\,C\,m^{-2}\,yr^{-1}$) *observed by moored sediment traps during the SEEP-I and -II experiments (Biscaye* et al., *1988; Biscaye & Anderson, 1994).*

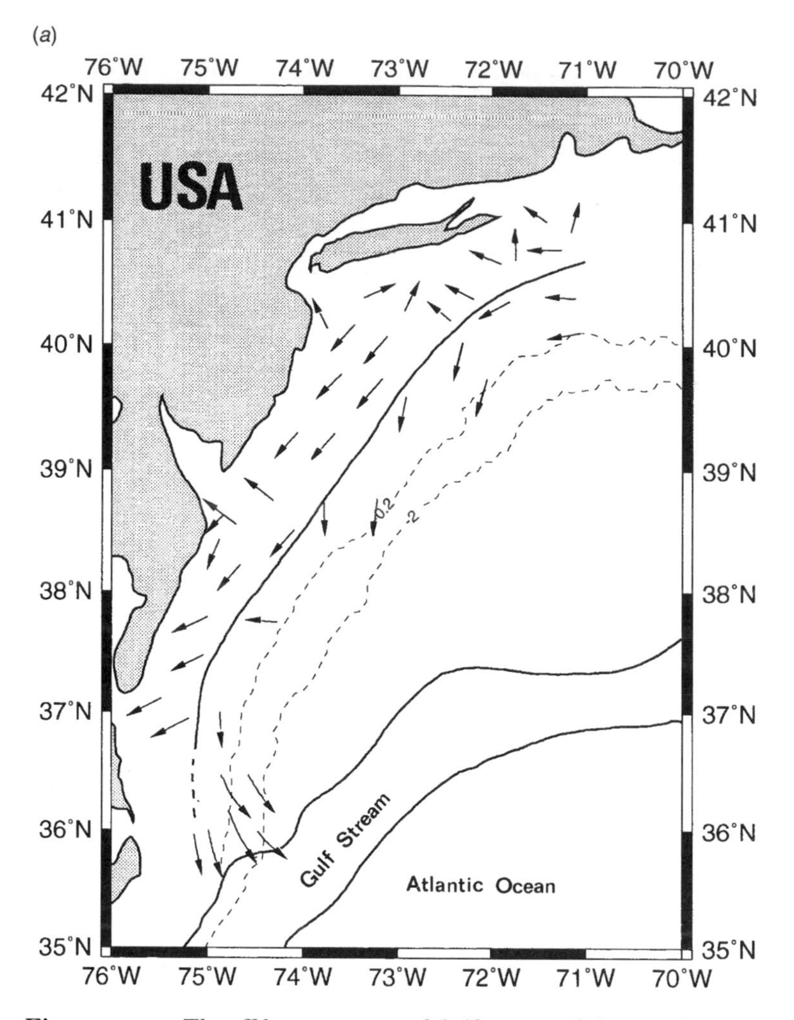

Figure 7.11 *The offshore transport of shelf water and the waterborne materials at the point of offshore deflection of the western boundary current for* (a) *the Gulf Stream in the Mid- and South Atlantic Bights, and* (b) *the Kuroshio in the southern East China Sea. The bottom currents on the shelf of the Mid-Atlantic Bight are shown by the arrows (in panel* a*), which are divided by the 'Bumpus Line' (Bumpus, 1973). Currents to the east of this line probably move sediments towards the slope off Cape Hatteras, forming the 'Hatteras funnel' (Rhoads & Hecker, 1994). The Changjiang runoff forms a coastal jet during the northeast monsoon (from September to May of the next year), which moves the sediments southward (Beardsley* et al., *1985; Sternberg* et al., *1985; Chao, 1991). The sediments may be transported across the Taiwan Strait northeast of Taiwan owing to blocking by the topographic high in the strait as well as by the Kuroshio branch current (Chao, 1991; Jan, 1995). A steady northward bottom current west of Taiwan and the cyclonic eddy northeast of Taiwan may work in series to move the sediments offshore (Chuang, 1985; Jan, 1995; T. Y. Tang, personal communication).*

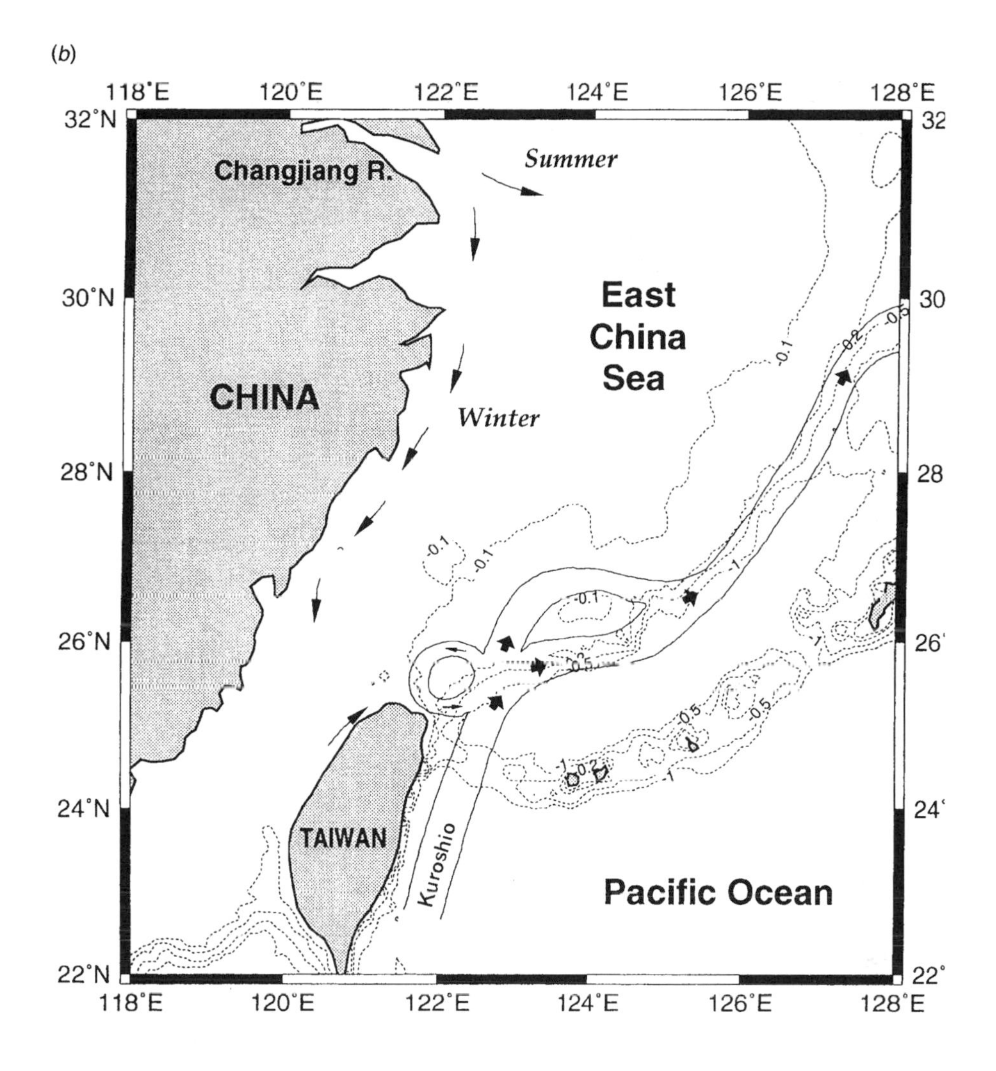

used such contrast as evidence of higher primary productivity in the Pacific Ocean. However, recent measurements of the re-mineralisation fluxes in the Atlantic Margin by Hales *et al.* (1994) and Blair *et al.* (1994) seem to contradict the previous findings. The recent values are mostly higher than the older values and follow the trend observed for the Pacific Margin (Jahnke, 1996).

In contrast to the low fluxes observed in the SEEP experiments, extremely high sedimentation rates (up to 1 cm yr^{-1}) have been observed in the marginal zone off Cape Hatteras (Fig. 7.9) during the Atlantic Continental Slope and Rise (ACSAR) project (DeMaster *et al.*, 1994). The corresponding burial rate of organic carbon is more than 100 g C m^{-2} yr^{-1}. To the south of this hot spot, the sedimentation rates decrease gradually back to the normal level of 7 cm kyr^{-1}. The hot spot seems to be the receiving end of the 'Hatteras funnel' (Fig. 7.11*a*).

Sediments are carried in by the southward-flowing bottom current over the outer shelf of the MAB seaward of the Bumpus line (Rhoads & Hecker, 1994), the sediment-carrying bottom water outwells towards Cape Hatteras. The SEEP-I and SEEP-II experiments (Fig. 7.10) also showed a southward increase of the POC fluxes in the MAB (Biscaye *et al.*, 1988; Biscaye & Anderson, 1994). Therefore, the advective transport of POC out of the shelf sea must be a more effective way of exporting POC than the diffusive transport.

Pacific margin of northeastern Asia

Along the Pacific Margin of northeastern Asia flows the Kuroshio Current, a strong western boundary current. It hugs the east coast of Taiwan before entering the Okinawa Trough, where it borders the shelf break of the East China Sea (Fig. 7.11*b*). This part of the Pacific margin is quite different from the Atlantic margin of the North America. Two of the largest rivers in the world, the Changjiang (Yangtze River) and the Huanghe (Yellow River), empty into the ocean at this margin. These two rivers discharge a large amount of runoff (1000 $km^3 yr^{-1}$) and sediments (1.5 $Gt yr^{-1}$) (Shen *et al.*, 1983; Qin & Li, 1983; Milliman *et al.*, 1985; Milliman, 1991). The Changjiang runoff carries rich loads of DOC (12 $Tg C yr^{-1}$, Gan *et al.*, 1983) and nitrate (0.8 $Tg N yr^{-1}$; Huang *et al.*, 1983). Just northeast of Taiwan, the Kuroshio Subsurface Water intrudes onto the shelf of the East China Sea as a cyclonic eddy providing additional nutrients (Wong *et al.*, 1991; Liu *et al.*, 1992*b*; Gong *et al.*, 1995; T. Y. Tang, personal communication). In this light, the shelf water of the East China Sea probably has a richer supply of nutrients than the MAB and SAB.

The primary production in the East China Sea shows strong seasonality as well as spatial variation (Guo, 1991). In summer, the Changjiang runoff bifurcates as it enters the shelf sea with a branch flowing to the northeast and the other branch flowing southward along the coast (Fig. 7.11*b*) (Beardsley *et al.*, 1985; Chao, 1991). Near the perimeter of the river plume, high primary productivity is induced by the high nutrient concentration (Edmond *et al.*, 1985; Guo, 1991; Gong *et al.*, 1995). In summer, the maximum primary production exceeds 1500 $mg C m^{-2} d^{-1}$, whereas it ranges from less than 100 $mg C m^{-2} d^{-1}$ up to 200 $mg C m^{-2} d^{-1}$ in the Kuroshio water off the shelf break (Guo, 1991). Most of the shelf waters have primary production values between the two extremes. In winter, the Changjiang Diluted Water forms a coastal jet flowing southward and turns cyclonically in the northern Taiwan Strait owing to blocking by a topographic high as well as by the Kuroshio northward branching current (Chao, 1991; Jan, 1995). The shelf water is well mixed vertically in winter by the prevailing northeast monsoon. Primary production in the shelf water (100–400 $mg C m^{-2} d^{-1}$) is much lower than in summer while that in the Kuroshio remains at the summer level (Guo, 1991).

The areal integration of Guo's (1991) atlas yields a primary production of 150 mg C m^{-2} d^{-1} in winter and 900 mg C m^{-2} d^{-1} in summer (J. Chang, personal communication). The annual mean is somewhere between 54 and 324 g C m^{-2} d^{-1} and lower than 300 g C m^{-2} d^{-1} for the MAB. This is puzzling in view of the richer supply of nutrients in the East China Sea. However, recent studies (Hama, 1995; Shiah *et al.*, 1995; Wen, 1995) show primary production values (Fig. 7.12) considerably higher than previous estimates, probably owing to the clean technique adopted for primary production measurement (Marra & Heinemann, 1987). The annual mean value of primary production in the Kuroshio should be in the range of 200–560 mg C m^{-2} d^{-1}, about twice the previous estimates. For coastal waters, primary production shows extreme variability, ranging from 80 mg C m^{-2} d^{-1} to 2000 mg C m^{-2} d^{-1}. The variability is probably controlled by light availability, which is in turn controlled by turbidity. In winter, the strong wind mixing causes very low productivity in the coastal waters. For mid-shelf waters, the recent measurements also showed primary production values between 500 and 1600 mg C m^{-2} d^{-1} (Shiah *et al.*, 1995; Hama, 1995) for spring and summer and 200–600 g C m^{-2} yr^{-1} in winter (Hama, 1995; Wen, 1995). These values are 30–100% higher than previous values measured in the same areas and the same seasons. For the upwelling region northeast of Taiwan, the primary production is quite high (1300–1600 mg C m^{-2} d^{-1}) throughout the year. In light of the discrepancies, better assessment of the mean primary production in the East China Sea is called for.

The export of particulate matter from the East China Sea shelf to the Okinawa Trough was observed in two experiments (see Fig. 7.1). The first observation was the Marginal Sea Flux Experiment (MASFLEX) in the northern part (Iseki *et al.*, 1995) and the other was the Kuroshio Edge Exchange Processes (KEEP) experiment in the southern part (Chen *et al.*, 1991; Chung & Chang, 1995). As observed at two MASFLEX stations, SST1 and SST2 (*c.* 1100 m), in the Okinawa Trough (Fig. 7.12), the particulate flux at 620 m was extremely low (less than 4 mg C m^{-2} d^{-1}) but increased markedly with depth (Fig. 7.13). A remarkable seasonality in particulate fluxes was also observed at both MASFLEX stations. The fluxes generally peaked from winter to early spring and were lowest during summer. Downwelling and upwelling cells induced by the northeast monsoon in winter and southwest monsoon in summer, respectively, may have caused the seasonality. Following this scenario, seaward bottom flow during northeast monsoon favours the offshore transport of sediments (Hu, 1994). At 50 m above the bottom, the POC flux at Station SST1 reached 40 mg C m^{-2} d^{-1} in May 1993. The yearly mean fluxes at 50 m above the bottom at the two stations were about 10–16 mg m^{-2} d^{-1} (4–6 g C m^{-2} yr^{-1}), comparable to the mean POC flux (5.3 g C m^{-2} yr^{-1}) observed

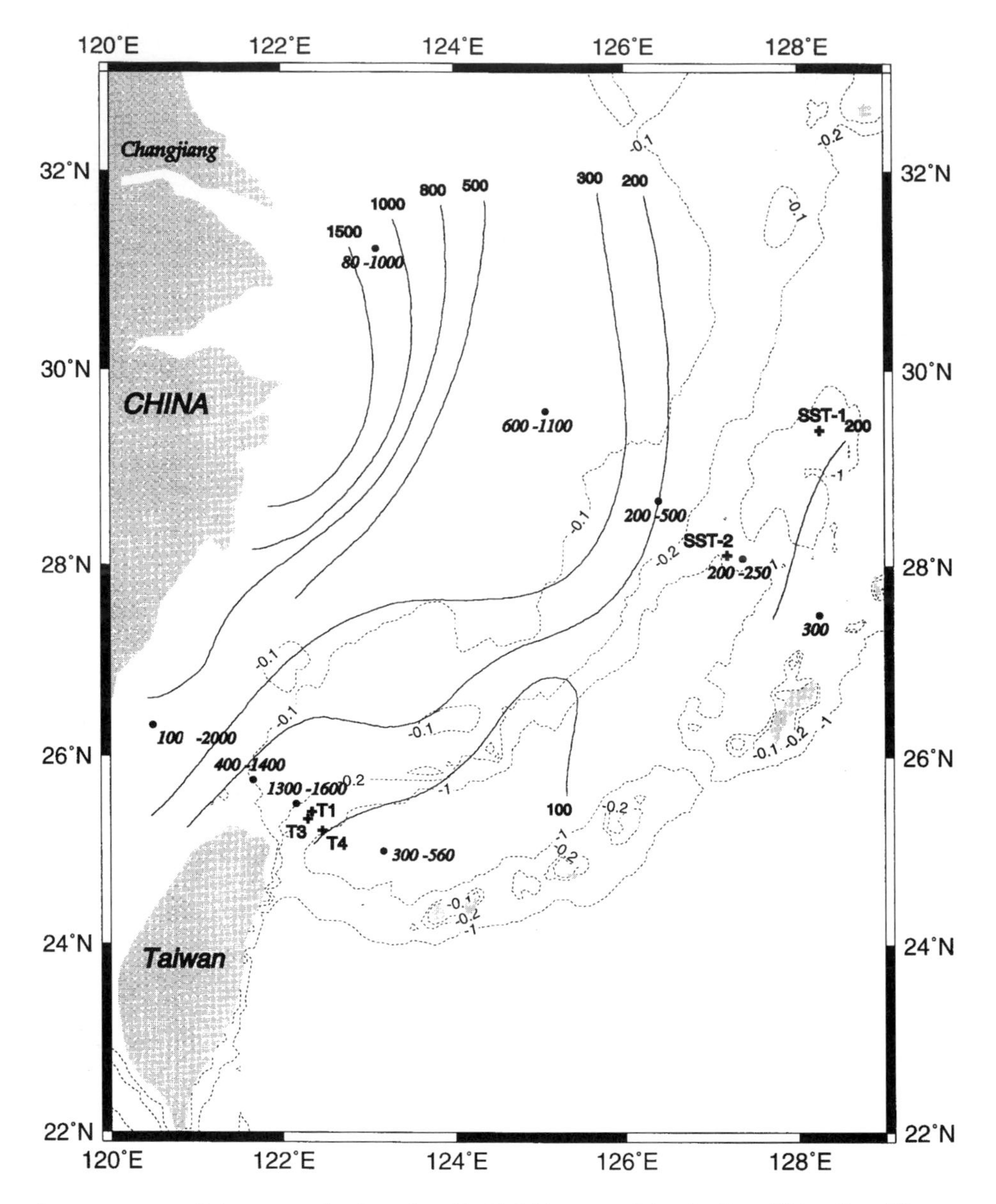

Figure 7.12 *Primary productions (mg C m^{-2} d^{-1}) reported for the East China Sea. The contours are the summer conditions reported by Fei* et al. *(1987) cited in Guo (1991). More recent measurements in the northern area (Hama, 1995) during the MASFLEX programme and in the southern area (Shiah* et al., *1995; Wen, 1995) during the KEEP programme are shown by the numerals. Also shown are the mooring locations for sediment traps: SST1 and SST2 in the north and T1, T3 and T4 in the south.*

over the slope in the SEEP-I experiment (Biscaye *et al.*, 1988). These features indicate that diffusive transport processes similar to the northern MAB also operate in the northern East China Sea. However, the strong seasonality in the

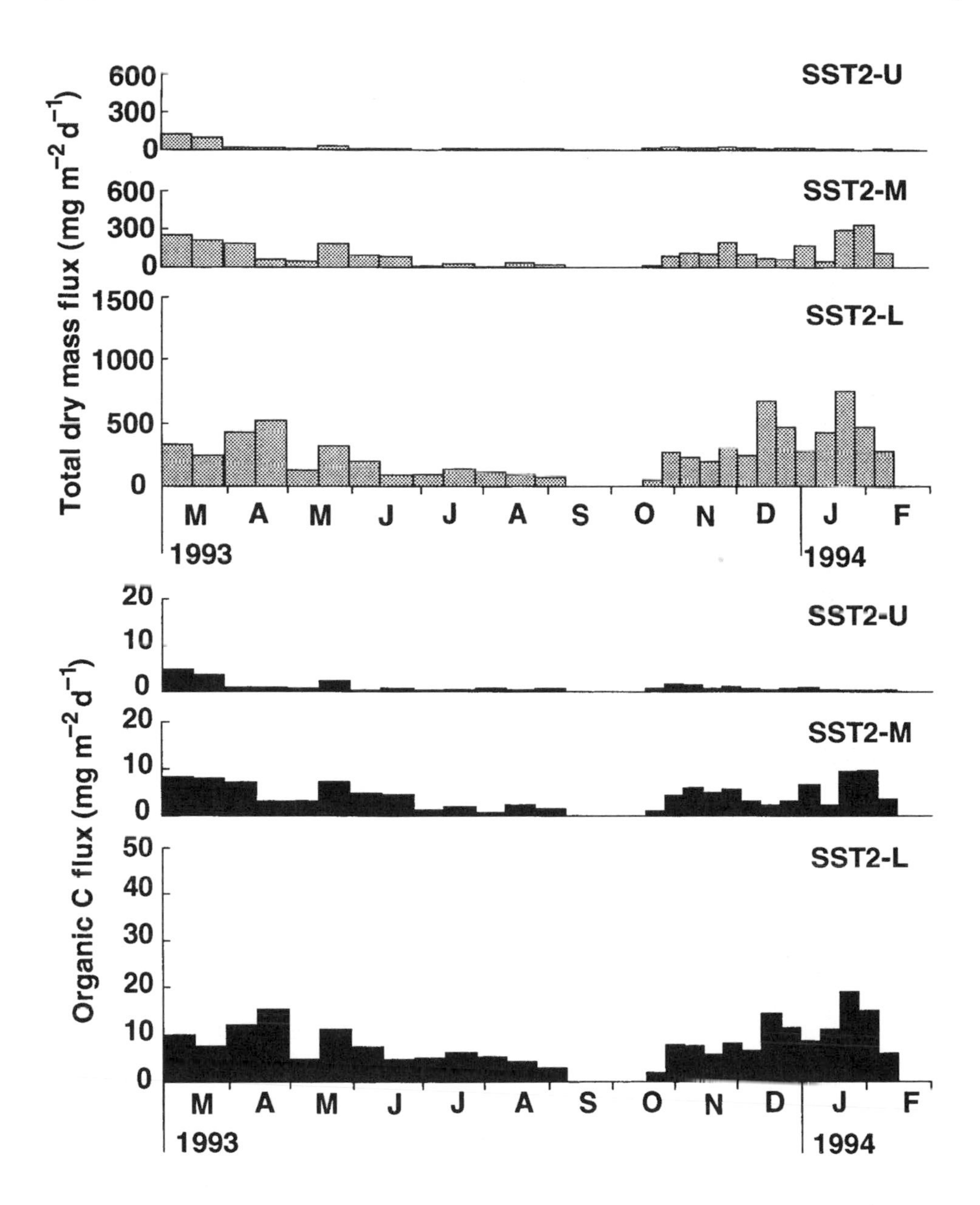

Figure 7.13 *Total mass fluxes and POC fluxes observed at SST2 (see Fig. 7.12) in the Okinawa Trough during the MASFLEX experiment in 1993–94 (Iseki* et al., *1995). The water depth was about 1100 m. The trap depths were 600 m (U), 800 m (M) and 50 m above bottom (L).*

East China Sea is mostly absent in the MAB. Further, the seasonal variation of the POC flux is out of phase with that of primary production, reflecting the transport dominated by monsoon-induced circulation.

In the southern East China Sea, a very high sedimentation rate (up to

0.9 cm yr^{-1}) has been observed in the slope sediments off northeastern Taiwan in the Okinawa Trough (Chung & Chang, 1995; Chen, 1995). The slope sediments are enriched in organic carbon relative to the adjacent zones on the shelf and in the basin (S. Lin *et al.*, 1992). The maximum POC accumulation rate reaches 56 g C m^{-2} yr^{-1} (Chen, 1995). An analogy can be drawn with the major depocentre on the slope off Cape Hatteras (DeMaster *et al.*, 1994). Both regions are characterised by offshore deflection of western boundary currents. The fast sedimentation rate in the southern East China Sea is accompanied by very large fluxes (3600–21 900 g m^{-2} yr^{-1}) of sinking particles (Y. C. Chung, personal communication). The mean POC flux observed at 80 m above the bottom on the upper slope (490 m) near a submarine canyon was as high as 100 g C m^{-2} yr^{-1} (J. J. Hung, personal communication). The elevated inventory of ^{210}Pb (300–900 dpm cm^{-2}) on the seafloor at the southern end of the Okinawa Trough northeast of Taiwan suggests 'sweeping' of ^{210}Pb from the shelf by horizontal transport (Chung & Chang, 1995). A rough estimate indicates that 10% of the ^{210}Pb from the East China Sea shelf may be deposited in a small region of 100 km × 100 km in the southern Okinawa Trough. For both the MAB and the East China Sea, the advective transport of particles seems to be a very significant mode of material transfer from the shelf to the deep ocean, and the 'funnels' point to the regions where western boundary currents are deflected offshore (Fig. 7.11).

Global estimations

Many authors (Table 7.2) have estimated the primary production of the coastal ocean. They all point out its importance but quantitative agreement among estimates is still wanting. High shelf productivity does not necessarily lead to a large capacity of the biological pump, which depends on the cross-shelf export of organic carbon to the subsurface ocean interior. Agreement among estimates of the export percentage of primary production is even worse (see Table 7.3). Factors contributing to the wide range of estimates are discussed below.

The estimates of global primary production fall in two groups: 27–31 and 45–51 Gt C yr^{-1}. Two of the higher values (Martin *et al.*, 1987; Smith & Hollibaugh, 1993) were based on extrapolation from measurements made with contamination-free incubation techniques (Marra & Heinemann, 1987). The recent estimation of global primary production based on a light-production model using CZCS images of near-surface chlorophyll yields a global primary production of 45–50 Gt C yr^{-1} (Longhurst *et al.*, 1995), which matches the higher range well. Antoine *et al.* (1996), using a similar approach but taking into consideration the active-to-total pigments ratio, obtained somewhat lower values (36.5–45.6 Gt C yr^{-1}). For this study, a value of 40 Gt C yr^{-1} is adopted for global primary production. The estimates of new production in the open ocean range from 1.9 to 6.5 Gt C yr^{-1} (Table 7.3). The mean value, 4 Gt C yr^{-1}, is adopted for this study.

The estimates of coastal primary production also fall in two ranges: 4.1–4.8 and 7.4–14.4 Gt C yr^{-1}. Diverse definitions of the coastal oceans contribute to the divergence of estimates. Following these definitions, the total area of the coastal ocean ranges from 29 to 58 × 10^{12} m^2, corresponding to 8–16% of the global ocean. The most conservative definition restricts the coastal ocean to the shelf areas less than 200 m deep (Smith & Hollibaugh, 1993); more generous definitions include adjacent water bodies, such as coastal upwelling systems, marginal seas and slope waters (up to 2000 m deep). Understandably, more inclusive definitions of the coastal ocean lead to higher estimates of primary production. How inclusive it should be is a matter yet to be defined. Ideally, the dynamic processes involved should dictate the definition of the coastal ocean. The coastal ocean is better defined as the region of fine-scale dynamic features that are poorly resolved or missed by global ocean models, yet important from a global perspective. At the present level of sophistication in global modelling, these neglected areas extend far beyond 200 m depth.

Even after the uncertainty in the area of coastal ocean is removed, the estimated coastal primary production per unit area still varies between 122 g C m^{-2} yr^{-1} and 385 g C m^{-2} yr^{-1}. The upper limit is derived from the CZCS chlorophyll data for the coastal region without consideration of the contribution of turbidity to the water-leaving radiation, and, therefore, must be an over-estimate (Longhurst *et al.*, 1995). Under the assumption that only 25% of radiation leaving the water represents chlorophyll, the value drops to 238 g C m^{-2} yr^{-1}. Quantification of chlorophyll concentrations in the turbid coastal water from ocean colour data remains highly uncertain (Antoine *et al.*, 1996). The rather high areal primary production (250 g C m^{-2} yr^{-1}) obtained by Martin *et al.* (1987) is an average of the coastal production values off Point Sur, California, under both upwelling and non-upwelling conditions. Such a value may be representative of the eastern boundary current system. Recent measurements in the East China Sea shelf adjacent to the Kuroshio suggest a mean primary production value around 180 g C m^{-2} yr^{-1} (Hama, 1995; Wen, 1995; G. C. Gong, unpublished data), which may be representative of the western boundary current systems. The average value (215 g C m^{-2} yr^{-1}) of the two representative types is adopted for this study. If the coastal ocean covers 10% of the global ocean, then global coastal ocean primary production is estimated to be 7.8 Gt C yr^{-1}, this value represents 20% of the global ocean primary production, in agreement with the mean value (Table 7.2).

Although primary production in the coastal ocean may account for 20% of the global ocean primary production, it is not clear what fraction of the primary production may contribute to the biological pump. Some estimates are listed in

Table 7.2. *Estimates of primary production of the global ocean and the coastal ocean*

	Global Ocean PP (Gt C yr^{-1})	Coastal Ocean Areal PP (g C m^{-2} yr^{-1})	Coastal Ocean Area (10^{12} m^2)	Coastal Ocean PP (Gt C yr^{-1})	Coastal ocean/Global Ocean Area	Coastal ocean/Global Ocean PP	References
1	31.1	122	33	4.1	9.1%	13.2%	Platt & Subba Rao (1975)
2[a]	23.7 (28.5)	124	39	4.8	10.8%	20.2% (16.8%)	Eppley & Peterson (1979)
3	29.7	128	58[b]	7.4	16.1%	24.9%	Walsh (1988)
4	51.0	250	36	9.0	9.9%	17.6%	Martin *et al.* (1987)
5	27.0	179	43	7.7	11.9%	28.5%	Berger (1989)
6	45.0	277	30	8.3[c]	8.3%	18.4%	Wollast (1991)
7	48.0	166	29	4.8	8.0%	10.0%	Smith & Hollibaugh(1993)
8	45–50	238–385	37.4	8.9–14.4	10.4%	19.9–28.7%	Longhurst *et al.* (1995)
Mean	38 ± 10	195 ± 75	38 ± 9	7.2 ± 2.5	10.6 ± 2.6%	19 ± 6%	
This study	40	215	36	7.8	10%	19.5%	Adopted values

Note: [a]Eppley & Peterson's (1979) original estimate for the global ocean PP does not include coastal ocean with depth < 200 m. The value in parenthesis is the original estimate plus the inshore PP. The coastal ocean is represented by the inshore waters; [b]including the slope waters; and [c] including macrophytes.

Table 7.3. Two types of organic carbon transfer are shown: new production (NP) and shelf export (X), corresponding to two modes of organic carbon transport in the coastal ocean. New production measured by nitrate uptake (Dugdale & Goering, 1967) represents the downward flux out of the euphotic zone, whereas the shelf export represents the lateral flux from the shelf to the slope. As discussed above under 'Systematics of fluxes and processes', both vertical and lateral processes are involved in the coastal ocean biological pump. Different processes prevail in different types of continental margins. At margins with strong coastal upwelling and filaments, new production is probably a good measurement of the biological pump because the POC carried by strong offshore advection sinks afterwards. On the other hand, margins with strong western boundary currents are usually not associated with strong coastal upwelling. Most of the cross-shelf POC transfers probably rely on lateral transport along the seafloor. Shelf export is a better measurement of the biological pump for the latter kind of margins.

For the coastal ocean new production, the estimates fall within two groups (Table 7.3): 1.5 and 3–4.7 Gt C yr^{-1}. The lower value comes from two entirely different approaches. Eppley & Peterson's (1979) estimate (for the inshore waters in their Table 1) is based on transforming the primary production map into a new production map, which is based on the empirical primary production – new production relation. The estimate by Martin *et al.* (1987) is based on extrapolation of observed POC fluxes at 100 m off Point Sur, California. The higher values were obtained with special consideration of very high *f* ratios encountered in highly productive coastal environments. Berger *et al.* (1989) used the following relation for the *f* ratio:

$$f = PP/400 - PP^2/340\,000. \qquad (2)$$

The unit of primary production (PP) is g C m^{-2} yr^{-1}.

The *f* ratio calculated from the above equation is 0.4 for the coastal ocean with a mean areal primary production of 215 g C m^{-2} yr^{-1}. Such high *f* ratios have been observed in the MAB, but most of the organic carbon was subsequently oxidised below the euphotic zone. When Eppley & Peterson (1979) made their estimate of global new production, they neglected so-called neritic waters because of the possible influence from river runoff and bottom regeneration. By doing so, they excluded the nearshore waters. Therefore, new production as defined by Dugdale & Goering (1967) is a measurement of the biological pump only in margins where the offshore advection can carry the POC off the shelf. In this scenario, the estimate by Martin *et al.* (1987) that 17% of the coastal primary production gets into the deep water should be representative of the eastern boundary current systems only.

For the global shelf export of organic carbon (Table 7.3), the estimates also

Table 7.3. *Estimates of new production (NP) or organic carbon export (X) from the coastal ocean and their significance relative to the primary production of the coastal ocean or to the global new production*

	Open ocean NP ($Gt\,C\,yr^{-1}$)	Coastal ocean PP ($Gt\,C\,yr^{-1}$)	Coastal ocean NP or X ($Gt\,C\,yr^{-1}$)	Coastal ocean NP/PP or X/PP	Global NP[a] ($Gt\,C\,yr^{-1}$)	Coastal NP/ Global NP or Coastal X/ Global NP	References
1	1.9 − 2.2	4.8	NP: 1.5	31%	3.4 − 3.7	40 − 44%	Eppley & Peterson (1979)
2	5.9	9.0	NP: 1.5	17%	7.4	20%	Martin *et al.* (1987)
3	3.0	7.7	NP: 3.0	25%	6.0	50%	Berger (1989)
4	3.0	13.7	NP: 4.7[b]	34%	7.7	64%	Bienfang & Ziemann (1992)
5		7.4	X: 1 − 1.5	14 − 20%			Walsh (1989)
6		6.9	X: 1.00	14%			Wollast (1991)
7		7.8[c]	X: 0.35	4.5%			Reimers, Jahnke & McCorkle (1992)
8	6.5	4.3	X: 0.21	4.9%	6.7	3.2%	Smith & Hollibaugh (1993)
9		7.8[c]	X: 0.44[d]	5.6%			Jahnke (1996)
Mean	4.1 ± 2.0	7.6 ± 2.3	NP: 2.7 ± 1.5 X: 0.7 ± 0.5	NP: 35 ± 23% X: 9 ± 6%	NP: 6.8 ± 2.5 X: 4.7 ± 2.0	NP: 40 ± 27& X: 14 ± 11%	
This study	4	7.8	0.9	11%	4.9	18%	Adopted values

Note: [a]The global new production is the sum of the open ocean NP and the coastal ocean NP or X. [b]Includes Arctic & Antarctic shelves. [c]The coastal ocean PP for this entry is the adopted value for this study. [d]This value is calculated from oxygen consumption fluxes over continental slope and rise at 1000 m (Jahnke, 1996) assuming a traditional Redfield ratio.

fall in two ranges: 1–1.5 and 0.21–0.44 Gt C yr^{-1}. The estimate by Walsh has been decreasing through the years from 3 Gt C yr^{-1} (Walsh *et al.*, 1981) to 2–2.5 Gt C yr^{-1} (Walsh *et al.*, 1985) to 1–1.5 Gt C yr^{-1} (Walsh, 1989). The last estimate was based on the POC deposition fluxes on the continental slope depocentres. However, his estimate of 0.3–0.5 burial flux on the slope was much higher than the 0.046 ± 0.012 Gt C yr^{-1} actually observed (Reimers *et al.*, 1992). On the other hand, the estimate of POC rain rates based on respiration measurements, 0.35 Gt C yr^{-1} by Reimers *et al.* (1992) or 0.44 Gt C yr^{-1} by Jahnke (1996) are definitely under-estimates of the shelf export. In these estimates, the POC oxidised in the overlying water column and the DOC fluxes out of the sediments were not included in the assessment. Besides, the sparse localities of their benthic survey missed some of the hot spots of POC deposition as mentioned above. According to the studies at both the Atlantic and Pacific margins of the North America (Martin & McCorkle, 1993; Bauer *et al.*, 1995), the DOC flux out of the slope sediments is as important as the DIC flux, if not more so. If this is a general case throughout the ocean, the total shelf export of POC could be more than double the estimated POC rain rates. The actual shelf edge export percentage could be somewhere between the upper limit (17%) obtained from upper water column sinking fluxes (Martin *et al.*, 1987) and the lower limit (5.6%) derived from respiration rates in the lower water column (Jahnke, 1996). The median value of 11% is adopted as the actual export fraction (f_x) for this study. Then, the POC export is estimated to be 0.9 Gt C yr^{-1}. Thus, the global ocean has a total biological pump capacity around 4.9 Gt C yr^{-1}, of which 18% is attributable to the coastal ocean.

The global estimations demonstrate the significance of coastal biological pumps but improved accuracy is needed. The coastal biological pumps are more vulnerable to human alteration than their counterparts in the open ocean. Their importance and susceptibility justify the employment of regional models to better our estimation and prognostic skill.

Strategy for synthesis and modelling

Extrapolation of biogeochemical fluxes for the global continental margin from limited observations is bound to be unreliable because of the variability of continental margin environments. Ideally, one should employ physical–biogeochemical models for extrapolation (Hofmann, 1991). Towards this goal, essential physical and biogeochemical processes have to be fully investigated first. Further, continental margins must be classified into different types, for which regional models need to be developed. The developed regional models may be embedded in, or provide boundary conditions for, the global ocean models of carbon cycle (Wroblewski & Hofmann, 1989).

Physical processes dictate much of the exchange of carbon between the continental margins and the open ocean. These processes include: (1) tides, (2) waves, (3) river runoff and buoyancy currents, (4) wind-driven coastal current, (5) coastal filaments, (6) coastal upwelling, (7) frontal eddies, (8) shelf break upwelling, (9) boundary current or slope water intrusions, (10) warm or cold core-ring impingement, (11) boundary current entrainment, (12) boundary countercurrents, (13) subduction along a slanted frontal boundary bordering the shelf, (14) sediment re-suspension, (15) turbidity currents, and (16) topographic effects. Brink (1987), Wroblewski & Hofmann (1989), Blanton (1991) and Huthnance (1995) have reviewed most of these processes. In addition, Wroblewski & Hofmann (1989) and Hofmann (1991) have reviewed important physical–biological models for the continental margins with emphases on trophic dynamics instead of carbon fluxes, although the two are closely related.

A quasi two-dimensional physical–biogeochemical model has been constructed for the study of carbon and nitrogen cycles in the Bering Sea (Walsh & Dieterle, 1994). The study provides a good description of some important processes involved in biogeochemical fluxes on continental margins. Some of the processes not covered in their study are discussed in the following. Although three-dimensional physical–biogeochemical models for the open ocean, such as the one for the North Atlantic (Sarmiento *et al.*, 1993), or for semi-enclosed seas, such as the one for the Baltic (Fennel & Neumann, 1996), have been developed, one for a continental margin has yet to be developed. Regional circulation models such as the ones for the East China Sea (Chao, 1991; Yanagi & Takahashi, 1993) do exist but lack biogeochemical components. Modelling efforts for both circulation and biogeochemical processes need to be stepped up.

Processes for modelling

Biogeochemical processes in the continental margin are numerous. More than 100 parameters have been used in a quasi-two-dimensional model for the carbon and nitrogen cycles in the Bering Sea shelf between the isobaths of 70 and 100 m (Walsh & Dieterle, 1994). Among all parameters, twelve are not considered in open-ocean models for the upper water column (Fasham *et al.*, 1990) but are critical to the continental margin environment (Table 7.4). These parameters are for the effects of bottom boundary layer (including sediments), tidal motion and ice cover. Other important processes and factors considered in Walsh & Dieterle's model are summarised in Table 7.4. Several important features pertaining to other types of continental margin are not considered in that model. The most important ones include river runoff, boundary current and

Table 7.4. *Parameters for the shelf environment used in a quasi-two-dimensional model for the carbon and nitrogen cycles in the Bering Sea*

Parameters unique to the shelf environment	Other factors and processes considered
Space domain	
Benthic boundary layer thickness	Vertical and horizontal dimensions
Sediment thickness	Surface mixed layer thickness
Sediment diffusive layer thickness	Coriolis parameter
Physical	
Bottom stress	Temperature and water density
Bottom friction	Wind driven circulation and mixing
Bioturbation coefficient	Lateral and vertical advection and mixing
Near bottom tidal velocity	Solar irradiation
Ice cover (%)	Air-sea heat exchange
Ice attenuation	Photosynthetic available radiation and light attenuation
Chemical	
	Salinity
	CO_2 chemistry
	Borate chemistry
	Lysed and excreted DOC, DON
	Photolysis of DOM
	pH
	Nitrate and ammonia
Biological	
Sedimentary denitrification rate	Phytoplankton biomass, growth (nitrogen uptake) and death
	Zooplankton grazing, respiration, assimilation, excretion, defecation (settling), etc.
	Bacterial biomass, growth (DOC uptake) and death,
	Nitrification
Geological	
Solubilisation rate of sedimentary POC	
C/N ratio of sedimentary POM	

Source: Walsh & Dieterle (1994).

shelf–slope exchange, and sediment transport. The effects of these processes on biogeochemical fluxes on continental margins are briefly discussed below.

River runoff

River runoff has strong influences on many aspects of the continental margin processes. Physically it drives a buoyancy current, which may dominate the shelf circulation. Biogeochemically, the materials carried in it may dictate chemistry and biology of the shelf water. Long-term accumulation of the riverine material often defines the geochemistry of the shelf sediments. The shelf sea is currently receiving an increasing amount of riverine nutrients (from 14 to 35 $Tg\,N\,yr^{-1}$), which may enhance the shelf productivity and increase organic matter export (Wollast, 1993). For example, the Changjiang River plume in the East China Sea shelf (Fig. 7.14) is characterised by high Chl *a* that is induced by the discharged nutrients (Gong *et al.*, 1995). The response of the marine ecosystem in the Gulf of Mexico to the riverine nutrients discharged from the Mississippi River has been studied in the Nutrient-Enhanced Coastal Ocean Productivity (NECOP) programme (Ortner & Dagg, 1995). Here, increased organic carbon concentrations have been observed in the sediments of the past 50 years on the shelf. DOC is produced at intermediate salinities of the river plume in all seasons. The fate of the POC and DOC is not clear. Radiochemical analysis of the sediments revealed re-suspension and re-distribution of the shelf sediments. If a significant fraction is transported offshore, it may amount to a significant organic carbon export. River-forced motion dictates much of the fate of the nutrient-enhanced primary production.

Large-scale rivers entering adjacent shelves are subject to the Coriolis deflection due to the Earth's rotation. In the absence of winds and ambient currents, the plume water turns anticyclonically until it impinges on the coast. Thereafter it protrudes away from the source along the coast, having a width scale of about one baroclinic Rossby radius of deformation (Chao & Boicourt, 1986). The transition zone between the plume and the coastal jet further down coast is characterised by strong downwelling, and may be a favourite site for deposition of waterborne materials. The front separating the riverine water from the ambient shelf water is a strong dynamic barrier against seaward dispersal of waterborne materials. It would take an external force of considerable strength to annihilate the front and disperse the plume water offshore.

An ambient current against the river-forced coastal jet serves the purpose well. If strong enough, it can force the separation of the nearshore river flow from the coast and causes it to disperse seaward. One notable example is the bimodal distribution of the Changjiang River plume over the East China Sea in summer months (Beardsley *et al.*, 1985). Forced by the southwest monsoon, the

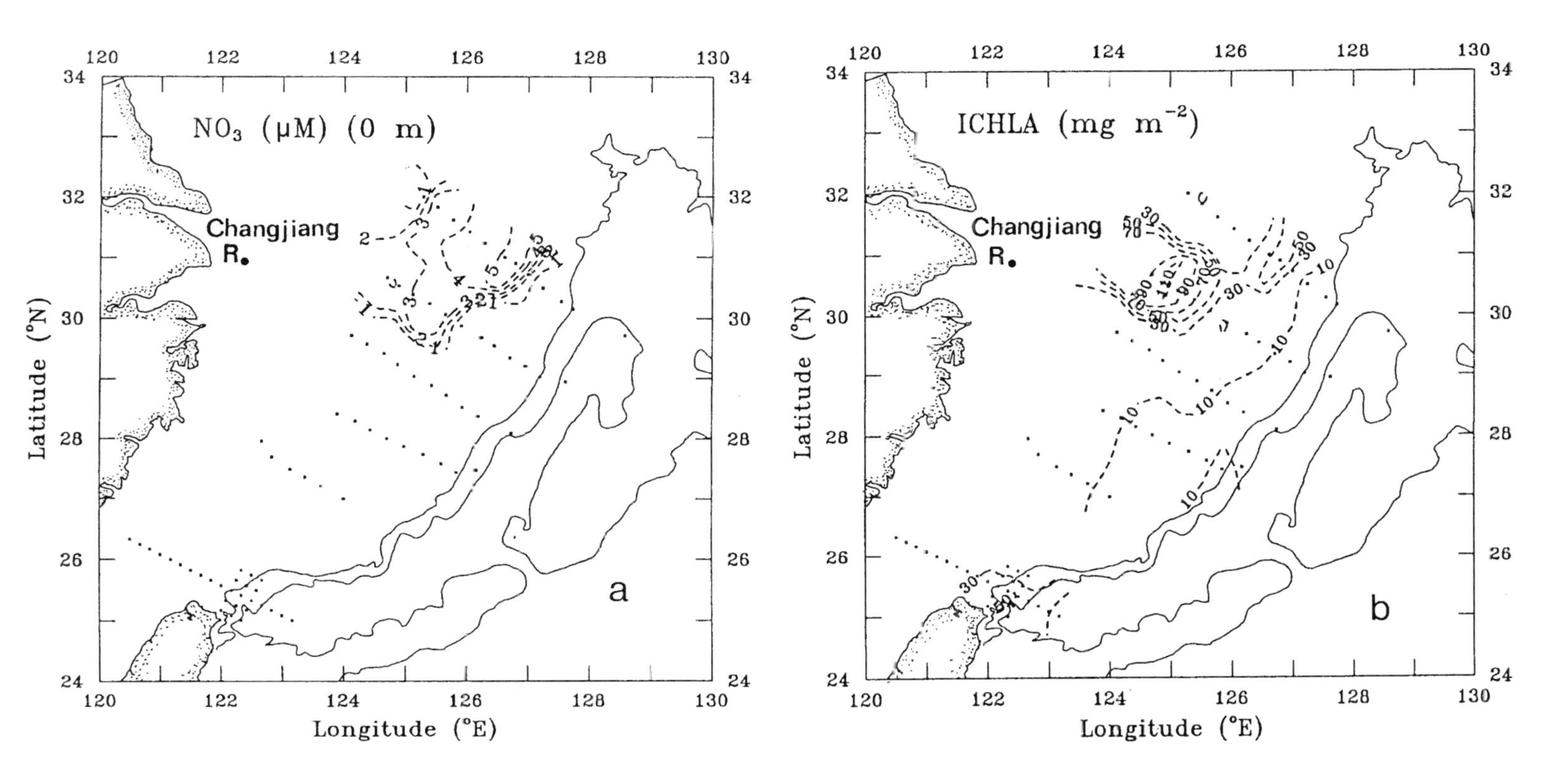

Figure 7.14 *The distribution of* (a) *nitrate concentration in the surface water and* (b) *integrated Chl* a *in the East China Sea observed during the KEEP experiment in July, 1992 (Gong* et al., *1995). The patch of high-nitrate water was the Changjiang Diluted Water with salinity less than 31 psu, which evidently induced a high Chl* a *standing stock.*

current in the marginal sea is essentially against the would-be southward coastal jet driven by the Changjiang freshwater discharge. The resulting plume becomes diffuse and disperses seaward. Offshore transport of waterborne materials is likely to be enhanced in the process.

In theory, fluctuating winds can also enhance the offshore dispersal. Though theoretically plausible, the type of dispersal has not been verified by carefully designed observational efforts. Consider a uniform wind field fluctuating in the longshore direction with some sub-tidal frequencies. Even if the mean wind after averaging over a substantial period is zero, offshore dispersal of river-forced coastal boundary layers is likely. The offshore dispersal is induced by the asymmetric response of the coastal boundary layer to upwelling-favourable and downwelling-favourable winds. Surface Ekman transport that is induced by upwelling-favourable winds forces the nearshore plume to expand seaward. After the wind relaxes, the expanded coastal boundary layer does not return to its original width. On the contrary, downwelling-favourable winds narrow the coastal boundary layer via a shoreward Ekman drift. The once narrowed coastal boundary layer more or less rebounds to its original width after wind relaxation (Chao, 1988). Conceivably, the asymmetry is likely to induce a seaward expansion of the plume water over long time scales.

Boundary current and shelf-slope exchange

The offshore advection associated with the eastern boundary current systems has been carefully examined in the Coastal Transition Zone Programme (Brink & Cowles, 1991). Its effect on cross-shelf transport is rather direct and efficient. By contrast, the cross-shelf transport processes in a western boundary current system are more complicated (Huthnance, 1995). Adjacent to western boundaries of the ocean, isobaths of the continental margin often vary smoothly in the meridional direction. For a western boundary current over the continental slope, exchange with shelf water often takes the form of frontal instabilities, barotropic and baroclinic alike. Along the South Atlantic Bight, the frontal waves and eddies associated with the Gulf Stream, commonly known as spin-off eddies, have wavelengths of the order of 100 km and propagate northward at speeds of tens of kilometres per day (Lee & Atkinson, 1983). Similar frontal waves and eddies have also been observed along the Kuroshio front off the East China Sea north of Taiwan (Qiu *et al.*, 1990; C. Y. Lin *et al.*, 1992). Such a mild form of lateral exchange is active only on the outer shelves. Inner and middle shelves remain largely unaffected.

In winter months, shelf waters can be sufficiently cooled by cold air outbreaks (Blanton *et al.*, 1987) to sink to considerable depths along the slanted frontal boundary of western boundary currents. Models suggest that the mechanism is particularly effective when aided by downwelling-favourable winds (Oey, 1986; Chao, 1992). The consequent downwelling cell over the shelf supplies cold bottom water for subduction. Cold air outbreaks over the South Atlantic Bight are not necessarily associated with downwelling-favourable winds. Over the East China Sea, the downwelling-favourable northeast monsoon is persistent and strong in winter. High particulate fluxes over the continental margin in winter (Fig. 7.13) have been attributed to the wind-induced downwelling cell (Hu, 1984, 1994). This is likely, as the sedimentation rates decrease markedly under the southwest monsoon in summer.

Sharper bends of isobaths allow the exchange between shelf waters and western boundary currents to take place in the longshore rather than the cross-shelf direction. Longshore exchange is in the preferred direction of shelf currents and is understandably more active than cross-shelf exchange. A mild example of longshore exchange takes place in the region north of the Cape Hatteras on the east coast of North America. From the South to the Mid-Atlantic Bight, isobaths of the continental slope region turn from essentially meridional to northeastward in the region north of Cape Hatteras, obstructing the would-be Gulf Stream path if it continues northward. The Gulf Stream turns offshore before interacting with the bending isobaths. The separation of the Gulf Stream near Cape Hatteras continues to be a source of raging debates. Whatever the reason, the separation appears to be related to the large-scale circulation of the North Atlantic. Upwelling of Gulf Steam water occurs slightly north of the Gulf Stream path in the southern extremity of the Mid-Atlantic Bight (Csanady & Hamilton, 1988). Moreover, the mean current is southward. Approaching the Gulf Stream, the resulting confluence presents the underlying physical basis for the 'funnel' hypothesis (Rhoads & Hecker, 1994) (Fig. 7.11*a*).

Similar processes in the southern end of the East China Sea are even more dramatic. Isobaths of the continental slope northeast of Taiwan are essentially zonal, perpendicular to the incoming Kuroshio path east of Taiwan (Fig. 7.11*b*). The direct impingement forces the major portion of the Kuroshio to turn northeastward. A cyclonic eddy is formed on the shoreward side of the Kuroshio (T. Y. Tang, personal communication). The resulting mushroom-like flow pattern (Fig. 7.11*b*), with a major branch on the seaward side and a minor branch on the shoreward side, is accompanied by upwelling of subsurface Kuroshio water (Chern *et al.*, 1990; Liu *et al.*, 1992*a*). As in the Mid-Atlantic Bight, mounting evidence suggests that this is the major depocentre of waterborne materials originating from the shelf (Chung & Chang, 1995).

Particle transport

As observed at the northeastern Atlantic margin during the ECOMARGE programme, the particles collected on the continental slope are fed by two sources: one in the upper water column and the other in the benthic re-suspension layer (Heussner, 1995). The former represents phytodetritus or faecal pellets produced in the upper water column. The latter represents mostly re-worked particles. Particles from the first source are mainly controlled by the cross-shelf dispersion of the shelf water and their sinking rate. Particles from the second source are fed by cross-shelf sediment transport along the bottom. More than 90% of total sedimentary organic carbon cannot be separated physically from its mineral matrix and probably exists as coatings on the mineral grains (Hedges & Keil, 1995). Therefore, POC export via sediment transport is likely to be substantial.

Suspended sediment transport is strongly influenced by erosion and deposition processes at the bottom boundary. Both depend on the intensity of gravity waves and currents. Suspension by storm-generated gravity waves is episodic. When averaged over a long time, transport by gravity waves alone is more isotropic. Nevertheless, gravity waves do enhance the amount of suspension available for transport by currents. For modelling purposes, it would be highly desirable to incorporate a sediment transport component into a hydrodynamic model, which includes both currents and gravity waves. Although such models do exist occasionally in one form or another, their validity still needs to be established scientifically. Further, the hybrid model is computationally demanding. A reasonable alternative for now would be to exclude explicit treatment of gravity waves in the hydrodynamic model while enhancing the sediment erosion rate to account for the missing physics. Sediment suspension thus modelled showed reasonable agreement with observations (Hamblin, 1989; Sanford *et al.*, 1991) and, when incorporated into a three-dimensional hydrodynamic model, provides a useful tool for process-oriented studies.

At the present level of understanding, the prudent strategy for modelling sediment transport is to parameterise sediment settling speeds, erosion and suspension rates as functions of bottom shear stress, but otherwise treat sediments as tracers in the water column. Models that regard erosion and deposition as concurrent processes generally produce more satisfactory results than models that treat the two processes as mutually exclusive (Sanford & Chang, 1997). Existing three-dimensional hydrodynamic models are poised to cope with such a sediment transport component numerically. Each marginal sea or shelf region deserves such a hybrid model to enhance the present level of understanding. Unfortunately, this remains as a distant goal at present.

Regional models

Types of continental margin

For global synthesis, classification of continental margins into different types is necessary for the development of regional models and for reliable extrapolation. That is, different physical–biogeochemical processes dominate carbon fluxes on different types of continental margin. For the Coastal Ocean Processes (CoOP) programme (Brink & Atkinson, 1995), continental shelves are categorised according to physical processes: (1) wind-driven transports (including coastal upwelling), (2) tidal-driven transports, (3) buoyancy-driven transports, (4) western boundary current shelves, and (5) ice-covered shelves.

An objective typology of the coastal units has been developed in the LOICZ programme (LOICZ, 1995). The properties considered in the typology are physical processes and morphology. The variables currently used are: (1) runoff, (2) tidal range, (3) shelf width, (4) June sea-surface temperature, (5) December sea-surface temperature and (6) coastal zone colour. Seven types of coastal sea have been identified with these variables. Additional parameters that may be important for the classification include: (1) major habitats, (2) sediment flux, (3) boundary-current strength, (4) marginal sea depth, (5) upwelling strength, and (6) ice cover. Further division of the coastal units into more types is inevitable.

The regional studies presented in this study represent only a small part of different coastal types. In fact, the completed or ongoing regional continental margin studies cover only very limited types of coastal unit (Fig. 7.1). Modelling is probably the only way to delineate the processes and environmental features that characterise each type of coastal unit. From these findings, better field programmes may be designed to obtain critical parameters for a reliable global synthesis in the future.

A practical approach to regional models

Global models lack the spatial resolution needed to simulate fine-scale features in the coastal ocean. Further, several important physical processes (such as tide and river discharge) are typically ignored in global models, and the biogeochemical processes are often simplified to include only the inorganic CO_2 chemistry (see, for example, Sarmiento & Orr, 1992). These are the primary justifications to develop regional models. With the rapid advance in massively powerful parallel computation technology, the resolution gap between global and regional models is narrowing, but the missing processes in global models will probably remain missing for a long time. In this light, regional models of high resolution are irreplaceable.

The development of hydrodynamic models has gone a long way in the past two decades. Enough experience has been gained, and a variety of three-dimensional models is available to simulate coastal ocean circulation realistically. These models are usually not computationally demanding; computation efficiency is no longer a good excuse to build biogeochemical modelling experience on idealised hydrodynamic models. To build coupled regional models, it seems wise to take advantage of realistic circulation models *ab initio*.

On the biogeochemical front, the modelling experience lags behind that in hydrodynamics. A one-dimensional physical–biogeochemical coupled model with seven compartments of plankton and nitrogen species has successfully simulated the annual cycle of plankton dynamics in the North Atlantic but has failed to predict realistic f ratios for new production (Fasham *et al.*, 1990). A ten-compartment biogeochemical model coupled with a two-dimensional flow field has been used to simulate the succession of trophic dynamics in the South Atlantic Bight during June–September 1981 (Hofmann, 1988). Aside from these limited cases, physical–biogeochemical coupled modelling has not been extensively employed for regional studies. Because hydrodynamic processes often drive biogeochemical processes, the numerical culture for the coupled approach needs to be transplanted and developed on a step-by-step basis. An all-inclusive approach from the very beginning is not in keeping with the current level of understanding and may lead to confusion. Along this line of approach, a realistic primitive-equation East China Sea circulation model has been partially developed (Y. Hsueh, personal communication). Upon its completion, a sediment transport component may be added to it. Further inclusion of biogeochemical components other than sediments will be the goal in the future. The strong seasonality of the monsoon system, the energetic Kuroshio Current System and the voluminous Changjiang runoff make the East China Sea an ideal marginal sea for process study. Co-ordinated efforts involving interaction between field observation and modelling may shed light on some of the critical features that dictate the organic carbon export from a marginal sea. Similar attempts in other types of continental margins are also highly desirable.

Summary

The significance of continental margins in the global carbon cycle is difficult to quantify in view of the complexity, variability and individuality of each shelf. Furthermore, they are most susceptible to anthropogenic changes. Nevertheless, we are certain that continental margins do serve as an important link in the global carbon cycle and contribute significantly to the biological pump of the ocean.

Previous studies show that the primary production in the coastal ocean represents 10–29% (with a mean of 19%) of global ocean production and its new production or shelf export represents 3–64% of the global new production. The variabilities are partly attributed to differences in definitions of the coastal ocean and the capacity of its biological pump. We propose to define the coastal ocean according to modelling needs, including coastal areas where physical as well as biogeochemical processes are not properly covered by either the global ocean model or the terrestrial ecosystem models. New production as defined by nitrate uptake (Dugdale & Goering, 1967) is a good measurement of the biological pump only for continental margins where offshore advection is effective. For margins where the longshore transport is dominant or the shelf is wide, the capacity of the biological pump is better represented by cross-shelf export of organic carbon to the sub-surface water beneath the thermocline of the ocean interior. The traditional new production over-estimates the biological pump capacity for the latter type of margins. A better definition of the f ratio for the coastal ocean is the fraction of coastal primary production, designated as f_x, exported to the sub-surface ocean interior. Based on previous observations, f_x is estimated to be 0.13, which yields 1 Gt C yr^{-1} for coastal biological pump capacity, accounting for 20% of the estimated global biological pump of 5 Gt C yr^{-1}.

Our understanding of the processes that are involved in carbon fluxes on continental margins is still inadequate for a more accurate estimate. In recent years, many regional programmes have carried out important continental margin studies (Fig. 7.1). These include the Benguela Ecosystem Study for the west coast of South Africa, CoOP (Coastal Ocean Processes) for the US coasts, ECOMARGE (Ecosystems of Continental Margins) for the French Atlantic and Mediterranean coasts, KEEP (Kuroshio Edge Exchange Processes) and MASFLEX (Marginal Sea Flux Experiment) for the East China Sea, LOIS (Land Ocean Interaction Study) for the British coasts, and OMEX (Ocean Margin Exchange) for the European coasts. These programmes form the backbone of the JGOFS–LOICZ continental margin studies. They are expected to shed new light on the physical as well as the biogeochemical processes in continental margins.

The cross-shelf export of organic carbon is difficult to measure directly. Mounting evidence indicates that diffusive transport of POC across the shelf in western boundary current systems is not very effective. On the other hand, offshore advective transport at certain regions, such as the point of sharp turning of the western boundary current, can be an effective mode of POC export. Such an intensified and localised cross-shelf transport mode may be easily missed by evenly distributed moorings of limited number, but drifters, drogues and buoys may help to show their localities. Such practice has been recently applied on the Hebridean shelf edge as a part of the LOIS programme and on the East China Sea shelf as a part of the KEEP programme. Other approaches, such as carbon budgeting, use of nutrient elements as proxies, or benthic flux measurement, should also be employed to constrain the lateral export. Moreover, previous studies of organic carbon export were mostly focused on POC fluxes. In the future, shelf export of DOC and DOC fluxes from sediments should also be assessed because some coastal waters are important sources of DOC and a significant fraction of sedimentary organic carbon is probably recycled as DOC.

The influence of river discharge on the coastal carbon cycle has only recently been carefully examined and definitely needs further research. River inputs are especially important for two reasons. First, carbon flux from rivers as a source of oceanic carbon dioxide must be better quantified before the net change in the air–sea exchange of carbon dioxide can be reliably determined. Second, anthropogenic nutrient fluxes discharged from rivers may enhance primary production over the shelves. Moreover, river discharge not only changes the chemistry of the receiving shelf water but also influences the shelf circulation, which may dictate the cross-shelf transport of organic carbon.

The coastal ocean may serve as both a source and a sink of atmospheric CO_2. Coastal upwelling may conceivably release carbon dioxide to the atmosphere by bringing carbon dioxide supersaturated water to the surface. This is partly offset by the downward flux of particulate organic carbon due to high productivity. However, intensive biological uptake may result in a net drawdown of atmospheric carbon dioxide because of the slow outgassing process in the initial stage. Therefore, the sign of this exchange process needs to be clarified. On the other hand, nutrients supplied by river discharge or onwelling of slope water in the frontal region could be offset by seaward transport of organic matter to balance the nutrient budget, which may result in a net drawdown of atmospheric CO_2.

Finally, simple linear extrapolation from limited and unevenly distributed observations to obtain global estimates of continental margin carbon fluxes severely compromises the reliability of the assessment. More sea-going observations may not be enough to resolve the complex processes and to better determine the carbon fluxes on continental margins, not to mention the understanding of their interannual fluctuations. Fine-scale regional models with coupled physical and biogeochemical processes must be employed to generalise the limited observations. Furthermore, new technologies in the fields of satellite remote sensing, continuous monitoring of sub-surface conditions by moored physical sensors and powerful parallel numerical computation will undoubtedly greatly assist the study of continental margins. With the aid of these new technologies and with more traditional sea-going observations, coupled physical–biogeochemical models may be developed for different types of continental margin so that we may have better grasp of the important processes and, eventually, achieve a reliable global synthesis.

Acknowledgements

This paper is a contribution of the Kuroshio Edge Exchange Processes programme. The authors are grateful for the suggestions from J. G. Field and J. M. Huthnance, and valuable information provided by many principal investigators of the KEEP programme, including J. Chang, Y. L. L. Chen, Y. C. Chung, G. C. Gong, J. J. Hung, S. Lin and T. Y. Tang. This study was supported by a grant (NSC 85–2611–M-002A-030) from the National Science Council of the Republic of China.

References

Anderson, R. F., Rowe, G. T., Kemp, P., Trumbore, S. & Biscaye, P. E. (1994). Carbon budget for the mid-slope depocenter of the Middle Atlantic Bight. *Deep-Sea Research*, Part II, **41**, 537–61.

Antoine, D., Andre, J.-M. & Morel, A. (1996). Oceanic primary production 2. Estimation at global scale from Satellite (coastal zone color scanner) chlorophyll. *Global Biogeochemical Cycles*, **10**, 57–69.

Baker, E. T. & Melburn, H. B. (1983). An instrument system for the investigation of particle fluxes. *Continental Shelf Research*, **1**, 425–35.

Bauer, J. E., Reimers, C. E., Druffel, E. R. M. & Williams, P. M. (1995). Isotopic constraints on carbon exchange between deep ocean sediments and sea water. *Nature*, **373**, 686–9.

Beardsley, R. C., Limeburner, R., Yu, H. & Cannon, G. A. (1985). Discharge of the Changjiang (Yangtze River) into the East China Sea. *Continental Shelf Research*, **4**, 57–76.

Berger, W. H. (1989). Global maps of ocean productivity. In *Productivity of the Ocean: Present and Past*, ed. W. H. Berger, V. S. Smetacek & G. Wefer, pp. 429–55. New York: Wiley.

Berger, W. H., Smetacek, V. S. & Wefer, G. (1989). Ocean productivity and paleoproductivity – an overview. In *Productivity of the Ocean: Present and Past*, ed. W. H. Berger, V. S. Smetacek & G. Wefer, pp. 1–34. New York: Wiley.

Berner, R. A. (1992). Comments on the role marine sediment burial as a repository for anthropogenic CO_2. *Global Biogeochemical Cycles*, **6**, 1–2.

Berner, R. A., Lasaga, A. C. & Garrels, R. M. (1983). The carbonate-silicate geochemical cycle and its effect on atmospheric carbon dioxide over the past 100 million years. *American Journal of Science*, **283**, 641–83.

Bienfang, P. K. & Ziemann, D. A. (1992). The role of coastal high latitude ecosystems in global export production. In *Primary Productivity and Biogeochemical Cycles in the Sea*, ed. P. G. Falkowski & A. D. Woodhead, pp. 285–97. New York: Plenum.

Billen, G., Lancelot, C. & Meybeck, M. (1991). N, P and Si retention along the aquatic continuum from land to ocean. In *Ocean Margin Processes in Global Change*, ed. R. F. C. Mantoura, J.-M. Martin & R. Wollast, pp. 19–44. New York: Wiley.

Biscaye, P. E. & Anderson, R. F. (1994). Fluxes of particulate matter on the slope of the southern Mid-Atlantic Bight: SEEP-II. *Deep-Sea Research*, Part II, **41**, 459–509.

Biscaye, P. E., Anderson, R. F. & Deck, B. L. (1988).Fluxes of particles and constituents to the eastern United States continental slope and rise: SEEP-I. *Continental Shelf Research*, **8**, 855–904.

Biscaye, P. E., Flagg, C. N. & Falkowski, P. G. (1994) The Shelf Edge Exchange Processes experiment, SEEP-II: an introduction to hypothesis, results and conclusions. *Deep-Sea Research*, Part II, **41**, 231–52.

Blair, N. E., Plaia, G. R., Boehme, S. E., DeMaster, D. J. & Levin, L. A. (1994). The remineralization of organic carbon on the North Carolina continental slope. *Deep-Sea Research*, Part II, **41**, 755–66.

Blanton, J. O. (1991). Circulation processes along oceanic margins in relation to material fluxes. In *Ocean Margin Processes in Global Change*, ed. R. F. C. Mantoura, J.-M. Martin & R. Wollast, pp. 145–63, New York: Wiley.

Blanton, J. O., Lee, T. N., Atkinson, L. P., Bane, J. M., Riordan, A. & Raman, S. (1987). Oceanographic studies during GALE. *Eos*, **68**, 1626–7, 1636–7.

Brink, K. H. (1987). Coastal ocean physical processes. *Review of Geophysics*, **25**, 204–16.

Brink, K. H. & Atkinson, L. P. (1995). A classification of U.S. continental shelves. *Newsletter of Coastal Ocean Processes*, No. **2** (April 1995), 1, 9.

Brink, K. H. & Cowles, T. J. (1991). The coastal transition zone program. *Journal of Geophysical Research*, **96**, 14637–47.

Broecker, W. S. & Peng, T. H. (1992). Interhemispheric transport of carbon dioxide by ocean circulation. *Nature*, **356**, 587–9.

Bumpus, D. (1973). A description of the circulation on the continental shelf of the east coast of the United States. *Progress in Oceanography*, **6**, 111–57.

Campbell, J. W. & O'Reilly, J. E. (1989). Role of satellites in estimating primary productivity on northwest Atlantic continental shelf. *Continental Shelf Research*, 8, 179–204.

Chao, S.-Y. (1988). Wind-driven motion of estuarine plumes. *Journal of Physical Oceanography*, **18**, 1144–66.

Chao, S.-Y. (1991). Circulation of the East China Sea, a numerical study. *Journal of the Oceanographic Society of Japan*, **46**, 273–95.

Chao, S.-Y. (1992). An air-sea interaction model for cold air outbreaks. *Journal of Physical Oceanography*, **22**, 821–42.

Chao, S.-Y. & Boicourt, W. C. (1986). Onset of estuarine plumes. *Journal of Physical Oceanography*, **16**, 2137–49.

Chavez, F. P., Barber, R. T., Kosro, P. M., Huyer, A., Ramp, S. R., Stanton, T. P. & de Mendiola, B. R. (1991). Horizontal transport and distribution of nutrients in the coastal transition zone off northern California: effect on primary production, phytoplankton biomass and species composition. *Journal of Geophysical Research*, **96**, 14833–48.

Chen, C. T. A. & Wang, S. L. (1997). Carbon and nutrient budgets on the East China Sea continental shelf. In *Biogeochemical Processes in the North Pacific*, ed. S. Tsunogai, pp. 169–86, Tokyo: Japan Marine Science Foundation.

Chen, C. T. A., Chuang, W. S. & Liu, K. K. (1991). The Kuroshio Edge Exchange Processes (KEEP) Program in Taiwan, R.O.C. *U.S. JGOFS Newsletter*, **3**(2), 7–8.

Chen, S. K. (1995). Sediment accumulation rates and organic carbon deposition in the East China Sea continental margin sediments. M.S. Thesis, National Taiwan University, Taipei, Taiwan. (In Chinese.)

Chern, C.-S., Wang, J. & Wang, D.-P. (1990). The exchange of Kuroshio and East China Sea shelf water. *Journal of Geophysical Research*, **95**, 16017–23.

Chuang, W. S. (1985). Dynamics of subtidal flow in the Taiwan Strait. *Journal of the Oceanographic Society of Japan*, **41**, 65–72.

Chung, Y. & Chang, W. C. (1995). Pb-210 fluxes and sedimentation rates on the lower slope between Taiwan and South Okinawa Trough. *Continental Shelf Research*, **15**, 149–64.

Csanady G. T. (1990). Physical basis of

coastal productivity – The SEEP and MASAR Experiments. *Eos*, **71**, 36, 1060–1, 1064–5.
Csanady, G. T. & Hamilton, P. (1988). Circulation of slope water. *Continental Shelf Research*, **8**, 565–624.
Degens, E. T., Kempe, S. & Richey, J. E. (1990). Summary: biogeochemistry of major world rivers. In *Biogeochemistry of Major World Rivers*, ed. E. T. Degens, S. Kempe & J. E. Richey, pp. 323–47. New York: Wiley.
DeMaster, D., Pope, R. H., Levin, L. A. & Blair, N. E. (1994). Biological mixing intensity and rates of organic carbon accumulation in North Carolina slope sediments. *Deep-Sea Research*, Part II, **41**, 735–54.
Ducklow, H. W. & Carlson, C. A. (1992). Oceanic bacterial production. In *Advanced Microbial Ecology*, ed. K. C. Marshall, vol. 12, pp. 131–81. New York. Plenum.
Dugdale, R. C. & Goering, J. J. (1967). Uptake of new and regenerated forms of nitrogen in primary production. *Limnology and Oceanography*, **12**, 196–206.
Edmond, J. M., Spivack, A., Grant, B. C., Hu, M.-H., Chen, Z., Chen, S. & Zeng, X. (1985). Chemical dynamics of the Changjiang estuary. *Continental Shelf Research*, **4**, 17–36.
Eppley, R. W. & Peterson, B. J. (1979). Particulate organic matter flux and planktonic new production in the deep ocean. *Nature*, **282**, 677–80.
Falkowski, P. G, Biscaye, P. E. & Sancetta, C. (1994). The lateral flux of biogenic particles from the eastern North American continental margin to the North Atlantic Ocean. *Deep-Sea Research*, Part II, **41**, 583–601.
Fasham, M. J. R., Ducklow, H. W. & McKelvie, S. M. (1990). A nitrogen-based model of plankton dynamics in the oceanic mixed layer. *Journal of Marine Research*, **48**, 591–639.
Fei, Z., Mao, X.-H., Lu, R.-H., Li, B., Guan, Y.-H., Li, B.-H., Zhang, X.-S. & Mou, D.-C. (1987). Distributional characteristics of chlorophyll *a* and primary productivity in the Kuroshio area of the East China Sea. In *Essays on the Investigation of Kuroshio*, ed. X. Sun, pp. 256–66. Beijing, China: Ocean Press. (In Chinese.)
Fennel, W. & Neumann, T. (1996). The mesoscale variability of nutrients and plankton as seen in a coupled model. *German Journal of Hydrology*, **48**, 49–71.
Friederich, G., Sakamoto, C.M., Pennington, J. T. & Chavez, F. P. (1995). On the direction of the air-sea flux of CO_2 in coastal upwelling systems. In *Global Fluxes of Carbon and its Related Substances in the Coastal Sea-Ocean-Atmosphere System*, ed. S. Tsunogai, K. Iseki, I. Koike & T. Oba, pp. 438–45. Yokohama, Japan: M. & J. International.
Galloway, J. N., Howarth, R. W., Michaels, A. F., Nixon, S. W., Prospero, J. M. & Dentener, F. J. (1996). Nitrogen and phosphorus budgets of the North Atlantic Ocean and its watershed. *Biogeochemistry*, **35**, 3–25.
Gan, W. B., Chen, H. M. & Han, Y. F. (1983). Carbon transport by the Yangtze (at Nanjing) and Huanghe (at Jinan) Rivers, People's Republic of China. In *Transport of Carbon and Minerals in World Major Rivers*, Part 2 (SCOPE/UNEP Soderband Heft 55), ed. E. T. Degens, S. Kempe & H. Soliman, pp. 459–70. Hamburg: University of Hamburg.
Gardner, W. D. (1980). Sediment trap dynamics and calibration: a laboratory evaluation. *Journal of Marine Research*, **38**, 17–39.
Gardner, W. D. (1985). The effect of tilt on sediment trap efficiency. *Deep-Sea Research*, **32**, 349–61.
Gong, G.-C., Liu, K.-K. & Pai, S.-C. (1995). Prediction of nitrate concentration from two end member mixing in the southern East China Sea. *Continental Shelf Research*, **15**, 827–42.
Gong, G. C., Chen, Y. L. L. & Liu, K.-K. (1996). Chemical hydrography and chlorophyll distribution in the East China Sea in summer: implications in nutrient dynamics. *Continental Shelf Research*, **16**, 1561–90.
Guo, Y. J. (1991). The Kuroshio. Part II.

Primary productivity and phytoplankton. *Oceanographic Marine Biology Annual Review*, **29**, 155–89.

Hales, B., Emerson, S. & Archer, D. (1994). Respiration and dissolution in the sediments of the western North Atlantic: estimates from models of *in situ* microelectrode measurements of porewater oxygen and pH. *Deep-Sea Research*, **41**, 695–719.

Hall, J. & Smith, S. (eds) (1995). *First Report of the JGOFS/LOICZ Continental Margins Task Team.* (LOICZ Reports and Studies No. 7/JGOFS Report No. 21.) Texel, The Netherlands: LOICZ.

Hall, J., Smith, S. V. & Boudreau, P. R. (eds) (1996). *Report on the International Workshop on Continental Shelf Fluxes of Carbon, Nitrogen and Phosphorus.* (LOICZ Reports and Studies 96–9.) Texel, The Netherlands: LOICZ. ii + 50 pp.

Hama, T. (1995) Seasonal change of the primary productivity in the East China Sea. In *Global Fluxes of Carbon and its Related Substances in the Coastal Sea-Ocean-Atmosphere System*, ed. S. Tsunogai, K. Iseki, I. Koike & T. Oba, pp. 74–9. Yokohama, Japan: M. & J. International.

Hamblin, P. F. (1989). Observations and models of sediment transport near the turbidity maximum of the upper Saint Lawrence Estuary. *Journal of Geophysical Research*, **94**, 14419–28.

Hedges, J. I. & Keil, R. G. (1995). Sedimentary organic matter preservation: an assessment and speculative synthesis. *Marine Chemistry*, **49**, 81–115.

Heussner, S. (1995). Sources and fate of settling particles on a northeastern Atlantic margin: recent issues from the ECOMARGE program (France-JGOFS). In *Global Fluxes of Carbon and its Related Substances in the Coastal Sea-Ocean-Atmosphere System*, ed. S. Tsunogai, K. Iseki, I. Koike & T. Oba, pp. 183–8. Yokohama, Japan: M. & J. International.

Heussner, S., Ratti, C. & Carbonne, J. (1990). The PPS 3 time-series sediment trap and the trap sample processing techniques used during the ECOMARGE experiment. *Continental Shelf Research*, **10**, 943–58.

Hinga, K. R. (1985). Evidence for a higher average primary productivity in the Pacific than in the Atlantic Ocean. *Deep-Sea Research*, **32**, 117–26.

Hofmann, E. E. (1988). Plankton dynamics on the outer southeastern U.S. continental shelf. Part III: a coupled physical-biological model. *Journal of Marine Research*, **46**, 919–46.

Hofmann, E. E. (1991). How do we generalise coastal models to global scale? In *Ocean Margin Processes in Global Change*, ed. R. F. C. Mantoura, J.-M. Martin & R. Wollast, pp. 401–17. New York: Wiley.

Honjo, S. & Doherty, K. W. (1988). Large aperture time-series sediment traps: design objectives, construction and application. *Deep-Sea Research*, **35**, 133–49.

Houghton, J. T., Jenkins, G. J. & Ephraums, J. J. (1990). *Climate Change, The IPCC Scientific Assessment.* Cambridge University Press.

Howarth, R. W., Billen, G., Swaney, D., Townsend, A., Jaworski, N., Lajtha, K., Downing, J. A., Elmgren, R., Caraco, N., Jordan, T., Berendse, F., Freney, J., Kudeyarov, V., Murdoch, P., Zhu, Z.-L. (1996). Regional nitrogen budgets and riverine N & P fluxes for the drainages to the North Atlantic Ocean: Natural and human influences. *Biogeochemistry*, **35**, 75–139.

Hu, D. X. (1984). Upwelling and sedimentation dynamics, I. The role of upwelling in Huanghai Sea and East China Sea. *Chinese Journal of Oceanology and Limnology*, **2**(1), 12–19.

Hu, D. X. (1994). Possible mechanism for material flux in the Yellow Sea and East China Sea from hydrographical viewpoint. In *Global Fluxes of Carbon and its Related Substances in the Coastal Sea-Ocean-Atmosphere System*, ed. S. Tsunogai, K. Iseki, I. Koike & T. Oba, p. 38. Yokohama, Japan: M. & J. International.

Huang, S., Yang, J., Ji, W., Yang, X. & Chen, G. (1983). Silicon, nitrogen and phosphorus in the Changjiang River Mouth water. In *Proceedings of International Symposium on Sedimentation on the Continental Shelf with Special Reference to the East China Sea, April 12–16, 1983, Hangzhou, China (Acta Oceanologica Sinica)*, pp. 220–8. Beijing, China: China Ocean Press.

Huthnance, J. M. (1995). Circulation, exchange and water masses at the ocean margin: the role of physical processes at the shelf edge. *Progress in Oceanography*, **35**, 353–431.

Iseki, K., Whitney, F. & Wang, C.S. (1980). Biochemical changes of sedimented matter in sediment trap in shallow coastal waters. *Bulletin of the Planktonic Society of Japan*, **27**, 27–36.

Iseki, K, Okamura, K. & Tsuchiya, Y. (1995). Seasonal variability in particle distributions and fluxes in the East China Sea. In *Global Fluxes of Carbon and its Related Substances in the Coastal Sea-Ocean-Atmosphere System*, ed. S. Tsunogai, K. Iseki, I. Koike & T. Oba, pp. 189–97. Yokohama, Japan: M. & J. International.

Jahnke, R. (1995). Assessing particulate organic carbon fluxes at continental margins by *in situ* flux chamber measurements. In *Global Fluxes of Carbon and its Related Substances in the Coastal Sea-Ocean-Atmosphere System*, ed. S. Tsunogai, K. Iseki, I. Koike & T. Oba, pp. 218–25. Yokohama, Japan: M. & J. International.

Jahnke, R. A. (1996). The global ocean flux of particulate organic carbon: areal distribution and magnitude. *Global Biogeochemical Cycles*, **10**, 71–88.

Jahnke, R. A. & Craven, D. B. (1995). Quantifying the role of heterotrophic bacteria in the carbon cycle: A need for respiration rate measurements. *Limnology and Oceanography*, **40**, 436–41.

Jahnke, R. A., Reimers, C. E. & Craven, D. B. (1990). Intensification of recycling of organic matter at sea floor near ocean margins. *Nature*, **348**, 50–4.

Jan, S. (1995). Seasonal variations of currents in the Taiwan Strait. Ph.D. Thesis, National Taiwan University, Taipei, Taiwan. 139 pp. (In Chinese.)

Kao, S.-J. & Liu, K.-K. (1996). Particulate organic carbon export from the watershed of a subtropical mountainous river (Lanyang Hsi) in Taiwan. *Limnology and Oceanography*, **41**, 1749–57.

Kao, S.-J. & Liu, K.-K. (1997). Fluxes of dissolved and non-fossil particulate organic carbon from an Oceania small river (Lanyang Hsi) in Taiwan. *Biogeochemistry*, **39**, 255–69.

Kemp, P. F., Falkowski, P. G., Flagg, C. N., Phoel, W., Smith, S. L., Wallace, D. W. R. & Wirick, C. D. (1994). Modelling vertical oxygen and carbon flux during stratified spring and summer conditions on the continental shelf, Middle Atlantic Bight, eastern USA. *Deep-Sea Research*, Part II, **41**, 629–55.

Kempe, S. & Pegler, K. (1991). Sinks and sources of CO_2 in coastal seas: the North Sea. *Tellus*, **43B**, 224–35.

Kennett, J. P. (1982). *Marine Geology*. Englewood Cliffs, NJ: Prentice-Hall.

Knauer, G. A., Martin, J. H. & Bruland, K. W. (1979). Fluxes of particulate carbon, nitrogen and phosphorus in the upper water column of the northeast Pacific. *Deep-Sea Research*, **26**, 97–108.

Lee, T. N. & Atkinson, L. P. (1983). Low-frequency current and temperature variability from Gulf Stream frontal eddies and atmospheric forcing along the southeast U.S. outer continental shelf. *Journal of Geophysical Research*, **88**, 4541–67.

Lin, C. Y., Shyu, C. Z. & Shih, W. H. (1992). The Kuroshio Fronts and Cold Eddies off Northeastern Taiwan Observed by NOAA-AVHRR Imageries. *Terrestrial, Atmospheric and Oceanic Sciences*, **3**, 225–42.

Lin, S., Liu, K.-K., Chen, M.-P., Chen, P. & Chang, F.-Y. (1992). Distribution of organic carbon in the KEEP area continental margin sediments. *Terrestrial, Atmospheric and Oceanic Sciences*, **3**, 365–77.

Liu, K.-K., Gong, G.-C., Lin, S., Yang, C.-Y., Wei, C.-L., Pai, S.-C. & Wu, C.-K. (1992*a*). The year-round upwelling at the shelf break near the northern tip of

Taiwan as evidenced by chemical hydrography. *Terrestrial, Atmospheric and Oceanic Sciences*, **3**, 243–75.

Liu, K.-K., Gong, G.-C., Shyu, C.-Z., Pai, S.-C., Wei, C.-L. & Chao, S.-Y. (1992*b*). Response of Kuroshio upwelling to the onset of northeast monsoon in the sea north of Taiwan: observations and a numerical simulation. *Journal of Geophysical Research*, **97**, 12511–26.

Liu, K.-K., Lai, Z.-L., Gong, G.-C. & Shiah, F.K. (1995). Distribution of particulate organic matter in the southern East China Sea: implications in production and transport. *Terrestrial, Atmospheric and Oceanic Sciences*, **6**, 27–45

LOICZ (1995). LOICZ Typology: Preliminary Version for Discussion. (LOICZ Reports & Studies No. 3.) Texel, The Netherlands: LOICZ.

Longhurst, A. R. & Harrison, W. G. (1989). The biological pump: profiles of plankton production and consumption in the upper ocean. *Progress in Oceanography*, **22**, 47–123.

Longhurst, A. R., Sathyendranath, S., Platt, T. & Caverhill, C. (1995). An estimation of global primary production in the ocean from satellite radiometer data. *Journal of Planktonic Research*, **17**, 1245–71.

Ludwig, W., Probst, J.-L. & Kempe, S. (1996). Predicting the oceanic input of organic carbon by continental erosion. *Global Biogeochemical Cycles*, **10**, 23–41.

Mantoura, R. F. C., Martin, J.-M. & Wollast, R. (edS) (1991). *Ocean Margin Processes in Global Change*. New York: Wiley.

Marra, J. & Heinemann, K. R. (1987). Primary production in the North Pacific Central Gyre: some new measurements based on ^{14}C. *Deep-Sea Research*, **34**, 1821–9.

Martin, J. H., Knauer, G. A., Karl, D. M. & Broenkow, W. W. (1987). VERTEX: carbon cycling in the north Pacific. *Deep-Sea Research*, **34**, 267–85.

Martin, W. R. & McCorkle, D. C. (1993). Dissolved organic carbon concentrations in marine pore waters determined by high-temperature oxidation. *Limnology and Oceanography*, **38**, 1464–80.

Matthews, E. (1994). Nitrogenous fertilizer: global distribution of consumption and associated emissions of nitrous oxide and ammonia. *Global Biogeochemical Cycles*, **8**, 411–39.

Meybeck, M. (1987). Global chemical weathering of surficial rocks estimated from river dissolved loads. *American Journal of Science*, **287**, 401–28.

Meybeck, M. & Jussieu, P. (1993). C, N, P and S in rivers: from sources to global inputs. In *Interactions of C, N, P and S Biogeochemical Cycles and Global Change*, ed. R. Wollast, F. T. Mackenzie & L. Chou, pp. 163–94. Berlin: Springer.

Milliman, J. D. (1991). Flux and fate of fluvial sediment and water in coastal seas. In *Ocean Margin Processes in Global Change*, ed. R. F. C. Mantoura, J.-M. Martin & R. Wollast, pp. 69–89. New York: Wiley.

Milliman, J. D. & Syvitski, J. P. M. (1992). Geomorphic/tectonic control of sediment discharge to the ocean: the importance of small mountainous rivers. *Journal of Geology*, **100**, 525–44.

Milliman, J. D., Xie, Q. & Yang, Z. (1984). Transfer of particulate organic carbon and nitrogen from the Yangtze River to the ocean. *American Journal of Science*, **284**, 824–34.

Milliman, J. D., Shen, H.-T., Yang, Z.-S. & Meade, R. H. (1985). Transport and deposition of river sediment in the Changjiang estuary and adjacent continental shelf. *Continental Shelf Research*, **4**, 37–45.

Monaco, A, Biscaye, P., Soyer, J., Pocklington, R. & Heussner, S. (1990*a*). Particle fluxes and ecosystem response on a continental margin: the 1985–1988 Mediterranean ECOMARGE experiment. *Continental Shelf Research*, **10**, 809–39.

Monaco, A, Courp, T., Heussner, S., Carbonne, J., Fowler, S. W. & Deniaux, B. (1990*b*). Seasonality and composition of particle fluxes during ECOMARGE-I, western Gulf of Lions. *Continental Shelf Research*, **10**, 959–87.

Oey, L. Y. (1986). The formation and maintenance of density fronts on the U.S.

southeastern continental shelf during winter. *Journal of Physical Oceanography*, **16**, 1121–35.
Ortner P. B. & Dagg, M. J. (1995). Nutrient-enhanced coastal ocean productivity explored in the Gulf of Mexico, *Eos*, **76**, 97–109.
Platt, T. & Subba Rao, D. V. (1975). Primary production of marine microphytes. In *Photosynthesis and Productivity in Different Environments*, ed. J. P. Cooper, pp. 249–80. London: Cambridge University Press.
Qin, Y. & Li, F. (1983). Study of influence of sediment loads discharged from the Huanghe River on sedimentation in the Bohai Sea and Huanghai Sea. In *Proceedings of International Symposium on Sedimentation on the Continental Shelf with Special Reference to the East China Sea, April 12–16, 1983, Hangzhou, China (Acta Oceanologica Sinica)*, pp. 83–92. Beijing, China: China Ocean Press.
Qiu, B., Tado, T. & Imasato, N. (1990). On the Kuroshio frontal fluctuations in the East China Sea using satellite and *in situ* observation data. *Journal of Geophysical Research*, **95**, 19 191–204.
Reimers, C. E., Jahnke, R. A. & McCorkle, D. C. (1992). Carbon fluxes and burial rates over the continental slope and rise off central California with implications for the global carbon cycle. *Global Biogeochemical Cycles*, **6**, 199–224.
Rhoads, D. D. & Hecker, B (1994). Processes on the continental slope off North Carolina with special reference to the Cape Hatteras region. *Deep-Sea Research*, Part II, **41**, 965–80.
Richey, J. E. & Victoria, R. L. (1993). C, P, N export dynamics in the Amazon River. In *Interactions of C, N, P and S Biogeochemical Cycles and Global Change*, ed. R. Wollast, F. T. Mackenzie & L. Chou, pp. 123–39. Berlin: Springer.
Rowe, G. T., Smith, S., Falkowski, P., Whitledge, T., Theroux, R., Phoel, W. & Ducklow, H. (1986). Do continental shelves export organic matter? *Nature*, **324**, 559–61.
Sanford, L. P. & Chang, M. L. (1997). The bottom boundary condition for suspended sediment deposition. *Journal of Coastal Research, Special Issue*, **25** 3 17.
Sanford, L. P., Panageotou, W. & Halka, J. P. (1991). Tidal resuspension of sediments in northern Chesapeake Bay. *Marine Geology*, **97**, 87–103.
Sarmiento, J. L. & Orr, J. C. (1992). A perturbation simulation of CO_2 uptake in an ocean circulation model. *Journal of Geophysical Research*, **97**, 3621–45.
Sarmiento, J. L. & Siegenthaler, U. (1992). New production and the global carbon cycle. In *Primary Productivity and Biogeochemical Cycles in the Sea*, ed. P. G. Falkowski & A. D. Woodhead, pp. 317–32. New York: Plenum.
Sarmiento, J. L. & Sundquist, E. T. (1992). Revised budget for the oceanic uptake of anthropogenic carbon dioxide. *Nature*, **356**, 589–93.
Sarmiento, J. L., Slater, R. D., Fasham, M. J. R., Ducklow, H. W., Toggweiler, J. R. & Evans, G. T. (1993). A seasonal three-dimensional ecosystem model of nitrogen cycling in the North Atlantic euphotic zone. *Global Biogeochemical Cycles*, **7**, 417–50.
SCOR (1990). *JGOFS Science Plan.* (JGOFS Report No. 4.) Halifax: Scientific Committee on Oceanic Research.
SCOR (1992). *Joint Global Ocean Flux Study Implementation Plan.* (JGOFS Report No. 9.) Baltimore: Scientific Committee on Oceanic Research.
SCOR (1994). *Report of the JGOFS/ LOICZ Task Team on continental Margin Studies.* (JGOFS Report No. 16.) Baltimore: Scientific Committee on Oceanic Research.
Shen, H., Li, J., Zhu, H., Han, M. & Zhou, F. (1983). Transport of the suspended sediments in the Changjiang Estuary. In *Proceedings of International Symposium on Sedimentation on the Continental Shelf with Special Reference to the East China Sea, April 12–16, 1983, Hangzhou, China (Acta Oceanologica Sinica)*, pp. 359–69. Beijing, China: China Ocean Press.
Shiah, F. K., Gong, G. C. & Liu, K. K. (1995). A preliminary survey on primary productivity measured by the ^{14}C assimilation method in the KEEP area.

Acta Oceanographica Taiwanica, **34**, 1–16.
Smetacek, V., Bathmann, U., Nolting, E.-M. & Scharek, R. (1991). Coastal eutrophication: causes and consequences. In *Ocean Margin Processes in Global Change*, ed. R. F. C. Mantoura, J.-M. Martin & R. Wollast, pp. 251–79. New York: Wiley.
Smith, K.L. Jr (1978). Benthic community respiration in the N.W. Atlantic Ocean: *in situ* measurements from 40 to 5200 m. *Marine Biology*, **47**, 337–47.
Smith, S. V. & Hollibaugh, J. T. (1993). Coastal metabolism and the oceanic carbon balance. *Review of Geophysics*, **31**, 75–89.
Spitzy, A. & Ittekkot, V. (1991). Dissolved and particulate organic matter in rivers. In *Ocean Margin Processes in Global Change*, ed. R. F. C. Mantoura, J.-M. Martin & R. Wollast, pp. 5–17. New York: Wiley.
Sternberg, R. W., Larson, L. H. & Miao, Y. T. (1985). Tidally driven sediment transport on the East China Sea continental shelf. *Continental Shelf Research*, **4**, 105–20.
Sverdrup, H. U., Johnson, M. W. & Fleming, R. H. (1942). *The Oceans: Their Physics, Chemistry and General Biology*. Englewood Cliffs, NJ: Prentice-Hall.
Tans, P. P., Fung, I. Y. & Takahashi, T. (1990). Observational constraints on the global atmospheric CO_2 budget. *Science*, **247**, 1431–8.
Walsh, J. J. (1981). A carbon budget for over fishing off Peru. *Nature*, **290**, 300–4.
Walsh, J. J. (1988). *On the Nature of Continental Shelves*. San Diego: Academic Press.
Walsh, J. J. (1989). How much shelf production reaches the deep sea? In *Productivity of the Ocean: Present and Past*, ed. W. H. Berger, V. S. Smetacek & G. Wefer, pp. 175–91. New York: Wiley.
Walsh, J. J. (1991). Importance of continental margins in the marine biogeochemical cycling of carbon and nitrogen. *Nature*, **350**, 53–5.
Walsh, J. J. (1994). Particle export at Cape Hatteras. *Deep-Sea Research*, Part II, **41**, 603–28.
Walsh, J. J. & Dieterle, D. A. (1994). CO_2 cycling in the coastal ocean. I – A numerical analysis of the southeastern Bering Sea with applications to the Chukchi Sea and the northern Gulf of Mexico. *Progress in Oceanography*, **34**, 335–92.
Walsh, J. J., Rowe, G. T., Iverson, R. L. & McRoy, C. P. (1981). Biological export of shelf carbon: a neglected sink of the global CO cycle. *Nature*, **291**, 196–201.
Walsh, J. J., Premuzic, E. T., Gaffney, J. S., Rowe, G. T., Harbottle, G., Stoenner, R. W., Balsam, W. L., Betzer, P. & Macko, S. A. (1985). Organic storage of CO_2 on the continental slope off the mid-Atlantic bight, the southeastern Bering Sea, and the Peru coast. *Deep-Sea Research*, **32**, 853–83.
Walsh, J. J., Biscaye, P. E., & Csanady, G. T. (1988). The 1983–1984 Shelf Edge Exchange Processes (SEEP)-I experiment: hypotheses and highlights. *Continental Shelf Research*, **8**, 435–56.
Wen, Y. H. (1995). A preliminary study of the seasonal variation of primary productivity and light/chlorophyll model in the sea off northern Taiwan. M.S. Thesis, National Taiwan Ocean University, Keelung, Taiwan. (In Chinese.)
Wollast, R. (1991). The coastal organic carbon cycle: fluxes, sources, and sinks. In *Ocean Margin Processes in Global Change*, ed. R. F. C. Mantoura, J.-M. Martin & R. Wollast, pp. 365–81. New York: Wiley.
Wollast, R. (1993). Interactions of carbon and nitrogen cycles in the coastal zone. In *Interactions of C, N, P and S Biogeochemical Cycles and Global Change*, ed. R. Wollast, F. T. Mackenzie & L. Chou, pp. 195–210. Berlin: Springer.
Wong, G. T.-F., Pai, S.-C., Liu, K.-K., Liu, C.-T. & Chen, C.-T. A. (1991). Variability of the chemical hydrography at the frontal region between the East China Sea and the Kuroshio northeast of Taiwan. *Estuarine Coastal Shelf Science*, **33**, 105–20.
Wood, A. M., Rai, H., Garnier, J., Kairesalo, T., Gresens, S., Orive, E. & Ravail, B.

(1992). Practical approaches to algal excretion. *Marine Microbial Food Webs*, **6**, 21–38.
Wroblewski, J. S. & Hofmann, E. E. (1989). U.S. interdisciplinary modelling of coastal–offshore exchange processes: Past and future. *Progress in Oceanography*, **23**, 65–99.
Yanagi, T. & Takahashi, S. (1993). Seasonal variation of circulations in the East China Sea and the Yellow Sea. *Journal of Oceanography*, **49**, 503–20.
Zeitzschel, B., Dickman, P. & Uhlmann, L. (1978). A new multisample sediment trap. *Marine Biology*, **45**, 285–8.
Zhang, J. (1996). Nutrient elements in large Chinese estuaries. *Continental Shelf Research*, **16**, 1023–45.

8 Sediment trap sampling in surface waters

W. D. Gardner

Keywords: sediment trap, vertical fluxes, carbon balance, measurements, particles, ^{234}Thorium, time-series stations.

Introduction

Vertical fluxes have been a central theme in the Joint Global Ocean Flux Study, (JGOFS), which relies heavily on fluxes measured with sediment traps. For the legacy of JGOFS and future flux studies, it is important that we understand what traps are measuring. As outlined in this report and in an earlier meeting on sediment trap technology (US GOFS Report no. 10, 1989), calibration schemes indicate that traps moored at depth in the ocean quantify the vertical flux of particulate matter reasonably accurately. However, concerns have been raised about the accuracy of fluxes measured with drifting sediment traps in the upper ocean (see, for example, Buesseler, 1991; Michaels *et al.*, 1994). The measurement in JGOFS of individual compounds such as CO_2 has been significantly improved with the establishment of, and adherence to, widely agreed protocols and standards thanks to the efforts of individuals such as Andrew Dickson and others (Dickson & Goyet, 1994). The measurement of particle fluxes, however, has no absolute standards because fluxes are the cumulative result of a series of complex hydrodynamic, biological and chemical processes rather than the precise analysis of a single compound. Carbon fixed into particulate form in the ocean surface can settle by gravity, be removed by vertically migrating organisms, be re-mineralised and mixed downward as DOC or across the air–sea interface as CO_2, or advected laterally. JGOFS studies have shown that *c.* 50–80% of the carbon flux occurs as gravitational settling. Therefore, quantification of the vertical flux and collection of samples of settling material for compositional analyses is of great importance in unravelling biogeochemical cycles, so we must improve our understanding of particle dynamics and settling and the tools used for particle collection.

During the First International JGOFS Symposium in Villefranche-sur-Mer in May 1995, a meeting was held to discuss the status of the use of sediment traps in the upper ocean for collecting and quantifying the flux of particles, especially organic carbon. During the 75 min meeting, we could scarcely do more than enumerate the potential problems in the use of traps in the open ocean. A draft of comments made at the meeting was circulated in August 1995 to the participants and to the invitees who were not able to attend and additional comments were solicited. Additional comments were solicited on a greatly expanded version of the report in November 1995 and posted on the World Wide Web at `http://www-ocean.tamu.edu/JGOFS/contents.html`. This final version has a few updates and is in a slightly different format from what appeared on the Web.

There is a wide range of possible trap designs and environmental conditions for their use. Protocols for the use of traps have been outlined twice (US GOFS, 1989; Knap *et al.*, 1996), but have not always been followed because of differing opinions and our inability to devise unequivocally accurate calibration schemes that can be routinely followed for surface waters. Thus, we have not progressed to the point of the CO_2 measurements in establishing protocols that are fixed, widely accepted and adhered to.

The objective of this chapter is to assess the present status of particle trapping in the upper water column and suggest plans on how to (1) estimate the magnitude of errors in trap and trap-related measurements, (2) resolve differences in fluxes measured by different methods in the upper 200 m, (3) establish which ancillary measurements are needed in future trapping experiments, and (4) verify which protocols should be used for traps.

To accomplish this objective, we list and discuss some of the potential causes of biases in using traps in surface waters and estimate the magnitude of their importance. We acknowledge disagreement in this assessment because we have not been able to adequately quantify the errors under all conditions. Although many of the issues in this chapter were discussed in the US GOFS Report no. 10 (1989), half of the references in this chapter were published after the GOFS report and they demonstrate substantial progress. This chapter is an update of the major issues with a narrow focus on using and calibrating traps in the upper 200 m of the water column.

Carbon imbalance and balance

As a unifying theme, the discussion centres around the potential impact of each parameter on the carbon imbalance reported by Michaels *et al.* (1994) at the Bermuda Atlantic Time-Series station (BATS), and the carbon balance at the Hawaii time-series station (HOT).

BATS carbon imbalance

At BATS a one-dimensional (vertical) mass balance was constructed for carbon during the April–December period when sediment traps were deeper than the mixed layer depth. The time course of total carbon standing stock (DIC + DOC + POC) was compared with the balance of all vertical fluxes (measured or estimated). The carbon changes in the water were three times greater than the balance of the fluxes. If the discrepancy were entirely due to under-trapping, the traps would have had to collect six times as much material. Alternatively, advection or vertical migration of zooplankton could account for some of the difference. Re-calculation and some new assumptions have reduced the BATS imbalance to a factor of 2.4 (Steinberg *et al.*, 1997).

HOT carbon balance

At HOT, two different 1-D models predict a carbon export from the upper ocean of approximately the same rate as the measured carbon flux in traps (without accounting for sample dissolution) (Emerson *et al.*, 1995; Quay & Anderson, 1996). Not included in this mass balance was an estimate of the carbon flux due to vertical migration of swimmers or the fluxes of DOC. Thus modifications of our sense of trap accuracy to bring the trap estimate more towards addressing the imbalance at BATS might create a disagreement between carbon fluxes and model estimates at HOT.

In constructing a carbon balance, we must remember that carbon can be removed from surface waters by several processes; gravitational settling of particles, vertical migration of zooplankton, vertical mixing of DOC, DIC and POC, advective transport that creates a horizontal gradient, and gas exchange of CO_2 with the atmosphere. Traps are intended to collect only the settling particles. Sources of errors in trap measurements include; swimmers, solubilisation of carbon within the trap, and hydrodynamic effects that include trap geometry, flow, wave-induced trap motion, tilt, and the effects of brine inside the trap.

The central question is whether differences in carbon budgets arise because we do not understand the behaviour of the instruments we are using (traps and *in situ* filtration systems) or because we do not understand the dynamics of the system (particle fluxes and dynamics in surface waters, radionuclide distributions and interactions with a wide range of particle types, spatial scales, etc.) or both.

For each issue presented there is a general discussion composed of comments made at the meeting or in response to the draft report, notes on the magnitude of the problem, effect on the balance or imbalance at the HOT and BATS sites and recommendations of what to do in the future. In addition to normal citations, names in parentheses indicate the person responsible for the comment.

Possible trap biases

Swimmers

Comments

Live organisms may swim or be swept into a trap and eat material collected, or die in the trap and contribute carbon to the sample. Protocols for picking swimmers have been established (Knap *et al.*, 1996), but keep in mind that some swimmers may be part of the vertical flux (Silver *et al.*, 1984). It would be preferable to have a swimmer-avoidance trap. Coale (1990) and Hansell & Newton (1994) have designed traps to separate active swimmers from passively settling material, but neither design is widely used. Peterson *et al.* (1993) designed a novel cylinder–cone trap with an indented rotating sphere (IRS) separating upper and lower trap segments. The device is designed to eliminate swimmers from lower regions of the trap and to isolate the sample. On average, the flux of material of diameter greater than 850 μm was reduced by 88% relative to identical control traps without the sphere. At the same time, the flux of material smaller than 850 μm decreased to 59–97% (mean 84%) of the flux in identically-shaped control traps without spheres. The C:N ratio of material in the IRS traps was about 10, whereas it was about 8 in the control traps, suggesting some differentiation in the type of material passing around the sphere. During the JGOFS EqPac programme the IRS traps collected much less material than the cylindrical traps of Murray *et al.* (1996) when they were surface tethered, but the moored IRS trap fluxes were more similar to the fluxes measured with other moored traps (Cindy Lee, personal communication).

In the Mediterranean, pteropods are regularly and abundantly found in the traps (Miquel *et al.*, 1994). In principle, shells with the animal inside do not belong to the trap flux whereas empty shells do contribute to the passive vertical flux (see Harbison & Gilmer, 1986).

Magnitude of the problem

The swimmer problem decreases with depth (Lee *et al.*, 1988; US GOFS, 1989). Even after manual removal of zooplankton under a strong dissecting microscope, much of the remaining carbon in very shallow traps is still swimmer or swimmer-derived particles (Michaels *et al.*, 1990). However, the problem decreases rapidly below 200 m. Karl & Knauer (1989) attempted to circumvent the swimmer problem by using combinations of screened and unscreened traps to calculate a swimmer-free flux. A preliminary comparison of picking methods between HOT and BATS found a 50–100% difference in total carbon. The screening technique used at HOT was employed on replicates of the BATS traps and the resulting carbon estimate with the HOT technique was higher than that with the traditional BATS technique.

Effect on BATS/HOT carbon imbalance/balance

Swimmer errors in poisoned traps usually bias traps towards higher fluxes. If more swimmers were picked, trap fluxes would be even lower and the BATS imbalance larger (Tony Michaels). However, removing swimmers also removes attached non-swimmer mass and so tends to under-estimate true fluxes. There is a crossover point somewhere: no swimmer removal over-estimates flux, complete swimmer removal and the attached particles under-estimates flux. The 'null' point will vary with size spectrum and, perhaps, species present (D. M. Karl). If HOT traps were processed like BATS traps, the flux estimate would probably be somewhat lower, creating a difference with the modelled fluxes (Tony Michaels).

If traps contain poison, it is not likely that swimmers will eat and leave (Gardner *et al.*, 1983). However, Lee *et al.* (1987) found a 43% loss of carbon in unpoisoned traps at 3 m in a shallow lake and attributed the loss to zooplankton feeding in the traps. The loss was only 1–3% at 8–10 m depth.

Recommendations

It is recommended that all investigators report how swimmers were removed and quantify their abundance. Inter-comparisons of swimmer removal techniques should be done.

Solubilisation of particulate matter in the trap

Comments

Organic carbon is lost with time (Lee & Cronin, 1982; Gardner *et al.*, 1983; Knauer *et al.*, 1984; Lee *et al.*, 1992), but decay components can be retained in the brine. Knauer *et al.* (1984) argued that the quantification of these components (e.g. phosphate) could be used to estimate carbon loss. Dennis Hansell pointed out, however, that this is true only if the decay products are unique to the sinking particles. If these decay products are also found in herniating swimmers, then the source cannot be identified uniquely. One would also have to know the percentage of DOC release per unit POC from swimmers and sinking particles.

Peterson & Dam (1990) demonstrated that the addition of brine to a trap will cause zooplankton to herniate. Hansell & Newton (1994) found a ten-fold difference in DOM accumulation between brined and brine-free trap solutions, which they attributed to herniation of swimmers. In deployments of a swimmer-segregating trap (1.7 d), the quantity of DOC released caused only a 7% decrease in the total POC flux compared with standard PITS traps (Hansell & Newton, 1994).

C. Lee *et al.* (unpublished data) measured DOC in their EqPac IRS traps that have reduced swimmer content. It appeared to make up 10–20% of the C flux. This may still be a problem, but probably more for those interested in specific organic compounds.

Measurements of DOC and DIC in sediment traps obviously have not found a wide acceptance and only very few examples of DOC measurements have been published (Hansell & Newton, 1994). DIC measurements in sediment trap samples have not been carried out or published at all to our knowledge. Wolfgang Koeve pointed out that, besides the additional work burden, theoretical considerations restrict the completeness of any DOC + DIC correction of POC flux measurements. Degradation of POC to DIC in sediment traps would increase the p_{CO_2}. This increase would give rise to loss of DIC from the sample before the DIC analysis. Excess DIC (above ambient values) would thus be a minimum estimate of carbon lost from the POC sample to the water of the sediment trap cups. The problem of the origin of this excess DIC (sinking particles vs. swimmers) will be similar to that of excess DOC discussed earlier. For routine measurements of DOC, additional problems arise for those samples poisoned with formaldehyde (the poison most recommended from the JGOFS protocols) (Knap *et al.*, 1996). The DOC introduced by formaldehyde appears to be a very large background signal that will not allow detection of excess DOC at a reliable level.

The 'Kiel Particle Flux Group' suggested furthermore that not only should the measurement of DOC and DIC in sediment trap cups be optimised, but one should also be aware of possible dissolution of other components. Excess phosphate has been observed frequently (Bodungen *et al.*, 1991) and should be monitored like other constituents of interest to the respective programme (e.g. trace metals, amino acids, fatty acids). If we believe that excess DIC is a significant source of error for our POC flux estimates, dissolved excess Ca also should be monitored to control our estimates of PIC fluxes. It will depend on the precautions carried out to control or even increase the buffer capacity of the seawater in the sediment trap cups whether or not dissolution of POC to DIC and the subsequent increase of the p_{CO_2} in the sample gives rise to the dissolution of PIC for a given sample. Lee *et al.* (1992) examined the effects of alternative poisons to buffered formaldehyde, e.g. $HgCl_2$, sodium azide, etc. Furthermore, without monitoring excess Ca in the sediment trap cups one would not be able to decide whether any excess DIC should be added to the POC or PIC flux estimate (Wolfgang Koeve).

Obviously, reliable detection of excess DOC and DIC in sediment trap samples relies on: (1) improving the swimmer avoidance; (2) using poisons other than formaldehyde; (3) controlling the whole carbonate system in the trap samples; (4) measuring and understanding the artificial losses of dissolved tracers from sediment trap cups; and (5) modelling the loss of DOC, DIC and other related dissolved components from sediment trap cups (Wolfgang Koeve).

Experimental laboratory and field tests carried out showed that losses of dissolved compounds are small for the 'Kiel Sediment Traps'. Investigation of losses of the supernatant sodium azide concentration, used as poison during these experiments, was carried out in a series of 11 sediment traps (up to 20 samples each) from moorings deployed over recent years in the northeast Atlantic. Deployment period was one year. On average, blank bottles, which were not exposed to the open funnel, showed losses of 7.5% of the initial value of the poison. Higher loss values (up to 20%) occurred in the sample bottles, which were exposed to the funnel for 8–28 d. These higher losses can be explained by reaction of the poison with particles, diffusion, swimmer activity, and probably turbulent mixing events (Lundgreen *et al.*, 1997). This source of uncertainty needs to be evaluated for other sediment trap designs. Because it will be highly dependent on the hydrodynamic environment of the sediment trap for a given experiment or field study, regular measurements are recommended. Tracers other than the poison sodium azide need to be discussed. Kremling *et al.* (1996) recently used ^{22}Na as a tracer in another study.

The VERTEX group found that dense NaCl brines cause $CaCO_3$ to dissolve (Honjo *et al.*, 1992), so they switched to a NaCl, $MgCl_2$, $CaCl_2$, KCl mix in making their brine. Brines also 'purge' interstitial fluids of particles that could contain substantial amounts of nutrients (Karl *et al.*, 1984) and biogenic gas (Karl & Tilbrook, 1994).

If C, N, P, etc., are lost owing to herniation of swimmers, maybe we should re-define the whole biogeochemical cycle in terms of Si, which has a much smaller magnitude problem with swimmers. If the carbon imbalance problem finally falls on swimmers, perhaps this is a viable solution, or at least a test of the hypothesis (Dave Karl).

Magnitude of the problem

The decay rate is a function of the length of deployment and what poisons and/or preservatives are used. In some earlier VERTEX studies, carbon fluxes were increased by a factor of 2.23 to 'correct' for the presumed dissolution in traps (Knauer *et al.*, 1990). This is no longer done (Tony Michaels). Lorenzen *et al.* (1981) reported a 7% loss per day for 5 d using sediment trap material. Iturriaga (1979) reported an 8% loss per day for zooplankton and a 3% loss per day for phytoplankton in water at 15 °C. In the upper water column, carbon losses were about 4% per day in 1.7 d (Hansell & Newton, 1994). Gardner *et al.* (1983) measured daily losses of 0.1–1% per day extended over 106 d in deep traps, and showed that if carbon losses were much greater than 1% per day for long-term, multiple-cup trap experiments, seasonal cycles would not be discernible.

Effect on BATS/HOT carbon imbalance/balance

The herniating of swimmers will increase carbon in the supernatant even if they are picked out. Hansell & Newton (1994) found only a 7% difference between total organic carbon and particulate organic carbon after a 1.7 d deployment in Monterey Bay. This is not large enough to solve the carbon imbalance (Tony Michaels).

Recommendations

Several measurements indicate that the carbon loss by solubilisation is at most a few per cent per day in unpoisoned traps. Unless swimmers are prevented from entering the trap, most of any excess DOC signal in trap supernatant water is derived from herniating swimmers. If trap samples are well picked for swimmers, the apparent magnitude of the solubilisation problem appears to be smaller than previously assumed.

Hydrodynamic biases

Comments

One of the major approaches in evaluating the efficiency of sediment traps has been to calibrate small models of traps in flumes or tanks where the conditions of sedimentation can be controlled and measured (Hargrave & Burns, 1979; Gardner, 1980*a*; Butman, 1986). The experimentally confirmed assumption has been that, in the absence of any current, cylindrical traps accurately intercept the material settling out of the water above the trap. In the presence of velocities up to about 10 cm s^{-1}, cylindrical traps still collect particles at the rate predicted ± 30–50%. The predicted rate is determined from the loss of particles above the trap (Hargrave & Burns, 1979; Gardner, 1980*a*) or the measured settling velocity of the particles used in the calibration (Butman, 1986). It is difficult to test traps in flumes at velocities much higher than about 10 cm s^{-1} because settled material is re-suspended from the flume bottom and has a second chance to enter the trap. In flume experiments conducted by Gust *et al.* (1996), VERTEX-style cylinders showed significant increases in flux between velocities of 5 and 10 cm s^{-1}. However, Gardner *et al.* (1997) argued that the methods (and therefore the results) of Gust *et al.* (1996) were flawed because they injected particles into the trap through a tube rather than allowing particles to be intercepted by the trap and collected naturally.

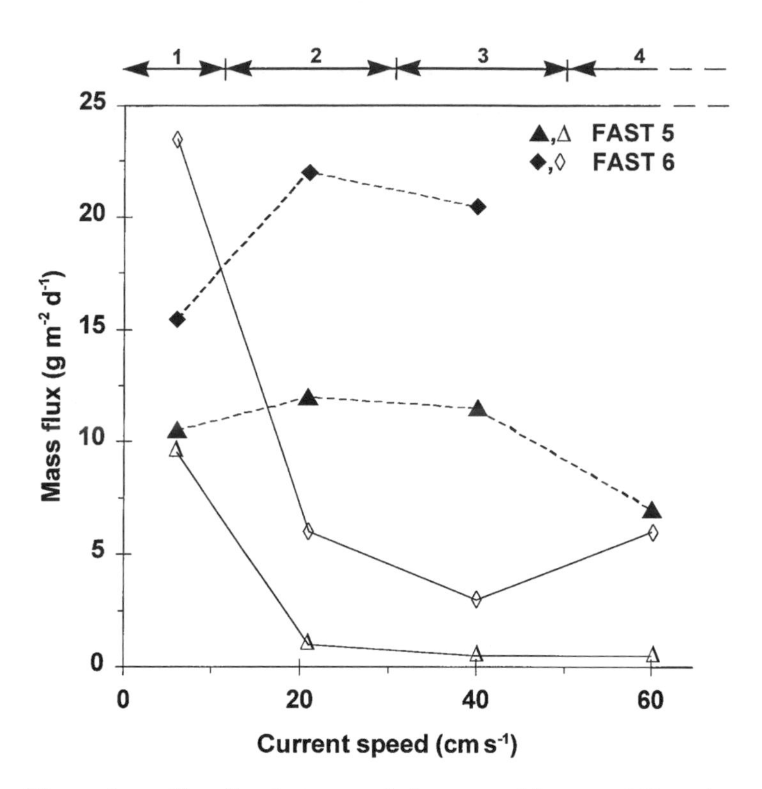

Figure 8.1 *Mass flux during two deployments of the moored Flow Actuated Sediment Trap (FAST) decreased sharply (open symbols) when flow increased from less than 12 cm s^{-1} to 12–30 cm s^{-1}, whereas the mean flux collected simultaneously by a drifting trap in the same area was fairly uniform (closed symbols). (Data from Baker* et al., *1988.)*

Moving to the field, one continues to make the reasonable assumption that cylindrical traps still accurately intercept the material settling out of the water above the trap if there is no movement past the trap, although independent verification of this assumption would certainly be desirable (see discussion below on independent measures of vertical flux). The effects of large-scale turbulence, internal waves, tilt and mooring-line motions that are not present in the flume have unknown effects on the comparison of flume and field data. Comparisons have been made between cylindrical traps and traps of other designs, such as funnels, that are deployed simultaneously (Honjo *et al.*, 1992). Comparisons of fluxes measured with drifting and moored traps of the same design at the same location can also be used to determine the efficiency of traps under different flow conditions. Baker *et al.* (1988) have done precisely that experiment.

Baker *et al.* (1988) showed that relative trap efficiency in moored, cylindrical traps with a steep inner funnel in the field dropped to 5–25% efficiency somewhere in the velocity range of 12–30 cm s^{-1} compared with the same design of trap when it was drifting with the water in the same place over the same time (Fig. 8.1). The corresponding trap Reynolds numbers (R_t = (velocity × trap diameter)/viscosity) for the 20 cm diameter Baker trap were 24 000–60 000. The R_t or range of R_t over which the flux decreases is unknown. The composition of particles was also very different at higher Reynolds numbers.

Field experiments by Gardner *et al.* (1997) showed no decrease in trapping efficiency up to a Reynolds number of 43 000 in cylindrical traps with an interior funnel moored at 4500 m (Fig. 8.2). If Reynolds number were the controlling parameter in trap efficiency, this suggests there should be no decrease in relative trapping efficiency for the 7.6 cm diameter BATS traps up to a velocity of 32–57 cm s^{-1}, which is faster than any of the velocities reported for those traps in the BATS area by Gust *et al.* (1992, 1994). However, based on personal observations of flow within model traps at just 20 cm s^{-1}, it is hard to believe that the collection efficiency of traps is not affected at velocities above 20 cm s^{-1} (W. D. Gardner). We must remind ourselves that, although we can scale the dynamics of flow between models and full-scale traps using dimensional analysis, we may not be able to predict the dynamics of natural marine particles within that turbulent flow. Furthermore, in the dynamic region of the upper ocean one must consider whether other factors such as large-scale flow, tilt and mooring dynamics are more important than Reynolds number.

Does the horizontal flux of particles past sediment traps influence the measurement of vertical flux within traps? Measurements by Gardner *et al.* (1997) of the horizontal flux past moored traps varied by a factor of 56, yet the quantity collected by the traps differed by only a factor of 2.7. Discounting two traps in the benthic boundary layer reduced the variation in apparent vertical flux to a factor of 1.4, and within the 95% confidence interval there is no significant correlation between the vertical and horizontal flux (Fig. 8.2).

Neutrally buoyant traps have been recommended (US GOFS, 1989) and tested (S. Honjo, personal communication, 1980; Diercks & Asper, 1994) but have not been widely used owing to engineering difficulties. Neutrally buoyant traps are not practicable for all environments, but they are well suited to near-surface deployments in the open ocean. Their use could significantly reduce questions about hydrodynamic effects on trapping efficiency. One must also have information about how tightly coupled a neutrally buoyant trap is to the surrounding water. Valdez & Price (1999) have constructed a neutrally buoyant sediment trap based on the RAFOS drifting float design and successfully tested it at BATS. During the periods of low flow and low flux that were tested, there was good agreement between fluxes measured with surface tethered and neutrally buoyant traps, but there were compositional differences

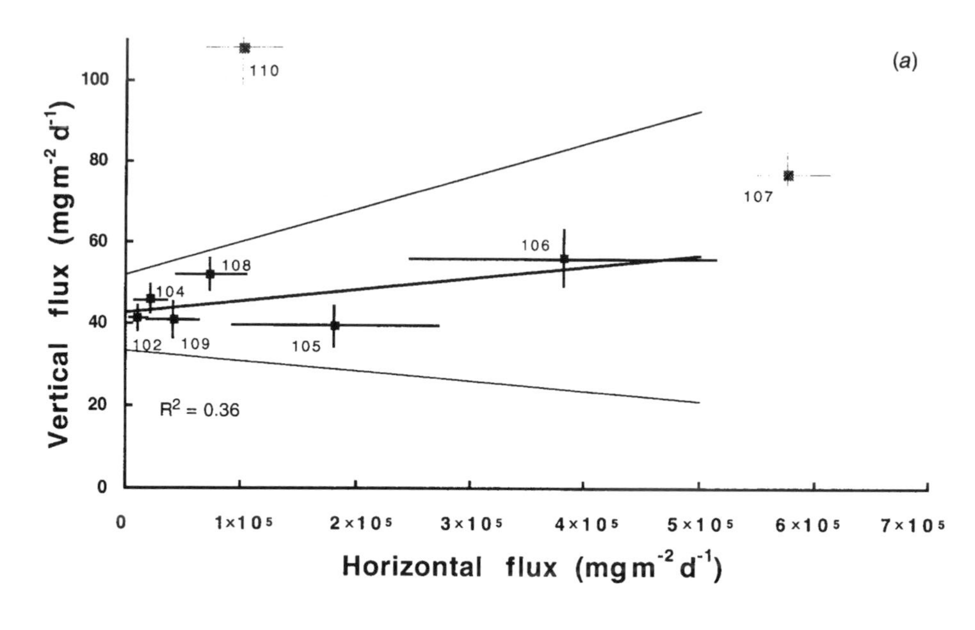

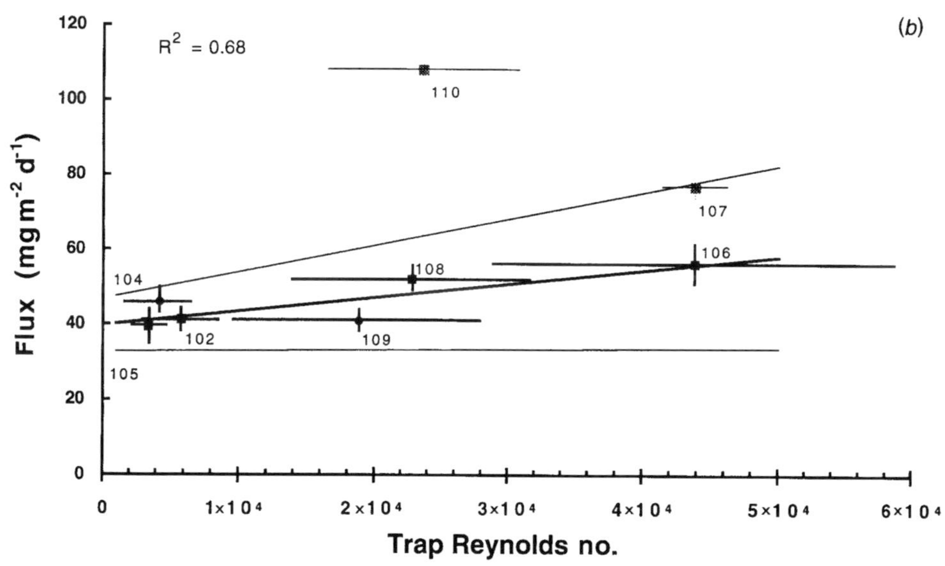

Figure 8.2 (a) *Total vertical flux measured by each cylindrical sediment trap versus the horizontal flux past the trap. Numbers are trap identifications. Traps 107 and 110 were in the benthic boundary layer (BBL) and are excluded from the statistics. Error bars are one standard deviation for the vertical and horizontal flux. The heavy line is the least-squares regression of the non-BBL traps. The upper and lower boundaries of the 95% confidence interval of the regression are given by the thin lines and show that statistically, the vertical flux could either increase or decrease with increasing horizontal flux.* (b) *Total vertical flux collected by each trap versus trap Reynolds number (R_t). Error bars are one standard deviation for the vertical flux and R_t.*

in the material colleced (Buesseler *et al.*, 1999). The neutrally buoyant traps contained significantly less swimmer mass. More tests are needed at periods when fluxes and lateral flow are larger. All traps should start with the same minimal amount of brine at the bottom of the trap for a controlled test. A neutrally buoyant trap should minimise the hydrodynamic questions, but leaves the solubilisation and migration transport questions to be quantified simultaneously.

For further discussion of trap dynamics see Hargrave & Burns (1979), Gardner (1980*a,b*, 1985), Bloesch & Burns (1980), Blomqvist & Kofoed (1981), Butman (1986), Butman *et al.* (1986), Hawley (1988), Gardner *et al.* (1997) and Gardner & Zhang (1997).

Gust *et al.* (1992) show a two-fold increase in flux with a doubling of velocity (in the 10–30 cm s^{-1} range) past large funnel traps during one-day deployments at the BATS site. It is possible that this is more than a velocity effect. Tilted cylinder traps collect more than upright cylinder traps: 25% at 5°, up to 200–250% at 30° (Gardner, 1985). Tilts measured by Gust *et al.* (1992) were less than 5°, but tilt effects on the efficiency of funnel-shaped traps have not been tested.

A. F. Michaels *et al.* (unpublished data) have measured velocities at the 150 m trap for nearly 4 years of the BATS programme and, for many of those samples, have looked in detail at the particle composition of the traps. When all the data are considered together, there is no significant trend with velocity (Fig. 8.3). However, if stations of high (over 350 mg C m^{-2} d^{-1}) and low (under 350 mg C m^{-2} d^{-1}) productivity are considered separately, there is an apparent pattern in collection of carbon with velocity. The collection differences are 2–3-fold higher with increased velocity over the range of approach velocities from 4 to 14 cm s^{-1}. In examining the major components of the carbon flux, there is no velocity pattern with the dominant particle type, marine snow (assuming this can be adequately identified in the trap). There is a strong velocity dependence for faecal pellets (10-fold differences in collection over the velocity range). Thus, the hydrodynamic effects on aggregates may be very different from those on particles that are more solid. Alternatively, faecal pellets in traps could be due to swimmers; swimmer collection might well be a function of approach velocity (and hence the amount of water that passes through the trap mouth).

One area that has received little or no attention is the behaviour and integrity of aggregates in the eddies and flow generated in and around traps. It is also necessary to study how aggregates cross a dense brine interface inside traps. One difficulty in this regard is in being able to identify aggregates once they are collected in a trap. Jannasch *et al.* (1980) attempted to preserve aggregates by adding a polyacrylamide gel in the sample collection area, so that aggregates

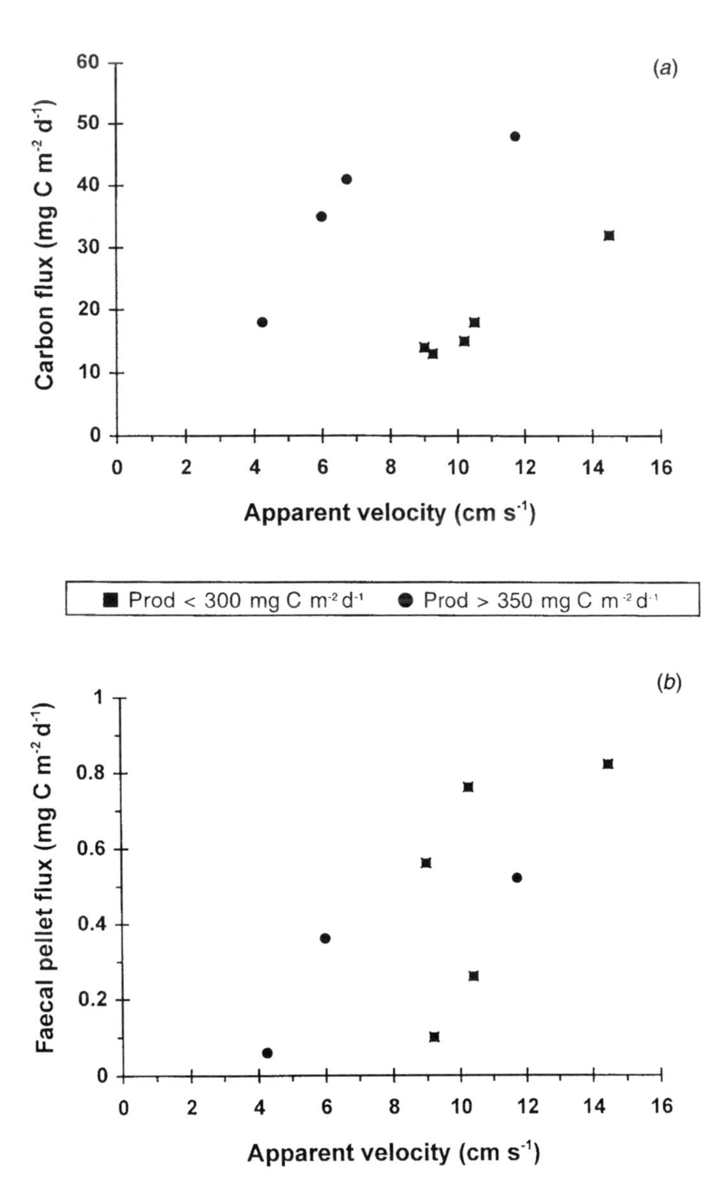

Figure 8.3 *Carbon flux* (a) *and faecal pellet flux* (b) *in traps at BATS versus apparent velocity at the trap mouth (based on distance/time travelled between deployment and recovery of traps). (Unpublished data from Tony Michaels.)*

could be thin-sectioned and studied. However, the viscosity of the gel was so large that aggregates rolled up like dust balls before they sank into the gel (H. W. Jannasch, personal communication, and observation by W. D. Gardner in 1979).

The VERTEX surface-tethered trap array has a stretchable member to dampen out waves. It is important to be sure that there is enough sub-surface flotation to prevent the stretchable member from ever becoming fully extended or the surface-wave energy and motion will be transmitted to the traps and may affect their efficiency (Vernon Asper, personal communication). Gust *et al.* (1994) measured significant vertical motion and wave energy on floating arrays of the VERTEX design that had a stretchable member, but it was not reported whether additional flotation was added to compensate for the weight of additional instrumentation on the array.

Vertical motion can be minimised by using a spar buoy as the first or only surface float for drifting traps. An even more effective solution is the VERTEX method of using a string of small floats on the surface, and it is easier to handle. The intent of a spar buoy is to minimise water displacement per unit of vertical displacement or to minimise the waterline area. The string of small floats has a small waterline area and functions like a 'flexible spar buoy' (Vernon Asper). The installation of a horizontal drag plate below the trap or at the bottom of the array can act as an effective sea anchor to reduce upward motion of floating traps (Gardner *et al.*, 1985). Most of the drag on a trap array is from the mooring line itself. This can be minimised by using very thin wire (Tony Michaels).

To minimise flow past the trap, position the floats below the Ekman layer to maximise the drag in the vicinity of the traps (Susanne Neuer). Traps at multiple depths make it difficult to minimise the flow past all traps. Deploying one trap per array would improve this dilemma.

Magnitude of the problem

Based on data from Baker *et al.* (1988), velocity effects could decrease the flux by a factor of 4–5, but potentially only at very high velocities (or Reynolds numbers, R_t). Based on unpublished field data there is potentially a 2–3-fold bias at lower velocities (Tony Michaels).

Effect on BATS/HOT carbon imbalance/balance

This could account for the entire imbalance of carbon at BATS, but there is still significant uncertainty about flow effects when floating traps are deployed. If the Reynolds number arguments are relevant in surface waters, then flow effects should be small except at high velocities. The data of A. F. Michaels *et al.* (unpublished) show a difference of a factor of 2–3 that may be attributed to velocity effects at lower velocities.

Approach velocities are likely to be in the same approximate range at HOT as at BATS based on similar drift patterns, array configurations and horizontal velocities. Thus, any *ad hoc* explanation for the BATS imbalance that involves a flow-based explanation should have the opposite impact on the comparison of the HOT data with the models at that site.

Recommendations

Develop a neutrally buoyant drifting sediment trap (US GOFS, 1989; Diercks & Asper, 1994).

More experiments of the type performed by Baker *et al.* (1988) should be made. The velocity bins need to be more narrowly limited, especially between 12 and 30 cm s^{-1} (R_t = 24 000–60 000) because that is where the trapping efficiency decreased markedly in their study. Reynolds numbers should be used when planning experiments and interpreting their results. One should also test funnel traps in this mode: they are one of the most widely used designs because they offer the advantage of large collection area and sample concentration.

Field experiments should be done to compare arrays drogued to have different approach velocities, to see whether there are collection differences in the configuration in which traps are actually used in the field. Although this will only establish a relative accuracy pattern, it will allow a determination of flow impacts in the regime in which these experiments are conducted.

To minimise the flow past the traps, deploy only a single trap per array. Use a buoy or several small floats to remove short-period waves from the array. Use a stretchable segment of mooring line above the trap and be sure it has a rapid response and is never fully extended. Put the maximum drag near the trap depth and use thin lines to minimise the drag at other portions of the array. Measure the velocity past traps on all deployments.

Effect of adding brine to traps

Comments

Brines with densities greater than that of seawater are often added to traps to isolate samples collected, prevent re-suspension of settled particles, retain poisons and preservatives, and retain dissolved or leached components of the sample. At the BATS and HOT time-series stations, traps are filled to the top with a 50 g l^{-1} excess brine before deployment. However, one of the important variables controlling trap efficiency is the aspect ratio (Hargrave & Burns, 1979; Gardner, 1980*a*,*b*; Bloesch & Burns, 1980; Blomqvist & Kofoed, 1981; Butman *et al.*, 1986). The addition of a brine introduces a hydrodynamic barrier that acts as a false bottom. Thus, brine changes the effective aspect ratio of a trap (the height dimension of the effective aspect ratio applies only to the brine-free region) and also excludes particles or aggregates whose density is less than the brine density.

The density difference between seawater and aggregates in the ocean has been determined to be as low as 10^{-4}–10^{-5} density units (Alldredge & Gotschalk, 1988). Macintyre *et al.* (1995) have observed aggregates accumulating at density gradients in the water column and report that it can take from hundreds of

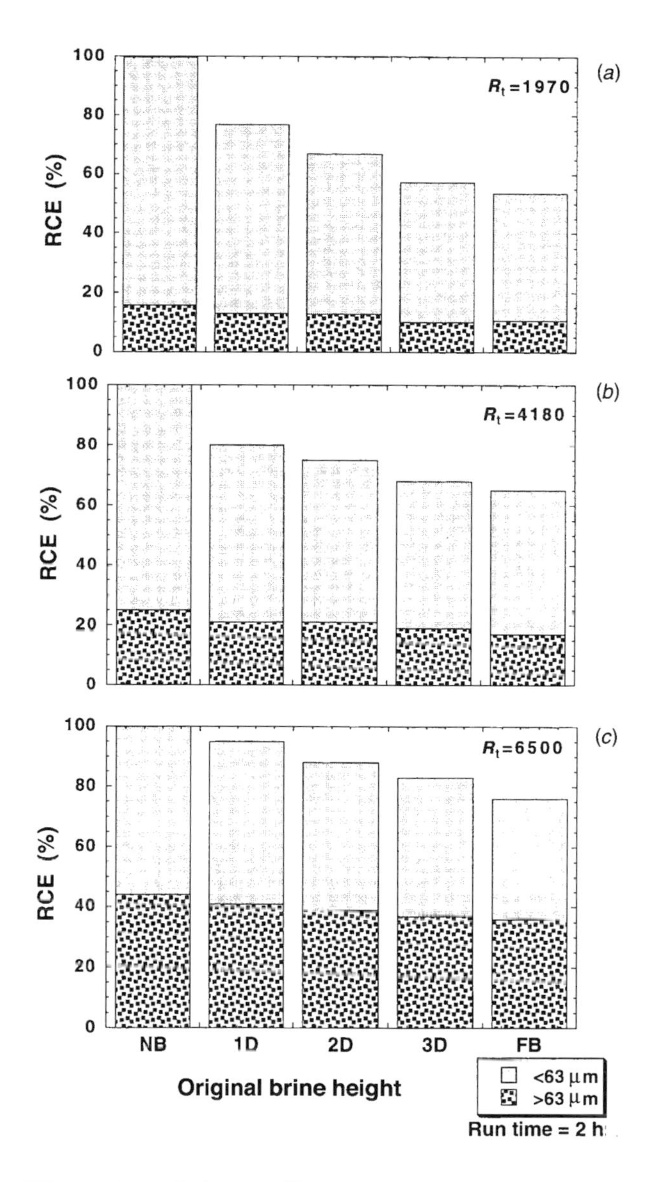

Figure 8.4 *Relative collection efficiency (RCE) of traps during experiments at three different velocities. Collection rates are normalised to the traps with no brine (NB). Samples were wet-sieved at 63 μm.* (a) *At the lowest water velocity (4.8 cm s⁻¹) the decreased trapping efficiency was largest: nearly 50%; in* (b) *water velocity was 10.2 cm s⁻¹.* (c) *At the highest velocity (15.9 cm s⁻¹), filling the trap with brine decreased trapping efficiency by the smallest amount (about 25%).*

seconds up to 3 h for the pore water to exchange in the aggregates. Flume experiments show that the addition of a 5 practical salinity units (psu) brine to traps decreased the collection rate (Gardner & Zhang, 1997). The trap efficiency was 54% at 5 cm s⁻¹ and 75% at 15 cm s⁻¹ (Fig. 8.4). Three BATS field

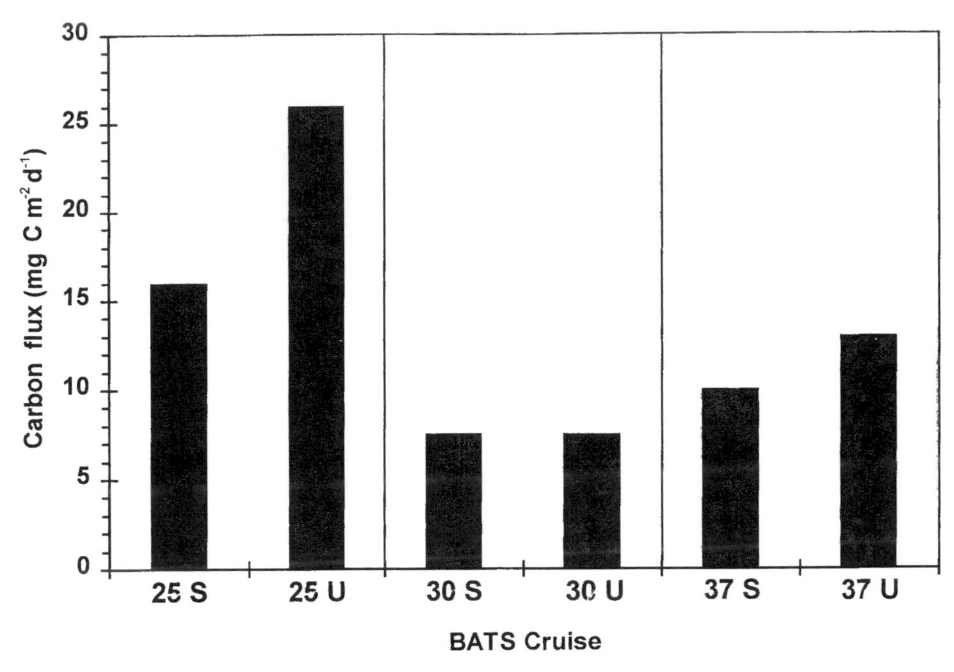

Figure 8.5 *Carbon flux in pairs of floating traps at BATS with (S) and without (U) brine added to the traps before deployment (unpublished data from Tony Michaels).*

experiments showed carbon fluxes 0, 25% and 60% higher in traps without brine (Fig. 8.5), but showed both increases and decreases in the flux of specific components of the flux (Tony Michaels). Scott Nodder (New Zealand; personal communication) tested cylindrical traps (ID = 9 cm, aspect ratio = 10.6) on frames 3 m above the harbour floor for 24 h filled with a 50 psu excess brine. He found that they collected 2–3 times less material than traps that were partly filled with the same brine (equivalent to 1 or 3 trap diameters of brine). This strongly argues that traps should not be filled with brine.

Tony Michaels argues that a change in any of the protocols at the JGOFS time-series station needs substantial justification that the new measurement will lead to an increase in the absolute accuracy, and it would require an extensive period of simultaneous measurements using both the old and new techniques (e.g. simultaneous brine and no-brine experiments overlapping for a year to test for seasonal differences in hydrography and particle types). Thus, operators of the time-series stations are reluctant to modify one facet of the method (brine) before there is an independent method to determine that the new collection techniques actually result in a flux estimate that is accurate on some absolute standard. It is not considered worth the risk to go through the extensive extra effort of switching one facet to find that the new method without brine is still very inaccurate but for a reason very different from the brine effect (e.g. flow, large-scale turbulence or something else that is not adequately considered now).

Two groups have discussed this issue (US GOFS, 1989; Knap *et al.*, 1996) and recommended that traps be deployed with no more than a 5 psu excess brine only in the sample-collection region of a trap (equivalent to one trap diameter in cylinders). Trapping programmes should adopt this and the other JGOFS recommendations.

Magnitude of the problem

Flume experiments with a 5 practical salinity units (psu) brine show up to a factor of 2 loss in the flux measured. Field tests with and without a 50 psu brine in cylindrical traps have shown decreases in flux anywhere from 0% to a factor of 2–3. The increases in carbon flux with floating traps in surface waters without brine have been 0–60% higher than with brine-filled traps.

Effect on BATS/HOT carbon imbalance/balance

A brine effect could increase trap fluxes at BATS but the limited brine experiments at the BATS site indicate that the effect would account for about 10% of the carbon imbalance, or 20% of the projected under-trapping. However, experiments in flumes and field experiments at other locations find brine effects of as much as a factor of 2–3. If these experiments are relevant to BATS, they could account for about half of the carbon imbalance, or up to all of the projected under-trapping. The magnitude of the brine effect decreases with increasing velocity and with increased exposure time. The HOT traps use the same protocol and they report no carbon imbalance in their measurements. Any effect of brine on the carbon budget at one site (BATS) must be applied to other sites (e.g. HOT) assuming hydrodynamic conditions are similar.

Recommendations

Follow the published protocols, which call for a 5 psu brine only in the bottom one diameter equivalent of a cylindrical trap.

Make necessary tests to convert the BATS/HOT trap protocols to brine only in the bottom. These comparisons need to be done in the context of experiments to determine the absolute accuracy of sediment traps if they are going to be useful for causing a change in protocols at a time-series station (Tony Michaels).

System dynamics questions

Vertical flux by zooplankton migration

Comments

The role of particle transport by vertically migrating organisms and respiration has been discussed for many years with little quantification because it is a difficult task (see Angel, 1989). However, traps are not designed to measure the effect of migrant transport. Other methods must be employed to quantify this process. The question of migrant transport does not directly affect the efficiency of traps, but it is an essential component when constructing mass balances in the upper water column for assessing the accuracy of floating traps. One must understand the dynamics of particles and cycling within the upper water column as well as the dynamics of sediment traps.

Migrant transport measured at the BATS site by Longhurst & Harrison (1988) was 8–28% of the carbon flux measured concurrently by floating traps. Recent measurements (Dam *et al.*, 1995) estimated the migrant transport at 18–70% of the trap flux. Walsh *et al.* (1988) noted a recurring deep (> 1000 m) particle flux maximum in MANOP annual sediment trap profiles. They concluded that as much as 50% of the flux measured in traps at 1500–1900 m bypassed or was produced below their shallower traps at 500–1000 m. This is well below the zone of interest in this discussion, but suggests that migrant transport is not isolated to surface waters. One would expect migrant transport to be largest in surface waters, where diel migration is well documented.

Effect on BATS/HOT carbon imbalance/balance

At BATS, the migrant flux of Longhurst & Harrison (1988) was used in the carbon balance (Michaels *et al.*, 1994). A re-calculation of the carbon balance using new migrant flux data decreased the BATS carbon imbalance to a factor of 2.4 (Steinberg *et al.*, 1997). At this site, they noticed that the larger the magnitude of migrant transport, the smaller the carbon imbalance.

Carbon imbalance was not implied by the comparison between the carbon budget and the particle fluxes at HOT, so external transport mechanisms such as migrant transport are not needed to balance carbon budgets. If vertical migrant fluxes occur at HOT, then that system is again out of balance with the carbon budget. Longhurst & Harrison (1988) estimated that migrant fluxes in the oligotrophic Pacific were 0.8–2.5 mg N m^{-2} d^{-1}, comparable to an annual carbon flux of 0.2–0.6 mol C m^{-2} yr^{-1}. This would increase the HOT annual flux by 22–66%. Is migrant transport less important at HOT than BATS, or are the measurements of migrant transport more likely to be on the low end of the range measured by Dam *et al.* (1995)? How does migrant transport vary seasonally?

Recommendation

The role of migrant transport requires further study, but should be included in an assessment of trap fluxes compared with other independent methods of calculating fluxes.

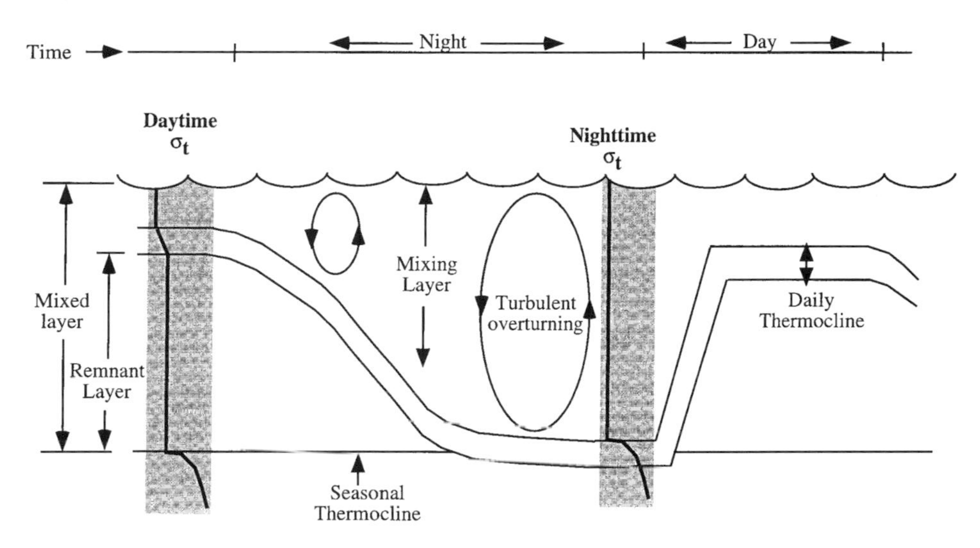

Figure 8.6. *Night-time cooling and wind mixing can deepen the mixing layer; daytime heating can stratify and thin the mixing layer. The shaded areas are density profiles used to define the depth of mixing and the mixed layer. Traps should always be below the mixed layer (modified from Brainerd & Gregg, 1995).*

Mixed-layer (ML) depth or mixed-layer pumping

Comments

Convective overturning increases the residence time of particles in the mixed layer (Kerr & Kuiper, 1997), so fluxes cannot be accurately measured with traps until you are below the mixed layer (Gardner & Richardson, 1992). Nocturnal increases in the mixed layer depth (Fig. 8.6) can move particles downward quickly where they are isolated and allowed to settle in non-turbulent flow when the mixed layer thins during the day (Woods & Onken, 1982; Gardner *et al.*, 1995). This is especially important for aggregates, which can rapidly settle below the depth of mixing by the following night and escape re-incorporation into the mixed layer. Conversely, nutrients, p_{CO_2} or any component whose concentration increases with depth in the zone of mixing will be mixed upward. As long as traps are deployed below the mixed layer, this should not affect the fluxes measured with traps.

Effect on BATS/HOT carbon imbalance/balance

Trap fluxes in BATS were examined only when ML depth was less than trap depth, so there should be no direct influence. At HOT the trap depths were always greater than the mixed-layer depths. Total carbon increases with depth, so ML pumping would increase carbon in the surface, not decrease it.

Recommendations

Traps should be deployed below the maximum mixed-layer depth during the time of deployment.

Spatial inhomogeneity

Comments

Particle fluxes have been measured in many parts of the world, but we have little information about the spatial inhomogeneity of vertical fluxes on a scale of kilometres to tens of kilometres. D. Lal suggested deploying multiple traps to test for homogeneity. Time-series traps average out some of the local inhomogeneity depending on length of deployment and advective rates. However, information about spatial homogeneity does not answer the question of accuracy of measured fluxes.

Michaels *et al.* (1994) analysed the effect of stochastic events on the carbon imbalance problem. Because of the large number of measurements, it is statistically very unlikely that rare, missed events could explain the imbalance at BATS. With the frequency of sampling, there is only a small number of events that could have occurred and not been seen in the BATS sampling. However, these events would have to account for the entire discrepancy and would require an unreasonably large flux (more than the total standing stock of POC every day) to make up for the imbalance. Since the HOT sampling is of similar frequency, the same conclusion can be drawn at that site. The traps at the time-series stations are not grossly inaccurate because they miss rare events.

For a regional study, we may not have enough measurements near BATS to determine whether the error lies in the traps or is a result of large-scale advection. The advective field is not known. Gradients in ^{234}Th and carbon are small. Do we need to know the 3-D flow field (Tony Michaels)? Are 3-D experiments necessary for every place in which traps are deployed? The question of the role of advection is much larger than the trap accuracy issue and hits to the heart of the interpretation of all JGOFS data (Tony Michaels).

As discussed below in the section on independent measures of vertical flux, mass balances based on oxygen production match the trap carbon fluxes at the HOT site, but not at the BATS site. Perhaps this results from differences in the degree of spatial inhomogeneity of the two sites.

Over what time scales does primary production prevail? Note that trap collections from short-term deployments at multiple depths each reflect the surface production history for different periods (Ian Walsh). Traps at multiple depths may also reflect the flux from different source regions (Siegel *et al.*, 1990).

Magnitude of the problem

Unknown.

Effect on BATS/HOT carbon imbalance/balance

Missing rare events probably cannot explain the annual patterns. Advection may play an important role at each station.

Recommendations

Consider deploying multiple trap arrays to test for homogeneity.
Do control-volume experiments as a follow-on to JGOFS.

Independent measures of vertical flux

Comments

What is the accuracy of the trap measurements? Invoking closure of mass balance cannot be just a convenience when testing for trap accuracy (Tony Michaels). Conservation of mass must be maintained. Closure of mass balances provides validity for other types of calibrations, but the time scales of each measurement must be comparable. Closure should be attempted whenever possible; when observations violate this condition, it is a powerful constraint on the interpretation of data.

When we see disagreements between traps and other measurements, other processes are often invoked as an explanation, but if agreement is seen then the same questions are not raised (Ken Buesseler). This is true of any endeavour in science. We always seek closure of balance and when it is achieved, we move on.

Thorium-modelled and -measured fluxes

Comments

Methodology for this technique is described in Moore *et al.* (1981), Coale & Bruland (1985, 1987), Eppley & Peterson (1979), Buesseler (1991) and Buesseler *et al.* (1994). It is based on the fact that ^{238}U is well mixed in the ocean, is closely correlated with salinity concentrations, and does not react with particles, whereas the daughter product, ^{234}Th, rapidly adheres to particles and settles out of surface waters with particles. One measures the ^{234}Th deficiency (relative to ^{238}U) in surface waters, integrating down to the depth of disequilibrium, which is usually between 70 and 100 m but can extend to 200 m (Fig. 8.7). Obviously, this method is restricted to predicting fluxes to this depth: other studies show that the flux of material decreases rapidly with depth below the upper 50–150 m (VERTEX) or at least that the abundance of particles known to contribute to the vertical flux decreases rapidly with depth (Bishop *et*

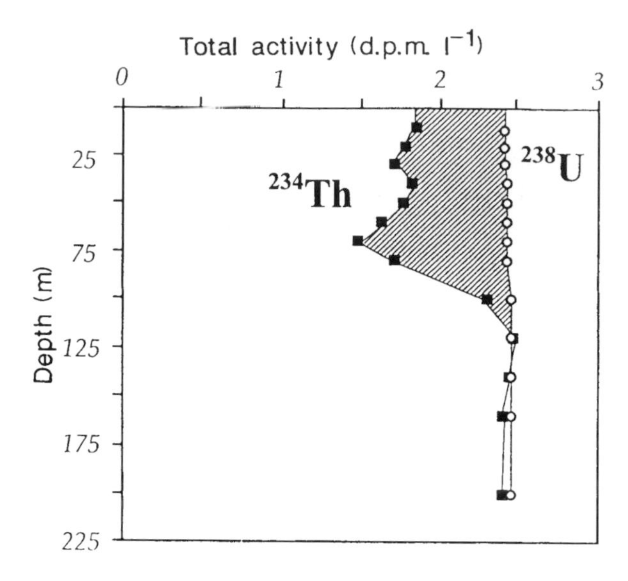

$$F_{Th} = \lambda_{Th} \int_0^z (U - Th)\, dz$$

$$F_C = F_{Th} \left(\frac{C}{Th} \right)_{\text{sinking particles}}$$

Figure 8.7 *Profiles of ^{234}Th and ^{238}U concentrations in surface waters of the Pacific during VERTEX 3 (data from Coale & Bruland, 1987). The ^{234}Th deficiency is integrated to calculate the flux of ^{234}Th from the zone of deficiency. The flux of organic carbon is obtained by multiplying the ^{234}Th flux by the POC: ^{234}Th ratio for settling particles.*

al., 1980). This method may not work along continental margins owing to scavenging and lateral advection and mixing of waters in those regions (Baskaran *et al.*, 1996).

From the ^{234}Th deficiency, and correcting for the half-life of ^{234}Th (24 d), one can predict the ^{234}Th flux (i.e. the loss) over the depth of disequilibrium. Given this and a ^{234}Th measurement in a trap, one can theoretically make an independent estimate of whether the trap is collecting ^{234}Th-bearing particles in a predictable (i.e. accurate) fashion. Fortunately, swimmers have low ^{234}Th concentrations, so their inclusion or exclusion should not affect the measurement (Jim Murray; Buesseler *et al.*, 1994). In an experiment at BATS, Buesseler *et al.* (1994) predicted a ^{234}Th flux from 27 water column profiles of

-30 ± 140 dpm m^{-2} d^{-1}. However, there was a wide range of both positive and negative values from day to day and over short spatial distances assuming steady-state conditions (-300 to $+237$ dpm m^{-2} d^{-1}). Buesseler *et al.* said that the positive values were due either to upwelling of deeper, ^{234}Th-rich water, which seemed unlikely, or to large errors in ^{234}Th activity at the low concentrations that result from low fluxes in an oligotrophic area like BATS. Thus, they assumed that the predicted ^{234}Th flux was essentially zero. Their trap ^{234}Th fluxes from two arrays during that time was significantly higher (290 ± 15 dpm m^{-2} d^{-1}). They explained the difference with a hypothesis that upper ocean traps at BATS may over-collect during low-flux periods and under-collect during high-flux periods based on these data and earlier BATS data. The Buesseler *et al.* (1994) data demonstrate that multiple profiles are needed in areas of low flux to obtain a statistically meaningful result. More data are needed in areas of higher flux.

Murray *et al.* (1996) measured the flux of organic carbon from the central equatorial Pacific during EqPac using floating traps at multiple depths and the ^{234}Th approach described above. In their calculations of ^{234}Th flux, they included terms to account for upwelling and meridional advection away from the equator. Zonal gradients in ^{234}Th were found to be small (Buesseler *et al.*, 1995) and were neglected. Comparison of the model ^{234}Th fluxes with the corrected ^{234}Th fluxes shows that the model fluxes shallower than 150 m were much less than the trap fluxes. The fluxes at 150 and 200 m agreed to within a factor of ± 2 from 12°N to 12°S. Murray *et al.* (1996) argue that all drifting sediment trap studies should be conducted as a function of depth and include ^{234}Th analyses. The recommendation to have only one trap per depth would require a large number of arrays deployed simultaneously. Error bars also need to be included in reported data, as in Buesseler *et al.* (1994).

One question about the method concerns the time scales (in terms of minimum and maximum fractions or multiples of half-lives) that are appropriate on which to make ^{234}Th or other radionuclide measurements. Ken Buesseler responded that if you have only a single radionuclide profile, you need to integrate particle fluxes over a comparable time scale appropriate to the tracer activity. If ^{234}Th activities are decreasing with time (i.e. particle fluxes are increasing), a single ^{234}Th profile would actually underestimate the net particle removal at the time you took your sample. If you take a non-steady-state approach (Buesseler *et al.*, 1994) and you measure ^{234}Th in a time-series (or better yet, 4-D) manner, then you can predict the ^{234}Th export flux within the measurement period. That is, a period of 2–5 d period is acceptable, as long as you have data to examine whether the ^{234}Th activity is varying with time during this same period.

The flux of ^{234}Th will be related to carbon fluxes later; the argument here is that if the trap is not collecting particles bearing ^{234}Th in a predictable way, then there is no confidence that they are collecting other particle types in an accurate fashion. However, surface area per unit mass is greatest on the smallest particles, and the small particles are not dominant in the vertical flux of mass. An equally important question is how the adsorption of ^{234}Th differs between particles of different composition.

Magnitude of the problem

Buesseler (1991) noted that fluxes predicted based on ^{234}Th-deficiency and trap-measured ^{234}Th fluxes often disagree by more than a factor of three, with some experiments showing over-trapping and others showing under-trapping by that amount. Time-series measurements of ^{234}Th deficits were made at BATS during trap deployments (Michaels *et al.*, 1994) and can account for much of the carbon imbalance based on the ^{234}Th calibration of the traps and the measured range of C: ^{234}Th ratios. Over the course of two years of simultaneous ^{234}Th profiles and ^{234}Th collections in traps, both under-trapping and over-trapping have been observed at the BATS site. Their explanation of the variable over- or under-trapping was that their traps over-collected during times of low flux and under-collected during times of high flux at BATS, thus dampening out the seasonal signal of the total flux cycle. Although this may occur at BATS, Rivkin *et al.* (1997) argued that this is not a global trend.

Effect on BATS/HOT carbon imbalance/balance

The difference between the predicted and measured ^{234}Th flux is of the same order as the carbon imbalance at BATS (Michaels *et al.*, 1994). If we appeal to advection to account for this ^{234}Th difference, the magnitude of advection would have to be large based on known gradients of ^{234}Th concentrations, and there would have to be a seasonal trend to account for the flux imbalance. One might ask why there should be a seasonal trend in ^{234}Th concentrations in the western North Atlantic. Vernon Asper noted that if the ^{234}Th deficiency is caused by scavenging and if scavenging is related to the production and flux of particles, how could there not be a seasonal trend in ^{234}Th concentrations?

^{234}Th measurements had not been made around HOT until recently (Dunne & Murray, 1997) because the scavenging of ^{234}Th is generally low in oligotrophic regions, making the ^{234}Th disequilibrium very small. When scavenging is low, this means that the errors can be large for this calculation (Ken Bruland). Dunne & Murray (1997) found that recycling of POC in the euphotic zone of the equatorial Pacific was 2–10 times faster than at sites previously studied. Their results also suggested that, assuming all particles re-mineralise at the same rate, ^{234}Th recycled 3–4 times between dissolved and particulate phases before being removed from the euphotic zone on particles. ^{234}Th deficiency profiles published by Coale & Bruland (1987) in oligotrophic waters show relationships with mixed layers and zones of particle production that suggest that ^{234}Th recycling and removal from surface waters is a complex process.

There are times when there is no ^{234}Th deficiency at BATS, which would imply no particle flux yet the traps collect particles (Buesseler *et al.*, 1994), suggesting we do not fully understand the system or the tools used.

Recommendation

See recommendations in following section.

^{234}Th-derived estimates of particulate organic C flux

Comments

The ^{234}Th deficiency method is used to calculate a carbon flux by multiplying the ^{234}Th flux by the C: ^{234}Th ratio in sinking particles as determined from trap samples or from *in situ* pump samples. One then compares the carbon flux measured in the trap with the carbon flux predicted based on the preceding calculation. This approach has its own uncertainties, including the time variability and transport terms (Buesseler *et al.*, 1992, 1994; Wei & Murray, 1992) and the C: ^{234}Th ratio in sinking particles (Michaels *et al.*, 1994; Buesseler *et al.*, 1995).

To estimate the accuracy of traps at collecting carbon by comparison with the ^{234}Th-derived calibration, however, requires the assumption that the ^{234}Th is distributed similarly to organic C among the different types of sinking particles, based on composition and settling velocity distribution, that are responsible for the vertical flux of carbon in the ocean. ^{234}Th adsorption is a function of surface area, and there is much greater surface area per unit mass for small particles that may not be sinking rapidly. How quickly do these small particles become incorporated into larger particles and sink? C: ^{234}Th ratios vary by particle type and size, but by how much? The biggest concern is the assumption that the particles collected with *in situ* pumps are representative of the local pool of sinking particles. Existing data demonstrate that the C: ^{234}Th ratio differs between trap and large-volume filtration samples; this result is not surprising since filtration extracts all particles whether they are sinking rapidly or slowly or are neutrally buoyant. The trap–filtration difference during the North Atlantic Bloom Experiment (NABE), however, was as much as a factor of 4 for the same depth and time interval (Table 8.1) and could significantly affect the predicted flux. The C: ^{234}Th ratio for both traps and *in situ* filters varied by a factor of 2 over 6 weeks in NABE and could vary between seasons (Buesseler *et al.*, 1992). Murray *et al.* (1996) found a very different C: ^{234}Th ratio during two different cruises across the equator. These changes must be adequately measured and incorporated into the models (Buesseler *et al.*, 1995). ^{234}Th on fragile, sinking

Table 8.1. *Ratio of particulate organic carbon to* ^{234}Th *from* in situ *filtration and from sediment traps deployed at two depths over three time periods in the North Atlantic Bloom Experiment*

Deployments	Depth (m)	Filter POC: ^{234}Th (μmol dpm^{-1})	Trap POC: ^{234}Th (μmol dpm^{-1})
1	150	15.0	3.9
	300	10.6	2.4
2	150	11.8	5.9
	300	8.0	3.6
3	150	8.9	7.4
	300	7.9	4.0

Source: Data from Buesseler *et al.* (1992).

aggregates may break up when they encounter a sediment trap and pieces broken up may not be retained in the trap, or the aggregate may not be able to penetrate the high-density brine used in some traps, thus decreasing the collection of both carbon and ^{234}Th.

There is an increasing awareness that the POC loss from surface waters due to physical transport and respiration of vertical migrators is important. The physical transport loss is probably included in the upper-ocean ^{234}Th balance, but the respiration loss is not. Neither portion is likely to be recorded in fluxes measured with sediment traps and this could account for some of the discrepancies between the two methods. Swimmers have low ^{234}Th relative to POC (Coale, 1990; Buesseler *et al.*, 1994), hence the POC: ^{234}Th ratios in traps may be elevated if swimmers are not adequately removed.

The same concerns expressed above for ^{234}Th fluxes alone apply here to POC flux, i.e. if the predicted ^{234}Th flux is incorrect, then the POC fluxes would also be in error (assuming the C: ^{234}Th ratios have been measured and modelled correctly).

Effect on BATS/HOT carbon imbalance/balance

Using the predicted ^{234}Th fluxes and measured POC: ^{234}Th on particles, the ^{234}Th-derived POC flux would account for up to 80% of the apparent carbon imbalance at BATS (Michaels *et al.*, 1994).

Recommendations

Further studies are essential to examine the range of C: ^{234}Th ratios in size-sorted sediment trap and size-fractionated filtered particle samples to determine whether the ^{234}Th-derived trap calibration can be directly applied to POC.

Studies of the U–Th system and ^{234}Th transfer between dissolved, colloidal, particulate, and aggregate states are also essential, including temporal and spatial variability. However, ^{234}Th is the most promising independent particle tracer we have. Some feel that any JGOFS trap study should have ^{234}Th measurements made at the same time (Jim Murray; Buesseler *et al.*, 1994).

In order to validate the ^{234}Th -deficiency method for estimates of carbon fluxes, a trap–^{234}Th calibration experiment should be designed and conducted in the highest-productivity, lowest-energy regime (i.e. low advection) that is practicable. Continental margins should be avoided because of the potential for thorium scavenging in the sediments of those regions. This experiment must include explicit consideration of non-steady-state and 3–D effects on the time scale of the experiment (Tony Michaels & Ken Buesseler). Measurements of migrant fluxes should be made simultaneously to account for system processes.

Oxygen and carbon mass balance

Comments

Spitzer & Jenkins (1989) calculated a carbon flux of 3 ± 1 mol C m^{-2} yr^{-1} in the western North Atlantic from an oxygen mass balance model. The carbon flux from floating traps at BATS is only 0.8 ± 0.2 mol C m^{-2} yr^{-1} (Michaels *et al.*, 1994), leaving a wide discrepancy between the two methods. Ducklow *et al.* (1995) calculated that on an annual basis the carbon flux via mixing of DOC equalled or exceeded the POC flux near BATS.

However, the carbon flux calculated from an O_2 mass balance model by Emerson *et al.* (1995) in the HOT area was 1 ± 0.5 mol C m^{-2} yr^{-1}; the carbon flux from floating traps was 0.9 ± 0.3 mol C m^{-2} yr^{-1} (Karl *et al.*, 1996), which is in remarkable agreement. Emerson *et al.* (1995) also calculated that about 25% of the carbon flux was carried as DOC. The carbon flux calculated from a mass balance of DIC carbon and ^{13}C-DIC in a 1-D model in the euphotic layer (100 m) by Paul Quay was 1.7 ± 1.0 mol C m^{-2} yr^{-1} (personal communication, 1997). This value is larger than the 1 ± 0.5 mol C m^{-2} yr^{-1} referred to in Quay & Anderson (1996) because the latter value did not take into account northward Ekman flow in the surface layer. This allows for some discrepancy between the trap measurements and models at HOT, but it is still within the error bars and is much smaller than the discrepancy at BATS.

Magnitude of error

The methods show more agreement near Hawaii than Bermuda. However, the comparisons do not include all of the same processes at each of the two sites. Changes in the BATS flux, as a result of a correction for an inferred source of error, tend to have the effect of creating an imbalance at HOT. At BATS there is a difference of a factor of three between trap, ^{234}Th and carbon budget methods (Michaels *et al.*, 1994), which has revised downwards to a difference of 2.4 (Steinberg *et al.*, 1997). The mass-balance budgets include all vertical processes, but not horizontal advection. At HOT the three methods (trap, oxygen and carbon isotope budgets) agree within the accuracy of the data. They do not include a number of other processes, which may export carbon vertically (DOC, migrant fluxes). The budgets do not include horizontal advection.

One might infer that the agreement at HOT means that traps are accurate at that site. Perhaps there are fewer environmental variables to contend with around Hawaii than Bermuda, e.g. fronts passing the area, winter over-turn, mode-water formation, proximity to a major current such as the Gulf Stream with its attendant rings, and general advection. So there is a better chance of reaching closure for budgets of carbon, oxygen and ^{234}Th for calibration with trap fluxes. However, it is also possible that the agreement between one form of carbon flux (traps) and the overall 1-D organic carbon budget means that the trap is over-collecting. The current comparison leaves no room for other processes of transport and error such as migrant fluxes, hydrodynamics and solubilisation of carbon.

Recommendations

Continue to make mass balances of this sort where possible whether or not other means of calibration are available. If possible, use multiple independent strategies for comparison with traps.

Comparison with sediment accumulation rates

Comments

Some trap fluxes have matched well with (1) accumulation rate of underlying sediments based on radionuclide dating (Pennington, 1974; Soutar *et al.*, 1977; Dymond *et al.*, 1981; Gardner *et al.*, 1985); (2) accumulation of radionuclides (Moore *et al.*, 1981; Anderson *et al.*, 1983; Bacon *et al.*, 1985; Biscaye *et al.*, 1988; Biscaye & Anderson, 1994; Colley *et al.*, 1995); (3) accumulation above a known sediment horizon in lakes (Pennington, 1974); and (4) varves (Soutar *et al.*, 1977; Brunskill, 1969 as discussed in Gardner, 1980*a*; Hay *et al.*, 1990).

It is very difficult to use accumulation rates as a calibration standard of the carbon flux for short-term near-surface traps in the open ocean because so much degradation occurs between the surface and the sea floor. Even in shallow lakes, the time scales can also be orders of magnitude different between trap deployments and accumulation rate measurements (decades for ^{210}Pb and 100–1000 years for ^{14}C).

The flux of inert components such as Al can be used as a calibration standard with the assumption that trapping efficiency for POC matches that for aluminosilicates. One must always be aware of the possible 'contamination' by lateral advection of material re-suspended from boundaries, both in traps and in the sediments.

The accumulation of short-lived radionuclides within the water column (e.g. ^{234}Th) can be measured on time scales close to those of trap deployments.

Seasonality

Trap fluxes have been shown to have a seasonal cycle (Deuser & Ross, 1980; Honjo, 1982; Deuser, 1986, 1987). However, a seasonal cycle can exist without one knowing the absolute flux because the entire cycle or parts of the cycle could be biased high or low depending on the dominant particle type or sinking speed and hydrodynamic conditions.

Buesseler *et al.* (1994) argued that at the BATS site floating traps over-collected during periods of low productivity and under-collected during times of high productivity, thus smoothing out the seasonal cycles, but resulting in an annual average under-collection. Rivkin *et al.* (1997) re-examined the data of Buesseler (1991) and agreed that there may be a slight under-collection by floating traps, but argued against the hypothesis of over-collection during periods of low productivity and under-collection during periods of high productivity.

Correlation with ocean colour or chlorophyll

Comments

B. G. Mitchell *et al.* (in preparation) have compiled floating trap data from numerous projects (RACER, ProMARE, NABE, HOT, CABS, BATS, EqPac) and plotted the fluxes against an algorithm-derived parameter based on sea-surface temperature and a blue: green water-leaving radiance ratio. The 48 points have an r^2 fit of 0.71 on a log–log plot. This is comparable to the r^2 fit of 0.76 for the same number of points fitting the data of Chl + phaeopigments versus the blue: green water-leaving radiance ratio for the same sites.

Baines *et al.* (1994) also found good correlations between the amount of chlorophyll in the surface waters and the flux of carbon measured with sediment traps in both lakes and the ocean, although a correlation of the data had a different slope in the two aquatic environments (Fig. 8.8).

One must recognise, however, that there could be a good correlation between flux and any algorithm and still the trap fluxes could all be too high or too low. There is also enough scatter that some of the values could be high or low by factors of 2–3, as seen in the data collated by Buesseler (1991). Care must be exercised in selecting data for this sort of comparison as some of the scatter may result from differences in the depths at which measurements are made, etc.

The important, encouraging point is that there appears to be a real correlation

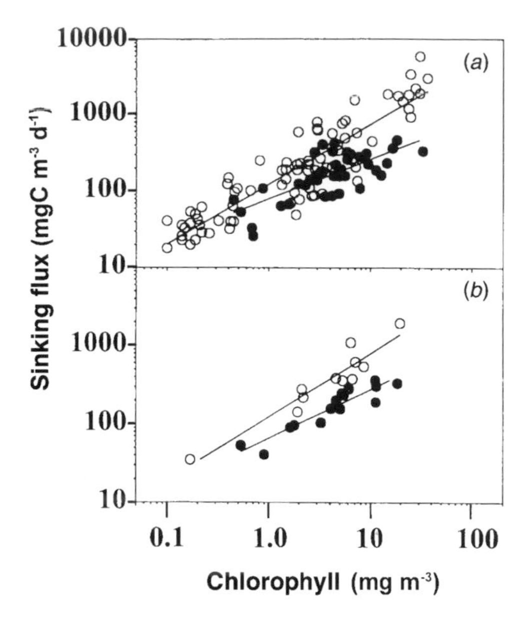

Figure 8.8 *Sediment trap fluxes of carbon and simultaneous measurements of integrated chlorophyll in lakes (filled circles) and oceans (open circles) (after a compilation by Baines* et al., *1994). All data for each study are shown in* (a)*; part* (b) *averaged all the data in each study. Trends are not obvious in a single study, but the data spanning two decades of fluxes and chlorophyll standing stocks from a variety of environments demonstrate definite correlations.*

between chlorophyll or ocean colour and particle flux. Such a relation may seem intuitively obvious to some, but others have questioned whether such a relation could be demonstrated. It is crucial to cover the entire dynamic range of oceanic conditions to establish this relation rather than examining only a small portion of the entire range at one geographic location, in which case the correlation might not be so obvious.

Possible trap errors and their magnitude

Comments

Traps are intended to collect only the particles settling under gravitational forces, not the material carried by vertically migrating organisms or re-mineralised and mixed vertically or advected laterally. It appears that most trap fluxes are within a factor of ±2 times the true flux, which is actually quite encouraging for a measurement that is the result of so many different biogeochemical and hydrodynamic processes. If trap methodology is not properly adhered to, errors can be larger. Table 8.2 provides a summary of possible errors associated with trap measurements plus system processes that

Table 8.2. *Summary of the magnitude of possible errors with traps and the environment*

Trap errors

Source of error	*Magnitude of error*
Swimmers	Up to 2-fold, depending on techniques
Solubilisation of carbon	
Poisoned	0.1% to 1% per day
Unpoisoned	3–7% per day near surface
Hydrodynamic effects that include	
Trap geometry	Extremes of several-fold for odd-shaped traps Funnels may be 0.7–1.0 × cylinders
Flow	× 0.2–2 depending on configuration and velocity
Wave-induced circulation	Not quantified
Deployment of traps in the ML	Not quantified
Tilt	× 1.25–2 for cylinders; not tested for funnels
Apparent tilt from internal waves	Not quantified
Effects of brine in the trap	× 1–3 (× 1.6 is maximum seen in surface water)

System dynamics resulting in apparent trap flux errors

Mechanism	*Magnitude of carbon export*
Vertical migration of zooplankton	8–70% of trap flux
Vertical mixing of DOC, DIC and POC	7–50% in three estimates
Advective transport	Undetermined
Gas exchange of CO_2 with atmosphere	2% in one estimate

may cause discrepancies between trap fluxes and balances in the carbon budget, and erroneously may be attributed to trap errors.

Summary comments

Sediment traps have opened a new era of investigation of biogeochemical cycles in the ocean. They provided the first proof that seasonal and episodic variations in surface-water productivity could result in variable fluxes at depth in the ocean, thus triggering many new questions to pursue. The collection of samples at various depths has allowed studies of the recycling of oceanic particles and helped to elucidate where many processes are occurring. In turn, this information is extremely important in making comparisons with the small residue of biogenic material that reaches the sea floor, and the even smaller residue that is preserved in the sediments. This is critical to accurately (albeit very imperfectly) interpret the palaeorecord from sediment cores. It is equally important for understanding correlations on short time scales between remotely sensed ocean colour data from satellites and processes that lead to the export of carbon from surface waters.

Traps have proven to be valuable tools. Particle cycling in the upper ocean is more complex than previously realised. The lack of mass balance in some studies based on trap fluxes is part of what has made us realise that complexity. That does not mean *a priori* that traps do not (or do) work. We need to make a more concerted effort to fully understand both traps and the particle dynamics in the environments in which they are used. Existing data clearly show that there are trap designs and hydrodynamic regimes in which the results from traps cannot be used in either quantitative (flux) or qualitative (compositional) analyses. On the other hand, there are studies using floating traps where the carbon fluxes match well the macro-scale carbon budgets derived from oxygen budgets determined by completely independent measurements (Emerson *et al.*, 1995; Karl *et al.*, 1996). Other studies show mismatches of as much as a factor of 2.4–3 (Michaels *et al.*, 1994; Steinberg *et al.*, 1997). Further studies are needed to understand why this discrepancy exists.

Standard measurements are needed that can verify or validate trap fluxes. Three basic approaches have been used to calibrate sediment traps. The first has been to calibrate specific trap designs under known conditions in a flume (e.g. velocity, Reynolds number, tilt) and then to use those traps in the field and accept only those data that are collected within physical parameters proven to be acceptable. A variation on this approach is to compare fluxes between a trap moving with the water and one that is not moving with the water to extend the limits of acceptable hydrodynamic conditions. Only a limited number of studies of this type have been made and the hydrographic constraints within which trap fluxes match still-water fluxes must be better defined. The development of neutrally buoyant traps would improve on this method significantly. A second approach is to compare trap fluxes with independent measures of fluxes such as accumulation rates of labile components in the sediments or the loss of a radionuclide from the water column and its accumulation in sediment traps. ^{234}Th is the best candidate for surface traps, but there are still some caveats in its use. A third calibration scheme is to develop carbon budgets independent of trap fluxes as a standard for trap carbon fluxes. Comprehensive measurements made in JGOFS and other programmes are making such carbon budgets possible.

Ultimately, one of the prime JGOFS goals is to develop accurate budgets of carbon, carbonate, silica, etc., not to determine a radionuclide flux or trap accuracy. Different environments and seasons have a different portion of material transported by vertical settling versus migrant transport, DOM mixing or other mechanisms (advection). From that viewpoint, it is a higher priority to develop a method that can be used to predict the removal of carbon (by all pathways combined) from the surface water than to determine the absolute accuracy of traps. However, an accurate calibration of traps would inform us of the portion carried by gravitational settling. If the ^{234}Th-deficiency method could be shown unequivocally to accurately predict the flux of carbon out of surface waters under all conditions, it would be a very important advancement for JGOFS. Presently, we assume that traps probably do not collect the material carried by vertical migrators and they certainly do not collect DOC or significant colloidal material. Unfortunately, a carbon flux via DOC is not likely to be apparent in the ^{234}Th-deficiency method either. Furthermore, biogeochemical studies require knowledge about more than just carbon cycling, and trap samples afford the opportunity to examine the composition of settling material if they are functioning in an unbiased manner. Present JGOFS studies suggest that *c*. 50–80% of the carbon flux occurs as settling particles (Carlson *et al.*, 1994; Gardner, 1997). That is what makes it imperative to calibrate traps.

It is important to remember that comparisons between the trap fluxes and ^{234}Th fluxes apply only to the depth over which particle scavenging creates a ^{234}Th depletion (the upper 70–150 m or so in the open ocean). Carbon re-mineralisation is rapid below the euphotic zone, so carbon fluxes decrease rapidly. Still, a proven calibration scheme for traps in this depth range is important for comparisons with short-term processes in the euphotic zone and with satellite data. With regard to long-term sequestration of carbon, fluxes to the deep ocean and to the sea floor are far more significant than carbon fluxes out of the upper 200 m. Agreement between longer-lived radionuclides and trap fluxes at greater depths has been much better, but that is beyond the scope of this report.

How can we reconcile the apparent agreement between traps and independent measurements at HOT and an apparent *c*. 3–fold disagreement at BATS? One suggestion is to conduct a 4-D-scale carbon and ^{234}Th calibration to verify the carbon and ^{234}Th budgets. If done at BATS, would this be a calibration of the trap methodology or a calibration of the carbon dynamics around the BATS region? ^{234}Th measurements have been made only recently around HOT because the scavenging of ^{234}Th is so low in oligotrophic regions that the errors are large for this calculation and the data are not yet fully analysed. K. Buesseler generally finds enough scavenging at BATS to make these measurements, although there are times when there is no disequilibrium, which means the predicted flux is zero.

What, then, is the calibration standard? Is it necessary to measure ^{234}Th depletion every time a trap measurements is made in the upper 200 m to determine whether it is accurate? If we conduct a 4-D-scale trap–^{234}Th experiment at a simple site, does that guarantee that similar trap measurements conducted elsewhere can use the same correction factor or scheme? Or do we take the hydrodynamics viewpoint and determine the conditions (especially velocity past the trap) under which traps collect particles without bias and then say that traps can be used under those conditions but that if the hydrodynamic conditions are not met the data must be discarded? There are strong proponents of both approaches.

We can make recommendations for JGOFS, but what is the penalty for non-compliance? What about all the trap measurements made outside JGOFS? They probably constitute the majority of trap measurements both now and in the future. What advice and legacy do we leave? Most people who use traps do not have the resources to measure either ^{234}Th or currents.

Before NASA sends an instrument into space (except for the Hubble space telescope), it is tested for responses in all conditions it might experience. Many calibration and comparison experiments have been made with sediment traps in both the laboratory and the field, but few calibration measurements have been made in the upper 200 m of the water column. The only calibration technique that has been suggested for calibrating traps on short time scales in this region is ^{234}Th. Therefore, a 4-D ^{234}Th calibration of traps should be made to answer the question of trap efficiency, and there should be an oversight committee to ensure that all parameters are sufficiently characterised and measured. Such a study would also have to include measurements of vertical migrant and DOM transport.

Despite the concern about fluxes measured with floating sediment traps, Baines *et al.* (1994) have demonstrated a correlation between integrated chlorophyll concentration and carbon flux when a wide range of values are measured. Similarly, B. G. Mitchell *et al.* (in preparation) have shown a correlation between ocean colour plus surface temperature and particle flux when measured over the global range of ocean colour. Although there is significant scatter in the data and the precision is not known, it provides encouragement that it is possible to develop even better algorithms to make truly global calculations for the Joint Global Ocean Flux Study by using floating sediment traps.

Major recommendations for trap sampling in surface waters

1. Design and conduct a trap–^{234}Th calibration experiment in the highest-productivity, lowest-energy regime that is practicable. There should be community input to the design of such an experiment even if only one or more groups conduct the work. The C: ^{234}Th ratio and ^{234}Th

cycling between sinking and non-sinking particle pools is one of the crucial points of such an experiment. Migrant fluxes should also be measured simultaneously.

2. Pending the outcome of the above experiment, measure the ^{234}Th deficiency during trap studies.
3. Make more experiments of the type of Baker *et al.* (1988) where the fluxes of moored traps are compared with fluxes in floating traps to test for hydrodynamic effects.
4. Develop and test neutrally buoyant traps.
5. Construct floating arrays to minimise the flow past traps, and measure the velocity past traps during deployment.
6. Decouple the trap from surface wave motion.
7. Measure flux at a single depth per array. Measurements at multiple depths are always desirable, but we must consider the importance of one measurement in which we have confidence compared with several numbers that might be compromised because of velocity effects.
8. Deploy traps with just 5 practical salinity units excess brine only in the bottom of the trap. Make the necessary tests to convert the BATS–HOT trap protocols to add brine only to a height of one trap diameter.
9. Carefully remove swimmers from samples.
10. Put the JGOFS protocols on the World Wide Web so they can be accessed easily.
11. Fully report methods and errors in all publications.

Acknowledgements

The input of all participants of the sediment trap workshop at the JGOFS Symposium is appreciated. Contributions from Tony Michaels, Ken Buesseler, Dave Karl and Wolfgang Koeve to the early drafts significantly honed and improved the concepts and text. Assistance from Roger Hanson and Sarah Searson is gratefully acknowledged. This is US JGOFS Contribution No. 362.

References

Alldredge, A. L. & Gotschalk, C. (1988). In situ settling behaviour of marine snow. *Limnology and Oceanography*, **33**, 339–51.

Anderson, R. F., Bacon, M. P. & Brewer, P. G. (1983). Removal of Th-230 and Pa-231 from the open ocean. *Earth and Planetary Science Letters*, **62**, 7–23.

Angel, M. V. (1989). Does mesopelagic biology affect the vertical flux? In *Productivity of the Ocean: Present and Past*, ed. W. H. Berger, V. S. Smetacek & G. Wefer, pp. 155–73. Chichester: John Wiley & Sons.

Bacon, M. P., Huh, C. A., Fleer, A. P. & Deuser, W. G. (1985). Seasonality in the flux of natural radionuclides and

Plutonium in the deep Sargasso Sea. *Deep-Sea Research*, **32**, 273–86.
Baines, S. B., Pace, M. L. & Karl, D. M. (1994). Why does the relationship between sinking flux and planktonic primary production differ between lakes and oceans? *Limnology and Oceanography*, **39**, 213–26.
Baker, E. T., Milburn, M. B. & Tennant, D. A. (1988). Field assessment ofsediment trap efficiency under varying flow conditions. *Journal of Marine Research*, **46**, 573–92.
Baskaran, M., Santschi, P. H., Laodong, G., Bianchi, T. S. & Lambert, C. (1996). 234Th: 238U disequilibria in the Gulf of Mexico: the importance of organic matter and particle concentration. *Continental Shelf Research*, **16**, 353–80.
Biscaye, P. E. & Anderson, R. F. (1994). Fluxes of particulate matter on the slope of the southern Middle Atlantic Bight: SEEP-II. *Deep-Sea Research*, Part II, **41**, 459–509.
Biscaye, P. E., Anderson, R. F. & Deck, B. L. (1988). Fluxes of particles and constituents to the eastern United States continental slope and rise: SEEP-I. *Continental Shelf Research*, 8 855–904.
Bishop, J. K. B., Collier, R. W., Ketten, D. R. & Edmond, J. M. (1980). The chemistry, biology, and vertical flux of particulate matter from the upper 1500 m of the Panama Basin in the Equatorial Pacific Ocean. *Deep-Sea Research*, **27A**, 615–40.
Bloesch, J. & Burns, N. M. (1980). A critical review of sedimentation trap technique. *Schweizerische Zeitschrift für Hydrologie*, **42**, 15–55.
Blomqvist, S. & Kofoed, C. (1981). Sediment trapping – a subaquatic in situ experiment. *Limnology and Oceanography*, **26**, 585–90.
Bodungen, V., Wunsch, B. M. & Fürderer, H. (1991). Sampling and analysis of suspended and sinking particles in the northern North Atlantic. *Geophysical Monographs*, **63**, 47–56.
Brainerd, K. E. & Gregg, M. C. (1995). Surface mixed layer and mixing layer depths. *Deep-Sea Research*, **42**, 1521–43.
Brunskill, G. J. (1969). Fayetteville Green Lake, New York, III. Precipitation and sedimentation of calcite in a meromictic lake with laminated sediments. *Limnology and Oceanography*, **14**, 830–47.
Buesseler, K. O. (1991). Do upper-ocean sediment traps provide an accurate record of particle flux? *Nature*, **353**, 420–3.
Buesseler, K. O., Cochran, J. K., Bacon, M. P. & Livingston, H. D. (1992). Carbon and nitrogen export during the JGOFS North Atlantic Bloom Experiment estimated from ^{234}Th : ^{238}U disequilibria. *Deep-Sea Research*, **39**, 1115–37
Buesseler, K. O., Michaels, A. F., Siegel, D. A. & Knap, A. H. (1994). A three-dimensional time-dependent approach to calibrating sediment trap fluxes. *Global Biogeochemical Cycles*, **8**, 179–93.
Buesseler, K. O., Andrews, J. A., Hartman, M. C., Belastock, R. & Chai, F. (1995). Regional estimates of the export flux of particulate organic carbon derived from thorium-234 during the JGOFS EqPac program. *Deep-Sea Research*, **42**, 777–804.
Buesseler, K. O., Steinberg, D. K., Michaels, A. F., Johnson, R., Andrews, J. E., Valdes, J. R. & Price, J. F. (1999). A comparison of the quantity and composition of material caught in a neutrally buoyant versus surface-tethered sediment trap. *Deep-Sea Research* (submitted).
Butman, C. A. (1986). Sediment trap biases in turbulent flows: results from a laboratory flume study. *Journal of Marine Research*, **44**, 645–93.
Butman, C. A., Grant, W. D. & Stolzenbach, K. D. (1986). Predictions of sediment trap biases in turbulent flows: A theoretical analysis based on observations from the literature. *Journal of Marine Research*, **44**, 601–44.
Carlson, C. A., Ducklow, H. W. & Michaels, A. F. (1994). Annual flux of dissolved organic carbon from the euphotic zone in the northwestern Sargasso Sea. *Nature*, **371**, 405–8.Coale, K. H. (1990). Labyrinth of doom: a device to minimise the 'swimmer' component in sediment

trap collections. *Limnology and Oceanography*, **35**, 1376–80.
Coale, K. H. & Bruland, K. W. (1985). $^{234}Th:^{238}U$ disequilibria within the California Current. *Limnology and Oceanography*, **30**, 22–33.
Coale, K. H. & Bruland, K. W. (1987). Oceanic stratified euphotic zone as elucidated by $^{234}Th:^{238}U$ disequilibria. *Limnology and Oceanography*, **32**, 189–200.
Colley, S., Thomson, J. & Newton, P. P. (1995). Detailed ^{230}Th, ^{232}Th and ^{210}Pb fluxes recorded by the 1989/90 BOFS sediment trap time-series at 48°N, 20°W. *Deep-Sea Research*, **42**, 833–48.
Dam, H. G., Roman, M. R. & Youngbluth, M. J. (1995). Downward export of respiratory carbon and dissolved inorganic nitrogen by diel-migrant mesozooplankton at the JGOFS Bermuda time-series station. *Deep-Sea Research*, **42**, 1187–97.
Deuser, W. G. (1986). Seasonal and interannual variations in deep-water particle fluxes in the Sargasso Sea. *Deep-Sea Research*, **33**, 225–47.
Deuser, W. G. (1987). Seasonal variations in isotopic composition and deep-water fluxes of the tests of perennially abundant Planktonic Foraminifera of the Sargasso sea: results from sediment-trap collections and their Paleoceanographic significance. *Journal of Foraminifera Research*, **17**, 14–27.
Deuser, W. G. & Ross, E. H. (1980). Seasonal change in the flux of organic carbon to the deep Sargasso Sea. *Nature*, **283**, 364–5.
Diercks, A. & Asper, V. (1994). Neutrally buoyant sediment traps: The first designs. *EOS, Transactions of the American Geophysical Union*, **75**, 22.
Dickson, A. G. & Goyet, C. (eds) (1994). *DOE Handbook of Methods for the Analysis of the Various Parameters of the Carbon Dioxide System in Sea Water.* Version 2. ORNL/CIDIAC-74.
Ducklow, H. W., Carlson, C. A., Bates, N. R., Knap, A. H. & Michaels, A. F. (1995). Dissolved organic carbon as a component of the biological pump in the North Atlantic Ocean. *Philosophical Transactions of the Royal Society of London* B **348**, 161–7.
Dunne, J. P. & Murray, J. W. (1997). ^{234}Th and particle cycling in the central equatorial Pacific. (ASLO Aquatic Sciences Meeting, Santa Fe, NM, February 1997.) *American Society of Limnology and Oceanography Abstracts*, p. 152.
Dymond, J., Fischer, K., Clauson, M., Cobler, R., Gardner, W., Richardson, M. J., Berger, W., Soutar, A. & Dunbar, R. (1981). A sediment trap intercomparison study in the Santa Barbara Basin. *Earth and Planetary Science Letters*, **53**, 409–18.
Emerson, S., Quay, P. D., Stump, C., Wilbur, D., & Schudlich, R. (1995). Chemical tracers of productivity and respiration in the subtropical Pacific Ocean. *Journal of Geophysical Research*, **100**, 15873–87.
Eppley, R. W. & Peterson, B. J. (1979). Particulate organic matter flux and planktonic organic matter in the surface layer of the ocean. *Deep-Sea Research*, **30**(A), 311–23.
Gardner, W. D. (1980*a*). Field calibration of sediment traps. *Journal of Marine Research*, **38**, 41–52.
Gardner, W. D. (1980*b*). Sediment trap dynamics and calibration: a laboratory evaluation. *Journal of Marine Research*, **38**, 17–39.
Gardner, W. D. (1985). The effect of tilt on sediment trap efficiency. *Deep-Sea Research*, **32**, 349–61.
Gardner, W. D. (1997). Mechanisms, Methods and Measurements of the Vertical Flux of Particles to the Deep Sea. *Abstracts, The Oceanography Society, 1997*, p. 24.
Gardner, W. D. & Richardson, M. J. (1992). Particle export and resuspension fluxes in the western North Atlantic. In *Deep-Sea Food Chains and the Global Carbon Cycle*, ed. G. T. Rowe & V. Pariente, pp. 339–64. Dordrecht: Kluwer Academic Publishers.
Gardner, W. D. & Zhang, Y. (1997). The effect of brine on the collection efficiency

of cylindrical sediment traps, *Journal of Marine Research*, **55**, 1029–48.
Gardner, W. D., Hinga, K. R. & Marra, J. (1983). Observations on the degradation of biogenic material in the deep ocean with implications on the accuracy of sediment trap fluxes. *Journal of Marine Research*, **41**, 195–214.
Gardner, W. D., Southard, J. B. & Hollister, C. D. (1985). Sedimentation and resuspension in the western North Atlantic. *Marine Geology*, **65**, 199–242.
Gardner, W. D., Chung, S. P., Richardson, M. J. & Walsh, I. D. (1995). The oceanic mixed-layer pump. *Deep-Sea Research*, Part II, **42**, 757–75.
Gardner, W. D., Biscaye, P. E. & Richardson, M. J. (1997). A sediment trap experiment in the Vema Channel to evaluate the effect of horizontal particle fluxes on measured vertical fluxes. *Journal of Marine Research*, **55**, 995–1028.
Gust, G., Byrne, R. H., Bernstein, R. E., Betzer, P. R. & Bowles, W. (1992). Particle fluxes and moving fluids: experience from synchronous trap collections in the Sargasso Sea. *Deep-Sea Research*, **39**, 1071–83.
Gust, G., Michaels, A. F., Johnson, R., Deuser, W. G. & Bowles W. (1994). Mooring line motions and sediment trap hydromechanics: in situ intercomparison of three common deployment designs. *Deep-Sea Research*, **41**, 831–57.
Gust, G., Bowles, W., Giordano, S. & Huettel, M. (1996). Particle accumulation in a cylindrical sediment trap under laminar and turbulent steady flow an experimental approach. *Aquatic Sciences*, **58**, 297–326.
Harbison, G. R. & Gilmer, R. W. (1986). Effects of animal behaviour on sediment trap collections: implications for the calculation of aragonite fluxes. *Deep-Sea Research*, **33**, 1017–24.
Hansell, D. A. & Newton, J. A. (1994). Design and evaluation of a 'swimmer'-segregating particle interceptor trap. *Limnology and Oceanography*, **39**, 1487–95.
Hargrave, B. T. & Burns, N. M. (1979). Assessment of sediment trap collection efficiency. *Limnology and Oceanography*, **24**, 1124–36.
Hawley, N (1988). Flow in cylindrical sediment traps. *Journal of Great Lakes Research*, **14**, 76–88.
Hay, B. J., Honjo, S., Kempe, S., Ittekkot, V. A., Degens, E. T., Konuk, T. & Izdar, E. (1990). Interannual variability in particle flux in the southwestern Black Sea. *Deep-Sea Research*, **37**, 911–28.
Honjo, S. (1982). Seasonality and interaction of biogenic and lithogenic particulate flux at the Panama Basin. *Science*, **218**, 883–4.
Honjo, S., Spencer, D. W. & Gardner, W. D. (1992). A sediment trap intercomparison experiment in the Panama Basin, 1979. *Deep-Sea Research*, **39**, 333–58.
Iturriaga, R. (1979). Bacterial activity related to sedimenting particulate matter. *Marine Biology*, **55**, 157–69.
Jannasch, H. W., Zafiriou, O. C. & Farrington, J. W. (1980). A sequencing sediment trap for time-series studies of fragile particles. *Limnology and Oceanography*, **25**, 939–43.
Karl, D. M. & Knauer, G. A. (1989). Swimmers: a recapitulation of the problem and a potential solution. *Oceanography*, **2**, 32–5.
Karl, D. M. & Tilbrook, B. D. (1994). Production and transport of methane in oceanic particulate organic matter. *Nature*, **368**, 732–4.
Karl, D. M., Knauer, G., Martin, J. & Ward, B. (1984). Bacterial chemolithotrophy in the ocean is associated with sinking particles. *Nature*, **309**, 54–6.
Karl, D. M. Christian, J. R., Dore, J. E. Hebel, D. V. Letelier, R. M. Tupas, L. M. & Winn, C. D. (1996). Seasonal and interannual variability in primary production and particle flux at Station ALOHA . *Deep-Sea Research*, **43**, 539–68.
Kerr, R. C. & Kuiper, G. S. (1997). Particle settling through a diffusive-type thermohaline staircase in the ocean. *Deep-Sea Research*, **44**, 399–412.
Knap, A., Michaels, A., Close, A., Ducklow, H. & Dickson, A. (eds) (1996). *Protocols for the Joint Global Ocean Flux Study (JGOFS) Core Measurements.* (JGOFS

Report No. 19, pp. vi + 170. Reprint of the IOC Manuals and Guides No. 29, UNESCO 1994.)
Knauer, G. A., Karl, D. M., Martin, J. H. & Hunter, C. N (1984). In situ effects of selected preservatives on total carbon, nitrogen and metals collected in sediment traps. *Journal of Marine Research*, **42**, 445–62.
Knauer, G. A., Redalje, D. G., Harrison, W. G. & Karl, D. M. (1990). New production at the VERTEX time-series site. *Deep-Sea Research*, **37**, 1121–34.
Kremling, K., Lentz, U., Zeitzschel, B., Schultz-Bull, D. E. & Duinker, J. C. (1996). New type of time-series sediment trap for the reliable collection of inorganic and trace chemical substances. *Review of Science Instrumentation*, **67**, 4360–3.
Lee, C. & Cronin, C. (1982). The vertical flux of particulate organic nitrogen in the sea: Decomposition of amino acids in the Peru upwelling area and the equatorial Atlantic. *Journal of Marine Research*, **40**, 227–51.
Lee, C., McKenzie, J. A. & Sturm, M. (1987). Carbon isotope fractionation and changes in the flux and composition of particulate matter resulting from biological activity during a sediment trap experiment in Lake Greifen, Switzerland. *Limnology and Oceanography*, **32**, 83–96.
Lee, C., Wakeham, S. G. & Hedges, J. I. (1988). The measurement of oceanic particles flux – are 'swimmers' a problem? (Review and Comment). *Oceanography*, **1**, 34–6.
Lee, C., Hedges, J. I., Wakeham, S. G. & Zhu, N. (1992). Effectiveness of various treatments in retarding microbial activity in sediment trap material and their effects on the collection of swimmers. *Limnology and Oceanography*, **37**, 117–30.
Longhurst, A. R. & Harrison, W. G. (1988). Vertical nitrogen flux from the oceanic photic zone by diel migrant zooplankton and nekton. *Deep-Sea Research*, **35**, 881–9.
Lorenzen, C. J., Shuman, F. R. & Bennett, J. T. (1981). In situ calibration of a sediment trap. *Limnology and Oceanography*, **26**, 580–5.
Lundgreen, U., Waniek, J., Schulz-Bull, D. E. & Duinker, J. C. (1997). Azide as a tool to evaluate sediment trap behaviour. *German Journal of Hydrography*, **49**, 57–69.
Macintyre, S., Alldredge, A. L. & Gotschalk, C. C. (1995). Accumulation of marine snow at density discontinuities in the water column. *Limnology and Oceanography*, **40**, 449–68.
Michaels, A. F., Silver, M. W. , Gowing, M. M. & Knauer, G. A. (1990). Cryptic zooplankton 'swimmers' in the upper ocean sediment traps. *Deep-Sea Research*, **37**, 1285–96.
Michaels, A. F., Bates, N. R., Buesseler, K. O., Carlson, C. A. & Knap, A. H. (1994). Carbon-cycle imbalances in the Sargasso Sea. *Nature*, **372**, 537–40.
Miquel, J. C., Fowler, S. W., La Rosa, J. & Buat-Menard, P. (1994). Dynamics of the downward flux of particles and carbon in the open northwestern Mediterranean Sea. *Deep-Sea Research*, **41**, 243–61.
Moore, W. S., Bruland, K. W. & Michel, J. (1981). Fluxes of uranium and thorium series isotopes in the Santa Barbara Basin. *Earth and Planetary Science Letters*, **53**, 391–9.
Murray, J. W., Young, J., Newton, J., Dunne, J., Chapin, T., Paul, B. & McCarthy, J. J. (1996). Export flux of particulate organic carbon from the central equatorial Pacific determined using a combined drifting trap-^{234}Th approach. *Deep-Sea Research*, Part II, **41**, 1095–32.
Pennington, W. (1974). Seston and sediment formation in five Lake District lakes. *Journal of Ecology*, **62**, 215–51.
Peterson, M. L., Hernes, P. J., Thoreson, D. S., Hedges, J. I., Lee, C. & Wakeham, S. G. (1993). Field evaluation of a valved sediment trap. *Limnology and Oceanography*, **38**, 1741–61.
Peterson, W. & Dam, H. G. (1990). The influence of copepod 'swimmers' on pigment fluxes in brine-filled vs. ambient seawater-filled sediment traps. *Limnology and Oceanography*, **35**, 448–55.
Quay, P. D. & Anderson, H. (1996). Organic carbon export rates in the subtropical N.

Pacific. *EOS, Transactions of the American Geophysical Union*, **76**, OS85.
Rivkin, R. B., Legendre, L., Deibel, D., Tremblay, J.-E., Klein, B., Crocker, K., Roy, S., Silverberg, N., Lovejoy, C., Mesple, F., Romero, N., Anderson, M. R., Matthews, P., Savenkoff, C., Vezina, A., Theriault, J.-C., Wessen, J., Berube, C. & Ingram, R. G. (1997). Measuring biogenic carbon flux in the ocean: Response. *Science*, **275**, 554–5.
Siegel, D. A., Granata, T. C., Michaels, A. F. & Dickey, T. D. (1990). Mesoscale eddy diffusion, particle sinking, and the interpretation of sediment trap data. *Journal of Geophysical Research*, **95**, 5305–11.
Silver, M. W., Gowing, M. M., Brownlee, D. C. & Corliss, J. O. (1984). Ciliated protozoa associated with oceanic sinking detritus. *Nature*, **309**, 246–8.
Soutar, A., Kling, S. A., Crill, P. A., Duffrin, E. & Bruland, K. W. (1977). Monitoring the marine environment through sedimentation. *Nature*, **266**, 136–9.
Spitzer, W. S. & Jenkins, W. J. (1989). Rates of vertical mixing, gas exchange and new production: estimates from seasonal gas cycles in the upper ocean near Bermuda. *Journal of Marine Research*, **47**, 169–96.
Steinberg, D. K., Carlson, C. A., Bates, N. R. & Michaels, A. F. (1997). Zooplankton vertical migrations and the active transport of dissolved organic and inorganic carbon in the Sargasso Sea. *American Society of Limnology and Oceanography, Program and Abstracts*, **1997**, 311.
US GOFS 1989. *Sediment Trap Technology and Sampling*. (US GOFS Report No. 10.) Available from US JGOFS Planning Office, Woods Hole Oceanographic Institution, Woods Hole, MA, USA. 94 pp.
Valdes, J. R. & Price, J. F. (1999). Development and testing of a neutrally buoyant sediment trap for studies of biogeochemical cycling in the upper ocean. *Journal of Atmospheres and Oceans* (submitted).
Walsh, I., Fischer, K., Murray, D. & Dymond, J. (1988). Evidence for resuspension of rebound particles from near-bottom sediment traps. *Deep-Sea Research*, **35**, 59–70.
Wei, C.-L. & Murray, J. W. (1992). Temporal variations of ^{234}Th activity in the water column of Dabob Bay: particle scavenging. *Limnology and Oceanography*, **37**, 296–314.
Woods, J. D. & Onken, R. (1982). Diurnal variation and primary production in the ocean – preliminary results of a Lagrangian ensemble model. *Journal of Plankton Research*, **4**, 735–56.

Attendees at the Villefranche trap meeting

*People who contributed material and comments after the first draft.

Nicholas Bates <nick@sargasso.bbsr.edu>
U. Bathmann <ubathmann@awi-bremerhaven.de>
*Ken Buesseler <kbuesseler@whoi.edu>
Craig Carlson <ccarlson@bbsr.edu>
Fei Chai <fchai@athena.umeoce.maine.edu>
Andrew Dickson <adickson@ucsd.edu>
Jan Duinker
*Steve Emerson <emerson@u.washington.edu>
*Wilford Gardner <wgardner@ocean.tamu.edu>
Julie Hall <hall@eco.cri.n2>

Nobuhiko Handa <h4496a@nucc.cc.nagoya-u.ac.jp>
*Dennis A. Hansell <dennis@bbsr.edu>
Roger Hanson <Roger.Hanson@jgofs.uib.no>
Andreas Irmisch, PTBEO Meeresforschung FAX +4938151509
Dale A. Kiefer <dale.kiefer@fao.org>
Tony Knap <knap@bbsr.edu>
*Wolfgang Koeve <wkoeve@ifm.uni-kiel.d400.de>
D. Lal <dlal@ucsd.edu>
Richard Lampitt <R.Lampitt@soc.soton.ac.uk>
K. K. Liu <kkliu@ccms.ntu.edu.tw>
*Ulrich Lundgreen, IfM Kiel, Germany
Jim McCarthy <james—j__mccarthy@harvard.edu>
Nick McCave <mccave@esc.cam.ac.uk>
Dennis McGillicuddy <mcgillic@epl.whoi.edu>
*Tony Michaels <tony@usc.edu>
J. Carlos Miquel <miquel@unice.fr>
*Jim Murray <jmurray@u.washington.edu>
Wajih Nagvi <naqvi@bcgoa.ernet.in>
Susanne Neuer <susanne@zfn.uni-bremen.de>
John Parslow <parslow@ml.csiro.au>
Don Rice <drice@nsf.gov>
Javier Ruiz <javier.ruiz@uca.es>
T. Saino <i45518a@nucc.cc.nagoya-u.ac.jp>
Jan Scholten <js@gpi.uni-keil.de>
Detlef Schulz-Bull, IfM Kiel, Germany
Ian Walsh <walsh@occ.orst.edu>, Rapporteur
Ning Xiuren, 2nd Institute of Oceanography, SOA, 310012 Hangzhou, China

People who were not present, but contributed material and comments after the first draft:

*Vernon Asper <vasper@whale.st.usm.edu>
*Cindy Lee <cindylee@ccmail.sunysb.edu>
*Dave Karl <dkarl@soest.hawaii.edu>
*Greg Mitchell <bgmitchell@ucsd.edu>
*Scott Nodder <nodder@greta.niwa.cri.nz>
*Juan Carlos Miquel <miquel@unice.fr>
*Kenneth Bruland <bruland@cats.ucsc.edu>
*Kenneth Coale <coale@mlml.calstate.edu>

PART THREE

REGIONAL-SCALE ANALYSIS AND INTEGRATION

9 Mixed-layer dynamics and primary production in the Arabian Sea

S. Sathyendranath and T. Platt

Keywords: phytoplankton blooms, mixed-layer depth, critical depth, solar radiation, bio-optics, phytoplankton production model

Introduction

For more than fifty years, it has been understood that the cycle of growth and decay of phytoplankton in the upper mixed layers of the oceans is controlled by the availability of light and nutrients in the layer (Gran & Braarud, 1935). Sverdrup (1953) provided a formal theoretical construct for this idea using the concept of critical depth: it is a concept that has dominated the thinking about phytoplankton blooms ever since.

Reduced to its most elementary level, the Sverdrup critical-depth criterion for phytoplankton growth is a statement of conservation of mass and energy: production depends on the energy available, and production must exceed loss if net growth is to occur. It is therefore a necessary condition for phytoplankton growth in the mixed layer. But the fact that Sverdrup was able to arrive at a plausible explanation for the occurrence of phytoplankton blooms assuming stable parameters for growth and loss also led to the common assumption that the dynamics of phytoplankton growth are determined largely by the light energy available in the mixed layer. Clearly, an increase in average light in the mixed layer would favour more photosynthesis and therefore increased growth. However, is it a sufficient condition? Is it always true that the phytoplankton biomass in the mixed layer merely reflects the prevailing physical conditions? On the other hand, is it possible that the adaptation of phytoplankton in response to the physical and chemical environment could also have a role as a regulatory mechanism? Is the biological–physical interaction always in one direction, or can it be that changes in phytoplankton concentration could, in turn, have an impact on ocean thermodynamics?

These are some of the questions that have been examined during the JGOFS years. In this chapter, we examine the bio-optical properties of phytoplankton as a means for studying the subtle inter-connections between phytoplankton and its environment. This has some obvious advantages. Optical tools are now available that can probe the oceans at a multitude of spatial and temporal scales, ranging from flow-cytometric studies that evaluate properties of single cells to remote-sensing techniques that are capable of global coverage. In recent years, these advances in optical techniques have been accompanied by significant improvements in modelling optical processes in the sea. Besides, optical properties of phytoplankton can be considered the interface between the solar forcing of the oceans and the ecological processes that are driven by it.

We begin with a brief review of the concept of critical depth, and then describe how it can be refined, using the Arabian Sea as a case study. One of the strategies of the JGOFS initiative has been to focus the research efforts on a few sites, selected for their contrasting characteristics. Whenever possible, we will use the North Atlantic as a contrasting point of reference, to gain more insight into the Arabian Sea. This is particularly useful since the North Atlantic is one of the areas where the Sverdrup concept has been applied with success in the past.

Concept of critical depth

The critical depth (Sverdrup, 1953) is defined as the lower boundary of a surface, mixed layer within which the total growth of phytoplankton is exactly matched by the total loss of phytoplankton. If $P_T(z)$ is the daily primary production at depth z, and $L_T(z)$ is the daily loss rate of phytoplankton at z, then we have:

$$\int_0^{Z_c} P_T(z)\,\mathrm{d}z = \int_0^{Z_c} L_T(z)\,\mathrm{d}z, \qquad (9.1)$$

where Z_c is the Sverdrup critical depth. Because production is driven by light, and light decreases exponentially with depth, $P_T(z)$ decreases non-linearly with z. On the other hand, $L_T(z)$ can often be treated as a function independent of depth, because it depends on the biomass, which is uniformly distributed within the mixed layer. A consequence of these differences in the depth distribution of the growth and loss terms is that production is likely to exceed loss in the mixed layer if the mixing depth Z_m is shallower than the critical depth Z_c, whereas losses would exceed production when $Z_c < Z_m$.

Sverdrup applied this concept to data from the North Atlantic, assuming that production was a linear function of available light, and showed that, when nutrients were non-limiting, relative changes in mixed-layer depth and critical depth were sufficient to explain the occurrence of phytoplankton blooms.

Although Sverdrup was aware of the complexity of the biological processes in the mixed layer, in his calculations he used an extremely simple biological model: a fraction m of the light available at depth z is converted into phytoplankton biomass; phytoplankton biomass is lost from any depth at a depth-independent rate n; the ratio of n to m is a constant ($n/m = I_c$, where I_c is the illumination at the 'compensation depth'). With these assumptions, critical depth is given by the equation:

$$Z_c = \frac{I(0)}{I_c} \frac{[1 - \exp(-KZ_c)]}{K}, \tag{9.2}$$

where $I(0)$ is the photosynthetically active irradiance at the sea surface, and K is the diffuse attenuation coefficient for this irradiance. (Note that, instead of $I(0)$ Sverdrup actually used an 'effective radiation' one metre below the sea surface, which he assumed was 20% of total incoming solar radiation at the sea surface, after correction for surface reflection losses; but the distinction between photosynthetically active radiation at the sea surface and effective radiation one metre below the surface is immaterial for our discussion, and so we have used the photosynthetically active irradiance at the surface, which is more commonly used today.) The term $(I(0)/K)[1 - \exp(-KZ_c)]$ represents the integral from the surface to Z_c of $I(z)$, the light available at z. The light available at the compensation depth (I_c) appears in the solution as a simple scaling factor for the surface light $I(0)$, and it is the only biological property in the solution.

In the rest of this chapter, we will examine how the concept of critical depth can be refined and applied to improve understanding of phytoplankton growth cycles. First, let us look at the role of the optical parameter K.

Role of diffuse attenuation coefficient in development of a bloom

The critical depth, as defined by Equation 9.2, decreases with increase in K. We know now that accumulation of biomass in the water causes significant increases in the optical attenuation coefficient. In fact, it is now recognised that the optical properties of open-ocean waters can be modelled well given information on the concentration of phytoplankton in the water. Therefore, we can be confident that the growth of phytoplankton in a layer would increase K, which would in turn decrease Z_c, thus arresting growth by the Sverdrup criterion, unless the mixed-layer depth decreased at the same time. One might conclude then that phytoplankton accumulation in the mixed layer is a self-regulatory process. However, the attenuation coefficient for solar radiation appears also in the computation of mixed-layer depth, and the question arises whether changes in critical depth would keep pace with the corresponding changes in mixed-layer depth, when there is an increase in K. We may anticipate that an increase in K would localise the solar heating nearer to the surface, which would ensure that the change in mixed-layer depth would be in a favourable sense.

In fact, using the solution of Denman (1973) for the depth of a shallowing mixed layer, we can make a simple analysis of how the mixed-layer depth and critical depth change relative to each other, when there is a change in the attenuation coefficient K. The depth Z_m of the mixed layer is given by:

$$\frac{2(G-D)}{\rho\alpha g} = Z_m\left(\frac{Q(0)}{\rho C_p} + \frac{I(0)}{\rho C_p} f_{Zm}\right), \tag{9.3}$$

where the $(G-D)$ term represents the balance between the generation and dissipation of turbulent kinetic energy; $Q(0)$ is the total, non-penetrative flux of heat at the sea surface, which includes latent heat, sensible heat, long-wave back radiation and the non-penetrative part of the solar radiation; $I(0)$ is the penetrative part of the solar radiation at the sea surface; ρ is the density of sea water; α is the coefficient of thermal expansion; g is the acceleration due to gravity; C_p is the specific heat of sea water at constant pressure; and f_{Zm} is a dimensionless function that determines the influence of K on Z_m, defined (Ravindran, 1996) as:

$$f_{Zm} = 1 - \frac{2}{KZ_m}(1 - \exp(-KZ_m)) + \exp(-KZ_m). \tag{9.4}$$

Let K^0 be the initial value of the attenuation coefficient at the time when a bloom begins to form, and let Z_m^0 be the corresponding mixed-layer depth. Let the attenuation coefficient increase from K^0 to K^1 because of bloom initiation, with a consequent change in mixed-layer depth from Z_m^0 to Z_m^1. If we assume that there is no change in the forcing fields during this period, then the relative change in mixed-layer depth can be estimated from Equation 9.3 (see also Ravindran, 1996):

$$\frac{(1 + [I(0)/Q(0)]f_{Zm}^0)}{(1 + [I(0)/Q(0)]f_{Zm}^1)} = \frac{Z_m^1}{Z_m^0}, \tag{9.5}$$

where the superscripts 0 and 1 on the function f_{Zm} are used to represent values of the function when K is equal to K^0 and K^1, respectively. If, during the same interval of time, the critical depth changed from Z_c^0 to Z_c^1, then the relative change in Z_c would be given by:

$$\frac{(1 - \exp(-K^1 Z_c^1))K^0}{(1 - \exp(-K^0 Z_c^0))K^1} = \frac{Z_c^1}{Z_c^0}. \tag{9.6}$$

From this solution we see that the critical depth will evolve such that

$Z_c^1K^1 = Z_c^0K^0$. As the mixed layer shallows, according to the Sverdrup criterion, a bloom will be initiated when Z_m becomes less than Z_c. Therefore, at the instant when the bloom began, the mixed-layer depth would be very close to the critical depth, so that we can put $Z_m^0 = Z_c^0$. Then taking the ratio of Equation 9.5 to Equation 9.6,

$$\frac{(1 + [I(0)/Q(0)]f^0_{Z_m})\,(1 - \exp(-K^0Z_c^0))K^1}{(1 + [I(0)/Q(0)]f^1_{Z_m})\,(1 - \exp(-K^1Z_c^1))K^0} = \frac{Z_m^1}{Z_c^1}. \tag{9.7}$$

This solution is worth some consideration, since the fate of the bloom would depend on how the mixed layer and the critical layer evolve relative to each other. According to this solution, the change in critical depth relative to the mixed-layer depth, due to an increment in K, would depend on the (constant) value of $I(0)/Q(0)$ prevailing. Because the typical values of this quantity in the temperate North Atlantic may not be similar to those in the tropical, monsoonal Arabian Sea, one may conclude that a bloom, once initiated, may evolve quite differently in these two regions.

This simple analysis also points to another role of the attenuation coefficient: changes in K could modify, in principle, the mixed-layer dynamics (see Equation 9.5, for example). In the next section, we examine this aspect of the problem in more detail.

Influence of the attenuation coefficient on mixed-layer dynamics

From Equations 9.3 and 9.4, it is clear that K has a role in determining mixed-layer depth. Because K determines the rate of penetration of solar radiation in the water column, an increase in its value would also have the consequence of localising light absorption, and therefore direct heating by the sun, closer to the sea surface. This could then lead to an increase in mixed-layer temperature.

It has been known for a long time that the results of mixed-layer models are sensitive to changes in the attenuation coefficient for light. In fact, Denman (1973) found that K had to be treated as a tuneable parameter in his model of mixed-layer dynamics at Station P in the North Pacific. However, in construction of mixed-layer models of the open ocean, it has been conventional to ignore naturally occurring variations in K. Instead, it has been customary to treat K as a geographic constant, invariant with time, in the belief that the range of variations in K for open-ocean waters was insufficient to be of significance for physical calculations. The advent of remote sensing of ocean colour has caused us to re-think these assumptions.

Images of the distribution of phytoplankton pigments derived from the ocean-colour signal became an important source of information in oceanography. Many of these images, particularly those of the Arabian Sea, showing the striking variations in phytoplankton concentrations with the reversal of the monsoons, provoked considerable debate. If the concentrations of plant pigments were so variable, then it was likely that variations in K, which would accompany the changes in pigment concentrations, were also important, even in open-ocean waters.

One way to explore the importance of variations in K for mixed-layer dynamics is to force a mixed-layer model with realistic variables and parameters for a given region, and then compare the model outputs with those obtained when the values of K are replaced by those of pure sea water. One such calculation for the Arabian Sea (Sathyendranath *et al.*, 1991) showed that the changes in phytoplankton biomass could in fact modify K by a factor of 5 in these waters. It also showed that the effect of these changes on the heating rates of the mixed-layer could often exceed 1 °C per month. Ravindran (1996) has carried out a systematic analysis of the role of K in mixed-layer dynamics, and shown that the models are sensitive to changes in K when the optical thickness of the layer (KZ_m) is small.

Collectively, these studies under-score the role of K as a link between dynamical and biological processes in the mixed layer. The depth of the mixed layer relative to the critical depth determines whether or not a bloom is possible; once a bloom is initiated, the resultant growth in phytoplankton modifies the attenuation coefficient; in turn, the change in K may modify the mixed-layer depth. Thus, the biological and dynamic processes in the mixed layer are linked through the optical properties of the layer, and these processes are best studied as coupled processes.

Analyses using mixed-layer models that were coupled to a phytoplankton growth model (Platt *et al.*, 1994; Sathyendranath & Platt, 1994) have also shown that, in such coupled models, the mixed-layer model would evolve differently from that in model runs in which the feedback between biological and physical processes was suppressed.

Practical applications of such coupled models would undoubtedly depend on how well the models are able to simulate phytoplankton growth. This brings us next to a closer examination of the phytoplankton growth term in the Sverdrup model, to see how it can be improved.

Modelling phytoplankton growth

The Sverdrup model laid a solid foundation for understanding how the growth of phytoplankton depends on their physical environment. In the implementation of the model for the North Atlantic, Sverdrup assumed that the one biological parameter in the model, I_c, had a constant value (0.15 gCal cm^{-2} h^{-1}). In spite of the simplicity of its biological component, this model was successful in explaining the North Atlantic spring bloom, which led to the commonly held idea that the cycle of phytoplankton biomass is determined solely by changes in the physical forcing (light energy available for production, kinetic energy available for mixing).

We now know that the phytoplankton growth model in Sverdrup (1953) was over-simplified. It is well understood, for example, that phytoplankton production is a non-linear function of available light. To represent this response, one needs at least two parameters: for example, the initial slope α^B that represents the rate of increase of photosynthesis at low light intensities, and the saturation parameter P^B_m that describes the production at high light intensities where production becomes independent of available light. (If light intensities are so high that photoinhibition sets in, a third parameter, say β^B, will be required to describe this effect.) Note that, in the notations used here, the superscript B indicates that the parameters are normalised to the phytoplankton biomass B. Given the non-linearity in the photosynthetic response of phytoplankton to available light, it would be appropriate, in computations of daily mixed-layer production, to allow for the diurnal variations in solar energy, rather than to force the model with a mean irradiance. Platt & Sathyendranath (1991; see also Platt *et al.*, 1991*a*) have proposed an analytical solution to daily mixed-layer production which includes these refinements. They also showed that the solution is precisely represented as a fifth-order polynomial:

$$P_{Z_{\mathrm{m}},T} = A \sum_{x=1}^{5} \Omega_x (I^{\mathrm{m}}_*)^x - A \sum_{x=1}^{5} \Omega_x M (I^{\mathrm{m}}_*)^x, \tag{9.8}$$

where $P_{Z_{\mathrm{m}},T}$ is the primary production integrated over time (the day length D) and over the mixed layer, A is the scale factor defined as $BP^B_{\mathrm{m}}D/\pi K$, Ω_x are the weighting functions ($x = 1, \ldots 5$), I^{m}_* is the maximum irradiance at local noon scaled to the photoadaptation parameter $I_{\mathrm{k}} = P^B_{\mathrm{m}}/\alpha^B$ and M is the transmittance through the mixed layer, defined as $M = \exp(-KZ_{\mathrm{m}})$. Based on dimensional considerations, Platt & Sathyendranath (1993) have also shown that all models of daily phytoplankton production in the mixed layer should have the general form

$$P_{Z_{\mathrm{m}},T} = Af(I^{\mathrm{m}}_*) - Af(MI^{\mathrm{m}}_*), \tag{9.9}$$

where f is a dimensionless function of irradiance.

Models based on available light, such as those represented by Equations 9.8 and 9.9, are easily transformed into absorbed-light models by making use of the equivalence $\alpha^B = \phi_{\mathrm{m}} a^B$, where ϕ_{m} is the maximum, realised quantum yield of photosynthesis, and a^B is the biomass-specific absorption coefficient of phytoplankton.

To determine the net accumulation of biomass, one also has to account for the loss terms. Because these encompass various processes such as respiration, excretion, grazing and sinking, four or more parameters are needed to represent these processes as biomass-normalised rates (Platt *et al.*, 1991*a*). This analysis can be used to derive a net growth rate for phytoplankton, the inverse of which would yield a time scale for phytoplankton growth, whose magnitude, compared with the mean time between passage of storms (note that storms break down the mixed-layer structure and dilute the phytoplankton in the layer), would determine whether or not a bloom would occur (Platt *et al.*, 1991*a*).

Phytoplankton growth and new production

Sathyendranath & Platt (1994) have also discussed how analyses of phytoplankton growth can be used to derive some minimum estimates of new production. Note that new production may be defined as that part of the production in excess of all the metabolic requirements of the entire community of organisms in the mixed layer (Platt *et al.*, 1992). Net growth of phytoplankton and zooplankton and sinking of organic material out of the mixed layer should therefore all be fuelled by new nutrients. Net growth of phytoplankton in the mixed layer, which can in fact be monitored through remote sensing, would therefore yield a minimum estimate of new production, if horizontal advection were negligible. Rate of growth of phytoplankton per unit biomass in the mixed layer can be written as (Sathyendranath & Platt, 1994):

$$\Delta = 2P^B_{\mathrm{m}} D I^{\mathrm{m}}_{*}(1 - M)/(\pi K \chi Z_{\mathrm{m}}) - 24L^B/\chi, \qquad (9.10)$$

where L^B is the biomass-specific loss rate which includes losses due to grazing, sinking, excretion and respiration, and χ is the carbon-to-chlorophyll ratio (Platt *et al.*, 1991*a*). The second term on the right-hand side of the equation would thus be equal to, or greater than, the regenerated production, or the production required to maintain the population in the mixed layer. It can, therefore be rewritten in terms of the *f*-ratio, the ratio of new production to total production. Methods are now under development (Longhurst *et al.*, 1995; Platt *et al.*, 1995; Sathyendranath *et al.*, 1995) for estimating the first term on the right-hand side of the equation, which represents total primary production, by using remotely sensed data. Thus, the potential exists of estimating the *f*-ratio, indirectly, through remote sensing. Note, however, that the new production estimated in this way would represent a 'local' estimate, rather than a long-term, seasonal, or annual estimate, as would be obtained through bulk-property techniques, such as apparent oxygen utilisation. The quality of such estimates will also depend on how well one is able to prescribe the relevant physiological parameters for the time and place concerned.

Parameterisation of bio-optical characteristics

From the early models in which phytoplankton growth was simulated by using a single parameter whose value was assumed to be a constant, modern concepts have evolved to the stage where the need for a number of parameters is recognised. This additional information is needed if we are to do justice to the complexities of phytoplankton ecology and physiology. We also recognise now that the variations in the parameter values with location and with season are of prime interest for understanding phenomena such as species succession. Langdon (1993) has argued, for example, that the low respiration rates of diatoms compared with those of other phytoplankton groups could be used to explain why diatoms generally dominate spring blooms in the North Atlantic. Banse (1994) has noted the need to incorporate variable photosynthesis-light parameters into models of primary production in the Arabian Sea to make them more realistic. Sathyendranath & Platt (1994) have also suggested that one has to invoke changes in the biological parameters that regulate phytoplankton growth in response to the availability of nutrients to explain the occurrence of some of the blooms in the Arabian Sea. This is in sharp contrast to the spring bloom in the North Atlantic, which can be explained, at least qualitatively, as just a biological manifestation of changing physical conditions (increase in incoming solar radiation, decrease in mixed-layer depth).

The importance of changes or differences in the physiological parameters of phytoplankton production and growth can also be demonstrated by examining in detail the absorption of light by phytoplankton in the mixed layer. For example, one can compute the light absorbed by phytoplankton in the mixed layer, given the satellite-derived information on the concentration of pigments, the depth of the mixed layer, and the amount of photosynthetically active radiation reaching the sea surface. In our calculations we used the mixed-layer fields for the Arabian Sea from Hastenrath & Greischar (1989), the monthly-composite phytoplankton distribution fields derived from the Coastal Zone Color Scanner data (Feldman *et al.*, 1989) and cloud cover information from the Max Planck Institute (Wright, 1988). The spectrally resolved irradiance at the sea surface was then computed as in Brock *et al.* (1993), using the clear-sky model of Bird (1984), combined with a cloud-cover correction as in Platt *et al.* (1991*b*). The light penetration through the mixed layer was then calculated by using a spectral, light-transmission model (Sathyendranath &

Platt, 1988) that allows for absorption and scattering by pure seawater, phytoplankton and dissolved organic matter. The computations were carried out from 400 to 700 nm at 5 nm intervals. For each waveband, the amount of solar radiation absorbed by phytoplankton in the mixed layer was also computed, and summed over the wavelength range to obtain the total, photosynthetically active radiation that was absorbed by the phytoplankton. This amount was divided by the mixed-layer depth to obtain the average light absorbed by phytoplankton per unit volume. These results were also divided by the concentration of phytoplankton, to obtain the light absorbed by phytoplankton per unit volume and per unit biomass.

An example of the computed light absorption by phytoplankton per unit volume in the mixed layer of the Arabian Sea is shown in Fig. 9.1*a*. This figure looks very much like the satellite images showing the phytoplankton distribution in the Arabian Sea, indicating that variations in the concentration of phytoplankton are indeed the primary cause of variations in the amount of light absorbed by phytoplankton in the mixed layer. The picture is remarkably different when we look at the amount of light absorbed by phytoplankton per unit volume, and per unit biomass (Fig. 9.1*b*). The variations seen in Fig. 9.1*b* reflect a number of factors. (1) The general light transmission model used in these computations allows for the commonly observed rule that phytoplankton in oligotrophic waters have a higher specific absorption coefficient than those in eutrophic waters. (2) The blue light at depth in oligotrophic waters is more favourable for absorption by phytoplankton than the greener light of phytoplankton-rich areas. (3) Finally, the distribution also reflects regional variations in the solar radiation reaching the sea surface, and in the mixed-layer depth. These differences in phytoplankton absorption are brought to the fore when the dominant effect of variations in biomass is removed by normalising the results to unit biomass. According to Fig. 9.1*b*, a given concentration of phytoplankton in the mixed layer of the central Arabian Sea would absorb more light, on average, than the same concentration of phytoplankton off the Arabian or Indian coastlines. If all the absorbed energy translates into primary production with equal efficiency everywhere, then one would predict that it is the central Arabian Sea that has more potential for blooms than the surrounding coastal waters, but this prediction would be entirely contrary to the actual observations of blooms in this area.

Such a simple analysis shows that, in the Arabian Sea, variations in the light environment of phytoplankton in the mixed layer, as computed here, are not in themselves sufficient to explain the blooms in this region. It leads to the conclusion that understanding phytoplankton blooms in the Arabian Sea requires detailed information on variations in the physiological parameters of photosynthesis, i.e., the parameters of the photosynthesis–light curve such as

P^B_m and α^B for use in models of available light, or parameters such as the quantum efficiency of production for use in models of absorbed light. Note that the modelling of light penetration in the Arabian Sea may also have to be improved to represent seasonal and regional variations in the optical characteristics of phytoplankton from the Arabian Sea. Therefore, one of the objectives of our JGOFS field programme was to make detailed measurements of the bio-optical characteristics of phytoplankton in the Arabian Sea. The results that are beginning to emerge from these field studies show a high variability in the optical and physiological characteristics of phytoplankton in the region; this variability appears to be related to changes in the population structure, as well as to variations in environmental conditions (Sathyendranath *et al.*, 1996, 1999).

Discussion

In this chapter we have discussed some aspects of interactions between biological and physical processes in the mixed layer, and highlighted some of the differences between the temperate North Atlantic and the tropical Arabian Sea. Based on a systematic study of CZCS images, Longhurst (1993) has also pointed out some of the differences in the physical mechanisms responsible for bloom initiations in the tropical regions of the Pacific, Atlantic and Indian Oceans. The picture that is emerging from all these studies suggests that, although the concept of critical depth provides the under-pinning for the study of blooms in general, it is the regional and seasonal differences in the biological and physical mechanisms in the mixed layer that merit close attention now, if indeed we are to arrive at quantitative estimates of primary production and new production at the global scale.

Although we have used only a simple, bulk model of the mixed layer (Denman, 1973) to illustrate the biological–physical feedbacks in the mixed layer, other studies based on more elaborate, turbulent-closure models (Dickey & Simpson, 1983; Stramska & Dickey, 1993) have also concluded that it is important to account for variations in K in mixed-layer models. Evidence from the Pacific Ocean (Lewis *et al.*, 1990; Siegel *et al.*, 1995) has also pointed to the need to account for optical variability in the water associated with phytoplankton, in computations of the heat budget of the layer. It thus appears that the need to see the biological and physical processes in the mixed layer as coupled processes is not restricted to the Arabian Sea or the North Atlantic.

Acknowledgements

We thank Heidi Maass for help with the computation and figures, and three

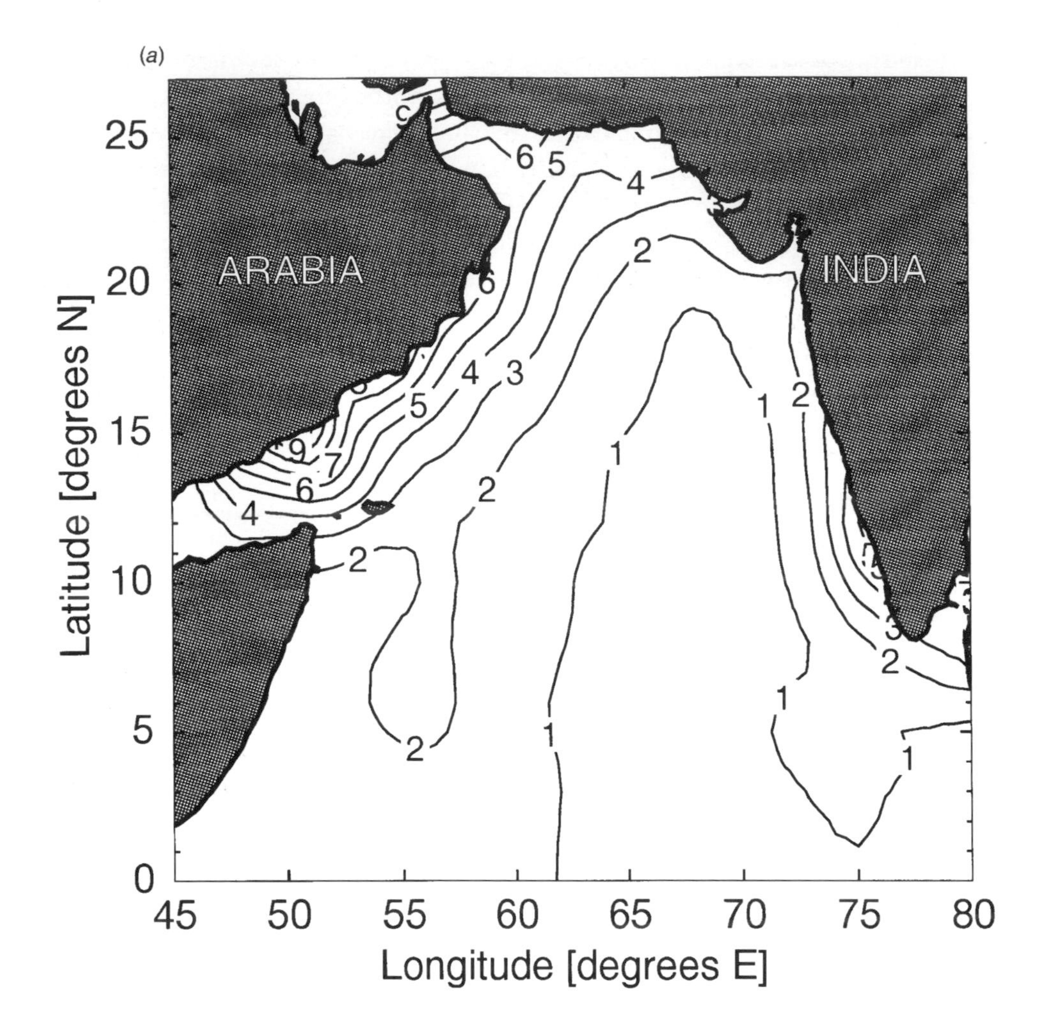

Figure 9.1 (a) *Amount of light absorbed by phytoplankton in the mixed layer of the Arabian Sea, per unit volume of water* ($W\,m^{-3}$)*, computed for September. The observed variations are due to (1) a non-linear relation between chlorophyll* a *concentration and absorption coefficient for phytoplankton (as in Sathyendranath & Platt, 1988); (2) regional variation in incoming solar radiation (computed as in Brock* et al.*, 1993; (3) variation in mixed-layer depth (from Hastenrath & Greischar, 1989), and (4) variation in phytoplankton concentration, from the NASA CZCS archives (Feldman* et al.*, 1989).* (b) *As* (a) *except that the results are normalised to the chlorophyll* a *concentration* ($W\,m^{-3}$ *(mg chl a)*$^{-1}$)*, to eliminate the dominating influence of variations in its concentration, and highlight the variability due to the other factors.*

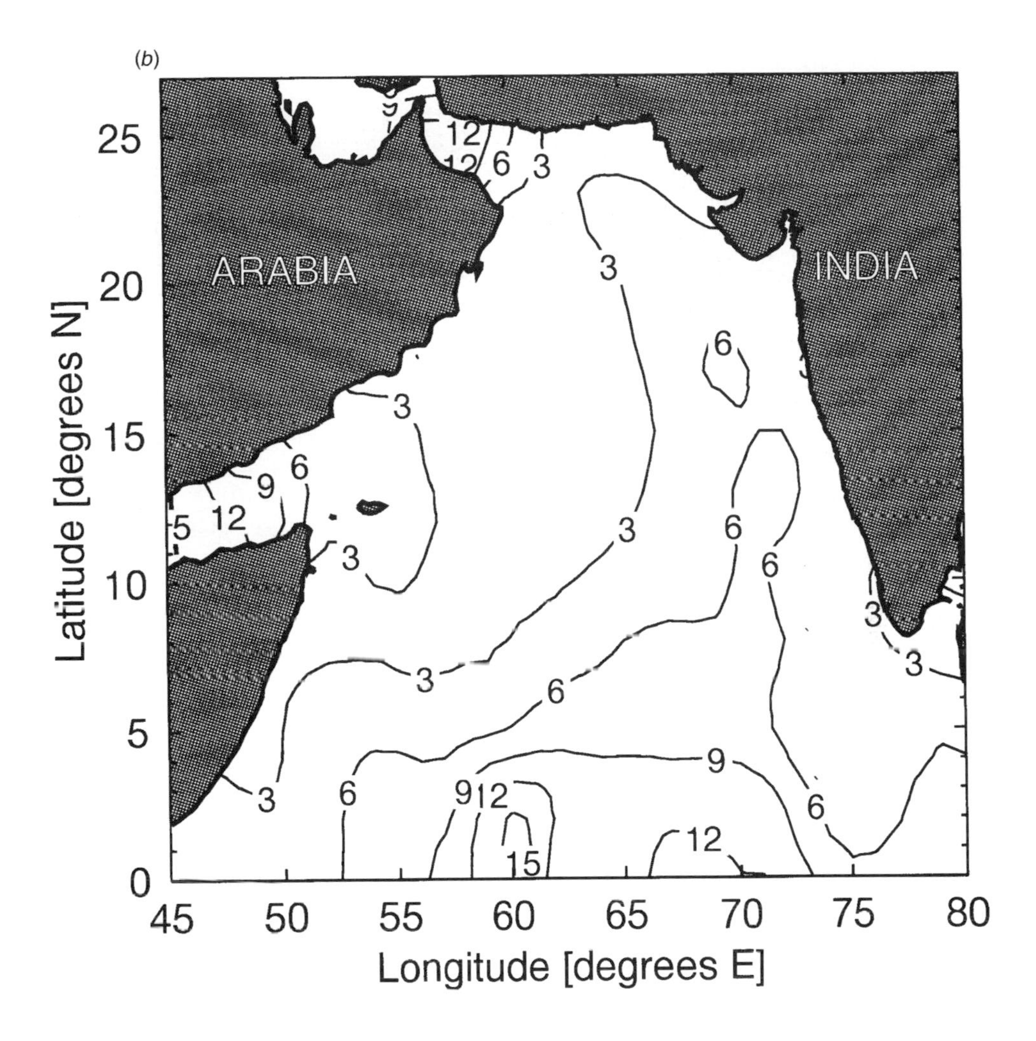

anonymous reviewers for their helpful comments. The Office of Naval Research, the National Aeronautics and Space Administration, and the Department of Fisheries and Oceans, Canada, supported the work presented in this chapter. The Natural Sciences and Engineering Research Council through research grants provided additional support to SS and TP. This work was carried out as part of the Canadian contribution to the Joint Global Ocean Flux Study.

References

Banse, K. (1994). On the coupling of hydrography, phytoplankton, zooplankton, and settling organic particles offshore in the Arabian Sea. *Proceedings of the Indian Academy of Science, Earth and Planetary Sciences*, **103**, 125–61.

Bird, R. E. (1984). A simple, solar spectral model for direct-normal and diffuse horizontal irradiance. *Solar Energy*, **32**, 461–71.

Brock, J., Sathyendranath, S. & Platt, T. (1993). Modeling the seasonality of subsurface light and primary production in the Arabian Sea. *Marine Ecology Progress Series* **101**, 209–21.

Denman, K. L. (1973). A time-dependent model of the upper ocean. *Journal of Physical Oceanography*, **3**, 173–84.

Dickey, T. D. & Simpson, J. J. (1983). The influence of optical water type on the diurnal response of the upper ocean. *Tellus*, **35B**, 142–54.

Feldman, G., Kuring, N., Ng, C., Esaias, W., McClain, C. R., Elrod, J., Maynard, N., Endres, D., Evans, R., Brown, J., Walsh, S., Carle, M. & Podesta, G. (1989). Ocean color. Availability of the global data set. *EOS, Transactions of the American Geophysical Union*, **70**, 634–5.

Gran, H. H. & Braarud, T. (1935). A quantitative study on the phytoplankton of the Bay of Fundy and the Gulf of Maine including observations on hydrography, chemistry and morbidity. *Journal of the Biological Board of Canada*, **1**, 219–467.

Hastenrath, S. & Greischar, L. L. (1989). *Climatic Atlas of the Indian Ocean*. Part III: *Upper-Ocean Structure*. Wisconsin: University of Wisconsin Press.

Langdon, C. (1993). The significance of respiration in production measurements based on oxygen. *ICES Marine Science Symposium*, **197**, 69–78.

Lewis, M. R., Carr, M.-E., Feldman, G. C., Esaias, W. & McClain, C. (1990). Influence of penetrating solar radiation on the heat budget of the equatorial Pacific Ocean. *Nature*, **347**, 543–5.

Longhurst, A. (1993). Seasonal cooling and blooming in tropical oceans. *Deep-Sea Research*, Part I, **40**, 2145–65.

Longhurst, A., Sathyendranath, S., Platt, T. & Caverhill, C. (1995). An estimate of global primary production in the ocean from satellite radiometer data. *Journal of Plankton Research*, **17**, 1245–71.

Platt, T. & Sathyendranath, S. (1991). Biological production models as elements of coupled, atmosphere-ocean models for climate research. *Journal of Geophysical Research*, **96**, 2585–92.

Platt, T. & Sathyendranath, S. (1993). Estimators of primary production for interpretation of remotely sensed data on ocean color. *Journal of Geophysical Research*, **98**, 14561–76.

Platt, T., Bird, D. F. & Sathyendranath, S. (1991*a*). Critical depth and marine primary production. *Proceedings of the Royal Society of London* B **246**, 205–17.

Platt, T., Caverhill, C. & Sathyendranath, S. (1991*b*). Basin-scale estimates of oceanic primary production by remote sensing: The North Atlantic. *Journal of Geophysical Research*, **96**, 15 147–59.

Platt, T., Jauhari, P. & Sathyendranath, S. (1992). The importance and measurement of new production. In *Primary Productivity and Biogeochemical Cycles in the Sea*, ed. P. G. Falkowski & A. D. Woodhead, pp. 273–84. New York: Plenum Press.

Platt, T., Woods, J. D., Sathyendranath, S. & Barkmann, W. (1994). Net primary production and stratification in the ocean. *Geophysical Monographs*, **85**, 247–54.

Platt, T., Sathyendranath, S. & Longhurst, A. (1995). Remote sensing of primary production in the ocean: Promise and fulfilment. *Philosophical Transactions of the Royal Society of London* B **348**, 191–202.

Ravindran, P. (1996). Physical-biological interactions and the thermodynamics of the upper ocean. Ph.D. thesis, Dalhousie University.

Sathyendranath, S. & Platt, T. (1988). The spectral irradiance field at the surface and in the interior of the ocean: A model for applications in oceanography and remote sensing. *Journal of Geophysical Research*, **93**, 9270–80.

Sathyendranath, S. & Platt, T. (1994). New production and mixed-layer physics. *Proceedings of the Indian Academy of Sciences, Earth and Planetary Sciences*, **103**, 177–88.

Sathyendranath, S., Gouveia, A. D., Shetye, S. R., Ravindran, P. & Platt, T. (1991). Biological control of surface temperature in the Arabian Sea. *Nature*, **349**, 54–6.

Sathyendranath, S., Longhurst, A., Caverhill, C. M. & Platt, T. (1995). Regionally and seasonally differentiated primary production in the North Atlantic. *Deep-Sea Research*, Part I, **42**, 1773–802.

Sathyendranath, S., Platt, T., Stuart, V., Irwin, B. D., Veldhuis, M. J. W., Kraay, G. W. & Harrison, W. G. (1996). Some bio-optical characteristics of phytoplankton in the N.W. Indian Ocean. *Marine Ecology Progress Series*, **132**, 299–311.

Sathyendranath, S., Stuart, V., Irwin, B. D., Maass, H., Savidge, G., Gilpin, L. & Platt, T. (1999). Seasonal variations in bio-optical properties of phytoplankton in the Arabian Sea. *Deep-Sea Research*. (In press.)

Siegel, D. A., Ohlmann, J. C., Washburn, L., Bidigare, R. R., Nosse, C. T., Fields, E. & Zhou, Y. (1995). Solar radiation, phytoplankton pigments and the radiant heating of the equatorial Pacific warm pool. *Journal of Geophysical Research*, **100**, 4885–91.

Stramska, M. & Dickey, T. D. (1993). Phytoplankton bloom and the vertical thermal structure of the upper ocean. *Journal of Marine Research*, **51**, 819–42.

Sverdrup, H. U. (1953). On conditions for the vernal blooming of phytoplankton. *Journal du Conseil*, **18**, 287–95.

Wright, P. B. (1988). *An atlas based on the 'COADS' data set: fields of mean wind, cloudiness and humidity at the surface of the global ocean.* Report No. 14. Hamburg: Max-Planck Institute of Meteorology.

⑩ Plankton ecology and biogeochemistry in the Southern Ocean: A review of the Southern Ocean JGOFS

U. Bathmann, J. Priddle, P. Tréguer, M. Lucas, J. Hall and J. Parslow

Keywords: carbon dioxide exchange, ocean carbon cycle, Southern Ocean, marginal ice zone, phytoplankton, bacterioplankton, zooplankton, frontal zones

Introduction

The Southern Ocean (SO) has been considered as an important sink for atmospheric CO_2 (Takahashi *et al.*, 1993) although it is now clear that the southern hemisphere alone cannot account for the 'missing CO_2' in flux budgets (Tans *et al.*, 1990). Moreover, mechanisms governing vertical stability and thermohaline circulation make the Southern Ocean sensitive to climate change, with uncertain consequences for CO_2 regulation. Current opinion recognises that the northern hemisphere is probably the major sink for atmospheric CO_2, although the northern terrestrial sink is probably much smaller than that calculated by Tans *et al.* (1990) because of improved regional estimates. Furthermore, recent shifts from equilibrium model simulations of CO_2 fluxes to improved time-dependent (transient) simulations indicate that warming over the Southern Ocean near Antarctica will be reduced by 60% or more at the time of CO_2 doubling.

Exchanges of CO_2 at the air–sea interface depend primarily on the relative atmospheric and surface water CO_2 values, the latter being strongly dependent on wind speed and water temperature. Because of this, the Southern Ocean has to be considered in calculating the physical solubility pump in the control of oceanic CO_2 fluxes. Seasonal sea-ice cover is also clearly important here through the inhibition of gaseous exchange (Harris & Stonehouse, 1991) while the development of sea-ice microbial communities may have important implications for carbon and nitrogen cycling (Garrison *et al.*, 1991). Indeed, sea-ice extent has a strong influence on the earth's radiative balance with consequent feedbacks on atmospheric CO_2 concentrations should sea-ice extent diminish in the future.

Phytoplankton biomass and photosynthesis provide the basis for transporting fixed carbon into the deep ocean. This is the 'biological pump' of CO_2 (Longhurst, 1991). It comprises the export of dissolved organic material; direct sedimentation of phytoplankton cells; and processes that transform phytoplankton into rapidly sedimenting particles such as faecal pellets and aggregates. Microphytoplankton, predominently diatoms, contribute much of the primary production in the Southern Ocean, especially in frontal and neritic areas and at the ice-edge. However, nano- and picoplankton dominate in some other areas and in some seasons. The Southern Ocean is the major high-nutrient – low-chlorophyll (HNLC) region in the global ocean, where low annual primary production occurs despite high nutrient concentrations. Explanations for this phenomenon include light limitation, strong grazing pressure, and the increasing probability that iron may limit algal growth (El-Sayed, 1984; Priddle *et al.*, 1986; Martin *et al.*, 1990; Smetacek *et al.*, 1990; de Baar *et al.*, 1995). Indeed, recent models suggest that, whereas iron fertilisation of the Pacific would have no long-term effects on CO_2 drawdown, this is not the case for the Southern Ocean (Cooper *et al.*, 1996).

Data for the Southern Ocean have historically been extremely sparse, and such that exist point to a patchwork of source and sink areas which created considerable uncertainty in ocean-atmosphere GCM models. It was of critical importance therefore that not only should better estimates of CO_2 flux be obtained for the Southern Ocean, but that in addition the physical and biological processes controlling carbon cycling needed to be properly understood (Bathmann, 1996). The development of the SO-JGOFS strategy was in response to these needs and now, five years later, we report on the accumulating evidence that certain areas of the SO are major sinks for CO_2, at least during the phytoplankton growing season. We cannot yet conclude, however, whether the SO as an entity functions as a source or a sink for CO_2 on a global scale.

The Southern Ocean-Joint Global Ocean Flux Study (SO-JGOFS) concentrated its activities within the four major sub-systems of the Southern Ocean identified by Tréguer & Jacques (1992): the permanently open ocean (POOZ), the frontal systems (e.g. PFZ), the region of seasonal sea-ice cover (SIZ), and the coastal and continental shelf zone (CCSZ). The scientific aims of SO-JGOFS were to investigate the following. What is the role of the Southern Ocean (SO) in contemporary carbon flux? What controls the magnitude and variability of primary production and the fate of particles? What are the major physical and chemical features of the essential biotic systems? What is the effect of sea ice on carbon flux? In addition, what was the role of the SO in this respect in the past and what will it be in the future?

JGOFS measurements conducted in the Southern Ocean included nearly all the core parameters specified in the JGOFS core measurement protocols. In

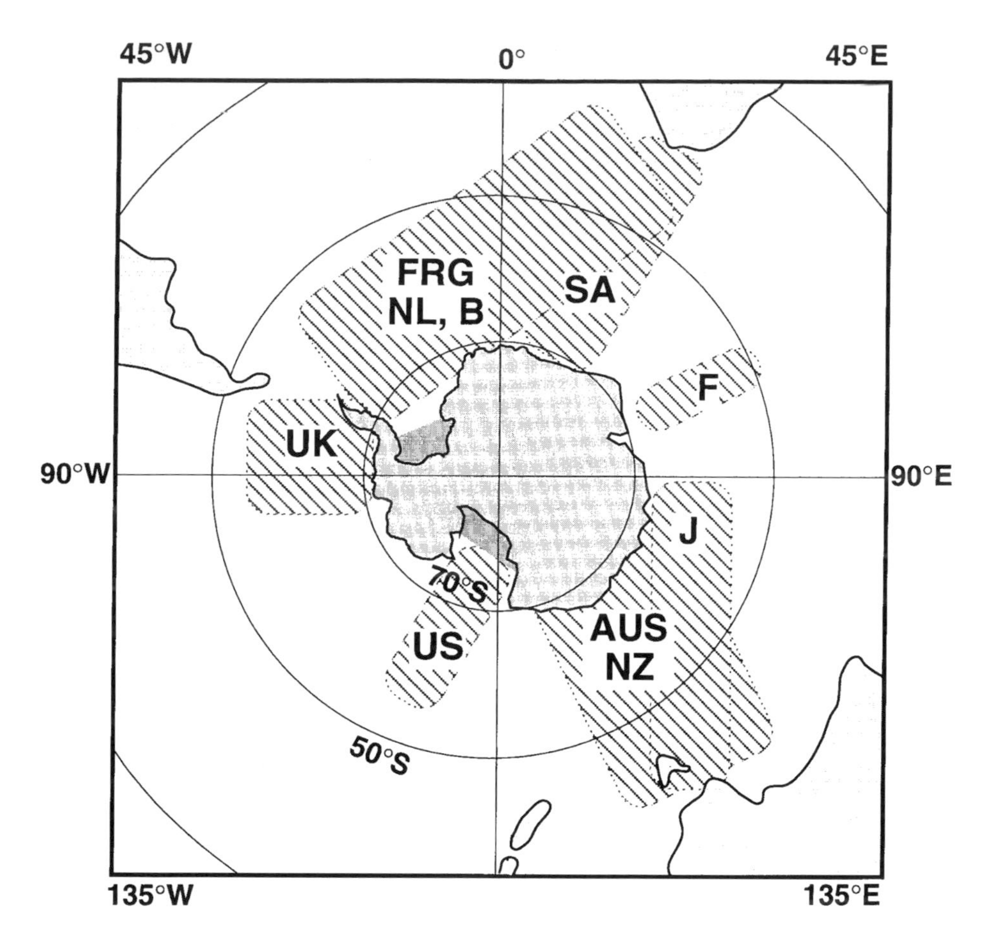

Figure 10.1 *Outline of national and regional JGOF-Studies in the Southern Ocean.*

addition, measurements of sea-ice distribution, ice biology and ice physics, silica cycling, dissolved iron – plankton interactions, dimethylsulfonium propionate (DMSP) production by algae, species composition and distribution of top predators, distribution of uranium and thorium in seawater, and geochemical and microbiological investigations of the sea floor were performed on many of the cruises. Particular emphasis was given to processes within the 'biological pump' of the upper water column and to studies on the role of the 'microbial loop'.

Within the first phase, SO-JGOFS (1990–95), about 15 cruises studied all of these Southern Ocean sub-systems to address the above questions. According to logistic requirements, fieldwork centred on various locations around the continent (Fig. 10.1), to allow regional comparisons to be made.

As not all sub-systems were investigated thoroughly at all sites, we present here a description of three systems, each with a scenario of its structure and function and its physical–chemical boundaries: the Subtropical Convergence (STC), the marginal sea-ice zone (MIZ) and the open ocean at the polar frontal zone (PFZ). A summary of the current knowledge of processes affecting fluxes of carbon, silicon and other biologically important elements in the Southern Ocean follows. Our chapter concludes by reflecting on how and where future activities should be concentrated in SO-JGOFS.

The Subtropical Convergence

The Subtropical Convergence region to the east of New Zealand is relatively narrow because of the physical constraints imposed by the Chatham Rise, which reaches within 150 m of the surface. In the studies reported here, replicate sites ranging from 170 to 210 nautical miles (n m) off the coast of New Zealand were sampled in the subantarctic convergence zone and subtropical waters (Table 10.1).

In subtropical waters east of New Zealand, the < 20 μm nanoplankton fraction dominated primary production in both winter and spring with the exception of Subtropical Convergence waters, where the 20–200 μm fraction dominated. Spring phytoplankton biomass (chlorophyll *a*) did not exceed 2.5 mg Chl *a* m^{-3} (Table 10.1). Production per unit chlorophyll biomass (PB) appeared to be large in subantarctic (SA) and subtropical waters (ST). At the Subtropical Convergence (STC) in October, PB was low, apparently because of low dissolved silicate (mean of 0.83 μm). In the < 2 μm size fraction the contribution to total phytoplankton biomass and production was very low in the STC (i.e. less than 15%) compared with the SA and ST waters from both seasons. Bacterial production and numbers increased in all water masses in spring, with the largest increase in production in the STC.

Microzooplankton biomass in spring was dominated by heterotrophic nanoflagellates in all water masses, whereas in winter ciliates were much more significant. Grazing of microzooplankton balanced phytoplankton growth in all water masses in winter for both chlorophyll *a* and picophytoplankton (Table 10.2). In subtropical waters in spring, total phytoplankton growth exceeded grazing although grazing exceeded growth for picophytoplankton populations. Grazing on bacteria balanced their growth in all water masses in spring.

Mesozooplankton biomass in all water masses was highest in spring and greatest in the > 1000 μm fraction followed by the 500–1000 μm and 200–500 μm fractions. Spring biomass in subtropical water was high (up to 101.4 mg m^{-3} dry mass) and was concentrated in the upper 100 m (Table 10.1).

Table 10. 1. *Distribution of biomass and production values in winter and spring at the STC east of New Zealand*
All integrated values to 100 m, except primary production to 1% light.

	Date	Latitude (°S)	Longitidue (°E)	Chl *a* ($mg\,m^{-2}$)	Primary production ($mg\,C\,m^{-2}\,d^{-1}$)	Bacteria ($mg\,C\,m^{-2}$)	Bacterial production ($mg\,C\,m^{-2}\,d^{-1}$)	Heterotrophic flagellates ($mg\,C\,m^{-2}$)	Ciliates ($mg\,C\,m^{-2}$)	Mesozooplankton ($mg\,C\,m^{-2}$)
Winter										
SA	22–24/6/93	46°10.0"	173°79.5'	11	45	935	108	27	59	305
STC	26–28/6/93	43°33.7"	176°29.4'	49	262	681	33	38	75	192
ST	1–2/6/93	41°28.9–	178°82.0'	19	149	600	135	71	74	177
Spring										
SA	10–13/10/93	46°06.1–	174°12.4'	14	251	1225	159	474	115	391
STC	14–16/10/93	43°31.4"	176°41.3'	108	986	1527	622	297	90	548
ST	17–18/10/93	41°02.2"	179°10.1'	47	971	2186	610	771	42	2895

Table 10.2. *Microzooplankton grazing as a percentage of phytoplankton biomass removed per day in winter and spring at the STC east of New Zealand*

	% Phytoplankton biomass removed d^{-1}	% Picophytoplankton biomass removed d^{-1}
Winter		
SA	38	76
STC	42	60
ST	86	25
Spring		
SA	58	57
STC	37	38
ST	43	51

Mesozooplankton biomass variability on a day–night basis often exceeded a factor of 2 and may be explained by sampling-net avoidance during daytime or by patchy distribution of the populations.

Grazing by the mesozooplankton community in spring was determined by gut fluorescence and did not show a strong diel pattern. Grazing was persistently low at many stations; perhaps this is the result of the relatively large proportion of primary production in the < 20 μm fraction. Total average ingestion by the mesozooplankton community was calculated to be 1–35 mg C m^{-2} d^{-1} with the largest grazing pressure occurring in subtropical waters and at the Subtropical Convergence. In winter and spring this represents grazing of 0.2 to 1.6% of the total integrated chlorophyll *a* standing stock. Thus, the amount of phytoplankton carbon ingested by the mesozooplankton did not meet basic metabolic requirements for some mesozooplankton size fractions in most water masses, particularly in subantarctic waters in either season. These data imply that mesozooplankton diets, especially for the two larger size fractions, were sometimes supplemented with non-phytoplankton sources of food.

Air–sea fluxes of CO_2 in austral spring 1993 in the surface waters east of New Zealand were generally negative and thus these waters acted as a sink for atmospheric CO_2 (p_{CO_2} from − 10 to − 110 μatm), particularly in the STC owing to increased biological productivity.

East of Tasmania, there is a complex frontal structure, with two weaker Subtropical Convergences (STC), characterised by surface salinities of *c.* 35.2 and 34.8, and the strong Subantarctic Front at 50°S, where salinity drops from 34.6 to 34.2. The latter forms the northern core of the Antarctic Circumpolar Current. There is a strong seasonal cycle in this zone, with deep winter mixed layers extending to over 500 m in the SAZ just north of the Subantarctic Front.

Shallow stratification occurs in the summer, with typical mixed layer depths of *c.* 50 m. There is a southward sub-surface intrusion of saline subtropical water at *c.* 100 m throughout the STC (north of the southern STC) in summer. Results from cruises, combined with satellite SST images, suggest strong inter-annual variation in the position of the STC, and mesoscale meanders in both the STC and SAF extending over several degrees of latitude.

Chlorophyll concentrations in the deep winter mixed layers were low (*c.* 0.1 mg m^{-3}), as might be expected, but P–E studies suggested cells that were healthy, and strongly low-light-adapted. Results from the continental slope show blooms in spring exceeding 3 mg m^{-3}, with estimated water-column primary production as high as 2000 mg C m^{-2} d^{-1}. Chlorophyll concentrations away from the slope in summer were typically 0.2–0.5 mg m^{-3}, with moderate levels of primary production (460–780 mg C m^{-2} d^{-1}). Nitrate was almost depleted (0.5–1 mmol m^{-3}) in the northern part of the STC, where results suggest that intermittent wind mixing may be important in sustaining summer production. Nitrate remained above 5 mmol m^{-3} in the SAZ, and this can be characterised as a high-nutrient, low-chlorophyll (HNLC) region.

The Subtropical Convergence (STC) to the south of Africa forms the poleward boundary of warm, salty, surface waters of the South Atlantic subtropical gyre. Lutjeharms & Valentine (1984) note that the location of the STC ranges from 40°35'S to 42°36'S with a mean position of 41°40'S, although there is considerable variability in its N–S latitudinal position. Whitworth & Nowlin (1987) describe the STC as a surface feature characterised by a southward temperature decrease from 14 °C to 10 °C and having a southward salinity gradient of 34.9 to 34.4. It defines the boundary between Subtropical Surface Water (SSW) and Subantarctic Surface Water (SAASW). Lutjeharms & Valentine (1984) give the mean central temperature of the STC as 14.2 °C within a cross-frontal range of 17.9 °C to 10.6 °C, a temperature decrease of 7.3 °C. Salinity decreases are variable but salinity falls by at least 0.5 in the range 35.5–35.6 to 34.3–34.6 (Lutjeharms *et al.*, 1993; Lutjeharms, 1985). South of Africa, the STC is a shallow feature of little more than 300 m in depth although its downstream streamlines are determined by the bottom topography (Weeks & Shillington, 1994). An extensive review of Southern Ocean fronts between the Greenwich Meridian and Tasmania appears in Belkin & Gordon (1996).

The optically clear, warm (18–25 °C) but nutrient-impoverished Agulhas Return Current (ARC), just to the north of the STC, lies between longitudes 13.5°E and 25°E and has a preferential retroflection at 20°E and a secondary retroflection at 16°E (Lutjeharms & van Ballegooyen, 1988). The Agulhas Front of the ARC has steeper density gradients than any other front in the Southern Ocean. With an observed southern limit of 40°01'S, it can almost merge with the STC over distances of ± 500 km, making these two features sometimes difficult to distinguish (Read & Pollard, 1993). Because of their mutual proximity, this region can be marked by extremely strong temperature gradients of up to 1 °C km^{-1}, (see Mantel *et al.*, 1999, for winter 1993 data). Cooling of the ARC both reduces thermohaline stability and increases the effectiveness of the CO_2 solubility pump.

Rings, eddies and filaments originating from the Agulhas Retroflection are the most energetic in the world and are responsible for a considerable transfer of heat, salt and energy from the South Indian Ocean into the South Atlantic (Perissinotto & Duncombe Rae, 1990; Van Ballegooyen *et al.*, 1994). On average, 7–9 warm-core eddies are shed annually from the Agulhas Retroflection Current into the South Atlantic. These eddies of approximately 200 km diameter and 750 m depth (as defined by the 10° isotherm) are often located at approximately 42.5°S 23°E and some appear to be topographically trapped by the Agulhas Plateau. Recent work (Schmitz, 1996; reviewed by Macdonald & Wunsch, 1996) suggests a direct link between Agulhas eddies and the intensity of global thermohaline circulation. Lutjeharms & de Ruijter (1996) have described the inter-basin leakage of heat and salt between the South Indian and South Atlantic oceans attributable to Agulhas current filaments, calculating that they contribute approximately 15% of the total inter-basin salt flux. The rate at which these rings and eddies cool also has significance for ocean–atmosphere exchanges of CO_2.

Remotely sensed data from AVHRR and CZCS (Weeks & Shillington, 1994) demonstrate that the surface expression of the STC is characterised by a relatively strong chlorophyll *a* signal of marked interannual variability. Spatial distribution patterns of chlorophyll *a* appear to be influenced by meridional excursions of the Rossby wave in the ARC and peak pigment concentrations occur in association with the 15 °C isotherm. In 1979, the mean chlorophyll *a* concentration was estimated to be 0.25 ± 0.12 mg m^{-3} (between 25° and 35°E) whereas in 1982 this was 0.57 ± 0.42 mg m^{-3}. Further east, between 35.8°E and 50°E, these values had diminished to 0.21 ± 0.09 mg m^{-3} (1979) and 0.44 ± 0.36 mg m^{-3} (1982) as Rossby wave patterns weakened easterly into the South Indian Ocean. The higher values were favoured when the ARC and STC were in close proximity.

By contrast, to the west in the mid-Atlantic sector, weak or ephemeral boundary conditions and a more variable latitudinal extent characterise the STC. The chlorophyll signal is weaker here also and cross-frontal exchanges of planktonic organisms are more in evidence (Barange *et al.*, 1999). The absence of the ARC to the west and its diminishing influence to the east provides the key to higher primary production and biomass immediately to the south of Africa. This appears to occur only when the ARC is in close and influential proximity to the STC or where temperature gradients are strongest.

Table 10.3. *Summary of chlorophyll concentrations, percentage phytoplankton size fractions, mean P–E parameters (n = 5) and rates of production for the three regions, April 1991*
Units and abbreviations: Chl *a* (mg m^{-3}); net- (20–200 μm), nano- (2–20 μm) and picoplankton (< 2.0 μm); PB_{max} (mg C mg^{-1} Chl *a* h^{-1}); α (mg C mg^{-1} Chl *a* h^{-1})/(μmol m^{-2} s^{-1}); E_k (μE m^{-2} s^{-1}); production (mg C m^{-2} d^{-1}) to 100 m depth.

Region	Z_m	Chl *a*	Net-	Nano-	Pico	PB_{max}	α	E_k	Production
Subantarctic	0	0.72	61	22	8	1.8	0.04	47.6	404 ± 78
	100	0.46	—	—	—	1.6	0.06	34.4	—
Eddy edge	0	0.92	62	21	9	1.5	0.06	31.3	453 ± 56
	100	0.53	—	—	—	1.4	0.09	22.0	—
Eddy core	0	0.36	41	47	9	1.4	0.06	35.0	287 ± 24
	100	0.37	—	—	—	1.9	0.07	29.1	—

Source: Dower & Lucas (1993).

Measurements of phytoplanktonic size-structure and primary production in the STC region south of Africa are summarised in Tables 10.3 – 10.7. During a late summer cruise in April 1991, Dower & Lucas (1993) measured primary production by using a photosynthetron. P–E parameters revealed the relation between the light environment, photoadaptation (E_k), the mixing depth (Z_{mix}) and production to test the hypothesis that if Z_{mix} exceeds the depth of E_k, then phytoplankton will experience a sub-optimal light regime, resulting in sub-optimal photosynthetic rates.

Appropriate P–E measurements were made within an anti-cyclonic warm-core eddy of ARC origin with further measurements extending into Subantarctic Surface waters. The eddy, of diameter ± 200 km, was situated at 42°11.8'S, 22°13.7'E and was characterised by the 10 °C isotherm extending to 750 m at its centre, resulting in deep mixing (Z_{mix} 140 m), which exceeded the 1.0% light euphotic depth range (Z_{eu}, 38–92 m). In subantarctic waters, Z_{mix} was 12–80 m and Z_{eu} was 40–75 m. NO_3 concentrations (1.7–2.1 mmol m^{-3}) were minimal within the eddy between 0 and 100 m. Elsewhere, NO_3 concentrations exceeded this value.

It is evident that there is enhanced biomass and production at the eddy edge. In the eddy core, biomass and production are much lower, whereas in subantarctic waters biomass and production are intermediary. Dower & Lucas (1993) conclude that it is the relationship between mixing depth (Z_{mix}), E_k depth (Z_{I_k}) and the euphotic depth (Z_{eu}), which exerts the strongest control on primary

Table 10.4. *The relation between mixing depth (Z_{mix}), E_k depth (Z_{E_k}) and the euphotic depth (Z_{eu}) for the three regions*

Region	Z (m)	Z_{E_k}/Z_{mix}	Z_{mix}/Z_{eu}
Eddy core	0	0.27	3.2
	100	0.27	—
Edge	0	1.10	0.72
	100	1.20	—
Subantarctic	0	0.68	1.1
	100	0.80	—

production. Increasing values of Z_{mix}/Z_{eu} and decreasing values of Z_{E_k}/Z_{mix} point to light limitation due to deep mixing below the euphotic zone or depth of optimal photosynthesis defined by E_k. A value of 1.0 or < 1.0 (Z_{mix}/Z_{eu}) implies optimal photosynthesis over the entire mixing depth, whereas values of > 1.0 imply sub-optimal light conditions for photosynthesis (Table 10.4).

The above analysis does show that the light environment within the eddy is the most unfavourable, owing to deep convective mixing. At the eddy edge, where warm water overlies cold subantarctic water, the mixing depth is reduced, so improving the light environment for photosynthesis. Perhaps this mechanism accounts for the observed zonal differences in biomass and production values in the STC region. Where warmer western boundary currents such as the Agulhas Retroflection impinge on the STC, satellite imagery (Weeks & Shillington, 1994) confirms elevated chlorophyll concentrations.

A central concern in JGOFS is that of estimating primary production at the ocean-basin scale. It is a scaling-up problem. One way to overcome this is to combine ocean colour satellite imagery with P–E parameters. An estimate of late summer and annual primary production for the STC region to the south of Africa has been calculated by using this approach (Table 10.5). The CZCS estimate of average surface Chl *a* for the region was 0.49 mg m^{-3} over a 7 yr period from 1979 to 1986 (Weeks & Shillington, 1994). The range for this period was 0.15–2.0 mg m^{-3}, representing a deviation of ± 69.4% from the mean on an inter-annual basis.

Summer productivity measurements at 30 nm intervals have been made along the WOCE A12/SR2 transect (1992–3) between Cape Town and Antarctica (Table 10.6) using simulated *in situ* on-deck ^{14}C incubations of 24 h and covering six depths to the 1% light depth. Primary production was highest at the MIZ, APF and STC frontal regions. However, netplankton dominated production only at the MIZ whereas at the STC nano- and picoplankton clearly dominated primary production on this occasion. These data and those of Dower

Table 10.5. *Synoptic primary production at the STC from P–E and CZCS imagery, April 1991*

Daily production	287–453 mg C m^{-2} d^{-1}
Ocean area	1.24 × 10^6 km^2
Annual production	1.33–2.09 × 10^{-1} Gt C yr^{-1}
Antarctic STC contribution to global production (%)[1]	0.50–0.80%

[1]Based on a combined open ocean of 3.1 × 10^8 km^2 and a continental margin of 0.6 × 10^8 km^2 (Walsh *et al.*, 1991).
Source: From Dower & Lucas (1993).

Table 10.6. *Summer (Jan.–Feb. 1993) primary production measurements (mg C m^{-2} d^{-1}) along a transect between South Africa and Antarctica*

Region	Total	% Netplankton	% Nanoplankton	% Picoplankton
MIZ	278.3	68.2	20.8	11.0
POOZ	105.6	24.6	49.4	26.0
APF	237.5	45.3	31.1	23.6
SAZ	125.8	14.3	50.0	35.7
STC	178.1	23.9	52.4	23.7
ARC	86.5	13.1	33.0	53.9

Source: Data from Laubscher *et al.* (1999).

& Lucas (1993) point to the considerable variability in phytoplankton abundance, community structure and productivity associated with the STC. Size-fractionated ^{15}N measurements by Mantel *et al.* (1999) of new and regenerated production on the same cruise during the summer of 1992–3 (Table 10.7) support the community structure observations. Furthermore, the maximum *f*-ratio value of 0.3 was observed at the MIZ whereas elsewhere the *f*-ratio was always less than 0.1, indicative of regeneration-based primary production where export production is insignificant.

Winter measurements of primary production at the STC and in an eddy at a similar position to that studied in 1991 (approx. 42°S; 22°E) were made in June 1993 (Dower & Lucas, 1999). Chlorophyll *a* concentrations seldom attained more than 0.7 mg m^{-3} except at the northern edge of the STC where patches of ± 1.0 mg m^{-3} were evident. As before, P–E parameters and light and chlorophyll profiles were used to estimate integrated primary production to the 1% light depth. Relative to April 1991, primary production was 30–50% lower in the winter and dominated by a nano- and picoplankton (61–83% by biomass) regeneration-based community utilising NH_4 and urea, rather than NO_3. NH_4 assimilation and regeneration were almost in balance, indicative of the close

Table 10.7. *Mean phytoplankton production (mg C m^{-2} d^{-1}) and* f*-ratios at the STC*

Region	Phytoplankton production		*f*-ratio
	June 1993	April 1991	
Eddy edge	140–332	453 ± 156	< 0.1
Eddy core	141–159	287 ± 124	< 0.05
STC	261–264	ND	< 0.1
Subantarctic	154–335	404 ± 178	< 0.1

Sources: Dower & Lucas (1999); Mantel *et al.* (1999).

Table 10.8. *Microzooplankton grazing at the STC (winter; June 1993)*

Region	Microzooplankton grazing	
	(% Chl d^{-1})	(% PP d^{-1})
Eddy edge	29–50	67–71
Eddy core	24–35	55–83
STC	41–53	56–67
Subantarctic	26–34	70–82

coupling between heterotrophic excretion of NH_4 and its photosynthetic assimilation (Mantel *et al.*, 1999).

Bacterial activity in the < 1.0 µm fraction is generally unimportant in NH_4 regeneration processes, whereas microzooplankton within the 2–20 µm size class are usually implicated. Table 10.8 shows information from Froneman *et al.* (1995) on microzooplankton grazing rates obtained from the same water as the P–E and ^{15}N studies of June 1993. Microzooplankton grazing removes between 55 and 82% of the primary production on a daily basis in all regions, confirming the close coupling between production, grazing and nutrient recycling in the nanoplanktonic community.

In winter, and quite frequently in summer, primary production at the STC is dominated by nanoplankton, is regeneration-based, and is coupled with microzooplankton grazing. These observations imply little biological drawdown of CO_2 at the STC. Instead, under-saturated p_{CO_2} gradients to the south of Africa are driven primarily by the 'physical solubility pump' associated with water cooling, or by subduction of Subantarctic Surface Water (SASW) at the STC. SASW finally appears as upwelled water in the southern Benguela upwelling region off the southwest coast of South Africa in the summer months. Ageing with CFC-113 : CFC-111 ratios indicate that upwelled water is probably less than ten years of age (Smythe-Wright *et al.*, 1996), so that its TCO_2 signal is relatively recent, implying that there may be no long-term storage of CO_2 in the STC region to the south of Africa.

The Polar Frontal Region

The Polar Frontal Region was investigated in detail in the Atlantic and Indo-Pacific sectors. One of these cruises was that of RV *Polarstern* (ANT X/6) during 6 weeks along 6°W on 11 latitudinal transects from 60°S (Weddell Gyre) to 47°S (Polar Front) in austral spring 1992. Australian studies were conducted on WOCE transects aboard *Aurora Australis* southwest of Tasmania along 140°E from 50°S to 60°S in austral late winter (October 1991), autumn (March 1993) and summer (January 1994, 1995).

The Joint Global Ocean Flux Study in the Atlantic Sector started with a cruise by RV *Polarstern* in 1991–92. The main goals of this cruise were the assessment of biological, chemical and physical processes governing plankton development and carbon dynamics in the ice-free areas and in the marginal ice zone. Investigations centred on the carbon cycle in the Southern Ocean, and one of the most important and much discussed questions was: 'What constrains phytoplankton from depleting the nutrient stocks in the Weddell Sea?' (Smetacek *et al.*, 1997). Hydrographic influences, grazing pressure and iron limitation were considered as potential limiting factors. The systems of the sea-ice, the upper mixed water column and the sea–sediment interface received special attention. Various experiments were conducted on board *Polarstern* to investigate specific biological processes of phytoplankton, zooplankton, bacteria and the microbial network.

The investigation area was chosen according to sea ice conditions observed from the ship and from NOAA-satellite ice charts. In mid-October, the pelagic system showed a typical late winter situation: biomass and productivity were low (Bathmann *et al.*, 1997), nutrient concentrations were high and aqueous carbon dioxide concentrations exceeded atmospheric concentrations (Bakker *et al.*, 1997). Only zooplankton, mainly consisting of krill and large copepods, were abundant and active under the ice and between ice floes (Dubischar & Bathmann, 1997; Fransz & Gonzalez, 1997).

One month later, in November, the situation had changed completely. The ice edge had retreated southwards, leaving only some icebergs behind (van Franeker *et al.*, 1997). Despite this massive ice retreat, no phytoplankton biomass enhancement was observed in the vicinity of the MIZ (Bathmann *et al.*, 1997) despite relatively high potential primary production (Jochem *et al.*, 1995) and very little grazing by mesozooplankton (Dubischar & Bathmann, 1997) (for comparison see also Marginal Ice Zone). At the MIZ and at the Polar Front, mesoscale eddies were recorded, which remained stable for about 4 weeks (Veth *et al.*, 1997). In the eddies at the Polar Front, phytoplankton accumulated in about 4 weeks, reaching biomass values of more than 4 mg Chl *a* m^{-3}.

Three spatially distinct phytoplankton assemblages were found in this high-biomass area, which extended to water depths of about 100 m (Bathmann *et al.*, 1997). Each phytoplankton bloom was dominated by a different species or species assemblage; the two most important ones were the bloom at 48°S dominated by the Antarctic diatom *Corethron* spp. undergoing sexual reproduction (Crawford, 1995), and another bloom dominated by *Fragilariopsis kerguelensis* (Crawford & Hinz, 1997) at 50°S. A total biomass of more than 280 mg Chl *a* m^{-2} (Bathmann *et al.*, 1997) in these areas was similar to those usually found in many shallow shelf seas of the world ocean. Primary productivity of up to 3 g C m^{-2} d^{-1}, recorded during this cruise (Jochem *et al.*, 1995), is only rarely found in open ocean waters of Antarctica. Thus, it seems that semi-permanent hydrographic stabilisation of the upper water column in the Polar Frontal region for some weeks gave favourable conditions for phytoplankton growth while growing conditions were probably influenced by chemical and biological factors as discussed below.

The control of primary production by the availability of biologically active iron is discussed elsewhere (de Baar *et al.*, 1995) but in summary, iron concentrations were high and co-varied with phytoplankton bloom occurrence. Other (macro-) nutrients were present in excess in all waters showing enhanced phytoplankton growth (except at the very northern limit of the transects) (Scharek *et al.*, 1997). Bacterial biomass and production followed phytoplankton development, and both were highest at the end of November in the Polar Frontal Zone. This close coupling between primary production and heterotrophic utilisation caused re-mineralisation of about one third of the freshly produced organic matter in the upper 200 m of the water column (Lochte *et al.*, 1997). Grazing impact on phytoplankton production by microzooplankton was in the same range in all regions (30% of primary production) with the exception of the MIZ, where microheterotroph grazing accounted for 10% of daily primary production (Klaas, 1997). Copepod grazing, by contrast, accounted for less than 3% of primary production in all regions investigated in this study (Table 10.9) (Dubischar & Bathmann, 1997).

The most important mesozooplankton grazers, however, were salps (*Salpa thompsoni*), which occurred in huge swarms in the southern Antarctic Circumpolar Current (ACC) south of the PFZ. The grazing impact of these salps exceeded primary production (Dubischar & Bathmann, 1997). Thus, bottom-up control of primary production (via availability of the micro-nutrient iron) and top-down control by grazing of salps concentrated in swarms occurred simultaneously in adjacent areas of the open Southern Ocean in spring.

Table 10.9. *Ingestion rates given for three dominant zooplankton species in the three areas investigated during the German SO-JGOFS cruise in 1992*

Area	Dominant species	Ingestion rates (% of daily primary production)
PFZ	*Rhincalanus gigas*	0.05–0.35
ACC	*Salpa thompsoni*	> 100
MIZ	*Calanus propinquus*	0.02–0.55

Phytoplankton accumulation was controlled by hydrographic conditions.

Export of particles (and associated biogenic matter such as organic carbon, silica, etc.) was high in the frontal region and is documented by depletion of ^{234}Thorium (Rutgers van der Loeff *et al.*, 1997). At the sea floor, the geochemical profile was influenced by the input of sinking organic material.

In conclusion, it is not always the MIZ, but frequently the PFZ, which is important in terms of phytoplankton biomass accumulation and subsequent sedimentation including drawdown of CO_2 from the ocean surface to the interior of the ocean (see below, Carbon dioxide exchange between the ocean and the atmosphere). Favouring the phytoplankton bloom were the hydrographical conditions of water column stabilisation in the upper 40–50 m induced by water mass overlayering. Relatively high concentrations (2–4 nmol l^{-1}) of dissolved iron in surface waters probably induced high phytoplankton growth rates, particularly of diatoms, but where total mesozooplankton grazing was only about 1–3% of primary production. Thus, the PFZ is one of the most important open-ocean sites where diatom growth is stimulated and where subsequent sinking of diatom frustules transports biogenic silica into the ocean interior.

The PFZ southwest of Tasmania along 141°E from *c.* 53°S to 58°S contrasts strongly with the zone north of the SAF in the same region. The PFZ in the Indian Ocean sector is characterised by relatively shallow winter mixed layers, with a permanent pycnocline at about 100–120 m. Shallower stratification occurs in the summer, with a seasonal mixed layer at *c.* 70 m. Surface p_{CO_2} data in this region suggest the zone is often a weak source of CO_2 to the atmosphere. Chlorophyll *a* concentrations are low and uniform in the winter. A surprising feature of this zone was the development of a sub-surface chlorophyll maximum in summer. This maximum appears to develop and deepen throughout the summer, possibly starting with a surface bloom associated with transient shallow stratification in December. The development of the sub-surface chlorophyll maximum appears to be associated with the depletion of silicic acid in the surface layer, whereas nitrate remains above 20 mmol m^{-3} year-round.

Estimated water-column primary production is low in this zone in winter, summer and autumn, but may be high in the spring bloom.

The Marginal Ice Zone

Sea ice is one of the dominant physical features of the Southern Ocean, covering an area of about 20 million km² at its maximum extent in winter. This accounts for 50% of the ocean area south of the Polar Front. Although only forming a layer no more than 1–2 m thick in most places, sea ice has significant physical influences through its modification of ocean–atmosphere fluxes of heat, momentum and materials, and it imposes constraints on biological processes, especially through its modification of underwater illumination.

Although the properties of ice as a feature of the ocean–atmosphere interface have undoubted biogeochemical significance, a second feature of the sea ice of the Southern Ocean is even more important. The large majority of the ice is seasonal, with only 4 million km² remaining in summer. This means that, for over half of the area of the Southern Ocean bounded by the Polar Frontal Zone, the seasonal pattern of biological and biogeochemical processes is dominated by the advance and retreat of sea ice. Whereas biogeochemical provinces in other parts of the World Ocean can be defined by correspondence to oceanographic regimes, the Southern Ocean ice-associated system transgresses most of the physical oceanographic features (see Tréguer & Jacques, 1992). The formation and melting of sea ice also gives rise to important physical effects. Cold saline brine is rejected during ice formation, giving strong vertical convection and being an essential element in the formation of Antarctic Bottom Water. When ice melts, corresponding low-salinity water results, and this may have an important influence on vertical stability of the water column. This stability may enhance phytoplankton productivity, giving rise to high nutrient utilisation rates by ice-edge blooms, which contrast strongly with the apparent low growth rates in the open waters of the Southern Ocean. The high albedo of ice, together with the high latent heat of melting compared with the specific heat of seawater, ensures that much of the solar energy incident at the surface of the Southern Ocean is either reflected or utilised in melting, rather than giving rise to seasonal warming.

Because of the importance of sea ice in the Southern Ocean biogeochemical system, this feature was given particular prominence in the plan for a JGOFS regional study in the area. Before JGOFS, studies of the production and dynamics of the ice-edge system had been relatively restricted in both their spatial extent and seasonal coverage. Crude extrapolation of these data had suggested that the Marginal Ice Zone (MIZ) might contribute as much as 40%

of the total photoautotrophic carbon fixation in the Southern Ocean. It was noted that the physical properties of the MIZ were dominated by the thermodynamics of the thin ice cover, and could thus be climatically sensitive.

A number of cruises by European groups and as part of the South African and Australian Southern Ocean programmes have extended the coverage of MIZ biogeochemical processes. The magnitudes and rates of biological and physical processes occurring at the ice edge and at the Antarctic Slope Front (ASF) in the Lazarev Sea were measured during two CTD grid surveys during a South African cruise (1995); several German cruises investigated the effect of sea-ice formation and melt on biological communities and processes.

On the Australian WOCE transects along 140°E, shallow stratification has been observed in the MIZ in both late winter (October) and summer, and phytoplankton were typically high-light-adapted. Chlorophyll concentrations were very low in winter, but a pronounced shallow sub-surface chlorophyll maximum, peaking at 1.2 mg Chl *a* m^{-3}, was observed in summer. However, estimated primary production levels were moderate.

The following data, however, derive predominantly from the UK study of the springtime evolution of the MIZ in the Bellingshausen Sea, on the Pacific Ocean side of the Antarctic Peninsula. The experiment had been designed specifically to study the development of the MIZ community as the ice retreated southwards in spring. The UK study involved two ships, one making detailed process measurements at a series of stations along a north–south transect, while the second vessel conducted larger-scale surveys using underway instrumentation. The site was selected on the basis of its relatively uncomplicated pattern of ice retreat.

The major feature of the study was a diatom bloom, which remained in approximately the same location during the entire study, despite a southward retreat of the ice edge by about 2° latitude. The location of the bloom was ice-covered at the start of the study, and its position was more or less constant rather than shifting with the ice edge. A further complication emerged in that the bloom appeared to be associated with a front, which corresponded to the southern branch of the Antarctic Polar Front. Thus the genesis and history of this bloom did not seem to correspond with what has become the classical picture of MIZ blooms, which is based on sea-ice meltwater stabilisation of the upper water column, thus enhancing phytoplankton production.

A large-scale survey using underway techniques, including ‘SeaSoar’, was used to map the horizontal distribution of properties relating to the carbon dioxide system and phytoplankton production in ice-free areas, while rate measurements were made at a limited number of stations in open water and in the ice.

Phytoplankton were dominated by large-celled diatoms in open water conditions, but small algae, predominantly flagellates, formed the bulk of the phytoplankton population under the ice. Phytoplankton biomass concentration exceeded 7 mg Chl *a* m^{-3} in the bloom, which comprised large centric diatoms together with *Phaeocystis*. Primary production measured in the open-water area north of the ice edge ranged from 36 to 495 mg C m^{-2} d^{-1} (Savidge *et al.*, 1995), but was in the region of 800 mg C m^{-2} d^{-1} within the open-water bloom region associated with the southern branch of the APF (Boyd *et al.*, 1995). Although primary production in open water was dominated by the larger-celled phytoplankton, smaller cells contributed more relative to their biomass, and close to the ice edge, the contribution of all size fractions was similar. Savidge *et al.* (1995) noted the similarity between the size fractionation of phytoplankton production in the Bellingshausen Sea study and that of the pre-bloom phase in the northeast Atlantic in spring.

Comparison can be made between ice production and the underlying water column. The production rates obtained for the ice were 18–174 mg C m^{-2} d^{-1} whereas column-integrated production was 87 mg C m^{-2} d^{-1} (0–70 m), or little over 1 mg C m^{-3} d^{-1} (cf. the value of 800 mg C m^{-2} d^{-1} quoted above is for the open water north of the ice.). Corresponding data for chlorophyll were 3.4–27.7 mg m^{-3} in ice and 2.8 mg m^{-2} (0–30 m). At all sites, phytoplankton were adapted to low light (Boyd *et al.*, 1995; Savidge *et al.*, 1995). These biological differences on an meridional line across the MIZ have strong impact on the CO_2 uptake and the relevant processes involved (see below, Carbon dioxide exchange between the ocean and the atmosphere, for further discussion).

Nutrient relationships in the Southern Ocean as a whole provide a complex paradox. In most other parts of the world ocean, phytoplankton growth in spring normally utilises (and is limited by) available inorganic nutrients in a seasonally well-stratified euphotic zone. Nutrients introduced into the mixed layer by winter mixing therefore provide the upper limit to carbon export into the deep ocean during the growing season. Further production will of necessity be based on recycled nutrients, and thus implicitly recycle carbon. However, in the Southern Ocean, inorganic nutrient concentrations in surface water remain high throughout the year, although there is a demonstrable decrease during summer. Nevertheless, reduced nitrogen utilisation by phytoplankton predominates, despite the fact that nitrate commonly represents more than 90% of the inorganic nitrogen pool.

During the Bellingshausen study, nitrate concentrations in near-surface water under the ice were approximately 33 mmol m^{-3} and were reduced to less than 22 mmol m^{-3} in the region of the bloom (Whitehouse *et al.*, 1995). This reduction in nitrate could be equated stoichiometrically with the apparent increase in phytoplankton biomass. However, ammonium and urea uptake measurements indicated that these regenerated and reduced nitrogen sources were markedly preferred (Waldron *et al.*, 1995). Under the ice, urea was the preferred substrate, with an uptake rate of 9.4 mg N m^{-2} d^{-1} whereas nitrate uptake was 1.3 mg N m^{-2} d^{-1} (Bury *et al.*, 1995). This probably reflected the relatively high concentration in the water column (1.4 mmol m^{-3} at the surface), and Bury *et al.* (1995) speculate that this may represent an allochthonous nitrogen source.

The grazing system at the MIZ was studied along the gradient from under the ice to open water, through the phytoplankton bloom. A range of grazers was examined, from protozoa to euphausiids. Under ice, grazing pressure was low. Protozoan concentrations – predominantly of oligotrich ciliates and dinoflagellates – were of the order of 60 mg C m^{-2} (Burkill *et al.*, 1995). Smaller zooplankton were dominated by copepods at all stations (Atkinson & Shreeve, 1995) and zooplankton biomass under ice was 300 mg C m^{-2} integrated to 600 m depth (Robins *et al.*, 1995). Copepod grazing rates were broadly comparable to those measured later in the year in northern parts of the Southern Ocean. However, copepod grazing accounted for a minute proportion of phytoplankton production (Atkinson & Shreeve, 1995), in contrast to microzooplankton grazing, which equated to 21% of the relatively low production under ice.

Within the phytoplankton bloom, microzooplankton standing stocks and grazing impacts were significantly higher, with phytoplankton removal rates of 3260 mg C m^{-2} d^{-1}, which is apparently more than could be sustained by *in situ* primary production (Burkill *et al.*, 1995). Metazooplankton biomass was slightly higher at this site, but only about double that seen under ice (Robins *et al.*, 1995) and this accounted for less than 10% of daily *in situ* primary production. Seasonal migrant copepods had ascended from depth to feed in the bloom.

Larger zooplankton, predominantly euphausiids, were detected acoustically. There were marked changes in biomass and distribution between surveys (Murray *et al.*, 1995). One very large acoustic target was interpreted as a krill swarm with a total biomass of 3.7×10^4 tonnes fresh mass. Murray *et al.* (1995) speculate that such a swarm could have produced 300 t of carbon in faecal pellets per day.

Much of the particulate material in the water column was of algal origin, as indicated by their biogenic hydrocarbon signature (Cripps, 1995). Material in surface sediment under the pack ice probably originated from deep-living copepod populations, whereas the algal bloom dominated the sediment hydrocarbon suite at more northerly stations. Radionuclide techniques showed that up to 250 mg C m^{-2} d^{-1} were exported from surface waters within the bloom, about 20% of the primary production (Shimmield *et al.*, 1995). Under ice, export was very much less, and represented a very small proportion of daily production.

The data obtained by this study indicate that some re-assessment of the basic biogeochemistry of the MIZ is needed. First, the physical system may be substantially less predictable than was previously thought. Interaction with the physical structure of the water column is undoubtedly a factor in determining the dynamics of productive systems (see, for example, Bianchi *et al.*, 1992), and the relationship between the ice edge and hydrographic features will be complex. The potential for establishing the supposedly typical MIZ production system based on the shallow low-salinity layer derived from ice melt must rest, pragmatically, on the conditions that determine whether this meltwater will be retained or dispersed. Although the Bellingshausen Sea site of the STERNA study was selected for its simple pattern of north-to-south ice retreat, it is also an area where the ice edge is exposed and where, during this study at least, the ice edge coincided with regions of strong easterly advection of surface water associated with the Southern APF.

The distribution and abundance of primary production is clearly variable and determined by a complex of physical and biological factors. In the STERNA study, the association of the main bloom with an eastward-flowing jet suggested that its dynamics might involve very much larger distance scales. The variation in p_{CO_2} clearly demonstrated that both biological and physical processes mediated in air–sea fluxes.

Nutrient uptake, particularly the partitioning of nitrogen, showed that it may not be possible to apply current paradigms to estimate Southern Ocean export production. As with other areas of the Southern Ocean, a substantial surplus of nitrate remains in bloom areas and phytoplankton production is fuelled primarily by regenerated nitrogen sources. Studies within the sea ice, where pulses of non-nitrate nitrogen may be released on melting, suggest that it is difficult to characterise allochthonous and autochthonous nutrient inputs by chemical speciation alone.

The fate of particulate material again showed some inconsistencies with the patterns considered typical for temperate oceans. Consider, for instance, a phytoplankton bloom in which nitrate concentrations are still 20 μM, grazing by microzooplankton alone appears to account for all of the daily production, and yet there is a clear signal of sedimented algal material at the seabed and export appears to be about 20% of daily primary production.

The ultimate paradox of the STERNA study can be summed up in two questions. First, was the bloom associated with the ice edge in any way, or did it arise in the front, which was co-located with it? Second, was the bloom self-sustaining, or was it topped up by material advected into the area from elsewhere?

The initial response to the first question is to suggest that the bloom was definitely not derived from the ice edge. To start with, it was not spatially correlated with the ice edge, but stayed in the same geographic location. At the start of the cruise, this was underneath pack ice and later an open area. The bloom also undoubtedly remained closely associated with the front. Yet we know that the ice edge had receded to the south of the bloom before the study, before extending northwards immediately before the arrival of the ships (Turner *et al.*, 1995). It is also clear that there was a developmental sequence of biological production from the ice-covered zone to open water. The extent to which the ice edge and the front is linked remains unclear, but it is unlikely that they are independent and it is possible that ice melt plays some part in determining the properties and possibly the position of the front.

The second question raises wider issues, which take us outside the dynamics of the MIZ *per se*. Read *et al.* (1995) indicate that the South Polar Front can be traced both up- and downcurrent for some distance. As Turner & Owens (1995) speculate, the source for the phytoplankton bloom could have been some distance upcurrent, which would not only have placed its origin further west but also a few hundred kilometres further north. Tracing the trajectory of the SPF backwards takes us close to Peter I Island: could the 'bloom' studied in the Bellingshausen Sea be the remains of a phytoplankton maximum associated with the island, which then ran the gauntlet of grazing and other loss processes as it was advected along the SPF? Several authors have suggested that the bloom could not be sustained by *in situ* production (Boyd *et al.*, 1995; Burkill *et al.*, 1995) and Whitehouse *et al.* (1995) suggest that the time taken to deplete nutrient concentrations to observed levels would have been longer than suggested by observations of ice retreat combined with plausible net growth rates.

Clearly, because of JGOFS studies, the role of the MIZ will need to be reassessed. It certainly cannot be treated as a system that is decoupled from the underlying water column, and transgression of fronts is likely to produce complex biogeochemical systems. Despite this, the MIZ remains an important facet of the Southern Ocean. A major component of the planned work by the JGOFS regional study will aim to test hypotheses on the generation and maintenance of MIZ systems.

In the Australian sector, there have been intensive studies of phytoplankton and microbial dynamics in Prydz Bay. A large seasonal drawdown of p_{CO_2} has been observed in Prydz Bay associated with intense phytoplankton blooms. Estimates of export production have been made only on the Tasmanian continental slope, using free-floating sediment traps. These indicate that export efficiencies through 200 m are low, being less than 10% of primary production, even during the spring bloom. The STC and SAZ show strong summer and annual mean drawdown of p_{CO_2}, and sinks of atmospheric CO_2. There are almost certainly both physical and biological contributions to these sinks, driven, respectively, by southward advection and cooling of low-DIC subtropical water, and northward Ekman transport of nitrate-rich, surface-equilibrated polar water. However, separating out these effects requires a better understanding of physical transport, including the location and magnitude of subduction and formation of Subantarctic Mode Water.

The fate of a phytoplankton bloom was followed in the Coastal Current (CC) in the Lazarev Sea (northeastern Weddell Sea) during an expedition with RV *Polarstern* in February and March 1991 (Bathmann *et al.*, 1992). At the beginning of the investigation, a maximum phytoplankton concentration of about 4 mg Chl *a* m^{-3} was measured, which was located above the pycnocline at about 40 m. Dominant phytoplankton species in the bloom were *Eucampia balaustium, Trichotoxon reinboldii, Chaetoceros dichaeta* and *Phaeocystis* spp. The bacterial biomass maximum was 17 mg C m^{-3} (2.6×10^5 cells ml^{-1}) which incorporated over 400 pmol thymidine l^{-1} d^{-1} (equivalent to a biomass production of approximately 11 mg C m^{-3} d^{-1}). The total amount of particulate organic carbon was about 450 mg POC m^{-3}. Nutrient concentrations during the bloom were 39 mmol m^{-3} silica, 16 mmol m^{-3} nitrate, 0.1 mmol m^{-3} nitrite and 1.2 mmol m^{-3} phosphate; these values are about half of the concentrations found below the pycnocline. Below the pycnocline, the biomass was considerably lower (0.02 mg Chl *a* m^{-3}, 2–9 mg C m^{-3} bacterial biomass and 50 mg POC m^{-3}).

After two days, the water column was mixed by a storm down to the shelf bottom at 250 m. As a result, all plankton biomass and other particulate and dissolved organic material was dispersed homogeneously in the entire water column, although integrated values of these parameters on a square metre basis remained rather constant. After another two days, Chl *a* and phytoplankton lipids were found in the upper 2 cm of the shelf sediments.

Calculations of particulate export from the euphotic layer due to the storm event based on organic carbon, chlorophyll biomass and thorium (^{234}Th) gave similar results. Given the bloom phytoplankton biomass as 100%, 71% sank out of the euphotic zone, of which 4.4% was incorporated into the sediments (nearly all of which was utilised within a week). We present evidence that, besides microbial breakdown of planktonic material, downslope transport of fresh phytodetritus is of importance. The calculations demonstrate that within the CC more than 13 g C m^{-2} of planktonic material was removed from the upper water column during the investigation. On an annual basis, about 50 g C m^{-2} may be exported from the upper 50 m of the CC in the Lazarev Sea.

Carbon dioxide exchange between the ocean and the atmosphere

The aim of Southern Ocean-JGOFS is to understand and to model the carbon cycle in a region that represents 20% of the World Ocean. Global atmospheric models predict that there is no net exchange of CO_2 between the ocean and the atmosphere. Large seasonal and spatial variations of CO_2 fluxes have been recorded in the different subsystems that compose the Southern Ocean (Tréguer & Jacques, 1992). Some examples of SO-JGOFS p_{CO_2} studies follow.

During the German SO-JGOFS 1992 cruise, Dutch investigators studied the temporal variability of the CO_2 flux at the atmosphere–water interface during spring, along a 6°W transect (Atlantic sector). In the Polar Front Zone, surface water was under-saturated with respect to atmospheric CO_2. Surface water p_{CO_2} decreased during the course of spring along with an increase in Chl *a* (Bakker *et al.*, 1997). In the Permanently Open Ocean Zone (POOZ), degassing of CO_2 was recorded owing to seasonal heating of the wind-mixed layer. The change from a slight source of CO_2 in early spring to a sink in austral summer (Takahashi *et al.*, 1993) could result from more active biological pumping during this season, although low Chl *a* concentrations and primary production have been recorded in this sub-system by many authors (Tréguer & Jacques, 1992). In the Seasonal Ice Zone (SIZ), Warm Deep Water upwelling beneath the sea ice explains CO_2 winter over-saturation of surface water parallel to oxygen under-saturation. In the wake of the retreating ice edge, during early spring, over-saturation was still measured in the remnant of Antarctic Winter Water, but during the course of spring, this CO_2 source changed into a CO_2 sink owing to phytoplankton growth.

An overall CO_2 balance was calculated during the one month study period by Bakker *et al.* (1997). For the whole 6°W transect from 47°S to about 60°S, the result clearly depends on wind speed estimates (local measurements or climatologic winds) and also on the exchange coefficient, which ranges between that of Liss & Merlivat (1986) and Wanninkhof (1992). A skin temperature effect can also affect the result. Accordingly, this balance may change from a small source (0.7 mmol CO_2 m^{-2} d^{-1}) to a strong sink (3.3 mmol CO_2 m^{-2} d^{-1}), including a null flux possibility, assuming a uniform 0.2 °C skin temperature difference.

The UK STERNA cruise in the Bellingshausen Sea (Pacific sector) was conducted almost at the same period. Classification of the meridional variation in properties allows an overview of the different processes at work in the system. Turner & Owens (1995) have defined a series of bands along the 85°W meridian, which passes through the middle of the Bellingshausen study area. In the north, there was under-saturation of p_{CO_2} with only moderate phytoplankton biomass. Turner & Owens (1995) interpreted this observation by analogy with summer post-bloom conditions in temperate waters, although clearly the fact that inorganic nutrient concentrations remained comparatively high showed that different factors controlled the inception and decline of the bloom. Within the frontal jet and to the immediate south of it, associated with the phytoplankton bloom, there was strong under-saturation in p_{CO_2} and high chlorophyll. A marked correlation between these two suggested that the CO_2 drawdown could be attributed to primary production. To the south of the front, p_{CO_2} was close to saturation and chlorophyll was low. Although similar to pre-bloom conditions in temperate waters, there was no indication that a bloom would have developed under the observed conditions.

Robertson & Watson (1995) have estimated the overall CO_2 uptake by a large region of the Southern Ocean during part of the austral summer, based on data from the Bellingshausen Sea and a second study at 80°E. They noted the likely presence of an overall sink for CO_2 over the entire Southern Ocean, and suggested that this may be a recent phenomenon, contrasting with the neutral picture of the region derived from early studies. These results are supported by measurements of p_{CO_2} by the Australian cruises in the marginal ice zone in the Indian sector of the Southern Ocean (Tilbrook *et al.*, 1995), which are providing evidence for large-scale zonal variability in the air–sea flux of carbon. The region of the SIZ between about 65°E and 110°E shows a strong decrease in p_{CO_2} apparently associated with the development of phytoplankton blooms over summer. The lowest values of p_{CO_2} have been measured in nearshore waters around Prydz Bay (p_{CO_2} of 109 µatm at about 65°E) during summer. Surface p_{CO_2} increased from the Prydz Bay region to about 110°E, where surface values are close to equilibrium with the atmosphere. The region between 110°E and 160°E has not been well sampled; however, the available data indicate that surface p_{CO_2} is close to atmospheric equilibrium (± 15 µatm) throughout much of the region. Flux estimates made using data collected south of 55°S (60°E to 150°E) indicate that the region is a weak net sink for atmospheric CO_2 of the order 0.03 Gt C between October and March. The large drawdown in p_{CO_2} observed in Prydz Bay and coastal waters was not associated with large net CO_2 fluxes because of their low areal extent and the low summer wind speeds that occur along the Antarctic coast.

During the SO-JGOFS Symposium held in Brest in August 1995, Tilbrook *et al.* (1995), reviewing the available data sets south of 55°S, concluded that the Southern Ocean is a net sink of CO_2 of 0.2–0.4 Gt C yr^{-1}. This is about 10–20% of the World Ocean total.

It is important to note this estimate excludes frontal areas. Numerous French cruises have covered this Southern Ocean sub-system in the Indian sector during Phase I of SO-JGOFS. Extrapolating the data to the whole Indian Ocean, Louanchi (1995) concluded that (1) the Indian Ocean CO_2 uptake is 0.3 Gt C yr^{-1}, i.e. 16% of the world ocean total CO_2 uptake, and (2) the latitudinal band between 20 and 50°S accounts for 0.07 Gt C yr^{-1}, i.e. 25% of the Indian Ocean total. No estimate is available at present for Southern Ocean

(*a*)

		Start	Finish
Minerve 25	Antares I	29/03/93	17/05/93
Minerve 28		25/07/93	8/08/93
Minerve 29	Antares II	28/01/94	23/02/94
Minerve 33	Antares III	28/09/95	20/10/95

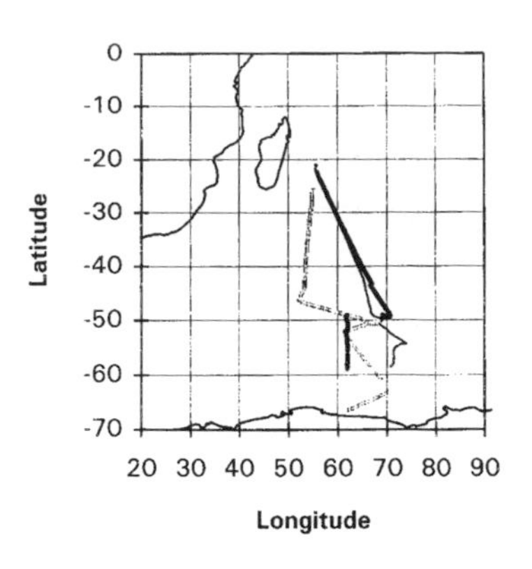

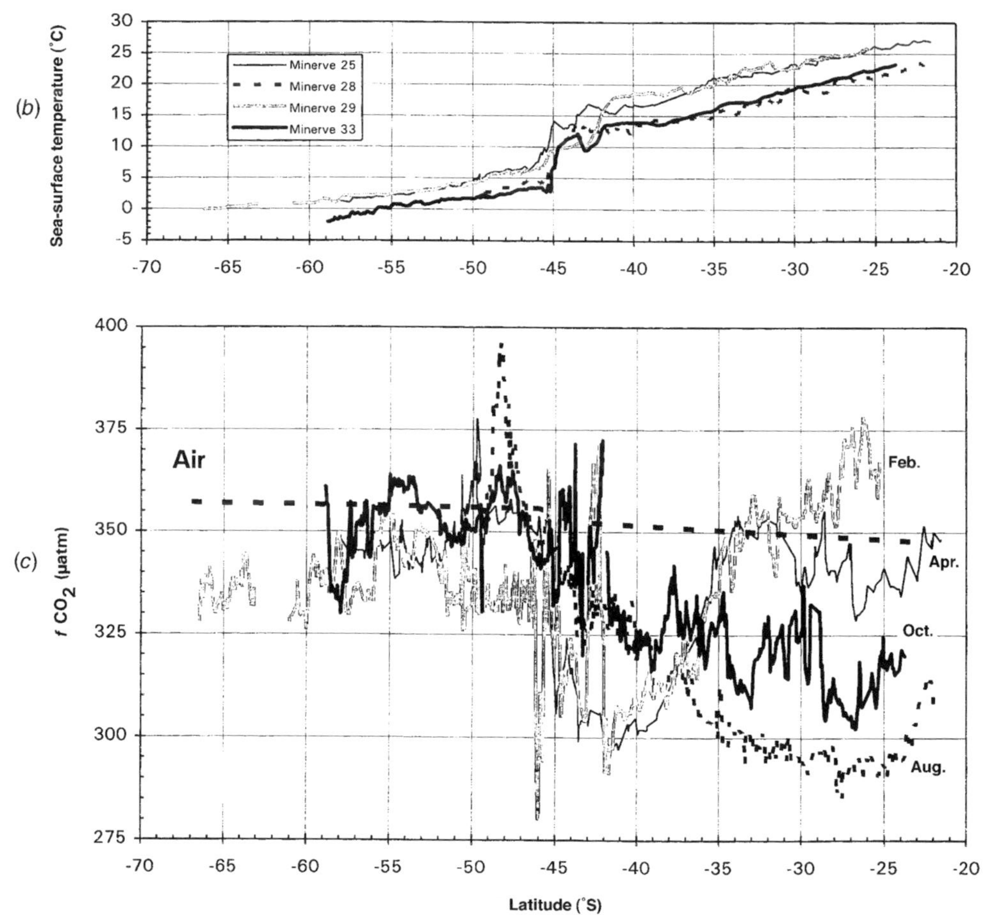

Figure 10.2 *Surface p_{CO_2} concentrations from La Réunion island (22°S) to ANTARES study area.* (a) *Cruise tracks of the MINERVA cruises;* (b) *sea-surface temperature vs. latitude, and* (c) *corresponding p_{CO_2} values vs. latitude.*

frontal zones *sensu stricto*. Our prediction is that their contribution to Southern Ocean CO_2 uptake might be important for two reasons. North of 45°S (i.e. north of the Subantarctic Front in the Crozet–Kerguelen zone) a permanent p_{CO_2} drawdown is observed, whatever the season (Fig. 10.2), related to Chl *a* concentrations (not shown in Fig. 10.2) and subduction at the STC. Transport can be as large as 50 μatm during summer and autumn. Because most frontal zones coincide with strong Westerlies, CO_2 flux is enhanced.

Data from SO-JGOFS Phase I confirm that the Southern Ocean is a high-nutrient low-chlorophyll (HNLC) ecosystem, with primary productivity typical of oligotrophic zones. This is especially true for the Permanently Open Ocean Zone (POOZ). Paradoxically the fluxes of biogenic matter at the sediment–water interface beneath the POOZ are more typical of mesotrophic than of oligotrophic ecosystems. Fluxes of biogenic matter in the POOZ have been documented both for surface waters (primary production and biogenic silica production) and at the sediment–water interface during the ANTARES/F-JGOFS cruises. For example, in the Indian sector, oxygen consumption at the sediment–water interface averaged 61 nmol cm^{-2} d^{-1} for the study area (ANTARES 1/F-JGOFS cruise). This is significantly higher than the mean value of 54 nmol cm^{-2} d^{-1} measured at the oligotrophic site of EUMELI/F-JGOFS (de Wit *et al.*, 1997). Bacterial abundance in the top centimetre for ANTARES 1 averaged 12×10^8 cells cm^{-3}, two times higher than for the oligotrophic NW African upwelling site of EUMELI. Incidentally, organic matter in the top centimetre was 4.3 mg C g^{-1} sediment for ANTARES 1, i.e. close to the value of 3.9 typical of the mesotrophic site of EUMELI.

Particulate organic carbon flux that reaches the sediment–water interface beneath the POOZ averaged 0.3–0.4 mol C m^{-2} yr^{-1} (Fig. 10.3*a*) (Rabouille *et al.*, 1997), which is 18–57% of the primary production calculated from the Geotop model. Buried carbon flux is estimated at 0.1 mol C m^{-2} yr^{-1} (Rabouille *et al.*, 1997), i.e. the preservation ratio ranges between 6 and 17% of the primary production. This is exceptionally high compared with the mean ratio for the World Ocean of about 0.4%. The rain rate of biogenic silica ranged between 1.4 and 2.5 mmol Si m^{-2} yr^{-1} (Fig. 10.3*b*) (Rabouille *et al.*, 1997) and the preservation ratio of biogenic silica for the POOZ sediments is estimated at about 20% (Nelson *et al.*, 1995), which again is exceptionally high compared with the 3% calculated for the World Ocean (Tréguer *et al.*, 1995).

The message is clear: the biogenic rain rate at the sediment–water interface is high in the Southern Ocean, despite its depth, compared with the rest of the World Ocean. Two lines of evidence indicate that fresh material can be rapidly transported from the surface layer to the seabed to account for these high flux rates.

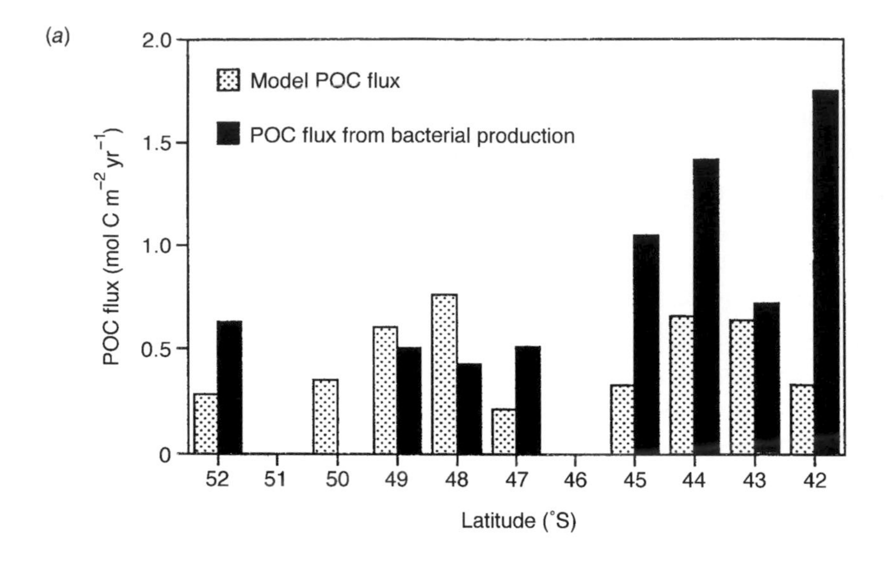

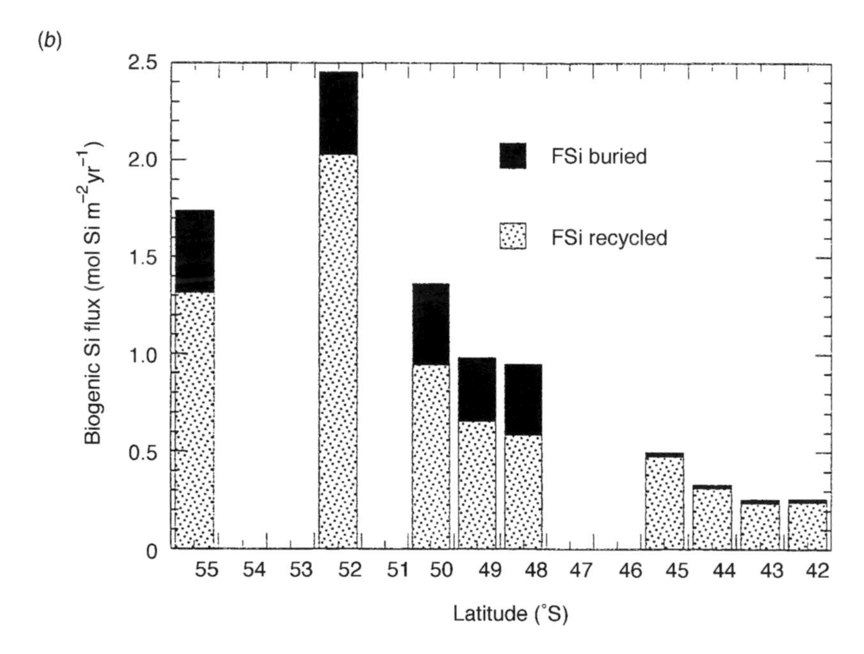

Figure 10.3 *ANTARES 1 cruise (autumn 1993). Flux of* (a) *particulate organic carbon (from model and from bacterial production), and* (b) *biogenic silica at the sediment–water interface beneath the Permanently Open Ocean Zone (south of 50°S) and beneath the frontal zone (north of 50°S) in the Indian sector of the Southern Ocean (58°E).*

First, during ANTARES 1 (austral autumn) the sediment–water interface consisted of 'fluffs' that were 4–5 cm thick. At 4000–4500 m depth, this fluff contained up to 13 mg m^{-3} of Chl *a* and up to 150 mg m^{-3} of phaeopigments (Riaux-Gobin *et al.*, 1997). It was mainly composed of well-preserved frustules of various diatom genera (of robust *Nitzschia*, and *Thalassiosira*, but also of delicate *Chaetoceros* and with *Corethron* and *Rhizosolenia*, which occurred still partly in chains). Also noticeable in the fluffs was the presence of well-preserved coccoliths and silicoflagellates. Revival tests (cultures of F/2 medium, initiated with samples from the sediment–water interface) were positive (Riaux-Gobin *et al.*, 1997), leading to diatom assemblages dominated by the genera *Chaetoceros* and *Nitzschia*, indicating that the deposited material was fresh.

Second, during ANTARES 2 (austral summer), salps were abundant (G. Champalbert, unpublished research) in the water column. These organisms are trophically important in many open-water areas of the Antarctic Circumpolar Current (see, for example, Bathmann *et al.*, 1997; Pakhomov & Perissinotto, 1997). Salps are known to produce faecal pellets that can sink from the photic zone to the abyssal seabed in 2–4 d (Fortier *et al.*, 1994), representing a rapid vertical conveyor for biogenic material from the surface to the sediments. *Salpa thompsoni*, the most abundant salp in the Southern Ocean, can graze a substantial fraction of the daily primary productivity (20 to over 100%), especially in oligotrophic waters.

Southern Ocean Science Symposium: future studies

Following the JGOFS evaluation meeting in Villefranche, the Southern Ocean JGOFS community held a major scientific symposium in Brest in August 1995. The symposium highlighted the priorities for future JGOFS studies in the region. Here we re-visit the SO-JGOFS research goals, and examine the main achievements and the direction for future research.

What role does the Southern Ocean play in contemporary global carbon flux?

Defining the sources and sinks for CO_2 in the Southern Ocean has been a major activity within Southern Ocean JGOFS. The initial perception of the region as a minor source has been re-evaluated, and it seems likely that the behaviour of the Southern Ocean has shifted from being neutral to a moderate CO_2 sink. The main effort of data collection has been concentrated in a few areas of the Southern Ocean. Taking the region as a whole, the annual variation in p_{CO_2} is poorly defined, although seasonal coverage is available for a few areas. The lack of winter data for most areas, and the lack of data from the Pacific Sector in all seasons, are the two major deficiencies, which need to be addressed, for example by means of unattended instrumentation. The US JGOFS programme will probably add significantly to the Pacific data set.

In conclusion, the description of the spatial and temporal variability of the air–sea flux of CO_2 in the Southern Ocean requires both the collection of more data and further analysis of existing information. An understanding of the apparent change in the behaviour of the Southern Ocean from being a neutral to a moderate sink needs to be explored within the context of studies of sub-decadal variability in the physical environment (see, for example, Murphy *et al.*, 1995; White & Peterson, 1996).

What controls the magnitude and variability of primary production and particle fate?

The Southern Ocean is the largest high-nutrient, low-chlorophyll (HNLC) region of the World Ocean. The so-called Antarctic Paradox, whereby the typically high ambient nutrient concentrations are not depleted by phytoplankton production, remains elusive despite two decades of research. The main factors thought to control production in the Southern Ocean are light and temperature. However, grazing impact is highly variable, and further information is likely to emerge from joint activities between the Southern Ocean studies in the JGOFS, GLOBEC and other programmes. Although there have been extensive studies of grazing impacts on production in the Southern Ocean, there remains a need to understand the interaction between patchiness in phytoplankton and grazers. This requires the collection of fine-scale data sets, and subsequent modelling studies. Overall, it was clear that identifying a single dominant limiting factor for the entire Southern Ocean was not a realistic goal, and that future modelling of the biogeochemistry of the Southern Ocean would require a multi-factoral approach. This has since been carried out.

Were iron fertilisation to be effective in increasing biological CO_2 drawdown in the Southern Ocean, this could potentially have a major impact on atmospheric CO_2 concentration (Cooper *et al.*, 1996). Because of the likely significance of iron limitation, it was felt that complementary studies should be encouraged. A first approach could be an IronEx type of experiment, where the effects of iron addition on a post-bloom situation are monitored by using a tracer-based Langrangian time-series study.

A second suggested approach involves the study of the natural spatial variability of iron concentrations and inputs, linking this variability to spatio-temporal patterns in phytoplankton production. It was emphasised that considerable effort needs to be invested in understanding the chemical and biological cycling of iron in the Southern Ocean.

Recent studies at the Antarctic Polar Front have shown that silicic acid can be exhausted by phytoplankton production. Given the essential role of silicic acid in the Southern Ocean biogeochemical system, it is clear that the effects of silicic acid exhaustion should be investigated. Such studies are underway.

The importance of physical forcing has already been demonstrated and may dominate at some spatial and temporal scales. More deployments of undulating instrumentation to describe the physical fields at appropriate scales for biological studies are required. During the 1996–7 austral summer, a German SO-JGOFS cruise undertook just such a study at the APF between 0 and 11°E.

The depletion of stratospheric ozone over Antarctica has led to increased ultraviolet (UV) flux to the surface waters of the Southern Ocean during summer. There is an increasing body of knowledge on the effects of increased UV radiation on a range of biological processes and species. Some effects should be included in modelling of biogeochemical processes in the Southern Ocean, although important long-term effects, such as community changes, cannot be estimated with any certainty. Three areas of UV-related research have priority within JGOFS. The first is the direct impact of UV radiation on biogeochemical processes, particularly impacts on primary production. Research should aim to formulate a UV contribution to photoinhibition as part of a standard photosynthesis–irradiance (P–E) model where measurements of the underwater and under-ice light fields should include the UV spectrum. This should include transmission through sea ice so that a comprehensive model of Southern Ocean primary production can be constructed. The second aim concerns UV effects on DOM and POM pools. Finally, possible longer-term effects should be investigated. Microplankton taxa differ in their response to UV flux, and it is likely that this will act as a selective pressure in the Southern Ocean over the next decade. The gross effects of such change on euphotic-zone carbon cycling should be assessed through ecological studies. Links are suggested with other programmes such as the ASPeCt and EASIZ programmes of SCAR (where UV-B research is already planned), GLOBEC (SCOR–IGBP) and SPARC (WCRP).

What are the major features of spatial and temporal variability in the physical and chemical environment, and key biotic systems?

The development of the Southern Ocean regional study of JGOFS has been based on an increasingly sophisticated portrayal of spatial and temporal variability. This is complicated by frontal systems, which may dominate export production, and by sea ice, which covers up to half of the Southern Ocean (south of the Subantarctic Front) and provides a uniquely mobile biogeochemical province.

The initial proposal for a JGOFS Southern Ocean regional study highlighted the need for an understanding of large-scale variability in the system. The apparent change in the flux of CO_2 to the Southern Ocean underlines this requirement, and recent research points to decadal and sub-decadal changes in the physical environment and ecosystem (Murphy *et al.*, 1995; White & Peterson, 1996).

What is the effect of sea ice on carbon fluxes in and to the Southern Ocean?

Before JGOFS, studies of the production and dynamics of the ice-edge system had been relatively restricted in both their spatial extent and their seasonal coverage. Extrapolation of these data had suggested that the Marginal Ice Zone (MIZ) might contribute a significant fraction of the total photoautotrophic carbon fixation in the Southern Ocean. It was noted that the thermodynamics of the thin ice cover could be climatically sensitive.

Studies during the JGOFS Southern Ocean programme have resulted in a better understanding of the MIZ biogeochemical system. Investigations at new study sites demonstrated interactions between the MIZ and hydrographic fronts, which give rise to different dynamics from the conventional picture of MIZ phytoplankton production. Following initial cruises (Germany, UK), these patterns have been confirmed by a South African cruise in the Lazarev Sea, which tracked an ice-edge system as it transgressed a front, which then 'adopted' the bloom.

There appear to be three outstanding problems: (1) we need to establish an operational protocol for using satellite remote sensing to map the MIZ as a biogeochemical province; (2) we need to understand the role of ice biota in the carbon cycle; and (3) we need to model the complex interactions between sea ice and hydrographic fronts.

How has the role of the Southern Ocean changed in the past and what will it be in the future?

The major problem associated with increasing our understanding of the role of the Southern Ocean in the past is the definition and calibration of proxies for contemporary processes. This was highlighted during the Brest symposium, when interpretations of past change based on different proxies often differed markedly. Because the Southern Ocean is a silica sink, the use of opal as a proxy is very promising, as highlighted in the OPALEO programme (Ragueneau *et al.*, 1996).

It was clear that a substantial modelling effort is being undertaken within SO-JGOFS, although much of this is directed towards smaller-scale modelling and there is little integration between the various national efforts. Modelling for JGOFS has two roles, which reflect the twin goals of the programme: to reduce uncertainties in our present understanding, and to develop a predictive capability. Both goals, but especially the second, imply large-scale modelling,

Table 10.10. *Overview of S–O-JGOFS field work. Updated information can be found at the JGOFS International Project Office URL*

National programmes	Areas	Number of cruises	Time period	Remarks
Australian	STC, SAZ	8 joint cruises	1990–95	Combined WOCE cruises
Belgium, Netherlands, Denmark, Italy	APF, STC, MIZ	several joint cruises	1990–97	in co-operation with other national programmes
France	POOZ	3	1990–95	plus time-series station
Germany	APF, POOZ, MIZ	2 + 4 joint cruises	1989–96	
Great Britain	APF, MIZ	2	1992–93	two-ships programme
New Zealand	STC	3 joint cruises	1992–95	
South Africa	STC, APF, MIZ	4 + 3 joint cruises	1991–96	WOCE-line A12/SR2
United States	APF, MIZ	3 years of cruises	1996–98	Ross Sea/ Pacific sector

either at the regional or at the global scale. Most of the small-scale biogeochemical models used in SO-JGOFS studies are too complex to interface with large-scale physical models, both because of the overheads in computing power and because the biogeochemical models are forced by environmental variables that simply do not appear in most large physical models. These smaller-scale models should therefore be used to optimise the design of the simpler models used in large-scale biogeochemical models.

Organisational aspects

Field work within SO-JGOFS was organised from about 15 special cruises and several more joint cruises from various countries. Specific information about these cruises may be obtained from the various national programmes. Table 10.10 provides an overview of these field activities.

Related Internet addresses

JGOFS International Project Office: http://ads.smr.uib.no/jgofs/jgofs.htm
SO-JGOFS page: http://ads.smr.uib.no/jgofs/pgtt.htm#SOUTHERN OCEAN PG

France link: http://www.obs-vlfr.fr/jgofs/html/bdjgofs.html
German link: http://www.ifm.uni-kiel.de/pl/dataman/dmpag1.html
UK link: http://www.pol.ac.uk/bodc/holdings/southern.html
US link: http://www1.whoi.edu/southern.html

Acknowledgements

On behalf of all participating scientists we thank all involved international and national research foundations for supporting Southern Ocean JGOFS activities. The conceptual and organisational framework was supported by SCAR and IGBP. Special thanks for providing national support for: Australia (Australian Southern Ocean Programme), France (Ministry of Universities, CNRS and IFRTP, publication number UMR 6539), Germany (Bundesministerium für Forschung und Technologie & Deutsche Forschungsgemeinschaft, AWI-publication number 1218), Great Britain (Natural Environment Research Council, through funding of the British Antarctic Survey and the BOFS programme); South Africa (South African National Antarctic Programme, SANAP) and New Zealand (Foundation for Research and Technology).

We also acknowledge the free access to data from numerous individual scientists working in the programmes, without it would have been impossible to draw the conclusions presented in the text. We would like to state that possible errors occurring from misinterpretation of the data are the full responsibility of the authors of this chapter.

This chapter was written to summarise the first phase of JGOFS activities in the Southern Ocean. Further synthesis of these data, and publication of results from later fieldwork, have added greatly to the value of the observations reported here. Landmarks include the output from the major symposium mentioned here (Le Févre and Tréguer, 1998) and the initial analysis of the results of the AESOPS cruise series undertaken by US JGOFS in the Ross Sea region (Anderson & Smith, 1998). *In situ* iron enrichment experiments, such as those already undertaken in the Pacific Ocean, have been planned and one, the multinational SOIREE experiment, was carried out in the 1998–99 austral summer.

References

Anderson, R. J. & Smith, W. O. (1998). AESOPS Data and Science Workshop 2. June 1998, Knoxville, Tennessee. US JGOFS Proceedings Report.

Atkinson, A. & Shreeve, R. S. (1995). Response of the copepod community to a spring bloom in the Bellingshausen Sea. *Deep-Sea Research*, Part II, **42**, 1291–311.

Bakker, D. C. E., de Baar, H. J. W. & Bathmann, U. V. (1997). Surface water CO_2 changes during spring in the Southern Ocean. *Deep-Sea Research*, Part II, **44**, 91–127.

Barange, M., Gibbons, M. J. & Carola, M. (1991). Diet and feeding of *Euphausia hanseni* and *Nematoscelis megalops*

(Euphausiacea) in the northern Benguela Current: ecological significance of vertical space partitioning. *Marine Ecology Progress Series*, **73**, 173–81.

Barange, M., Pakhomov, E. A., Perissinotto, R., Froneman, P. W., Verheye, H. M., Taunton-Clarke, J. & Lucas, M. I. (1999). Pelagic community structure of the subtropical Convergence region south of Africa and in the mid-Atlantic Ocean. *Deep-Sea Research*, Part II (in press).

Bathmann, U. V. (1996). Abiotic and biotic forcing on vertical particle flux in the Southern Ocean. In *Particle Flux in the Ocean*, ed. V. Ittekkot, P. Schäfer, S. Honjo & P. J. Depetris, pp. 243–9. Chichester: J. Wiley & Sons.

Bathmann, U. V., Schulz-Baldes, M., Fahrbach, E., Smetacek, V. & Hubberten, H.-W. (1992). The expeditions ANTARKTIS IX/1–4 of the research vessel POLARSTERN 1990/91. *Reports on Polar Research*, **100**, 1–403.

Bathmann, U., Scharek, R., Klaas, C., Dubischar, C. D. & Smetacek, V. (1997). Spring development of phytoplankton biomass and composition in major water masses of the Atlantic Sector of the Southern Ocean. *Deep-Sea Research*, Part II, **44**, 51–67.

Belkin, I. M. & Gordon, A. L. (1996). Southern Ocean fronts from the Greenwich meridian to Tasmania. *Journal of Geophysical Research*, **101**, 3675–96.

Bianchi, F., Boldrin, A., Cioce, F., Dieckmann, G., Kuosa, H., Larsson, A.-M., Nöthig, E.-M., Sehlstedt, P.-I., Socal, G. & Syvertsen, E. E. (1992). Phytoplankton distribution in relation to sea ice, hydrography and nutrients in the northwestern Weddell Sea in early spring 1988 during EPOS. *Polar Biology*, **12**, 225–35.

Boyd, P. W., Robinson, C., Savidge, G. & Williams, P. J. LeB. (1995). Water column and sea-ice primary productivity during austral spring in the Bellingshausen Sea. *Deep-Sea Research*, Part II, **42**, 1177–200.

Burkill, P. H., Edwards, E. S. & Sleigh, M. A. (1995). Microzooplankton and their role in controlling phytoplankton growth in the marginal ice zone of the Bellingshausen Sea. *Deep-Sea Research*, Part II, **42**, 1277–90.

Bury, S. J., Owens, N. J. P. & Preston, T. (1995). ^{13}C and ^{15}N uptake by phytoplankton in the marginal ice zone of the Bellingshausen Sea. *Deep-Sea Research*, Part II, **42**, 1225–52.

Cooper, D. J., Watson, A. J. & Nightingale, P. D. 1996. Large decrease in ocean-surface CO_2 fugacity in response to *in situ* iron fertilisation. *Nature*, **383**, 511–13.

Crawford, R. (1995). The role of sex in the sedimentation of a marine diatom bloom. *Limnology and Oceanography*, **40**, 200–4.

Crawford, R. M. & Hinz, F. (1997). Spatial and temporal distribution of assemblages of the diatom *Corethron criophilum* in the Polar Frontal region of the South Atlantic. *Deep-Sea Research*, Part II, **44**, 479–96.

Cripps, G. C. (1995). Biogenic hydrocarbons in the particulate material of the water column of the Bellingshausen Sea, Antarctica, in the region of the marginal ice zones. *Deep-Sea Research*, Part II, **42**, 1123–35.

de Baar, H. J. W, de Jong, J. T. M., Bakker, D. C. E., Löscher, B. M., Veth, C., Bathmann, U. & Smetacek, V. (1995). Importance of iron for plankton blooms and carbon dioxide drawdown in the Southern Ocean. *Nature*, **373**, 412–15.

Dower, K. M. & Lucas, M. I. (1993). Photosynthesis–irradiance relationships and production associated with a warm-core ring shed from the Agulhas Retroflection south of Africa. *Marine Ecology Progress Series*, **5**, 141–54.

Dower, K. M & Lucas, M. I. (1997). Phytoplankton biomass, P–I relationships and primary production at the STC in winter. *Polar Biology*, **16**, 41–9.

Dower, K. M & Lucas, M. I. (1999). Primary production at the Subtropical Convergence south of Africa during austral winter. *Deep-Sea Research*, Part II (in press).

Dower, K. M., Lucas, M. I., Phillips, R., Dieckmann, G. & Robinson, D. H.

(1996). Phytoplankton biomass, P–I relationships and primary production in the Weddell Sea, Antarctica, during the austral autumn. *Polar Biology*, **16**, 41–53.

Dubischar, C. & Bathmann, U. (1997). Mesozooplankton grazing across frontal systems in the Southern Ocean. *Deep-Sea Research*, Part II, **44**, 415–33.

Duncombe-Rae, R. C. M., Garzoli, S. L. & Gordon, A. L. (1996). The eddy field of the southeast Atlantic Ocean: A statistical census from the Benguela Sources and Transports Project. *Journal of Geophysical Research*, **101**, 11 949.

El-Sayed, S. Z. (1984). Productivity in the Antarctic waters – A reappraisal. In *Marine Phytoplankton and Productivity*, ed. O. Holm-Hansen, L. Bolis & R. Giles, pp. 19–34. Berlin: Springer-Verlag.

Fortier, L., Le Fèvre, J. & Lengendre, L. (1994). Export of biogenic carbon to fish and to the deep ocean: the role of large planktonic microphages. *Journal of Plankton Research*, **16**, 809–39.

Fransz, G. & Gonzalez, S. (1997). Vertical and latitudinal trends in mesozooplankton abundance, biomass, species composition, and population structure in the Circumpolar Current of the Atlantic sector of the Southern Ocean during spring 1992. *Deep-Sea Research*, Part II, **44**, 395–414.

Froneman, P. W., Perissinotto, R. & McQuaid, C. D. (1995). Microzooplankton grazing and community structure in the Southern Ocean: Seasonal variations and implications for carbon cycling. In *Carbon Fluxes and Dynamic Processes in the Southern Ocean: Present and Past* (symposium abstracts). Brest, France, 28–31 August 1995, p. 15.

Garrison, D. L., Buck, K. R. & Gowing, M. M. (1991). Plankton assemblages in the ice-edge zone of the Weddell Sea during the austral winter. *Journal of Marine Systems*, **2**, 123–30.

Harris, C. M. & Stonehouse, B. (1991). *Antarctica and Global Change*. London: Bellhaven Press. (198 pp.)

Jochem, F. J., Mathot, S. & Quéguiner, B. (1995). Size-fractionated primary production in the open Southern Ocean in austral spring. *Polar Biology*, **15**, 381–92.

Klaas, C. (1997). Microprotozooplankton distribution and grazing impact in the open waters of the Antarctic Circumpolar Current. *Deep-Sea Research*, Part II, **44**, 375–93.

Laubscher, R. K., McQuaid, C. D. & Dower, K. M. (1999). Size-fractionated primary production in the Atlantic sector of the Southern Ocean: implications for the food web. *Deep-Sea Research*, Part II (in press).

Le Fèvre, J. & Tréguer, P. (Editors) (1998). *Journal of Marine Systems*, **17** (parts 1–2). (39 papers from the 1995 Brest symposium.)

Liss, P. S. & Merlivat, L. (1986). Air-sea exchange rates: introduction and synthesis. In *The Role of Air–Sea Exchange in Geochemical Cycling*, ed. P. Buat-Menard, pp. 113–27. Dordrecht: D. Reidel.

Lochte, K., Bjørnsen, P., Becquevort, S., Giesenhagen, H. & Weber, A. (1997). Bacterial standing stock and production and its relation to phytoplankton biomass and productivity. *Deep-Sea Research*, Part II, **44**, 321–40.

Löscher, B., de Jong, J. T. M., de Baar, H. & Veth, C. (1997). Distributions of Fe in the Antarctic Circumpolar Current. *Deep-Sea Research*, Part II, **44**, 143–88.

Longhurst, A. R. (1991). Role of the marine biosphere in the global carbon cycle. *Limnology and Oceanography*, **36**, 1507–26.

Louanchi, F. (1995). Etude des variabilités spatio-temporelles de p_{CO_2} à la surface de l'Océan Indien. Processus et quantification. Thesis, Université P. et M. Curie, 290 pp.

Lucas, M. I., Lutjeharms, J. R. E., Field J. G. & McQuaid (1993). A new South African research programme in the Southern Ocean. *South African Journal*, **83**, 61–7.

Lutjeharms, J. R. E. (1985). Location of frontal systems between Africa and Antarctica: some preliminary results.

Deep-Sea Research, **32**, 1499–509.

Lutjeharms, J. R. E. & Valentine, H. R. (1984). Southern Ocean thermal fronts south of Africa. *Deep-Sea Research*, **31**, 1461–75.

Lutjeharms, J. R. E. & van Ballegooyen, R. C. (1988). The Agulhas retroflection. *Journal of Physical Oceanography*, **18**, 1570–83.

Lutjeharms, J. R. E. & Ruijter, W. P. M. de (1996). The influence of the Agulhas Current on the adjacent coastal ocean: possible impacts of climate change. *Journal of Marine Systems*, **7**, 321.

Lutjeharms, J. R. E., Valentine, H. R. & van Ballegooyen, R. C. (1993). On the Subtropical Convergence in the South Atlantic Ocean. *South African Journal*, **89**, 552–9.

Macdonald, A. M. & Wunsch, C. (1996). An estimate of global ocean circulation and heat fluxes. *Nature*, **382**, 436.

Mantel, J. D. & Lucas, M. I. (1999). The biophysical environment across the Subtropical Convergence and an associated warm eddy south of Africa, during austral winter 1993. *Deep-Sea Research*, Part II (in press).

Mantel, J. D., Lucas, M. I. & Probyn, T. (1999). New production across the Subtropical Convergence and an associated warm eddy south of Africa, during austral winter 1993. *Deep-Sea Research*, Part II (in press).

Martin, J. H., Gordon, R. M. & Fitzwater, S. E. (1990). Iron in Antarctic waters. *Nature*, **345**, 156–8.

Murphy, E. J., Clarke, A., Symon, C. & Priddle, J. (1995). Temporal variation in Antarctic sea ice: analysis of a long-term fast-ice record from the South Orkney Islands. *Deep-Sea Research*, **42**, 1045–62.

Murray, A. W. A., Watkins, J. L. & Bone, D. G. (1995). A biological acoustic survey in the marginal ice-edge zone of the Bellingshausen Sea. *Deep-Sea Research*, Part II, **42**, 1159–75.

Nelson, D. M., Tréguer, P., Brzezinski, M. A., Leynaert & Quéguiner, B. (1995). Production and dissolution of biogenic silica in the ocean: revised global estimates, comparison with regional data and relationship to biogenic sedimentation. *Global Biogeochemical Cycles*, **19**, 359–72.

Pakhomov, E. A. & Perissinotto, R. (1997). Mesozooplankton community structure and grazing impact in the region of the Subtropical Convergence south of Africa. *Journal of Plankton Research*, **6**, 675–91.

Perissinotto, R. & Duncombe Rae, C. M. (1990). Occurrence of anticyclonic eddies on the Prince Edward Plateau (Southern Ocean): effects on phytoplankton biomass and production. *Deep-Sea Research*, **37**, 777–93.

Priddle, J., Hawes, I., Ellis-Evans, J. C. & Smith, T. J. (1986). Antarctic aquatic ecosystems as habitats for phytoplankton. *Biology Reviews*, **61**, 199–238.

Rabouille, C., Gaillard, J. F., Tréguer, P. & Vincendeau, M. A. (1997). Biogenic silica recycling in surficial sediments across the Polar Front of the Southern Ocean (Indian sector). *Deep-Sea Research*, Part II, **44**, 1151–76.

Ragueneau, O., Leynaert, P., DeMaster, D. J. & Anderson, R. F. (1996). Opal studies as a marker of paleoproductivity. *EOS*, **77**, 491–3.

Read, J. F. & Pollard, R. T. (1993). Structure and transport of the Antarctic Circumpolar Current and Agulhas Return Current at 40°E. *Journal of Geophysical Research*, **98**, 12 281–95.

Read, J. F., Pollard, R. T., Morrison, A. I. & Symon, C. (1995). On the southerly extent of the Antarctic Circumpolar Current in the southeast Pacific. *Deep-Sea Research*, Part II, **42**, 933–54.

Riaux-Gobin, C., Hargraves, P. E., Neveux, J., Oriol, L. & Vètion, G. (1997). Microphyte pigments and resting spores at the water-sediment interface in the Subantarctic deep sea (Indian sector of the Southern Ocean). *Deep-Sea Research*, Part II, **44**, 1033–51.

Robertson, J. E. & Watson, A. J. (1995). A summer-time sink for atmospheric carbon dioxide in the Southern Ocean between 88°W and 80°E. *Deep-Sea Research*, Part II, **42**, 1081–91.

Robins, D. B., Harris, R. P., Bedo, A. W., Fernandez, E., Fileman, T. W., Harbour,

D. S. & Head, R. N. (1995). The relationship between suspended particulate material, phytoplankton, and zooplankton during the retreat of the marginal ice zone in the Bellingshausen Sea. *Deep-Sea Research*, Part II, **42**, 1137–58.

Rutgers van der Loeff, M., Friedrichs, J. & Bathmann, U. (1997). Carbon export during the spring bloom quantified with the natural tracer ^{234}Th. *Deep-Sea Research*, Part II, **44**, 457–78.

Savidge, G., Harbour, D., Gilpin, L. C. & Boyd, P. W. (1995). Phytoplankton distributions and production in the Bellingshausen Sea, austral spring 1992. *Deep-Sea Research*, **42**, 1201–24.

Scharek, R., van Leeuwe, M. & de Baar, H. (1997). Responses of Southern Ocean phytoplankton to the addition of trace metals. *Deep-Sea Research*, Part II, **44**, 209–27.

Schmitz, W. J. Jr (1996). On the eddy field in the Agulhas Retroflection, with some global considerations. *Journal of Geophysical Research*, **C101**, 16 259.

Shimmield, G. B., Ritchie, G. D. & Fileman, T. W. (1995). The impact of marginal ice zone processes on the distribution of ^{210}Pb, ^{210}Po and ^{234}Th and implications for new production in the Bellingshausen Sea, Antarctica. *Deep-Sea Research*, Part II, **42**, 1313–35.

Smetacek, V., Scharek, R. & Nöthig, E.-M. (1990). Seasonal and regional variation in the pelagial and its relationship to the life history cycle of krill. In *Antarctic Ecosystems. Ecological Change and Conservation*, ed. K. R. Kerry & G. Hempel, pp. 103–14. Berlin: Springer-Verlag.

Smetacek, V. S., de Baar, H. J. W., Bathmann, U. V., Lochte, K. & Rutgers van der Loeff, M. M. (1997). Ecology and biogeochemistry of the Antarctic Circumpolar Current during austral spring: A summary of Southern Ocean JGOFS cruise ANT X/6 of RV 'Polarstern'. *Deep-Sea Research*, Part II, **44**, 1–21.

Smythe-Wright, D., Gordon, A. L. & Jones, M. S. (1996). CFC-113 shows Brazil Eddy crossing the South Atlantic to the Agulhas Retroflection region. *Journal of Geophysical Research*, **C101**, 885.

Takahashi, T., Olafson, J., Goddard, J. G., Chipman, D. W. & Sutherland, S. C. (1993). Seasonal variations of CO_2 and nutrients in the high-latitude surface oceans: a comparative study. *Global Biogeochemical Cycles*, **7**, 843–78.

Tans, P. P., Fung, I. Y. & Takahashi, T. (1990). Observational constraints on the global atmospheric CO_2 budget. *Science*, **247**, 1431–8.

Tilbrook, B., Beggs, H., Wright, S. & Pretty, M. (1995). Southern Ocean sources and sinks for atmospheric CO_2. International SO-JGOFS symposium. *Carbon Fluxes and Dynamic Processes in the Southern Ocean: Present and Past.* Brest, France, 28–31 August 1995. Oral communication 170.

Tréguer, P. & Jacques, G. (1992). Review: dynamics of nutrients and phytoplankton, and fluxes of carbon, nitrogen, and silicon in the Antarctic Ocean. *Polar Biology*, **12**, 149–62.

Tréguer, P., Nelson, D. M., van Bennekom, A. J., DeMaster, D. J., Leynaert, A. & Quèguiner, B. (1995). The silica balance in the world ocean: a re-estimate. *Science*, **268**, 375–9.

Turner, D. R. & Owens, N. J. P. (1995). A biogeochemical study in the Bellingshausen Sea: overview of the STERNA 1992 expedition. *Deep-Sea Research*, Part II, **42**, 907–32.

Turner, D. R., Owens, N. J. P. & Priddle, J. (1995). Southern Ocean JGOFS: The U.K. 'STERNA' study in the Bellingshausen Sea. *Deep-Sea Research*, Part II, **42**, 905–6.

van Ballegooyen, R. C., Grundlingh, M. L. & Lutjeharms, J. R. E. (1994). Eddy fluxes of heat and salt from the southwest Indian Ocean into the southeast Atlantic Ocean: A case study. *Journal of Geophysical Research*, **99**, 14 053.

van Franeker, J. A., Bathmann, U. V. & Mathot, S. (1997). Carbon flux to Antarctic top predators. *Deep-Sea Research*, Part II, **44**, 435–55.

Veth, C., Peeken, I. & Scharek, R. (1997).

Physical anatomy of fronts and surface waters in the ACC near the 6°W meridian during austral spring 1992. *Deep-Sea Research*, Part II, **44**, 23–50.

Waldron, H. N., Attwood, C. G., Probyn, T. A. & Lucas, M. I. (1995). Nitrogen dynamics in the Bellingshausen Sea during austral spring 1992. *Deep-Sea Research*, Part II, **42**, 1253–76.

Walsh, J. J., Dieterle, D. A. & Priddle, J. R. (1991). Organic debris on the continental margins: a simulation analysis of source and fate. *Deep-Sea Research*, **38**, 805–28.

Wanninkhof, R. (1992). Relationship between wind speed and gas exchange over the ocean. *Journal of Geophysical Research*, **97**, 7373–82.

Weeks, S. J. & Shillington, F. A. (1994). Interannual scales of variation of pigment concentrations from coastal zone color scanner data in the Benguela Upwelling system and the Subtropical Convergence zone south of Africa. *Journal of Geophysical Research*, **C99**, 7385–99.

de Wit, R., Relexans, J. C., Bouvier, T. & Moriarty, D. J. W. (1997). Microbial respiration and diffusive oxygen uptake of deep-sea sediments in the Southern Ocean (ANTARES-I-cruise). *Deep-Sea Research*, **44**, 1053–68.

White, W. B. & Peterson, R. G. (1996). An Antarctic circumpolar wave in surface pressure, wind, temperature and sea-ice extent. *Nature*, **380**, 699–702.

Whitehouse, M. J., Priddle, J. & Woodward, E. M. S. (1995). Spatial variability of inorganic nutrients in the marginal ice zone of the Bellingshausen Sea during the austral spring. *Deep-Sea Research*, Part II, **42**, 1047–58.

Whitworth, T. & Nowlin, W. D. (1987). Water masses and currents of the Southern Ocean at Greenwich meridian. *Journal of Geophysical Research*, **92**, 6462–76.

11 Process studies in eutrophic, mesotrophic and oligotrophic oceanic regimes within the tropical northeast Atlantic

A. Morel

Keywords: pelagic biomasses, primary production, benthic fauna, water-column carbon flux, time-series station, southeast North Atlantic, ecosystem models

Introduction

The aim of this review is chiefly to indicate how the EUMELI–JGOFS programme was set up and on what previous knowledge it was based, and then how it was carried out and for what specific JGOFS objectives. It is intended to provide, in a synthetic way, a general survey of the results already obtained and presently available. Another purpose is to inform the international JGOFS community, and its modellers in particular, of the parameters that were (successfully, or less successfully) monitored, and also to provide appropriate references to detailed studies that are already published or expected for the near future, and have resulted from the EUMELI programme.

If the present chapter fixes the general interdisciplinary frame of the programme, it does not intend to be exhaustive with regard to all of its aspects, nor definitely conclusive. Indeed at this stage of completion, drawing final conclusions for the whole four-year programme, and while many activities are still ongoing, is out of reach. For instance, the preparation of the sediment-trap samples for further analyses was only completed in January 1997, so that many analyses remain to be made or are yet to be validated. Thus, only partial results for the fluxes and related parameters are displayed in the tables in this chapter, and at present they do not encompass the entire period of data acquisition. Therefore, the presentation below represents an intermediate and provisional state of the work, before a global synthesis can be made. There is no explicit presentation of materials and methods in what follows, as specific papers already published, and referred to, provide this information.

Historical background and objectives

The earlier root of the EUMELI programme (acronym for EUtrophic, MEsotrophic and oLIgotrophic regimes) is to be found in the International Programme of Co-operative Investigation of the northern part of the Eastern Central Atlantic (CINECA). This programme, born and then co-ordinated under the auspices of CIESM, COI and FAO, consisted of multidisciplinary studies of coastal upwelling events associated with the western boundary currents. Its main goals were to analyse the physical, chemical and biological processes within and around a coastal upwelling system, and to determine the extent and effect of such a system upon the adjacent zones. The multi-nation, multi-ship surveys lasted about ten years, ending with a Symposium, held in Las Palmas in 1978, entitled *The Canary Current, Studies of an Upwelling System*. Under the same title, a series of papers were published in 1982, under the editorship of G. Hempel. The French group that contributed to this programme (the 'Mediprod Group') had still not been disbanded in 1985, about ten years after the last cruise (CINECA-5). Under the guidance of H. J. Minas, this group elaborated a new project based on one of the conclusions drawn at the issue of the Las Palmas Symposium: 'Information on the near-shore area (along the African coast) is much better than that for the open ocean' (Hempel, 1982). This new project, initially termed 'upwelling dissipation', was aimed at studying the progressive changes in the biogeochemical processes and the gradual decline of biological activity when the upwelled waters drift from their sources toward the open ocean. It quickly became the seed for a participation in the emerging international JGOFS Programme, and its initial objectives were accordingly amended.

The restructuring was made in accordance with the basic elements, as succinctly defined in the SCOR–JGOFS Report of the International Scientific Planning and Co-ordination Meeting, and held in Paris (1987). More precisely, two basic elements, 'Process studies' and 'Modelling', were identified as the main driving ideas for a revised programme. The statement in this Report, about the future need for satellite observation of ocean colour to achieve the JGOFS goals, was also taken into consideration. The adopted strategy for the EUMELI programme essentially consisted of:

1. Carrying out detailed studies of the main processes governing the particle flux, and particularly the carbon flux, from the upper oceanic layers where photosynthesis occurs, throughout the water column, to ultimate deposition on the sea floor, and finally to incorporation into the sediments;
2. performing such studies in three typical trophic regimes, differing as distinctly as possible in their nutrient availability, and therefore in their

primary production rate, in intensity of vertical transport, and ultimately in deposition rate;
3. developing a series of generic models allowing biogeochemical fluxes toward the interior, and then to the bottom of the ocean, to be related to near-surface properties, in particular those detectable through remote sensing techniques, such as chlorophyll concentration, sea-surface temperature, incident irradiation and wind;
4. validating these models (operated in uncoupled or coupled modes) in various regimes by using field data.

Ab initio, it was admitted that localised time series would be more appropriate for validation exercises, rather than successive cruises. A quasi-permanent sediment-trap deployment was planned and seen as an essential component of the EUMELI programme. Satellite data were also expected for the period of field activities. With these tools, it was believed that at least a certain continuity in observation could be preserved.

Initially, six regularly spaced cruises were planned (each about 7–8 months apart). This interval was optimal for collecting the captured sediments and re-deploying the traps, current-meters and benthic modules left on the bottom. Because of the multi-disciplinary approach, the diversity of experiments, and as a consequence the number of people involved, it was necessary to specialise and diversify the cruises. Emphasis was thus successively put on bathymetry and sedimentology, studies of benthic communities and processes, then on water-column chemistry and pelagic studies, including primary production and photophysiology. Apart from the first cruise, hydrography and sampling for the JGOFS core parameters were systematically performed during all cruises, whatever their main topics. A significant atmospheric sampling programme (for aerosols and sulphur compounds; see Putaud *et al.*, 1993) was also implemented during cruises 3 and 4. Based on free-drifting sediment-trap deployments (at 200 m depth), the downward flux of detrital particulate dimethylsulphonio-propionate (DMSPp) was also studied (Corn *et al.*, 1994).

The programme began with the first cruise (EUM#1) in July 1989 (Table 11.1). Because of funding problems, in particular for purchasing a sufficient number of traps, and because of limited ship time in the selected zone, the programme after the first cruise was delayed until January 1991 (EUM#2). It was then regularly pursued until cruise 5. The sixth cruise was cancelled for logistic reasons, and replaced by a study of another oligotrophic regime in the Pacific Ocean (during a cruise called OLIPAC). Nevertheless, the deployment of the sediment traps and current meters, ending with cruise 5, lasted about two years over three consecutive periods. Only the Oligotrophic and Mesotrophic sites were equipped with moorings. For various security reasons, the Eutrophic

Table 11.1. *Location, bottom depth of the EUMELI stations, and chronology of visits*

Site	Z (m)	N	W	Cruise	Dates
EUtrophic	2030	20°32′	18°36′	EUM#1	19–25 VII 1989
				EUM#2	03–07 II 1991
				EUM#4	05–07 VI 1992
					13–15 VI 1992
				EUM#5	09–12 XII 1992
EM-04	3075	20°08′	19°03′	EUM#4	05 VI 1992
				EUM#5	12 XII 1992
EM-03	3245	19°43′	19°32′	EUM#4	04 VI 1992
				EUM#5	12 XII 1992
EM-02	3330	19°18′	20°00′	EUM#4	13 XII 1992
EM-01	3280	18°53′	20°30′	EUM#4	04 VI 1992
				EUM#5	12 XII 1992
MEsotrophic	3090	18°30′	21°00′	EUM#1	14–17 VII 1989
				EUM#2	22–30 I 1991
					08–12 II 1991
				EUM#3	16–18 IX 1991
					06–10 X 1991
				EUM#4	31 V–04 VI 1992
					15–19 VI 1992
				EUM#5	13–17 XII 1992
MO-03	3935	19°05′	23°30′	EUM#4	30 V 1992
MO-03				EUM#5	17 XII 1992
MO-02	4500	19°42′	25°52′	EUM#4	30 V 1992
MO-02(IMO)				EUM#3	11 X 1991
				EUM#5	18 XII 1992
MO-01	4850	20°25′	28°35′	EUM#4	29 V 1992
MO-01				EUM#5	19 XII 1992
oLIgotrophic	4600	21°00′	31°00′	EUM#1	06–09 VII 1989
				EUM#2	14–19 I 1991
					14–18 II 1991
				EUM#3	22–24 IX 1991
					13–20 X 1991
				EUM#4	24–28 V 1992
					21–26 VI 1992
				EUM#5	20–23 XII 1992
T1	4570	20°55′	30°47′	EUM#3	12 X 1991
T2	4620	20°55′	31°29′	EUM#3	13 X 1991
T3	4590	21°25′	31°08′	EUM#3	21 X 1991
				EUM#5	24 XII 1992

site was never instrumented; it was not visited at all during EUMELI#3, owing to its inaccessibility during the Gulf war.

Selection of the sites

A pre-selection of the three sites was made under several constraints. Besides the practical constraint of minimising the distances between the three locations, a pre-requisite was to identify and locate in the tropical Northeast Atlantic the three typical trophic situations that were hoped for, with well-established regimes and sufficiently steady characteristics. The Mauritanian upwelling area and open-ocean stations at the same latitude were selected *a priori* on the basis of climatic considerations (see below), as well as on the examination of satellite ocean imagery interpreted in terms of chlorophyll concentration (by using the data from the Coastal Zone Color Scanner, CZCS).

At such latitudes (about 20°N), the amplitude of the winter deep convection is known to be considerably reduced, compared with that occurring at higher latitudes (Levitus, 1982). The subsequent occurrence of vernal bloom is unlikely, and the seasonality is reduced (Hastenrath & Lamb, 1977). The inter-tropical convergence zone does not reach these latitudes in summer, so that rather regular trade winds ensure a relative steadiness in external forcing and expected biological responses. More precisely, upwelling along the Northwest African coast experiences a seasonal oscillation in latitude with, nevertheless, a permanent activity around 20°N, near Cape Blanc (Speth & Detlefsen, 1982; Mittelstaedt, 1982). Consequently, coastal upwelling is a permanent feature at this location, in spite of a fluctuating intensity, as already revealed by the CZCS scenes analysed by Bricaud *et al.* (1987). The eutrophic site was thus positioned not too far from the source of upwelled waters (actually at about 140 km from Cape Blanc; see Fig. 11.1). However, it was far enough away to avoid coastal turbid waters (Morel, 1982), as well as the influence of Banc d'Arguin waters (Peters, 1976), and far enough to reach deep waters and bottom depths exceeding 2000 m. The chlorophyll (Chl) concentration at this site is always above 1 mg m^{-3}, as recorded in the CZCS archive (see Morel, 1996).

This archive was also very helpful when selecting the position of the oligotrophic site. Indeed, the colour imagery showed that a permanent oligotrophic situation (with Chl below 0.1 mg m^{-2}, and no seasonal signal) could only be observed offshore, at a minimal distance of about 1500 km from the coast. Such a location, at the periphery of the North Atlantic gyre, and northward of the North Equatorial current, is in a rather quiet zone. Meandering of this current and eddy formation generated by baroclinic

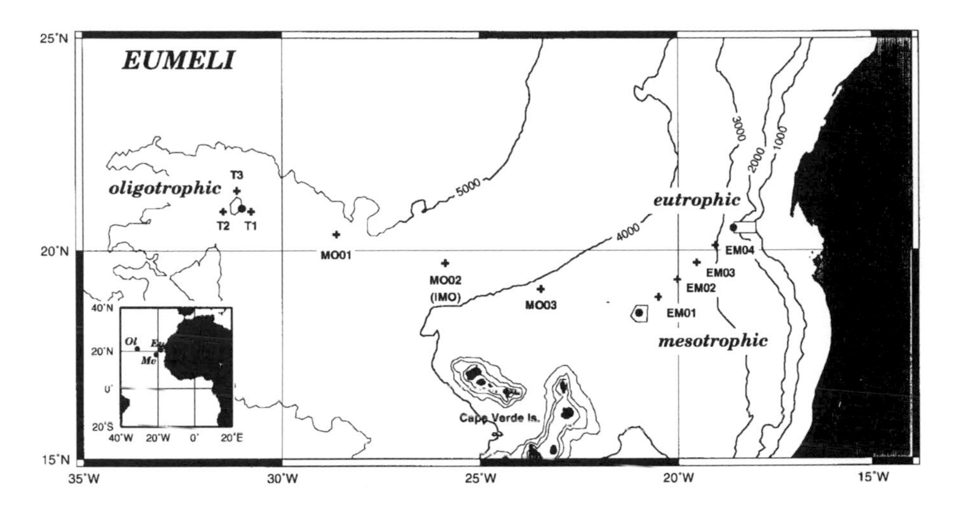

Figure 11.1 *Location of the EUMELI stations, including the three main sites E, M, and O, the intermediate stations, denoted EM and MO, visited during the cruises EUM#4 and 5, and the three stations forming a triangle around the O site, T1, T2, and T3 (see also Table 11.1). The shaded areas around the stations E, M, and O represent the boxes where intensive CTD casts (40) were performed during EUM#2.*

instabilities might induce some mesoscale variability near this area (Dadou *et al.*, 1996).

The choice of a site expected to be representative of mesotrophic conditions was less easy, as such regimes seem to be essentially transient and zonally migrating within the Mauritanian zone. The general circulation in this area (the Canary Current offshore, and a Northward coastal countercurrent), as well as the presence of eddies and filaments detaching from the upwelling system (Bricaud *et al.*, 1987; van Camp *et al.*, 1991), leads one to think that complex situations and mesoscale structures are likely to occur (see Mittelstaedt, 1982). On average, however, mesotrophic conditions seem to prevail in a zone midway between the eutrophic coastal region and the Cape Verde islands, with chlorophyll concentrations remaining moderate and varying between 0.3 and 1.3 $mg\,m^{-3}$, approximately, according to CZCS data.

The bottom topography, the sediment stability, and the deposition rate were also considered before deciding on the final position of the sites. For this purpose, the first cruise was devoted to a detailed bathymetric survey around the sites. It also provided the opportunity for a preliminary study of the benthic community and activity. Around each site, the bottom topography was accurately mapped by using side-scan sonar, then visually analysed by using sea-floor acoustic imaging devices. Such precise knowledge was essential for the deployment of benthic instrumentation and mooring lines (Auffret *et al.*, 1992; Sibuet *et al.*, 1993). It was also necessary to verify that no significant perturbation occurred near the selected sites, such as volcanic and hydrothermal activity, or turbidity events near the continental slope. A flat and extensive terrace, 2000 m deep, was selected for the eutrophic site. The mesotrophic site was located on the Cape Verde Terrace at an average depth of 3000 m, and the oligotrophic site on the abyssal plain (4600 m) far from the Cape Verde Islands rise, and from the mid-Atlantic ridge and scattered guyots. The sedimentation rates were found to be regular over the last post-glacial period and apparently correlated with the biological activity expected in the upper layers at the three sites (J.-L. Reyss, personal communication, 1991). The sites, as they were definitely selected after the first cruise, are shown in Fig. 11.1, as well as additional (intermediate) stations. Their geographical positions are given in Table 11.1. In what follows, the capitals O, M and E will be used to represent the oligo-, meso-, and eutrophic situations or sites.

Prevailing conditions at the three sites

The general climatology and main oceanographic features at these sites are summarised in Table 11.2. The seasonal variations to be expected at these latitudes can be found in the climatology of Hastenrath & Lamb (1977) (see also other estimates in Dadou & Garçon, 1993). The heat budget at the O site is in equilibrium over the year, whereas at the M and E sites a net oceanic heat gain (+ 40 and + 70 W m^{-2}, respectively) results from increasing solar radiation and decreasing evaporation when approaching the coast. The amplitude in the annual cycle of the heat budget also increases eastward: the amplitude values are 120, 160, and 170 W m^{-2} at the O, M, and E sites, respectively. The period of negative values (heat loss for the ocean) extends over 5 months (November–March) at the O site, 3 months (November–January) at the M site and only 1 month (December) at the E site. This site is almost permanently exporting energy through advection, whereas at the O site there is no net annual export. In spite of a rather long winter at the O site, the annual course of the upper-layer temperature (see Fig. 11.2) remains rather smooth. Winter values are not below 22.2 °C, and the narrow range of annual variation (by 3 °C) was fully confirmed by successive measurements. The temperature evolution at the M site, taken from Hastenrath & Lamb (1977), as well as the abundant oceanographic database for the CINECA region (Smed, 1982), spans a wider range.

The expected 5 °C change between early spring and fall was also confirmed during the successive cruises at this M site. At the E site, the annual amplitude is again of about 5 °C, but all values are shifted by 2 °C toward colder temperatures. The abrupt warming that starts in July probably results from the

Table 11.2. *Climatological and other relevant information for the three sites*

	Oligotrophic	Mesotrophic	Eutrophic
*Short wave radiation year average ($W\,m^{-2}$)	180	195	218
*Cloudiness index (tenth) min–max	5–7	4–6	3–5
*Net (short–long-wave) radiation ($W\,m^{-2}$)	+110	+130	+145
Latent heat flux year average ($W\,m^{-2}$)	−110	−90	−75
Ocean gain ($W\,m^{-2}$)	0	+40	+70
†Mixed-layer depth (m) min–max	35–70	20–35	20–35
*Surface temperature (°C) min–max	22.2–25.6	20.2–25.0	18.8–24.0
‡ Temperature at 100 dbar (°C) max–min	22.0–20.5	18.0–16.0	17.0–16.0
‡Surface Chl ($mg\,m^{-3}$) min–max	0.05–0.10	0.15–1.4	0.9–3.0
¶Primary production annual values ($g\,C\,m^{-2}$)	110	260	535
Average per day ($mg\,C\,m^{-2}$)	300	710	1500

Sources and symbols: * according to Hastenrath & Lamb (1977); † according to Levitus (1982); ‡ according to Maillard (1986); note that these temperatures are in opposite phase compared to those at surface, and thus are listed as max–min values, in correspondence with the min–max values for the surface; ‡ according to the CZCS-NASA climatological monthly maps (Feldman *et al.*, 1989); ¶ predicted from CZCS surface-chlorophyll (Morel, 1996).

weakening of the trade winds and thus of the upwelling activity. Such a warming, observed during CINECA cruises (Smed, 1982), has not been documented during EUMELI.

At the depth corresponding to 100 dbar (see Maillard, 1986, and Table 11.2), the annual temperature variation is small. Interestingly, at the O site in March, minimal temperature, almost constant within the water column (about 22 °C), suggests that the mixed layer extends down to at least 100 m, and thus deeper than indicated in the Levitus (1982) atlas. At the M and E sites, even in winter, the surface temperature always exceeds that at 100 m by about 2 °C, and indicates that the stratification is maintained throughout the year.

The satellite-derived chlorophyll values leave no doubt about the permanent oligotrophic regime, and the seasonally modulated eutrophic regime, which

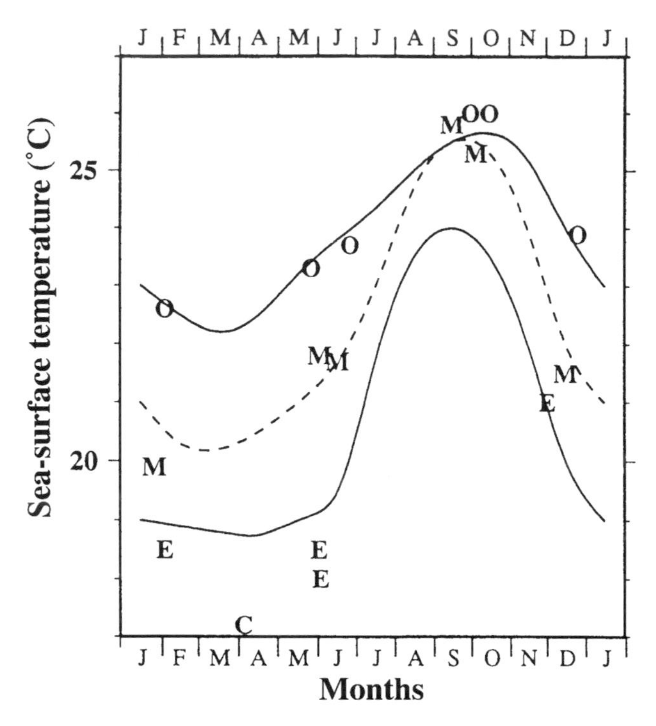

Figure 11.2 *Annual course of the surface temperature at the three sites taken from Hastenrath & Lamb (1977). The symbols O, M, and E represent mean values resulting from measurements at these stations made during successive cruises; the symbol C is for measurements made during an earlier cruise (CINECA 5, 1974), approximately at the position of the E site (see also Table 11.3).*

prevail at the corresponding selected sites. As anticipated for the M site, the algal biomass is much more variable, by a factor approaching 10, with erratic rather than well-defined seasonal patterns (see Fig. 2 in Morel, 1996). It can only be asserted that this regime is, on average, mesotrophic in terms of annual mean Chl content and of predicted primary production (Berthon, 1992).

Results

Water masses at the three sites

In contrast to the oligotrophic zone, which is rather poorly documented (see hydrographic stations compiled in Maillard, 1986, and in Lozier *et al.*, 1995), the E and M sites were intensively studied. These sites were intensively studied during the CINECA era (Tomczak, 1981; Fiuza & Halpern, 1982; Minas *et al.*, 1982; Manriquez & Fraga, 1982; Barton & Hughes, 1982; Mittelstaedt, 1982), and during several cruises afterwards (see, for example, Barton, 1987; Zenk *et al.*, 1991; Ahran *et al.*, 1994). Early observations (Fraga, 1974) showed that at latitudes 21–23°N a thermohaline front, without an appreciable density gradient, develops off Mauritania, with a complex meandering structure (see also Barton, 1987). Below the mixed layer, and down to about 800 m, the North Atlantic Central Water and the South Atlantic Central Water (NACW and SACW) are in contact. They interact along this front (named 'Cape Verde Frontal Zone' by Zenk *et al.*, 1991) through isopycnal mixing, often resulting in interleaved layers (Barton & Hughes, 1982).

The E and M sites are expected to be mainly influenced by the northward intrusion of SACW. With a general NE–SW orientation of the central-water discontinuity, these sites might be exposed to fluctuations resulting from the frontal-zone meandering (see Fig. 2 in Zenk *et al.*, 1991). The first intensive hydrographic survey (during the EUM#2 cruise; Vangriesheim *et al.*, 1993) included 16 CTD casts made within a $30 \times 60\,km^2$ box around the E site, and 13 others in a $30 \times 45\,km^2$ box containing the M site. These measurements confirmed the predominance of SACW at both sites; however, spatial (actually east–west) nuances were detected within the boxes around the E and M sites (Pierre *et al.*, 1994). The mixing proportions of SACW and NACW appeared more variable for the upper part of the water column (for $z < 220$ m and $\sigma_\theta < 26.8$), whereas for the deep part (down to 800 m and $\sigma_\theta = 27.3$) the θ–S diagrams (potential temperature against salinity) were found to be steadily superimposed on that of 'pure' SACW.

The θ–S diagrams for all cruises are separately shown in Fig. 11.3*a–c* for the three sites O, M and E. The seasonally changing upper-layer density profiles are displayed in Fig. 11.4*a–c*. Except for their upper parts, the θ–S patterns at the O site were found to be remarkably stable during the two years of this study (Fig. 11.3*a*). The Antarctic Bottom Water, with a potential temperature of 2 °C or slightly less (Fig. 11.5), was present below 4000 m and progressively diluted upward with the North Atlantic Deep Water, which is characterised by a salinity maximum (with $S > 35.05$ practical salinity units, psu). This maximum occurred at 1500, 1350, and 1250 m at the O, M, and E sites, respectively (see also the Salinity section in Fig. 11.6*a*). Above, namely at a depth of 925 m (O site), or 800 m (M and E sites), the relative salinity minima that were observed are the print of the Antarctic Intermediate Water, which forms a layer about 250 m thick (if demarcated by $S < 35.00$ psu).

Central waters usually extend from $\sigma_\theta = 27.3$ up to about 26.4, a value typical for the bottom of the mixed layer (see also Fig. 11.6*b*). At the O site, the central water was permanently of the NACW type, as expected. At the E site, a mixture, approximately 50–50%, of NA- and SACW was regularly present. At the M site, more complicated θ–S properties were observed. Although generally dominated by SACW, various mixtures were encountered in the upper layers of this water mass (for $z < 380$ m, and $\sigma_\theta < 27.0$). In contrast, the SACW was

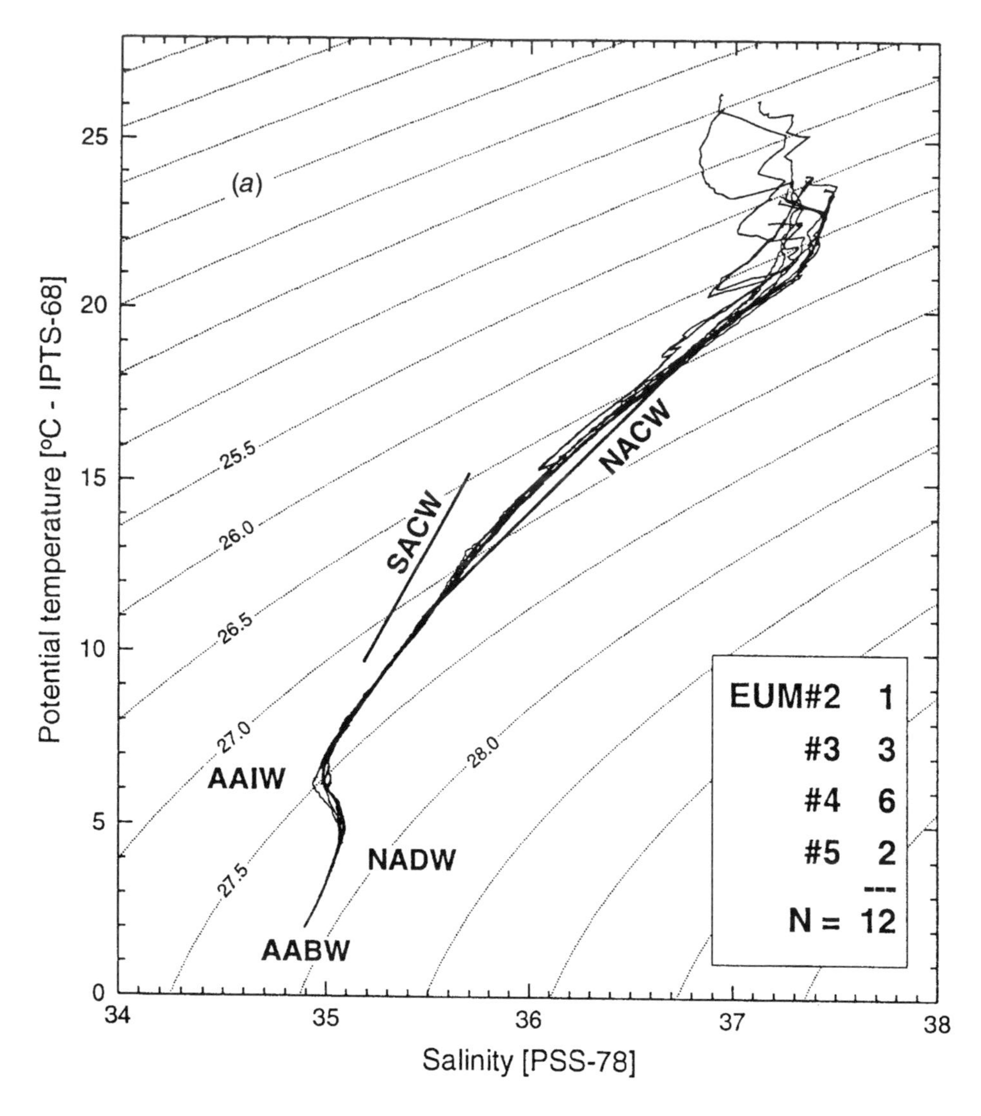

Figure 11.3 *Potential temperature – salinity (θ–S) diagrams for the three sites O* (a), *M (*b*) and E (*c*). All the deep CTD casts, down to the bottom (but only these) made during the cruises 2, 3, 4, and 5 are represented. The two straight lines represent the θ–S relations for the North Atlantic Central Water and the South Atlantic Central Water, as defined by Tomczak (1981). Note that, at the M site, the two superimposed curves (arrow) which differ from the others come from cruise 5, in December 1992.*

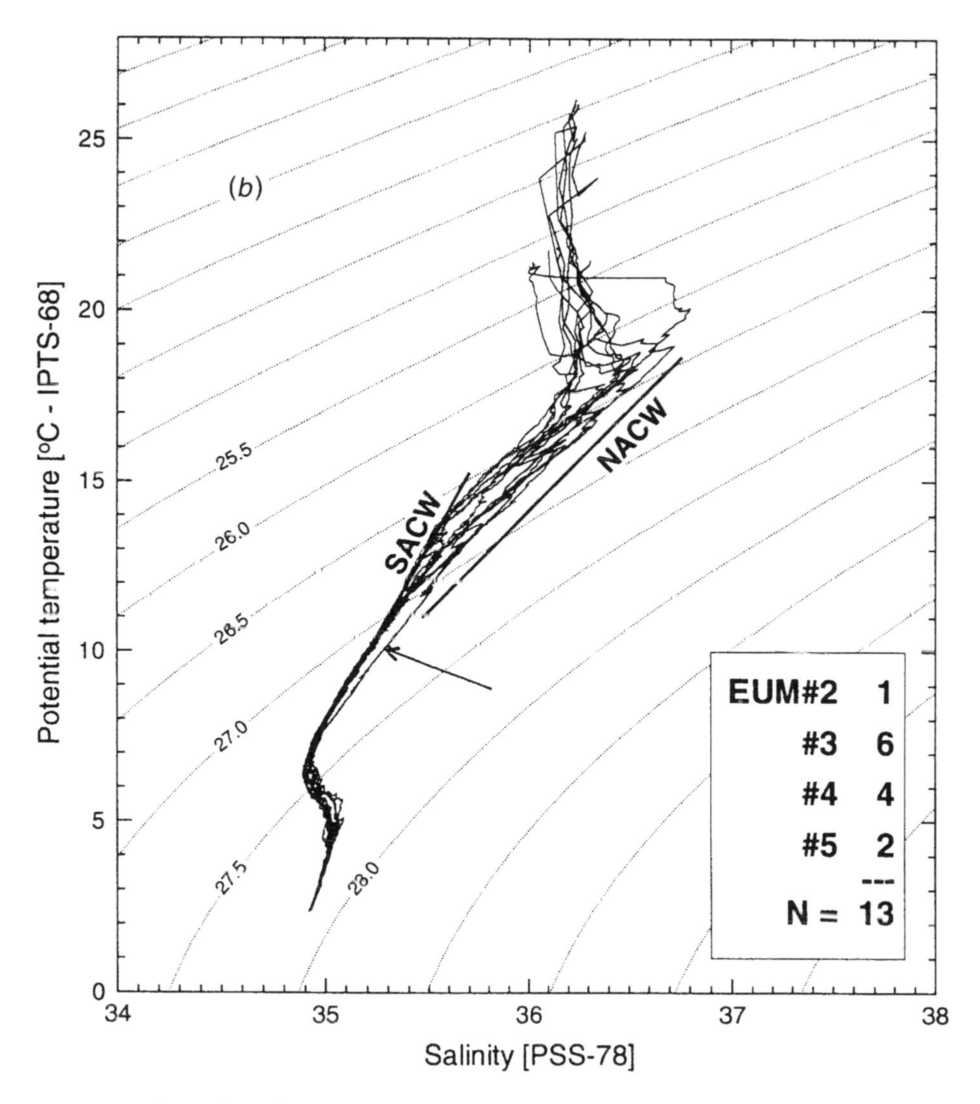

Figure 11.3b *For legend see opposite.*

almost pure within the deeper layers, down to 730 m, where σ_θ reaches 27.3; in December 1991, however, (during EUM#5), the deep SACW was slightly mixed with NACW. Above the central waters, the top layer, examined later, exhibited variable characteristics (Table 11.3) according to the period of the year (the successive cruises).

Time-averaged values, or range of variations, for the main properties at the three sites and various levels in the water column are given in Table 11.4. In relation to the water masses described above, in particular to the Central Waters, other parameters can be used as descriptors. As is well known (see, for example, Minas *et al.*, 1982; Table 2 in Zenk *et al.*, 1991), compared with

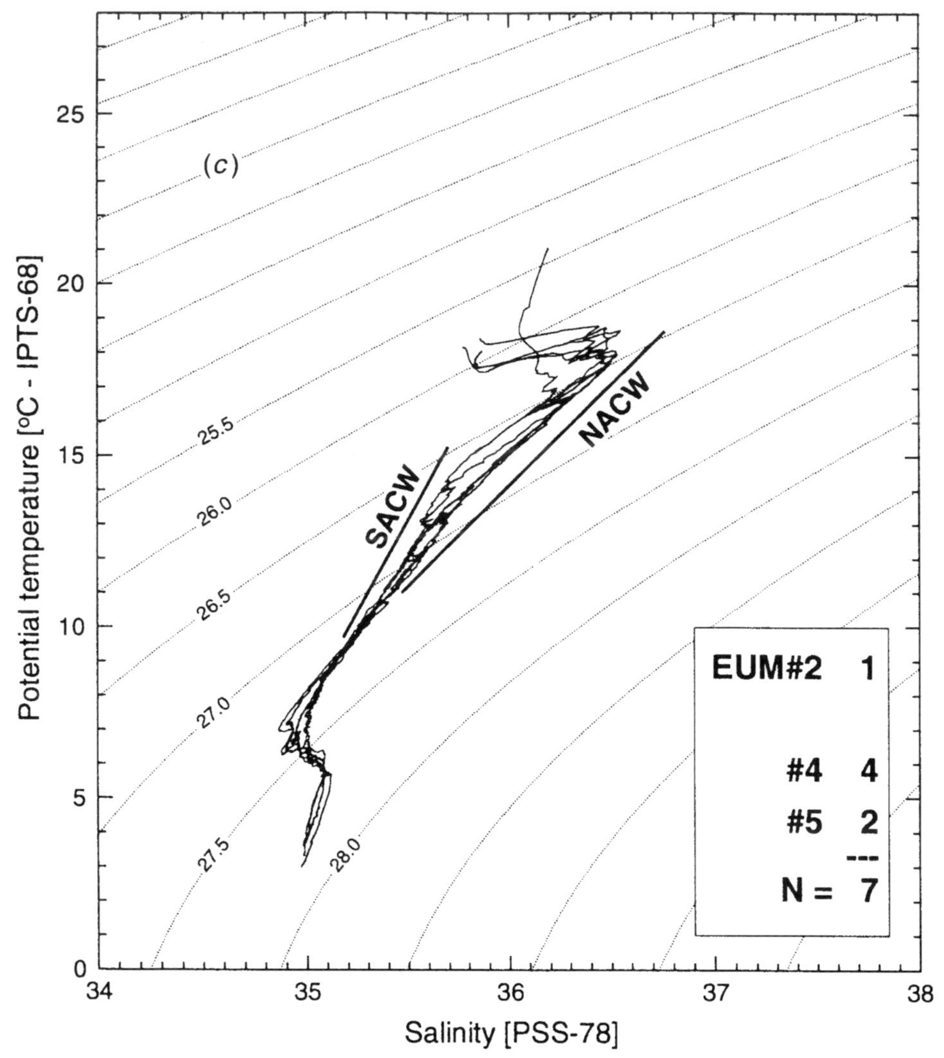

Figure 11.3c *For legend see p. 348.*

NACW, SACW is characterised by higher nutrient contents and a more pronounced oxygen minimum. These features are clearly seen in Table 11.4. At the O site, the nitrogen maximum (occurring at about 800 m with $\sigma_\theta = 27.42$) is typical of the lower end of the NACW. In contrast, at the M and E sites, the N-maximum values are those of pure SACW; they were located around 600–650 m, with $\sigma_\theta = 27.2$, in correspondence with the lower end of this water mass. The oxygen minima were found 150 m above the N-maxima. The minimal concentration at the O site was close to 2.6 ml l^{-1} in NACW, and thus almost twice that found in SACW, at both the M and E sites. A relative minimum in nitrogen (about 22.5 μM) was observed in the two sites (O and M) in the 2000–2500 m layer.

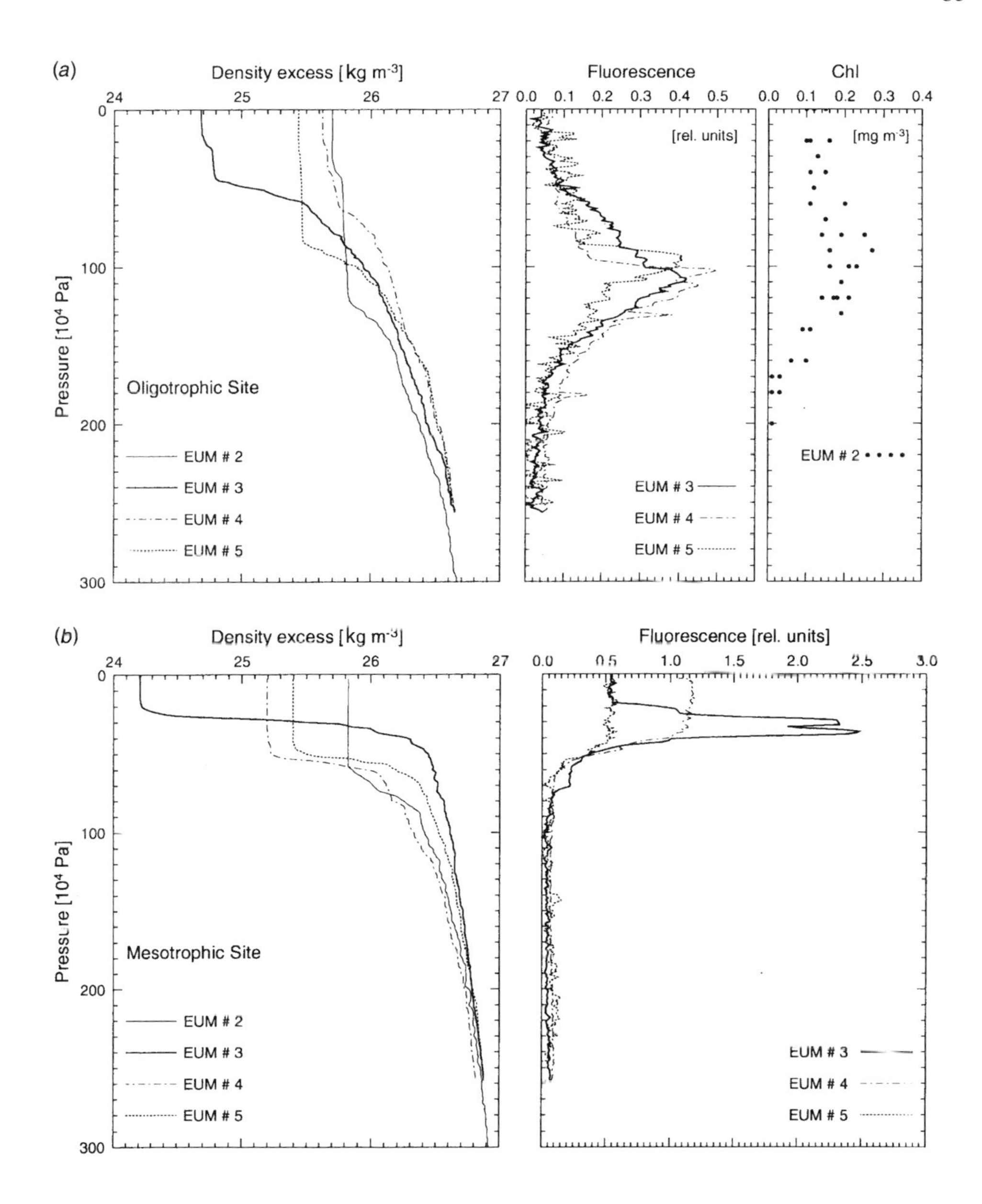

Figure 11.4 *Typical examples of vertical profiles of potential density (density excess in kg m^{-3}) and of algal chlorophyll fluorescence (relative units) at each of the three sites and at various seasons (successive cruises). Dates are as follows (months identified by roman numbers).* (a) *O site: 14 II 91, 14 X 91, 24 VI 92, and 22 XII 92; note that the chlorophyll concentration (absolute units, mg m^{-3}, Pujo-Pay & Raimbault, 1994), determined on discrete samples during EUM#2 (17, 18, and 19 I 91), are given in inset.* (b) *M site: 09 II 91, 09 X 91, 16 VI 92, and 15 XII 92.* (c) *E site: 05 II 91, 13 VI 92, and 12 XII 92; the results for discrete chlorophyll determinations (station 50, April 3, 1974, CINECA 5 cruise) at about the same location are also shown. The chlorophyll concentrations (as mg m^{-3}) can be read with the same numerical scale used for fluorescence.*

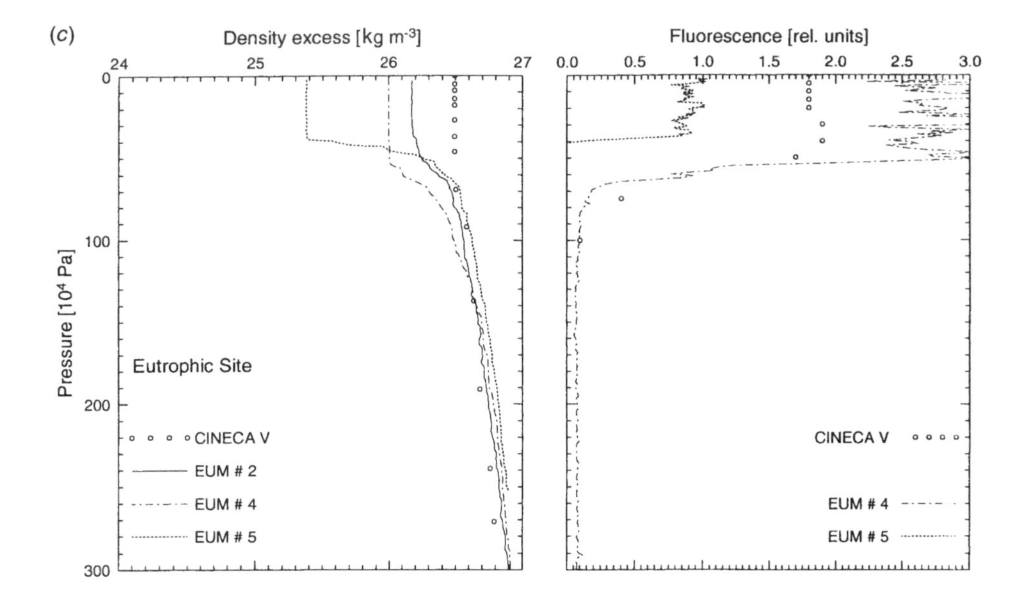

Figure 11.4c *For legend see* *p. 351.*

Hydrographic section between the three sites

From west to east, the replacement of NACW by SACW (or more precisely by NACW–SACW mixtures) is not progressive. A first salinity jump can be detected (Fig. 11.6*a*, showing EUMELI♯5 data) between stations MO-02 and MO-03. A second one, more abrupt, occurs near EM-03. These jumps in salinity have only a reduced impact on the essentially flat disposition of the isopycnals throughout the section (Fig. 11.6*b*). As already pointed out (Barton, 1987), when NACW replaces SACW, the concomitant changes in salinity and temperature are in opposite directions and thus leave the density practically unchanged. Very similar salinity and density fields were found during EUM♯4 (not shown), except perhaps for a steeper rise in salinity near the M site.

Concerning the large-scale circulation through the section displayed in Figs 11.6(*a*,*b*), the geostrophic currents computed in reference to the 3000 dbar level were found to be insignificant. The only exception was found between the M site and station MO-03, where velocities reaching 5 cm s^{-1} (towards S–SW) were derived within the upper 100 m layer. The orientation of the segments forming the section is admittedly not favourable for capturing the main circulation, which is essentially oriented southwest or westward in this zone. In addition, the distances between stations (about 280 km) are not appropriate for a detailed description of the velocity field at the mesoscale (Dadou *et al.*, 1996).

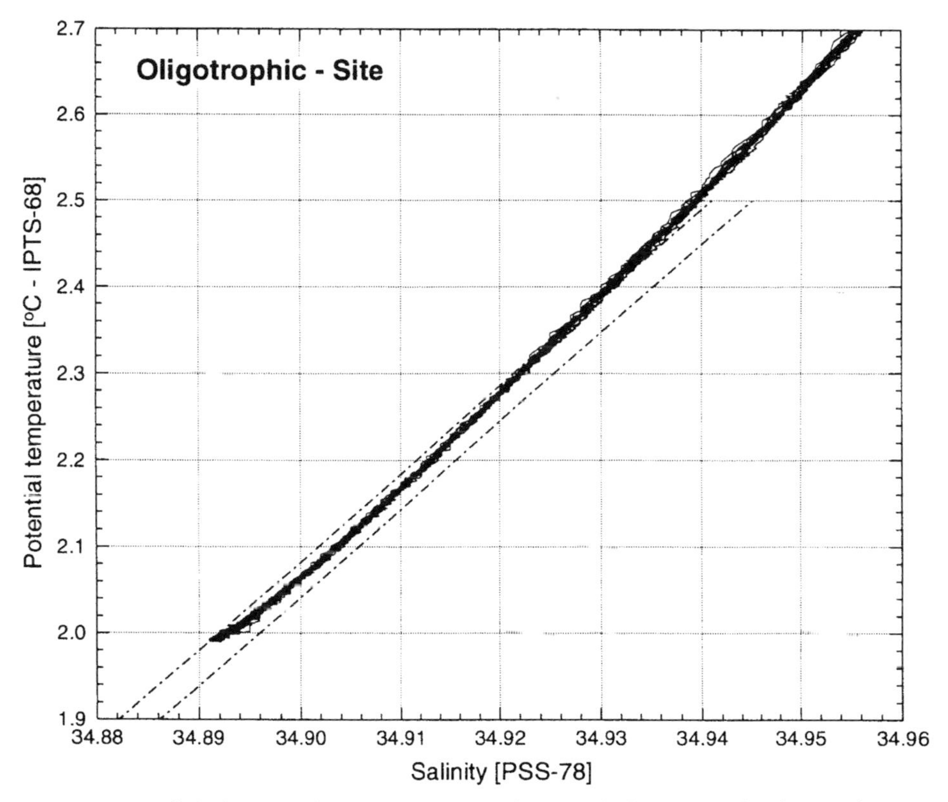

Figure 11.5 *θ-S diagrams for the deep waters, down to the bottom, at the oligotrophic site, as determined during the four cruises, from EUM#2 to 5. These curves show the internal consistency of measurements and calibrations (Tailliez, 1993); they slightly differ from the linear relation between S and θ proposed by Saunders (1986) for these waters. The two dashed lines are shifted by ± 0.002 psu, with respect to the Saunders reference line.*

From the CTD measurements made at the triangular array enclosing the O site (namely at the stations T1, T2, and T3, separated by 65 km), a very slow, SW-oriented, geostrophic motion could be computed between 200 and 1000 m (or deeper). Significant speeds, up to 10 cm s^{-1}, were found in the upper layer. The corresponding fluxes, which cannot be balanced within the triangle, suggest mesoscale activity near the surface. The intermittent occurrence of lower salinities reinforces the possibility of such local drifts of fresher waters at the very surface.

Associated with some of the sediment traps, current meters (Aanderaa) were deployed at 250, 1000, and 2500 m, at the O site and the M site (never at the E site). With three consecutive periods (from EUM#2 to EUM#5), the total records lasted 665 and 630 d at the O and M site, respectively. At 250 m, predominantly SW-oriented currents were recorded at both sites (Bournot *et al.*, 1995) with mean values over the entire period of 7.8 and 11.5 cm s^{-1}. Peak values were 38 and 50 cm s^{-1}, respectively, at the O site and the M site, even

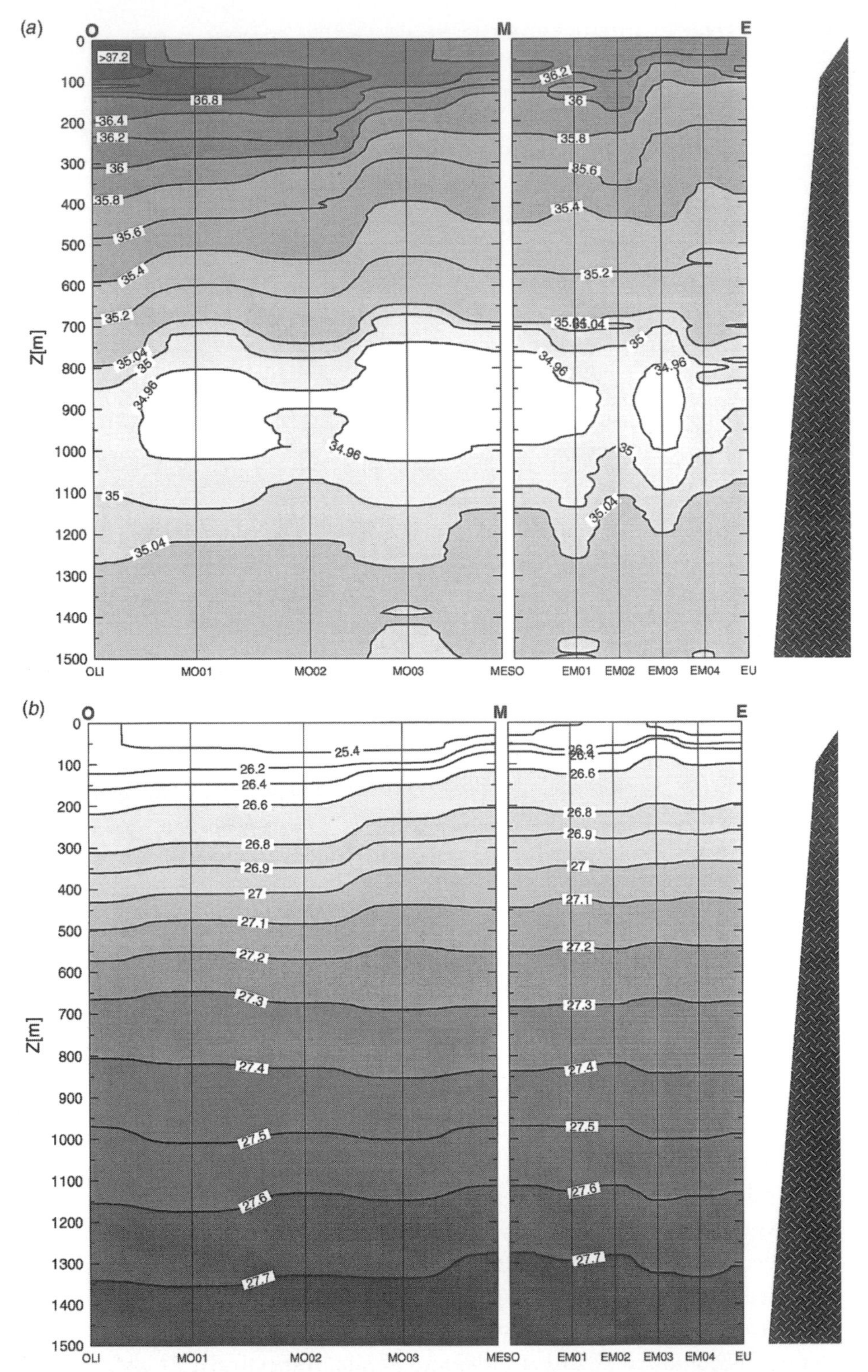

Figure 11.6 *(a) Vertical distribution of salinity, psu, and (b) potential density, density excess in kg m^{-3}, along the two contiguous sections, from the O site to the M site, and from the M site to the E site. The continental slope is schematically shown near the E site. Results from the cruise EUM♯5.*

with an exceptional event at the M site (reaching 98 cm s^{-1} and lasting a few days). With such episodic current velocities, and values often exceeding 15 cm s^{-1}, the collection efficiency of the sediment traps at this depth is questionable (and is not yet clear). The energy spectra show pronounced maxima at frequencies corresponding to the semi-diurnal and diurnal tides, and to the inertial periods (32.5 and 37.8 h, at the latitudes of the O and M sites, respectively). At 1000 m, the currents were generally flowing westward with rather low mean speeds of 4.8 and 4.5 cm s^{-1}, and decreasing at 2500 m to 3.0 and 2.3 cm s^{-1}, at the O site and the M site, respectively, without clear orientations.

Physical and biogeochemical properties within the upper layer at the three sites

Typical examples of the density structure within the upper (300 m thick) layer are displayed at the three sites (Fig. 11.4*a–c*). Fluorescence vertical profiles are also plotted to show the close correlation between the vertical distribution of the algal biomass and the thickness of the mixed layer as well as the density gradient within the pycnocline. Biochemical information for these layers, and time variability detected by the successive cruises, are also provided in Table 11.3 for the three sites.

At the O site, internal waves of wide amplitude (at least ± 10 m) were always observed, resulting in vertical oscillations of the bottom of the mixed layer (ML), as well as of the deep chlorophyll maximum (DCM). Examples are provided by Partensky *et al.* (their Fig. 7, 1996), Lazzara *et al.* (their Fig. 1, 1996) and Morel *et al.* (their Fig. 7, 1996). Therefore the depths of the mixed layer and of the euphotic layer (denoted Z_{ML} and Z_{eu}, respectively), which are given in Table 11.3, must be seen as representative mean values. The frequent occurrence of less saline waters drifting at the surface (probably owing to rains), often gave rise to two-step density profiles, in which the deepest one is the most significant (Fig. 11.4*a–c*). In mid-February (during EUM#2), about two weeks before the net heat budget returns to positive values, the winter deep convection was probably near its maximum and the mixed layer extended down to 120 m. This value greatly exceeds that given in Levitus (1982) (70 m). The minimal thickness of this layer (45 m), in coincidence with the warmest waters in October, agreed with the figure in Levitus (see also discussion and modelling in Dadou *et al.*, 1993). With regard to the algal biomass, this O site was characterised by a permanent deep chlorophyll maximum, generally below 100 m, and actually at a level below the euphotic depth, when defined in reference to the 1% level of the PAR (photosynthetically active radiation) surface irradiance. Vertically integrated phytoplankton biomass was rather constant throughout the year (Table 11.3). Not surprisingly, the less structured

Table 11.3. *Upper ocean layer characteristics*

Cruise no.	Mixed layer[1] Z_{ML} (m)	T (°C)	σ_θ kg m^{-3}	[N] (μM)	Depth[2], Z_{eu} (m)	Chlorophyll a[3] ML (mg m^{-3})	DCM	at Z (m)	Column Chl a (mg m^{-2})	Daily PP[4] (g C m^{-2})	F_p[5]	Z/C[6]	Bacteria[7] (0–200 m) (mg C m^{-2})	Zooplankton[8] (0–200 m) (mg C m^{-2})
Oligotrophic site														
2	120	22.6	25.7	0.00	97	0.12	0.25	100	21.3					
3 leg 1	45	26	24.5			0.06	0.28	110	24.9	0.29	0.055	0.23		205
3 leg 2	55(25)[10]	26	24.7	< 0.003[9]	95	0.08	0.3	110	23.1	0.33	0.061	0.24	783	
4 leg 1	60(25)[10]	23.3	25.6			0.06	0.25	120	24.8	0.34	0.06	0.29	729	
4 leg 2	70	23.7	25.7	< 0.003	102	0.05	0.28	120	23.6	0.35	0.07	0.3		
5	85	23.9	25.4	< 0.003	95	0.17	0.45	95						370
Mesotrophic site														
2	55	19.9	25.8	0.04		0.2	0.6	(50)	29.5					
3 leg 1	25	25.8	23.9		50	0.2	0.4	(40)	22.5	0.52	0.18	0.11		1148
3 leg 2	30	25.3	24.3	0.01	40	0.3	1.3	(45)	30.1	0.52	0.42	0.11	1385	
4 leg 1	35	21.8	25.1			0.45	(H)[12]		27.1	0.95	0.16	0.2	2001	
4 leg 2	40	21.7	25.2	0.7	25	1.25	(H)		44.2	1.5	0.56	0.2		
5	40	21.5	25.4	0.03	30	0.8	(H)							814

Eutrophic site														
2	55	18.5	26.2											
4 leg 1	40	18.5	25.8			1.5	(H)		59				3436	
4 leg 2	45	18.1	26	10	23	3.5	(H)		162	2	0.92	0.02		
5	40	21	25.4	0.25	32	1.2	(H)							1151
C-5[11]	70	17.1	26.5	9	30	1.2	(H)		65	1.24				

[1]Mixed layer depth (Z_{ML}), temperature, density excess (σ_θ), nitrogen ($NO_3 + NO_2$) concentration (Pujo-Pay & Raimbault, 1994)
[2]Depth of the euphotic layer, Z_{eu} (depth where PAR is reduced to 1% of its surface value)
[3]Chlorophyll *a* concentration within the mixed layer (ML) at the deep chlorophyll maximum (DCM), and depth of this maximum; column-integrated chlorophyll (Claustre & Marty, 1994; H. Claustre, personal communication, 1996)
[4]Daily primary production (Morel *et al.*, 1996)
[5]Pigment ratio, F_p, according to Claustre (1994)
[6]Zeaxanthin : chlorophyll ratio (H. Claustre, personal communication, 1996)
[7]Integrated (0–200 m) heterotrophic bacterial biomass (EUM#4: Dufour & Torréton, 1996; M. Bianchi, personal communication, 1994; EUM#3: M. Fernandez, F. Van Wambeke & J. Garnier, personal communication, 1992)
[8]Integrated (0–200 m) mesozooplankton biomass (WP-2 plankton net; S. Razouls & G. Gorsky, personal communication, 1996)
[9]Below detection level (3 nM) using nanomolar method (Raimbault *et al.*, 1990)
[10]Depth of a first minute step in the density profile (see Fig. 11.4*a*)
[11]April 3, 1974; station no. 50 at 25 nm in the NE of the Eutrophic site (CINECA-5, Cruise Report, 1976)
[12]H, homogeneous vertical distribution, without deep maximum

Table 11.4. *Main characteristics and fluxes at different levels for each site*

Asterisks, relative numbers (n ×)* used to compare the quantity in question (Sites M and E) with that observed at the reference site (Site O). Analyses of sediment trap collections are not presently completed (i.e. discrepancies remain between carbon flux and dry-mass flux measurements); daggers, cumulative deployment days for sediment trap collection.

	Site O	Site M	Site E	EUM Cruise no. (reference)
General information				
Bottom depth (m) and potential temperature (°C)	4600 (1.98)	3090 (2.38)	2030 (3.31)	
Oxygen minimum (μM kg^{-1})	115	60	70	
(at Z, m)	(700)	(450)	(280–360)	3, 4, 5 (1)
(at σ_θ, kg m^{-3})	(27.33)	(27.05)	(26.9–26.95)	
N maximum ($NO_3 + NO_2$, μM)	31	34.1	33.7	2, 3, 4 (2)
(at Z, m)	(800)	(600)	(650)	
(at σ_θ, kg m^{-3})	(27.40)	(27.22)	(27.25)	
pH minimum (at 25 °C)	7.595	7.506		3 (3)
(at Z, m)	(800)	(600)		
Mixed-layer depth (range) m	45–120[+]	25–55	40–70	2–5 (1, 4)
Upper layer (0–200 m depth)				
N ($NO_3 + NO_2$, μM), ML range	0	0.01–0.7	0.2–10	(2)
Euphotic depth (m)	95–105	25–50	23–32	(5)
Near-surface Chl (mg m^{-3}) (relative number)*	0.08 (1 ×)	0.50 (2 ×)	2 (25 ×)	(5)
Chl *a* (mg m^{-2})	23.6 (1 ×)	31.0 (1.3 ×)	95 (4.0 ×)	(5)
Primary production (mg C m^{-2} d^{-1})	330 (1 ×)	870 (2.6 ×)	1380 (4.2 ×)	(5)
Particulate organic carbon (mg C m^{-2})	3600 (1 ×)	7560 (2.1 ×)	12 550 (3.5 ×)	4 (2)
Heterotrophic bacteria (mg C m^{-2})	756 (1 ×)	1693 (2.2 ×)	3436 (4.5 ×)	3, 5 (5)
Zooplankton (mg C m^{-2})	287 (1 ×)	981 (3.4 ×)	1151 (4.0 ×)	3, 5 (5)

Water column (0–1000 m depth)				
Zooplankton (mg C m^{-2}) (relative numbers)*	644 (1 ×)	2303 (3.6 ×)	4380 (6.8 ×)	5 (6)
Macro-zooplankton*	1 ×	4.6 ×	7.8 ×	4 (7)
Heterotrophic bacteria (mg C m^{-2})	3900 (1 ×)	9100 (2.3 ×)	9800 (2.5 ×)	4 (8)
Particulate organic carbon (mg C m^{-2})	27 500 (1 ×)	26 000 (0.9 ×)	31 400 (1.1 ×)	4 (2)
Mean vertical fluxes				
Dry mass flux (mg m^{-2} d^{-1}) at 2500 m	36.32 (1 ×)	178 (5.1 ×)	—	
(cumulative deployment days)†	(604)	(586)	—	(9)
Dry mass flux (mg m^{-2} d^{-1}) at 200 m above bottom	30.25 (1 ×)	249 (8.2 ×)	—	
(cumulative deployment days)†	(210)	(120)	—	(9)
Carbon flux (mg m^{-2} d^{-1}) at 2500 m	1.5 (1 ×)	29 (19 ×)		
(cumulative deployment days)†	(240)	(240)		(10)
Carbon flux (mg m^{-2} d^{-1}) at 200 m above bottom	1.1 (1 ×)	5.5 (5.0 ×)		
(cumulative deployment days)†	(210)	(120)		(10)
Sediments (0–10 cm)				
Bacteria (mg C m^{-2})	175 (1 ×)	645 (3.7 ×)		(11)
Meiofauna (mg C m^{-2})	11.2 (1 ×)	43.4 (3.9 ×)	96.3 (8.6 ×)	
Total (bacterial C + meiofauna C m^{-2})	186 (1 ×)	688 (3.7 ×)		
Total particulate organic carbon (g C m^{-2})	159 (1 ×)	257 (1.6 ×)		(11)
Sedimentation rate	5 (1 ×)	15 (3 ×)	44 (9 ×)	1 (12)

Sources: (1) Tailliez *et al.* (1993); (2) Pujo-Pay & Raimbault (1994); (3) C. Copin, personal communication (1992); (4) Vangriesheim *et al.* (1994); (5) references given in Table 11.3; (6) G. Gorsky, personal communication (1997); (7) Andersen *et al.* (1997); (8) Dufour & Torréton (1996); (9) N. Leblond, personal communication (1997); (10) Legeleux *et al.* (1996); (11) Relexans *et al.* (1996); (12) J. L. Reyss, personal communication (1991).

vertical Chl profile, with slightly higher values in the upper layer (0.15 instead of less than 0.10 mg Chl m^{-3}), and exhibiting a rather diffuse DCM, was observed at the moment of the deepest vertical convection (Fig. 11.4*a*).

The temporal change of the mixed layer thickness at the E site was rather weak. For the sake of completeness, a density profile typical of April (CINECA 5 cruise) has been added in Fig. 11.4*c*. It shows denser waters near the surface and a deeper extension of the ML (down to 60–70 m) at this period of the year. In all circumstances, including that of April 1974, the Chl profile was found to be featureless at this site and the algal biomass homogeneously dispersed throughout the mixed layer. It is worth noting that the depth of the mixed layer at this site always exceeded that of the euphotic layer.

At the M site, which was also affected by internal waves (see Fig. 1 in Lazarra *et al.*, 1996), the relatively shallow thermocline was found to be permanent. It was extremely pronounced in October, with a thin ML (25 m); in this case only, a sharp DCM developed below the mixed layer, with short-term variations in chlorophyll concentration. Otherwise, as observed during the other cruises and seasons, a uniform algal biomass was maintained within the entire mixed layer, extending deeper than the euphotic level, as in the E site.

In summary, the O site and the E site were found typical of the two opposite situations that were hoped for, because of the highly contrasted nutrient levels and vertical distribution (Table 11.3). At the M site, the vertical distribution of the phytoplanktonic biomass tends, most of the time, to resemble that of the eutrophic regime, yet with lower pigment concentrations (and change in species composition, see below). The qualifier ‘mesotrophic’ is thus appropriate. During the period of the strongest stratification at this M site, which results in nutrient depletion within the mixed layer, the vertical algal profile conversely tends to resemble that of an oligotrophic regime. However, chlorophyll values are distinctly higher and the DCM is a rather ‘shallow’, compared with the situation in the O site. The regime remains of the mesotrophic type, even if its mode of functioning is different during such stable stratification periods.

Biological structures in the upper layer at the three sites

The vertical structure of the phytoplanktonic assemblages at the EUMELI stations has been thoroughly analysed by using flow cytometry and spectrofluorometry (Partensky *et al.*, 1996; Lazzara *et al.*, 1996). A spectrofluorimetric method was developed and used to quantify phycoerythrins (Lantoine & Neveux, 1997). The trophic status of the three main sites and their algal populations has been investigated by using a pigment biomarker approach (Claustre, 1994; Claustre & Marty, 1995; Lantoine & Neveux, 1997). Not surprisingly (Malone, 1980; Chisholm, 1992), large species, including diatoms in particular, predominate at the E site, with relatively high Chl *c* and fucoxanthin concentrations. The pigment ratio, F_p, was found to be close to 0.9 (defined in Claustre, 1994). Note that F_p is the ratio of the sum of fucoxanthin and peridinin pigments to the sum of all pigments. In brief, it expresses the relative contribution of diatoms and dinoflagellates to any algal assemblage, which in addition may comprise nano- and green flagellates, cryptophytes, cyanobacteria and prochlorophytes. Such F_p values are typical of populations of diatoms and dinoflagellates, and indicate a predominance within the algal assemblage of algae directly involved in new production (Claustre, 1994). In agreement with the F_p values, the contribution of small-sized cells was found to be minor, as revealed by flow cytometry and by the low value of the zeaxanthin: chlorophyll ratio, denoted Z/C (see Table 11.3). Conversely, the lowest F_p values and highest Z/C values were observed at the O site. Cell counting consistently demonstrated the almost complete dominance of tiny species, essentially *Prochlorococcus, Synechococcus* and picoeukaryotes. They showed non-random vertical distributions, maintained through the seasons, which are similar to those found in the subtropical Pacific (Campbell & Vaulot, 1993; Campbell *et al.*, 1994). At the M site, a conspicuous feature is the relative importance of cyanobacteria, in the deep maximum when it existed (in October), or within the entire mixed layer, as it was observed in June and again in December. Exceptionally dense *Synechococcus* populations were observed at this location, as well as in its neighbourhood (station EM-02; see Partensky *et al.*, 1996). This predominance had important consequences for the optical properties of the water (Morel, 1997) as well as for the productivity at this site (Morel *et al.*, 1996). Prochlorophytes, also present at the M site, provided a minor contribution to the algal standing stock, so that the high Z/C values were entirely due to the zeaxanthin-bearing cyanobacteria.

Somewhat surprisingly, a comparison of the seasonal values (Table 11.3) or mean values (as reported in Table 11.4) shows that the algal stocks are not markedly different at the M and the O sites. Indeed, the ratio of the vertically integrated Chl biomass at the M site to that at the O site is on average 1.3. This relative closeness mainly originates from the permanent presence at the O site of a deep chlorophyll maximum with algal populations often extending down to 200 m. In contrast, chlorophyll concentration was generally insignificant beyond 70–80 m at the M site. At the E site, the algal biomass is about 4 times larger than that observed at the O site. With relative numbers 1, 2.6, and 4.2 (Table 11.4), the mean primary production values are arranged as expected for oligotrophic, mesotrophic, and eutrophic regimes; in this way they differ from the biomass values themselves. The reason might be found in the photophysiological responses of the algal assemblages typical of the three sites. The photosynthetic parameters of algae have been studied in relation to the thickness of the mixed layer, the nitrogen availability, and the pigment composition (Babin *et al.*, 1996), and in effect, the photophysiological responses strongly differ in the three trophic situations. Before summarising these results, a first global approach is also informative.

In terms of photosynthetic capacity, the algal population at the M site is definitely more competent than those at the two other sites. This is made clear simply by forming the ratio of the integrated primary production to the integrated chlorophyll biomass, which turns out to be 14 g C g^{-1} Chl d^{-1} at both the O and E sites, whereas it is 28 g C g^{-1} Chl d^{-1} at the M site. A more accurate comparison must account for the variation in the incident PAR irradiation, from site to site and season to season. It consists of computing ψ^*, the bulk cross section for photosynthesis per unit of Chl (Morel, 1978, 1991). For the O site, it had already been shown (Claustre & Marty, 1995) that ψ^* seems to be seasonally constant, with a value around 0.06 m^2 (g Chl)$^{-1}$ at the summer solstice as well as near the autumn equinox. The same value actually is found for the E site in June. In contrast, ψ^* reaches 0.11 m^2 (g Chl)$^{-1}$ at the M site, a high, albeit common, value (Platt, 1986). These differences in bulk efficiency, and in particular the enhanced photosynthetic capacities at the M site, originate from the differences in photophysiology. Actually, it also reflects a weakness when photosynthetic efficiency is discussed based on chlorophyll only, as soon as other photosynthetically active pigments are abundantly represented, as is the case at the M site, with abundant phycoerythrin-bearing *Synechococcus* cells. The corresponding characteristics, namely the photosynthetic rate in the light-limited regime (α), the maximum quantum yield for C-fixation ($\phi_{C_{max}}$), and the maximum rate at saturating irradiance ($P_{B_{max}}$), were determined through incubation experiments and fluorescence studies (Babin *et al.*, 1996). In particular, the maximum quantum yield for carbon fixation has been found to decrease between the E site, the M site and the O site, with values about 0.05, 0.03 and 0.005 (mol C mol^{-1} quanta absorbed). The latter is for the mixed layer at the oligotrophic site, where high values of 0.06 were also observed but only in the deep algal population. Part of the variation in quantum yield is clearly to be attributed to the variable contribution of non-photosynthetic pigments (such as zeaxanthin) to total pigments. Another part originates from changes in the fraction of functional PSII reaction centres, measured by using fast repetition rate fluorometry, and is to be related to nitrate availability. The $P_{B_{max}}$ values (Fig. 5 in Babin *et al.*, 1996) at the M site are at least twice the values at the two other sites.

These photophysiological parameters were used for *a posteriori* computations of the vertical profiles of primary production via a modelling approach (Morel *et al.*, 1996), and in this way, they allowed the differences in photosynthetic performances at the three sites to be understood. Only during cruise 3 and at the oligotrophic site were new and regenerated production determinations performed. By using the ^{15}N tracer technique, a comparison of nitrate and ammonium assimilation rates allows the *f* ratio to be estimated (ratio of new to total production) (Dugdale & Goering, 1967; Eppley & Peterson, 1979). In the upper nutrient depleted layer in this site, the *f* ratio was about 0.10, whereas it reached 0.15–0.18 at the bottom of the euphotic layer (P. Raimbault, personal communication, 1997).

Interestingly, the other (0–200 m) integrated biomass values, namely heterotrophic bacteria and mesozooplankton biomass, and even the particulate organic carbon standing stocks, are all increased by the same factor between the O and the E site, as already found for integrated chlorophyll (i.e. about 4). Such a likeness does not hold true, however, when the M site is compared to the O site. Although the factor for areal Chl content is 1.3, as stated before, those for the other trophic compartments and POC stock are definitely higher (from 2.1 to 3.4; see Table 11.4). If, instead of considering the Chl biomass, the primary production daily rates are compared, a factor of 2.6 is found between the M and O situations, which agrees well with the other factors. Inasmuch as the heterotrophic communities (or the detrital compartment) depend in a direct (or indirect) way on the organic matter synthesised by the primary producers, there is some logic in observing a better correlation with the algal productivity rather than with the algal standing stock itself. However, in the absence of knowledge about the carbon to chlorophyll ratio (not determined) within the phytoplanktonic assemblages investigated, it is difficult to go further into the above explanation supplied as a heuristic hypothesis.

In a global view, however, Dufour & Torréton (1996) found significant log–log regressions of bacterial biomass (expressed as carbon) on chlorophyll concentrations, when all the data for the three sites and depths were pooled together. Notwithstanding the above remark concerning the C: Chl ratio, if a tentative value of 50 is adopted for this ratio (as in Redalje, 1983), the bacterial C to algal C ratio would be 0.64, 1.09 and 0.72 within the 200 m thick upper layer at the O, M, and E sites, respectively. If the same computations are repeated for the water columns inhabited by phytoplankton (namely for the 200, 70, and 50 m thick layers, at the O, M, and E sites, respectively), the above ratios become 0.64, 0.38 and 0.18. Despite its approximate character, such a calculation clearly shows the regular increase of bacterial carbon biomass relative to algal carbon, when the trophic regime changes from eutrophic to oligotrophic (as already described; see, for example, Cole *et al.*, 1988). It is also worth remarking that the ratio of bacterial carbon to POC for the (200 m thick) upper layer is almost constant, with a mean value of 24% (± 3%) at the three sites.

The heterotrophic bacterial production also differs at the three sites and would be significantly related to the photoautotrophic production (Dufour &

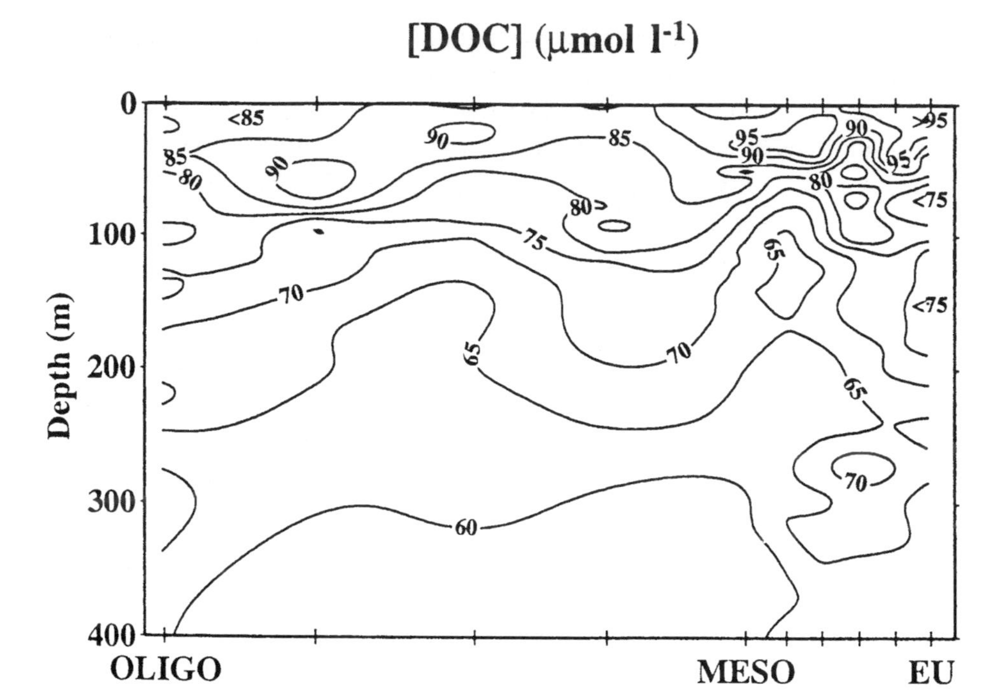

Figure 11.7 *Dissolved organic carbon (DOC) concentration determined during EUM 4 at all stations along the two contiguous sections from the E site to the O site (re-drawn from Avril, 1995). Below 400 m (not shown), all the values are between 60 and 55 μgmol l⁻¹.*

Torréton, 1996). Within the upper layer, the measured production rates (via the tritium–thymidine incorporation method) lead to similar turnover times for heterotrophic bacteria (biomass/production) at the E and M site (2–6 d). These turnover times were distinctly increased at the O site, with values between about 8 and 20 d (Dufour & Torréton, 1996). Bacterial utilisation of glucose is also discussed in relation to bacterial biomass and production in the three sites (Bianchi *et al.*, 1998).

Within the surface waters, a gradient in the partial pressure of CO_2 (p_{CO_2}) occurs along the section between the three sites (Copin-Montégut & Avril, 1995). In the recently upwelled waters spreading at the E site, p_{CO_2} remains high, with values (computed for 25 °C) ranging between 550 and 600 μatm (compared with 353 in air). At the M site, outgassing and net biological consumption reduce the supersaturation, and p_{CO_2} values (at 25 °C) of 410 and 405 μatm were measured during the two cruises EUM#3 and 4. At the O site, p_{CO_2} values were found to be slightly above equilibrium (370 ± 5 μatm) in September–October, as well as in June.

Depth profiles of the dissolved organic carbon (DOC) were systematically determined with the high temperature catalytic oxidation method (Avril, 1995). In contrast to all the properties briefly examined above, the DOC distribution appeared to be rather uniform (Fig. 11.7). In this vertical section, an increase by only 10 μmol l^{-1} can be observed between the O and E sites; it appears to be weak compared with the gradient in productivity. This increase, related to the freshly produced, and presumably labile, DOC fraction, is confined to the near-surface layer ($Z < 50$ m). Once integrated over a layer 200 m thick, the DOC contents do not differ significantly between the three sites and amount to 190 g C m^{-2} (mean concentration 0.95 g C m^{-3}). Such a conservative behaviour tends to demonstrate that the biological production of DOC, related to phytoplankton abundance, is balanced by consumptive processes, in which bacterioplankton (see below) play a major role. These upper layer values exceed those uniformly found below 500 m and down to the bottom, steadily around 55 μM (or 0.67 g C m^{-3}) at all stations of the E–M–O section. Changes in DOC concentration were insignificant and barely discernible between cruises 3, 4, and 5 (Avril, 1995). The above DOC concentrations are similar to those determined in the northwestern Mediterranean Sea (Copin-Montégut & Avril, 1993), and distinctly higher than those observed in the equatorial Pacific Ocean (Carlson & Ducklow, 1995). Similar conclusions can be drawn from a preliminary study (Pujo-Pay, 1995) of the dissolved organic nitrogen (DON) distribution. The concentrations (about 5–6 μM) in the upper layers were similar, whatever the site, and exceed those uniformly found (about 3 μM) at depth.

Water column characteristics and biomass at the three sites

The abundance, community structure, and vertical distribution of the macroplanktonic and micronektonic groups, as determined from oblique (0–1000 m) hauls, have been analysed at the three sites (Andersen *et al.*, 1997). The diversity index and vertical migration amplitude both increase from the E site to the O site. Zooplanktonic C biomasses, integrated over the 0–1000 m water column, are distributed at the three sites according to the following numbers: 1, 3.6 and 6.8 (see Table 11.4) These values span an interval slightly wider than that characteristic of the upper (0–200 m) layer for parameters such as primary production rates, POC content, and even 0–200 m integrated zooplankton. This widening could indicate an increasing efficiency in the C transfer between the primary and the secondary producers, when changing from an oligotrophic to a eutrophic system. It is worth remarking that the zooplanktonic biomass, more abundant at the E site, is also more regularly spread out within the water column. The ratios of the 0–1000 m biomasses to the 0–200 m biomasses are 2.24, 2.34, and 3.80 at the O, M and E sites, respectively, and are to be compared with the value of 5 that would correspond to a uniform distribution. The diel variations and the effect of vertical migrations (G. Gorsky, personal communication, 1997) are smoothed out in these numbers, by averaging all the plankton net catches.

Heterotrophic bacterial carbon values, when integrated from surface to bottom, are almost equal at the E and M sites, and on average 2.4 times higher than that found at the O site. In terms of mean concentration, computed by dividing the biomasses by the bottom depths, the relative numbers (namely 1, 3.42, and 5.66) are more contrasted between the sites and arranged in the usual manner. Indeed, when the effect of the water column height is removed, these relative numbers fall into the range found for other relative numbers; in particular, they are close to those related to the 0–1000 m mesozooplankton compartment (amounting to 1, 3.6, and 6.8, respectively).

After the overwhelming DOC pool, particulate matter (POC) represents the second most important carbon pool. These particles, which include small living organisms (essentially bacteria), remain dominated by various organic detritus and small-sized, slowly sinking debris. The whole water-column POC contents are roughly the same (from 26 to 31 $g\,C\,m^{-2}$) for the three sites. Once converted to mean concentrations, as was done for bacterial C, the relative contrast between sites is restored; indeed, the relative numbers for the O, M and E sites are 1, 1.41, and 2.58, respectively. This inter-site contrast, however, is noticeably weaker than for other trophic compartments. This narrowing could be consistent with the hypothesis, made for oligotrophic environments, of the presence of smaller debris, generated by an ecological system based on picoplanktonic species and bacteria. In such a regime, the residence time within the water column of these tiny suspended particles would be increased accordingly. It is worth recalling that the POC pool represents a very small fraction of the total (POC + DOC) organic carbon pool. In the present case, the POC to DOC ratios are only 0.9, 1.2 and 2.2%, for the O, M and E sites, respectively.

Fluxes and benthic responses

As stated in the Introduction, the sedimenting material, sequentially collected by traps, has not yet been fully analysed for the entire deployment period and for all parameters. Nevertheless, detailed specific studies are published (Legeleux *et al.*, 1994*a,b*, 1995, 1996; Relexans *et al.*, 1996; Tachikawa *et al.*, 1997; Hamelin *et al.*, 1997). It is therefore unsafe and premature to relate the corresponding numbers or annual means to those pertaining to the upper layer or water column quantities.

With the available results, however, it can be noted that the mass fluxes, determined at 2500 m at the M and O sites (in a ratio of 5.1 to 1; Table 11.4), merely reflect the general tendency observed for other parameters. As a preliminary comment about the magnitude of these particle fluxes, it is worth remarking that they are well arranged within the general trend to be related to biogeochemical provinces with their peculiar planktonic ecosystems (Jickells *et al.*, 1996). These authors analysed the sediment trap data obtained along a N–S transect from 48° to 19°N, at about 20°W. Three sites of this transect, located off Mauritania, represent typical oligotrophic subtropical conditions, namely the sites studied by Lampitt (1992), by Kremling & Streu (1993) and by Jickells *et al.* (1996). The mass fluxes amounted to 33.4, 51.0, and 27.4 mg C m^{-2} d^{-1}, respectively. Over two years, a very similar mean flux is observed at the O site, namely 36.2 mg C m^{-2} d^{-1}. In addition, over a shorter period (120 d, analysed by Legeleux *et al.*, 1996), the organic carbon fraction (4.7%) and the calcium carbonate fraction (66.7%) are practically confounded with the corresponding values given by Jickells *et al.* (1996; their Table I) for the three sites referred to above. For the M site, the present results can also be compared with those of Wefer & Fischer (1993). Their results were collected at nearby sites CB1 and CB2, located about 300 km off the Mauritanian coast at the latitude of the E site, in a regime presumably similar to that of the M site. The two successive time-series provided mean mass fluxes of 183 mg m^{-2} d^{-1} (during CB1, trap depth 2195 m) and 148 mg m^{-2} d^{-1} (during CB2, trap depth 3502 m); the fluxes recorded at the M site (2500 m) consistently result in a mean value of 178 mg m^{-2} d^{-1}. Over a restricted period (120 d; see Legeleux *et al.*, 1996), the organic carbon fraction at the M site amounted on average to 9.2%, with important fluctuations, compared with mean values of 4.1 and 3.0% for CB1 and CB2. The $CaCO_3$ fraction remained steadily around 56%, compared with 42.9% and 51.7% at CB1 and CB2. The lithogenic fraction varied between 11% and 35% at the M site, whereas it was 40.8% at CB1 and 37.9% at CB2. Another, less direct, comparison is also possible for the site studied by Jickells *et al.* (1996), located at 19°N 20°W, which practically coincides with the EM-02 station (Table 11.1), intermediate between the E and M sites. As expected, the mass flux at this site, 245 mg m^{-2} d^{-1}, at 2190 m as well as the organic carbon fraction, 12.6%, are both distinctly higher than the corresponding values at the M site.

The biomass of micro- and meiobenthos were measured and the benthic metabolic activities studied in parallel with determinations of the organic content and composition of the upper part of the sediments (see Relexans *et al.*, 1996). According to the conclusions of these authors (see also Table 11.4), the benthic response to food supply from the upper layers is rather linear, to the extent that the relative numbers for benthic microbial biomass and primary production in the upper layers (at the O and M sites) are relatively close. In the same reference, it is also suggested that micro- and meiobenthos could be more efficient utilisers of the downwelling carbon supply than their counterparts in the eutrophic site.

The macrofauna community structure, the spatial patterns, and scale of the organisms' aggregation were studied during three cruises (2, 3, and 4); 53 box core samples were obtained by using a modified USNEL sampler (Cosson *et al.*, 1997). The mean densities of the total benthic macrofauna, everywhere dominated by the Polychaeta, decreased significantly from the E site, to the M and O sites, as did the number of taxonomic groups (Sibuet *et al.*, 1993; Cosson *et al.*, 1997). The benthic response to primary productivity within the upper layers depends on both the organic matter flux exported downward and the bottom depth, or the distance travelled before deposition. It is worth recalling that for the EUMELI sites, the decrease in biological activity within the upper layers parallels the increase in bottom depth. A positive linear relation between the macrofaunal density at the three sites and the calculated organic carbon flux at the bottom level has been obtained (Rabouille *et al.*, 1993). The organic carbon burial flux would represent an extremely small fraction of the flux that reaches the bottom, which means that this energy supply is efficiently exploited by the benthic organisms (Khripounoff *et al.*, 1998; see also Table 7 in Cosson *et al.*, 1997).

Perspective and conclusion

Studies are still ongoing in various fields, in particular those related to sedimentation or to the exchanges and partitioning of chemical species between the dissolved fraction, the minute particles (collected by *in situ* pumping), and the sinking larger particles (collected in the drifting and moored traps). Results have recently been presented about the distribution of rare earth and neodymium isotopic composition in the suspended matter (Tachikawa *et al.*, 1997), and regarding the use of lead isotopes as tracers of exchange between the particulate and dissolved matter (Hamelin *et al.*, 1997). The composition of the sedimentary organic matter and of the sedimenting material collected in the traps, and the relative importance of the lithogenic, opal, and calcite fluxes still deserve further studies. The susceptibility of biogenic elements to re-mineralisation during settling, and in general all the processes within the water column, are not yet analysed as completely as those occurring in the upper layer. The same can be said about the modelling of the complex and intricate processes involving various compartments and the whole water column.

In effect, with a series of papers published in a special section of *Deep-Sea Research*, Part I (vol. 43, no. 8), the algal and heterotrophic bacterial compartments in the upper layer are well documented and described in detail. The specific studies of the phytoplankton bio-optical and physiological properties (Babin *et al.*, 1996) have also allowed a validation of the spectral primary production model to be performed (Morel *et al.*, 1996). To the extent that the community structures and photophysiological responses of algae are very different at the three sites, a meaningful assessment of the predictive skill of the carbon uptake model was attainable. The outputs of the model have been compared with *in situ* determinations at all levels, as well as with the resulting column-integrated primary productions. Because of this comparison, a 'version 2' of Morel's (1991) model was proposed. It differs from the previous version by a slight change within the set of standard photophysiological parameters, without modifying its general parameterisation. This version, adapted for general use (Antoine & Morel, 1996), was applied at a global scale (Antoine *et al.*, 1996) to the monthly maps of the satellite chlorophyll concentration, as derived from the coastal zone color scanner data archive (Feldman *et al.*, 1989). This model represents the initial segment of a series of models, dealing with the carbon flux from the surface down to the sediment. They are currently in progress and will be presented in their first versions at the JGOFS modelling symposium.

In reference to the initial strategy for EUMELI and to its main objectives, as summarised in the introductory section, several comments can be made. The general patterns expected for the zone and the three selected sites have been verified during the successive campaigns. Submitted to similar climatic conditions, the differing trophic regimes were characterised by mean primary production rate spanning a factor of 4 approximately. The other non-algal compartments and stocks related to the initial formation of particulate organic matter through photosynthesis are arranged roughly in the same way, as are the various fluxes within the water column and down to the bottom. This still preliminary conclusion tends to support the validity of hypotheses (necessarily made, but always questionable) underlying such a one-dimensional approach. Future detailed studies will probably discover some nuances. With abundant documentation in contrasted situations, and a databank available soon, the framework for further process studies, or for development and validation of models and use of future satellite data, is prepared.

Acknowledgements

Financial support for the EUMELI Programme has been provided by the French Agencies funding the JGOFS-France activities, particularly by CNRS-INSU and IFREMER, as the main contributors, and also by ORSTOM and CNES. The international scientific committee of JGOFS-France has been helpful during the execution of the programme, by its continuous advice and encouragement, as well as by its recommendations when funding problems have arisen. Over one hundred technicians and scientists were involved in EUMELI, at sea or in laboratory activities, which for some of them are still ongoing. The author of this chapter thanks the many participants for their effort in collecting and analysing the data, and particularly those who have made available, as personal communications, their unpublished results for the completion of the various tables of this chapter. Special thanks are due to D. Tailliez, who carefully checked the hydrographic data and prepared the figures from the data bank. The officers and crews of the hydrographic vessels *Jean Charcot*, l'Atalante and *le Suroît* are warmly thanked for their efficient technical assistance and companionship at sea.

References

Ahran, M., Colin de Verdière, A. & Mémery, L. (1994). The Eastern boundary of the subtropical North Atlantic. *Journal of Physical Oceanography*, **24**, 1295–301.

Andersen, V., Sardou, J. & Gasser, B. (1997). Macroplankton and micronekton in the northeast tropical Atlantic: abundance, community composition and vertical distribution in relation to different trophic environments. *Deep-Sea Research*, Part I, **44**, 193–222.

Antoine, D. & Morel, A. (1996). Oceanic primary production, 1: Adaptation of a spectral light-photosynthesis model in view of application to satellite chlorophyll observations. *Global Biogeochemical Cycles*, **10**, 43–55.

Antoine, D., André, J.-M. & Morel, A. (1996). Oceanic primary production, 2 : Estimation at global scale from satellite (Coastal Zone Color Scanner) chlorophyll. *Global Biogeochemical Cycles*, **10**, 57–69.

Auffret, G. A. *et al.* (1992). Caracterisation sédimentologique et biologique préliminaire des sites du projet EUMELI. *Comptes Rendus de l'Académie des Sciences, Paris*, **314**, 187–94.

Avril, B. (1995). Le carbone organique dissous en milieu marin. Thèse, Université Pierre et Marie Curie, Paris, 250 pp.

Babin, M., Morel, A., Claustre, H., Bricaud, A., Kolber, Z. & Falkowski, P. G. (1996). Nitrogen- and irradiance-dependent variations of the maximum quantum yield of carbon fixation in eutrophic, mesotrophic, and oligotrophic marine systems. *Deep-Sea Research*, Part I, **43**, 1241–72.

Barton, E. D. (1987). Meanders, eddies and intrusions in the thermohaline front off Northwest Africa. *Oceanologica Acta*, **10**, 267–83.

Barton, E. D. & Hughes, P. (1982). Variability of water mass interleaving off NW Africa. *Journal of Marine Research*, **40**, 963–84.

Berthon, J.-F. (1992). Evaluation de la biomasse phytoplanctonique et de la production primaire associée, à partir de données satellitaires 'couleur de l'océan': application à la zone tropicale au large de la Mauritanie. Thèse de l'Université Pierre et Marie Curie, 160 pp.

Bianchi, A., Van Wambeke, F. & Garcin, J. (1998). Bacterial utilization of glucose in the water column from eutrophic to oligotrophic pelagic areas in the eastern North Atlantic Ocean. *Journal of Marine Systems*, **14**, 45–55.

Bournot, C., Vangriesheim, A. & Vigot, A. (1995). *Campagnes EUMELI 2 à 5. Rapport des sonnées de courant des mouillages des pièges, 250 m, 1000 m et 2500 m.* IFREMER Report DRO/EP 95/202–AV. 453 pp.

Bricaud, A., Morel, A. & André, J.-M. (1987). Spatial-temporal variability of algal biomass and potential productivity in the Mauritanian upwelling zone, as estimated from CZCS data. *Advances in Space Research*, **100(C7)**, 13321–32.

Campbell, L. & Vaulot, D. (1993). Photosynthetic picoplankton community structure in the subtropical North Pacific Ocean near Hawaii (Station Aloha). *Deep-Sea Research*, Part I, **40**, 2043–60.

Campbell, L., Nolla, H. A. & Vaulot, D. (1994). The importance of *Prochlorococcus* to community structure in the central North Pacific Ocean. *Limnology and Oceanography*, **39**, 954–61.

Carlson, C. A. & Ducklow, H. W. (1995). Dissolved organic carbon in the upper ocean of the central equatorial Pacific Ocean, 1992: Daily and fine scale vertical variations. *Deep-Sea Research*, Part II, **42**, 639–56.

Chisholm, S. W. (1992). Phytoplankton size. In *Primary Productivity and Biogeochemical Cycles in the Sea*, ed. P. G. Falkowski & A. D. Woodheads, pp. 213–37. New York: Plenum Press.

Claustre, H. (1994). The trophic status of various oceanic provinces as revealed by phytoplankton pigment signatures. *Limnology and Oceanography*, **39**, 1206–10.

Claustre, H. & Marty, J.-C. (1995). Specific phytoplankton biomasses and their relation to primary production in the tropical North Atlantic. *Deep-Sea Research*, Part I, **42**, 1475–93.

Cole, J. J., Findlay, S. & Pace, M. P. (1988). Bacterial production in fresh and saltwater ecosystems: a cross-system overview. *Marine Ecology Progress Series*, **43**, 1–10.

Copin-Montégut, G. & Avril, B. (1993). Vertical distribution and temporal variation of dissolved organic carbon in the northwestern Mediterranean Sea. *Deep-Sea Research*, Part I, **40**, 1963–72.

Copin-Montégut, C. & Avril, B. (1995). Continuous p_{CO_2} measurements in surface water of the northeastern tropical Atlantic. *Tellus*, **47B**, 86–92.

Corn, M., Belviso, S., Nival, P., Vigot, A. & Buat-Menard, P. (1994). Downward flux of particulate dimethylsulphonio-propionate (DMSPp) in the tropical ocean. *Oceanologica Acta*, **17**, 233–6.

Cosson, N., Sibuet, M. & Galeron, J. (1997). Community structure and spatial heterogeneity of the deep-sea macrofauna at three contrasting stations in the tropical northeast Atlantic. *Deep-Sea Research*, Part I, **44**, 247–70.

Dadou, I. & Garçon, V. (1993). EUMELI oligotrophic site: response of an upper ocean model to climatological and ECMWF atmospheric forcing. *Journal of Marine Systems*, **4**, 371–90.

Dadou, I., Garçon, V., Andersen, V., Flierl, G. L. & Davis, C. S. (1996). Impact of the North Equatorial current meandering on a pelagic ecosystem: a modelling approach. *Journal of Marine Research*, **54**, 311–41.

Dufour, P. & Torréton, J.-P. (1996). Bottom-up and top-down control of bacterioplankton from eutrophic to oligotrophic sites in the tropical northeastern Atlantic Ocean. *Deep-Sea Research*, Part I, **43**, 1305–20.

Dugdale, R. C. & Goering, J. J. (1967). Uptake of new and regenerated forms of nitrogen in primary productivity. *Limnology and Oceanography*, **12**, 196–206.

Eppley, R. W. & Peterson, B. J. (1979). Particulate organic matter flux and phytoplanktonic new production in the deep ocean. *Nature*, **282**, 677–80.

Feldman, G., Kuring, N., Ng, C., Esaias, W., McClain, C., Elrod, J., Maynard, N., Endres, D., Evans, R., Brown, J., Walsh, S., Carle, M. & Podesta, G. (1989). Ocean Color: Availability of the global data set. *EOS Transactions, American Geophysical Union*, **70**, 634.

Fiuza, A. F. G. & Halpern, D. (1982). Hydrographic observations of the Canary Current between 21°N and 25°N in March–April 1974. *Rapport et Procès-Verbaux des Réunions, Conseil International pour l' Exploration de la Mer*, **180**, 58–64.

Fraga, F. (1974). Distribution des masses d'eau dans l'upwelling de Mauritanie. *Tethys*, **6**, 5–10.

Hamelin, B., Ferrand, J. L., Alleman, L., Nicolas, E. & Veron, A. (1997). Isotopic evidence of pollutant lead transport by the subtropical North Atlantic gyre. *Geochimica et Cosmochimica Acta*, **20**, 4423–8.

Hastenrath, S. & Lamb, P. J. (1977). *Climatic*

Atlas of the Tropical Atlantic and Eastern Pacific Oceans. 105 pp. University of Wisconsin Press.

Hempel, G. (1982). The Canary Current: Studies of an upwelling system; Introduction. *Rapport et Procès-Verbaux des Réunions, Conseil International pour l' Exploration de la Mer*, **180**, 7–8.

Hughes, P. & Barton, E. D. (1974). Stratification and water mass structure in the upwelling area off Northwest Africa in April–May 1969. *Deep-Sea Research*, **21**, 611–28.

Jacques, G. (1993). Observations on the pelagic system in the tropical Atlantic at 20°N (The EUMELI Program). *Annales de l'Institut Océanographique, Paris*, **69**, 9–14.

JGOFS (1987). *The Joint Global Ocean Flux Study; Background, Goals, Organization, and Next Steps*. Report of the International Scientific Planning and Coordination Meeting for Global Ocean Flux Studies. SCOR Editor, Baltimore: The Johns Hopkins University.

Jickells, T. D., Newton, P. P., King, P., Lampitt, R. S. & Boutle, C. (1996). A comparison of sediment trap records of particle fluxes from 19 to 48°N in the northeast Atlantic and their relation to surface water productivity. *Deep-Sea Research*, Part I, **43**, 971–86.

Khripounoff, A., Vangriesheim, A. & Crassous, P. (1998). Vertical and temporal variations of particle fluxes in the deep tropical Atlantic. *Deep-Sea Research*, Part I, **45**, 193–216.

Kremling, K. & Streu, P. (1993). Saharan dust influenced trace elements fluxes in deep North Atlantic subtropical waters. *Deep-Sea Research*, Part I, **40**, 1155–68.

Lampitt, R. S. (1992). The contribution of deep-sea Macroplankton to organic remineralisation: results from sediment trap and zooplankton studies over the Madeira Abyssal Plain. *Deep-Sea Research*, **39**, 221–33.

Lantoine, F. & Neveux, J. (1997). Spatial and seasonal variations in abundance and spectral characteristics of phycoerythrins in the tropical northeastern Atlantic Ocean. *Deep-Sea Research*, Part I, **44**, 223–46.

Lazzara, L., Bricaud, A. & Claustre, H. (1996). Spectral absorption and fluorescence excitation properties of phytoplanktonic populations at a mesotrophic and an oligotrophic site in the tropical North Atlantic (EUMELI Program). *Deep-Sea Research*, Part I, **43**, 1215–40.

Legeleux, F., Reyss, J.-L., Bonte, P. & Organo, C. (1994*a*). Concomitant enrichments of uranium, molybdenum and arsenic in suboxic continental margin sediments. *Oceanologica Acta*, **17**, 417–29.

Legeleux, F., Reyss, J.-L. & Schmidt, S. (1994*b*). Particle mixing rates in sediments of the northeast tropical Atlantic: evidence from ^{210}Pb, ^{137}Cs, $^{228}Th_{xs}$, and $^{234}Th_{xs}$ downcore distributions. *Earth and Planetary Science Letters*, **128**, 545–62.

Legeleux, F., Reyss, J.-L. & Floris, S. (1995). Entrainement des métaux vers les sédiments sur les marges continentales de l'Atlantique Est. *Comptes Rendus de l'Académie des Sciences, Paris*, **320**, 1195–202.

Legeleux, F., Reyss, J.-L., Etcheber, H. & Khripounoff, A. (1996). Fluxes and balance of ^{210}Pb in the tropical northeast Atlantic. *Deep-Sea Research*, Part I, **43**, 1321–41.

Levitus, S. (1982). *Climatological Atlas of the World Ocean*. (NOAA Professional Paper 13.) Washington, DC: US Government Printing Office. 173 pp.

Lozier, M. S., Owens, W. B. & Curry, R. G. (1995). The climatology of the North Atlantic. *Progress in Oceanography*, **36**, 1–44.

Maillard, C. (1986). *Atlas Hydrologique de l'Atlantique Nord-Est*. Service de la Documentation et des Publications de l'IFREMER, Brest. 299 pp.

Malone, T. C. (1980). Algal size. In *The Physiological Ecology of Phytoplankton*, vol. 1, ed. I. Morris, pp. 433–63. University of California.

Manriquez, M. & Fraga, F. (1982). The distribution of water masses in the upwelling region off Northwest Africa in November. *Rapport et Procès-Verbaux des*

Réunions, Conseil International pour l' Exploration de la Mer, **180**, 39–47.

Minas, H. J., Codispoti, L. A. & Dugdale, R. C. (1982). Nutrients and primary production in the upwelling region off Northwest Africa. *Rapport et Procès-Verbaux des Réunions, Conseil International pour l' Exploration de la Mer*, **180**, 148–83.

Mittelstaedt, E. (1982). The upwelling area off Northwest Africa: A description of phenomena related to coastal upwelling. *Progress in Oceanography*, **12**, 307–31.

Morel, A. (1978). Available, usable, and stored radiant energy in relation to marine photosynthesis. *Deep-Sea Research*, **25**, 673–88.

Morel, A. (1982). Optical properties and radiant energy in the waters of the Guinea dome and the Mauritanian upwelling area in relation to primary production. *Rapport et Procès-Verbaux des Réunions, Conseil International pour l' Exploration de la Mer*, **180**, 94–107.

Morel, A. (1991). Light and marine photosynthesis: a spectral model with geochemical and climatological implications. *Progress in Oceanography*, **26**, 263–306.

Morel, A. (1996). An ocean flux study in eutrophic, mesotrophic and oligotrophic situations: the EUMELI program. *Deep-Sea Research*, Part I, **43**, 1185–90.

Morel, A. (1997). Consequences of a *Synechococcus* bloom upon the optical properties of oceanic Case 1 waters. *Limnology and Oceanography*, **42**, 1746–54.

Morel A., Antoine, D., Babin, M. & Dandonneau, Y. (1996). Measured and modelled primary production in the Northeast Atlantic (EUMELI-JGOFS program); the impact of natural variations in photosynthetic parameters on model predictive skill. *Deep-Sea Research*, Part I, **43**, 1273–304.

Partensky, F., Blanchot, J., Lantoine, F., Neveux, J. & Marie, D. (1996). Vertical structure of picophytoplankton at different trophic sites of the tropical northeastern Atlantic Ocean. *Deep-Sea Research*, Part I, **43**, 1191–213.

Peters, H. (1976). The spreading of the water masses of the Banc d'Arguin in the upwelling area off the northern Mauritanian coast. 'Meteor' *Forschungs-Ergebnisse*, A **18**, 78–100.

Pierre, C., Vangriesheim, A. & Laube-Lenfant, E. (1994). Variability of water masses and of organic production-regeneration systems as related to eutrophic, mesotrophic and oligotrophic conditions in the Northeast Atlantic Ocean. *Journal of Marine Systems*, **5**, 159–70.

Platt, T. (1986). Primary production of the ocean water column as a function of surface light intensity: algorithms for remote sensing. *Deep-Sea Research*, **33**, 149–63.

Pujo-Pay, M. (1995). L'azote et le phosphore en milieu marin; importance des formes organiques en milieu océanique du large. Thèse, Université de la Méditerranée, Marseille. 164 pp.

Pujo-Pay, M. & Raimbault, P. (1994). Observations chimiques et biomasse dans l'océan Atlantique tropical nord; campagnes EUMELI. Résultats des mesures. Report, 65 pp. Centre Océanologique de Marseille.

Putaud, J.-P., Belviso, S., Nguyen, B. C. & Mihalopoulos, N. (1993). Dimethylsulfide, aerosols and condensation nuclei over the tropical northeastern Atlantic Ocean. *Journal of Geophysical Research*, **98**, D8, 14 863–71.

Rabouille, C., Gaillard, J.-F., Sibuet, M., Beaucaire, C. & Bonte, P. E. (1993). Sediment geochemistry in the three EUMELI sites in the northeast tropical Atlantic: General presentation and first results. *Annales de l'Institut Océanographique, Paris*, **69**, 35–42.

Raimbault, P., Slawyk, G., Coste, B. & Fry, J. (1990). Feasibility of using an automatic procedure for the determination of seawater nitrate in the 0–100 nM range: examples from field and culture. *Marine Biology*, **104**, 347–51.

Redalje, D. G. (1983). Phytoplankton carbon biomass and specific gross rates determined with the labelled chlorophyll *a* technique. *Marine Ecology Progress*

Series, **11**, 217–25.
Relexans, J.-C., Deming, J., Dinet, A., Gaillard, J.-F. & Sibuet, M. (1996). Sedimentary organic matter and micro-meiobenthos with relation to trophic conditions in the tropical northeast Atlantic. *Deep-Sea Research*, Part I, **43**, 1343–68.
Saunders, P. M. (1986). The accuracy of measurement of salinity, oxygen, and temperature in the deep ocean. *Journal of Physical Oceanography*, **16**, 184–95.
Sibuet, M., Albert, P., Charmasson, S., Deming, J., Dinet, A., Galeron, J., Guidi-Guilvard, L.& Mahaut, M.-L. (1993). The benthic ecosystem in the three EUMELI sites in the northeast tropical Atlantic Ocean: General perspectives and initial results on biological abundance and activities. *Annales de l'Institut Océanographique, Paris*, **69**, 21–33.
Smed, J. (1982). The oceanographic data base for the CINECA region. *Rapport et Procès-Verbaux des Réunions, Conseil International pour l' Exploration de la Mer*, **180**, 11–28.
Speth, P. & Detlefsen, H. (1982). Meteorological influences on upwelling off Northwest Africa. *Rapport et Procès-Verbaux des Réunions, Conseil International pour l' Exploration de la Mer*, **180**, 29–34.
Tachikawa, K., Jeandel, C. & Dupré, B. (1997). Distribution of Rare Earth elements and Neodymium isotopes in settling particulate material of the tropical Atlantic Ocean (EUMELI sites). *Deep-Sea Research*, **44**, 1769–92.
Tailliez, D. (1993). Campagnes EUMELI 3, 4 et 5: Présentation des données hydrologiques et calibration des capteurs. Internal report, 66 pp. Villefranche-sur-Mer.
Tailliez, D., Bournot, C., Leblond, N., Lefevre, D., Minas, M., Palazzoli, I., Pierre, C., Raunet, J. & Vangriesheim, A. (1993). Données hydrologiques; campagnes EUMELI 3, 4, and 5; JGOFS Program. Internal reports, 850 pp. Villefranche-sur-Mer.
Tomczak, M. Jr (1981). An analysis of mixing in the frontal zone of South and North Atlantic Central Water off northwest Africa. *Progress in Oceanography*, **10**, 173–92.
Van Camp, L., Nykjaer, L., Mittelstaedt, E. & Schlittenhardt, P. (1991). Upwelling and boundary circulation off Northwest Africa as depicted by infrared and visible satellite observations. *Progress in Oceanography*, **26**, 357–402.
Vangriesheim, A., Pierre, C. & Laube, E. (1993). Hydrological conditions in the EUMELI area in the NE tropical Atlantic: water masses, variability of productivity/regeneration and of particle load. *Annales de l'Institut Océanographique*, **69**, 15–20.
Vangriesheim, A., Sibuet, M., Gouillou, J.-P., Pelle, H., Courau, P., Billant, A., Branellec, P. & Lagadec, C. (1994). Campagne EUMELI 2 – leg 2, Rapport de données CTD. Report DRO/EP 94/01, 99 pp. Brest: Ifremer.
Wefer, G. & Fischer, G. (1993). Seasonal patterns of vertical flux in equatorial and coastal upwelling areas of the eastern Atlantic. *Deep-Sea Research*, Part I, 1613–45.
Zenk, W., Klein, B. & Schroder, M. (1991). Cape Verde Frontal Zone. *Deep-Sea Research*, **38**, Supplement 1, S505–30.

12 The North Atlantic carbon cycle: new perspectives from JGOFS and WOCE

S. C. Doney, D. W. R. Wallace and H. W. Ducklow

Keywords: ocean CO_2 carbon cycle, atmospheric CO_2 sinks and sources, anthropogenic CO_2 input, carbon transport mechanisms, ocean carbon budgets and models

Introduction

A central objective of the Joint Global Ocean Flux Study (JGOFS) is to improve our estimates and understanding of the flux of carbon both within the ocean and across the ocean–atmosphere and ocean–sediment interfaces. One rationale for focusing on the North Atlantic carbon system in particular revolves around the high-latitude formation of deep and intermediate water masses and the subsequent important role of the North Atlantic in the natural carbon cycle and uptake of anthropogenic CO_2 (Sarmiento *et al.*, 1995). A second, equally compelling, argument rests on using the diverse range of biogeographical and hydrographic regimes (Longhurst *et al.*, 1995) across the basin to explore the interaction among air–sea exchange, circulation and biological processes that determine the surface distribution of dissolved inorganic carbon. The knowledge gained by JGOFS is directly applicable to predicting the future uptake of CO_2 by the ocean, and because of its unique traits the North Atlantic is a key location for potential physical and biological feedbacks between climate and the global carbon cycle.

The first major JGOFS field effort focused on the spring bloom in the eastern North Atlantic (North Atlantic Bloom Experiment, NABE) (Ducklow & Harris, 1993), and a long-term biogeochemical time series was initiated in the Sargasso Sea in 1988 (Bermuda Atlantic Time-Series Study, BATS) (Michaels & Knap, 1996). The large-scale distribution of ocean carbon parameters for the basin is also being re-evaluated as part of the JGOFS global CO_2 survey in conjunction with the World Ocean Circulation Experiment (WOCE) hydrographic programme (Fig. 12.1). Together with a wealth of historical data sets (e.g. TTO) (Brewer *et al.*, 1985; Takahashi *et al.*, 1995), the data collected

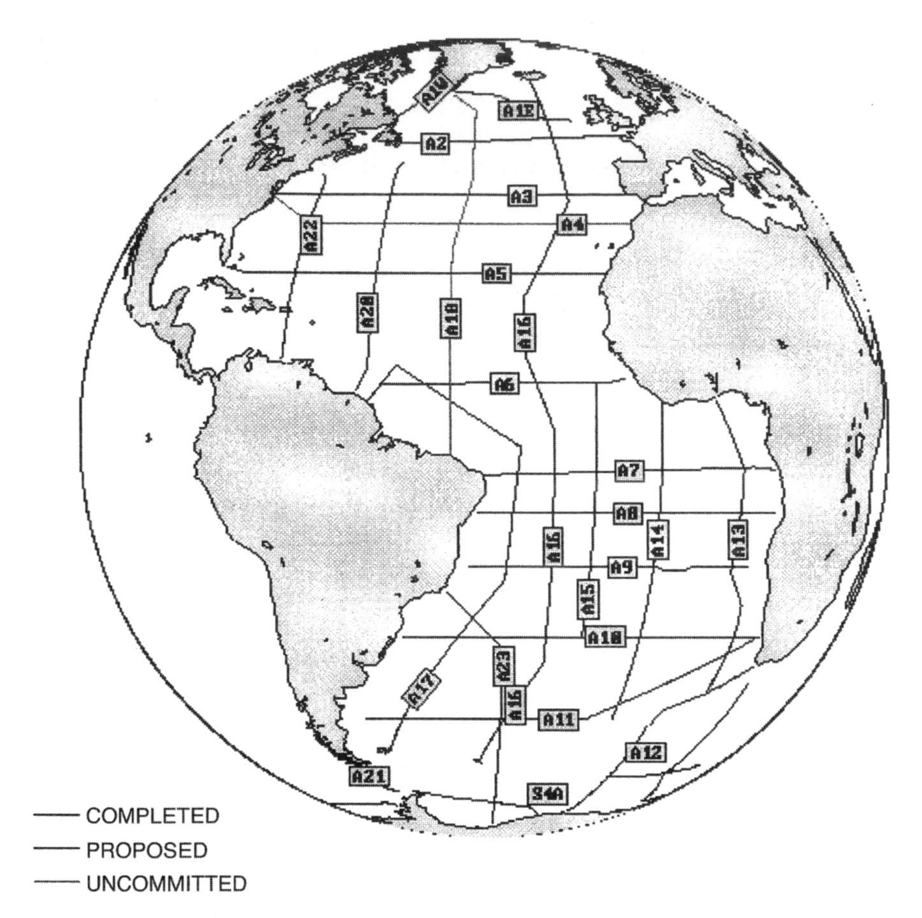

Figure 12.1 *Map of the JGOFS–WOCE one-time CO_2 survey sections for the Atlantic Ocean (WOCE Hydrographic Programme Office, http://whpo.whoi.edu/).*

over the last decade in the North Atlantic provide perhaps our best opportunity to constrain the carbon budget for an ocean basin.

The marine carbon system is a challenging subject, requiring a detailed knowledge of the interplay between ocean biogeochemistry and physics. The ocean carbon cycle can also be perturbed by climate variability on time scales ranging from the interannual ENSO and North Atlantic Oscillation to millennia and glacial–interglacial transitions. The oceanic uptake of anthropogenic CO_2 adds further complications. Traditional time-series data or oceanographic cruise data are limited in spatial and temporal extent, and a variety of approaches incorporating ship of opportunity, moorings and satellite data, data-assimilation models, and novel measurement systems are needed to address such issues on the regional or basin scale.

Preliminary efforts for a basin-wide synthesis of the North Atlantic carbon budget are underway, and both the vitality of the debate and the present level of uncertainty are highlighted by a series of papers presented at the Royal Society North Atlantic Carbon Cycle meeting in the autumn of 1994 (Eglinton *et al.*, 1995). Here, we present one view of the problem, in particular stressing the components of the carbon budget that require further work.

The North Atlantic's role in the global carbon cycle

A schematic of the global carbon cycle (Fig. 12.2) indicates the approximate present size and annual exchange rates among the various active marine, terrestrial, and atmospheric carbon reservoirs (Schimel *et al.*, 1995). Because carbon dioxide (CO_2) reacts with seawater to form carbonate species (H_2CO_3, HCO_3^- and CO_3^{2-}), the oceanic reservoir of dissolved inorganic carbon (DIC) is about 50 times that in the atmosphere. Carbon dioxide is actively exchanged between the atmosphere and ocean, and the ocean plays an important role in modulating atmospheric CO_2 concentrations on time scales ranging from years to millennia.

Despite the low biomass in the surface ocean, the marine and terrestrial global primary production rates are comparable. The particulate matter in the surface ocean turns over rapidly, however, and only a fraction of the fixed carbon is exported from the surface layer by gravitational sinking of large particles and advection of suspended particles and dissolved organic carbon (DOC). New production or export flux, which is supported mostly by upward vertical nutrient fluxes, acts to sequester carbon in the thermocline and deep ocean for periods of several years to centuries and contributes about two thirds of the observed vertical DIC gradient between the surface and deep ocean. The remaining third comes from solubility effects because CO_2 is more soluble in cold water than in warm.

Over the last two centuries, the natural carbon cycle has undergone major perturbations because of fossil-fuel burning and land-use change. Much of the current interest in the global carbon cycle is motivated by the measured anthropogenic rise in atmospheric CO_2, an important greenhouse warming gas for the Earth's climate system, from pre-industrial levels of 280 μatm to more than 355 μatm by 1995. If the surface ocean were to keep track with the atmosphere, the corresponding increase in upper ocean DIC would be about 2.5% over the same period.

During the decade of the 1980s, fossil-fuel burning and land-use change released on average an estimated 7.1 ± 1.1 Gt C yr^{-1} (1 Gt = 10^{15} g) into the atmosphere, of which about 45% (3.2 ± 2 Gt C yr^{-1}) stayed in the atmosphere with the remainder removed by marine (2.0 ± 0.8 Gt C yr^{-1}) and terrestrial (1.9 ± 1.6 Gt C yr^{-1}) sinks (Schimel *et al.*, 1995). The fate of anthropogenic

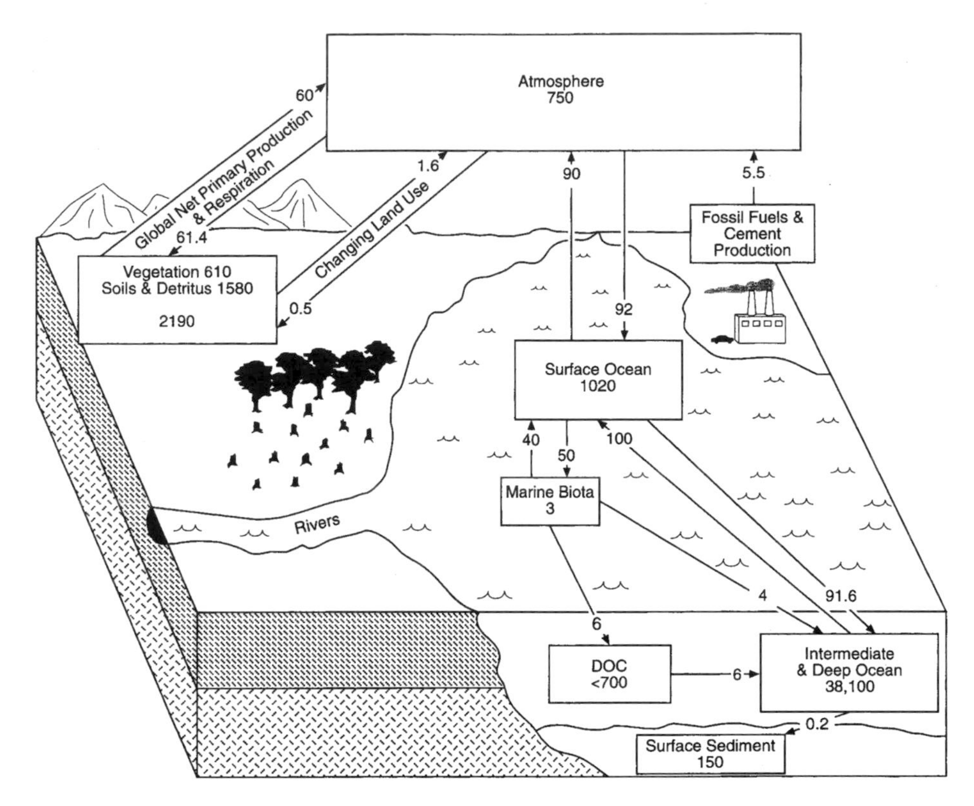

Figure 12.2 *Schematic of global carbon cycle for the 1980s in Gt C yr^{-1} (based on 1994 IPCC assessment; Schimel* et al., *1995).*

carbon cannot be studied in isolation from the large, background natural carbon cycle, however, and the explicit controls and regional distribution for these marine and terrestrial carbon sinks are not well determined at present, leading to significant uncertainties in projected future atmospheric CO_2 levels.

Beyond simply quantifying the present state, a critical question for future projects is whether there are substantial climate-related feedbacks governing the partitioning of CO_2 between the atmosphere and the ocean. Depending on their sense, such feedbacks could either partly mitigate or exacerbate potential climate-change effects and are important for assessing the effectiveness of the CO_2-mitigation strategies under consideration by the global community. A reduction in the North Atlantic thermohaline overturning rate owing to CO_2-induced warming (Manabe & Stouffer, 1993), for example, could limit future ocean uptake, thus amplifying climate sensitivity (Sarmiento & Le Quere, 1996). The large changes in atmospheric CO_2 from glacial to interglacial periods deduced from the ice core record suggest that such feedbacks do exist for the natural system, but the causal mechanisms underlying the climate–CO_2 relationship remain poorly understood.

Anthropogenic carbon is released primarily in the northern hemisphere, and the resulting latitudinal CO_2 gradient can be used along with atmospheric transport models to diagnose the location (latitude) of the major carbon sources and sinks. One influential study by Tans *et al.* (1990) found that the bulk of the anthropogenic CO_2 removal occurs in the northern hemisphere. Based on the fact that most of the ocean surface lies in the southern hemisphere and on a limited surface ocean CO_2 partial pressure (p_{CO_2}) data set, Tans *et al.* (1990) went on to suggest a relatively minor oceanic contribution to the inferred CO_2 sink. The terrestrial 'missing sink' of carbon was attributed to processes such as forest re-growth and nitrogen and CO_2 fertilisation at temperate latitudes.

An alternative interpretation of the atmospheric data involves a large, natural northern hemisphere sink of carbon thus generating an opposing, background south–north atmospheric CO_2 gradient (Keeling *et al.*, 1989). Outgassing of CO_2 from the equatorial upwelling regions and uptake at the high latitudes marks the large-scale distribution of air–sea CO_2 flux. An asymmetry in this general pattern can arise because of the extensive formation of deep waters in the North Atlantic. Cold deep waters can carry more DIC than warm surface waters, resulting in a net uptake and southward ocean carbon transport associated with the North Atlantic branch of the global thermohaline circulation (Broecker & Peng, 1992). The Keeling *et al.* (1989) hypothesis, however, requires that the northern hemisphere surface waters be substantially under-saturated in p_{CO_2} in order to support both the natural and anthropogenic CO_2 drawdowns.

The controversy surrounding the missing carbon sink has spurred the development over the past decade of a number of new methods for constraining the carbon cycle, two more notable examples based on the measurement of $^{13}C:^{12}C$ and $O_2:N_2$ ratios in the atmosphere. The carbon isotopes ^{13}C and ^{12}C are strongly fractionated during photosynthesis but not as much by air–sea exchange, and thus the combination of the $\delta^{13}C$ and CO_2 signals in atmosphere can be used to separate marine and terrestrial processes. One recent analysis by Ciais *et al.* (1995) points towards large net carbon sinks from both the northern hemisphere ocean and land, -2.4 and -2.2 Gt C yr^{-1}, respectively, for 1992.

The burning of fossil fuels consumes a corresponding amount of oxygen, reflected in an observed decrease in the atmospheric $O_2:N_2$ ratio over time (Keeling & Shertz, 1992). Unlike the case for CO_2, the ocean does not strongly interact with the anthropogenic O_2 perturbation because of the low oxygen solubility in seawater, and the relative change in atmospheric O_2 to CO_2 provides another method for partitioning carbon uptake between land and ocean reservoirs (Keeling *et al.*, 1996). Keeling *et al.* (1996) also estimate the approximate strengths of the northern and southern hemisphere carbon sinks required to match both the CO_2 and O_2 hemispheric gradients, finding an intermediate net northern hemisphere ocean uptake of about 1.0–1.3 Gt C yr^{-1}, of which 0.4 Gt C yr^{-1} is driven by the North Atlantic thermohaline circulation.

Basin-scale carbon fluxes

The major terms for the North Atlantic carbon budget, shown schematically in Fig. 12.3, include: the net air–sea CO_2 flux, the meridional fluxes of dissolved inorganic and organic carbon (DOC) at the northern and southern boundaries, vertical re-distribution through surface new production, downward particle and DOC export and subsequent re-mineralisation, riverine input and sediment burial, and temporal changes in the DIC inventory related to uptake of anthropogenic CO_2. The magnitudes of these fluxes can be estimated to varying levels of certainty, either directly, as we discuss below, or indirectly from atmospheric data, as above.

Two quantities are needed to compute the air–sea CO_2 flux: the ocean–atmosphere partial pressure difference (Δp_{CO_2}) and the gas exchange rate, which is generally parameterised as a function of wind speed. Surface water p_{CO_2} increases with DIC and temperature and decreases with alkalinity and less strongly with salinity. Biological DIC uptake for the production of organic matter typically dominates over biological calcification. The 'biological' and 'solubility' pumps (Volk & Hoffert, 1985) are often out of phase over the seasonal cycle with temperature-dominated, summer outgassing (positive Δp_{CO_2}) in oligotrophic regions (see, for example, Bates *et al.*, 1996*a*), and biology-dominated, summer drawdown (negative p_{CO_2}) at the more productive, higher latitudes (Chipman *et al.*, 1993; Takahashi *et al.*, 1993).

Although a fairly extensive historical surface p_{CO_2} data set exists for the North Atlantic (see, for example, Takahashi *et al.*, 1995), the coverage is insufficient to adequately represent the large seasonal and spatial variability in the surface water Δp_{CO_2} ($-$ 120 μatm to + 100 μatm), and some form of space–time extrapolation must be applied to the data (see, for example, Takahashi *et al.*, 1995; Lefevre, 1995). Advances have also been made in constraining gas exchange rates by using dual-tracer release experiments, but questions remain regarding how to scale such results to CO_2 (Wanninkhof, 1992; Watson *et al.*, 1995); at present, air–sea gas exchange rates are known to no better than a factor of two.

Based on a historical reconstruction of the seasonal p_{CO_2} field, Takahashi *et al.* (1995) estimate that the annual mean p_{CO_2} difference with the atmosphere is about $-$ 8.6 μatm over the entire study region (18°S–78°N). Large under-saturations of $-$ 20 to $-$ 60 μatm are found for the sub-polar gyre north of 42°N, with moderate under-saturation and super-saturation for surface water

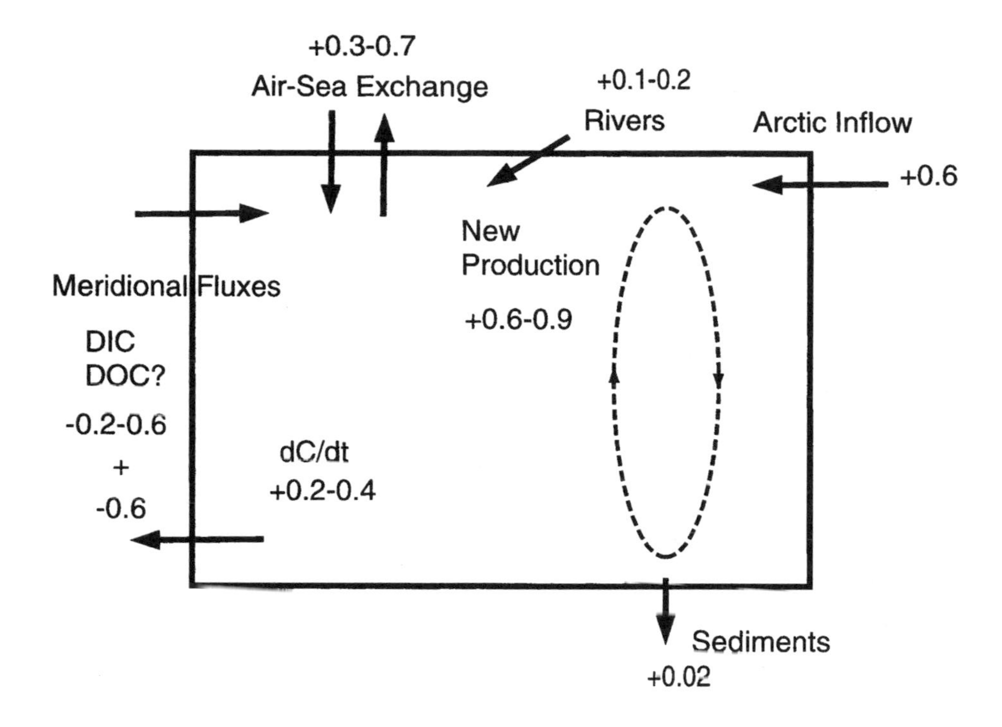

Figure 12.3 *Modern, basin-scale carbon flux estimates for the North Atlantic Ocean in Gt C yr⁻¹ (see text for details).*

p_{CO_2} in the subtropical and tropical Atlantic, respectively. Focusing on the extratropics north of 18°N, the net ocean CO_2 uptake for the basin is between 0.33 and 0.70 Gt C yr^{-1}, depending on the choice of gas exchange parameterisation.

Further refinement of the estimates of the net North Atlantic air–sea flux is difficult, considering the spatial and temporal data requirements and size of the Δp_{CO_2} signal. The available surface p_{CO_2} database is a composite over time, with only sparse coverage in any particular year, and may be aliased by inter-annual to decadal variability especially at high-latitude sites of deep water formation (see, for example, Wallace & Lazier, 1988; Schlosser *et al.*, 1991). A 2.5 μatm change in mean p_{CO_2} accounts for a flux of about 0.1 Gt C yr^{-1} over the basin compared with current measurement uncertainties of about 1.0 μatm. Barring a dramatic increase in the amount of available p_{CO_2} data, one approach for improving basin-scale p_{CO_2} maps would be take advantage of correlations of p_{CO_2} with properties such as SST, nutrients and ocean colour (Watson *et al.*, 1991; Takahashi *et al.*, 1993; Stephens *et al.*, 1995) that can either be measured by remote sensing or estimated from the broader hydrographic databases.

On the basin scale, the net meridional transports of DIC due to thermohaline overturning, wind-driven gyres and surface Ekman flow become increasingly important. The northward inflow of warm surface water and export of cold deep water sets up a natural CO_2 solubility pump with net air–sea uptake at high latitudes and southward DIC outflow at depth. The thermohaline component is closely linked to the overturning transport ($10–20 \times 10^6$ $m^3 s^{-1}$) and northward ocean heat transport, and a number of diagnostic calculations give a range of pre-industrial carbon fluxes out of the North Atlantic: 0.6 Gt C yr^{-1} (Broecker & Peng, 1992); 0.5 ± 0.1 Gt C yr^{-1} (Watson *et al.*, 1995); 0.33 ± 0.15 Gt C yr^{-1} (Keeling & Peng, 1995).

Meridional CO_2 transports can alternatively be calculated by combining zonal DIC section data with hydrographic section velocity estimates, preferably derived from an inverse procedure. The resulting CO_2 fluxes include the thermohaline component as well as fluxes due to a geostrophic surface Ekman flow and wind-driven gyre circulation. The calculations are complicated by uncertainty in the surface wind stress and depth-averaged or barotropic velocity field and by the poor DIC coverage in space and time. The computed net carbon transport estimates for the North Atlantic vary with investigator and latitude, in part reflecting the net surface flux patterns discussed above, but are generally southward with a range of 0.4–0.8 Gt C yr^{-1} (0.4 Gt C yr^{-1} at 25°N, Brewer *et al.*, 1989; 0.4–0.8 Gt C yr^{-1} over 20°–60°N, Martel & Wunsch, 1993; 0.5 Gt C yr^{-1} at about 20°S, Robbins, 1994; *c.* 0.6 Gt C yr^{-1} over 1°S–30°S, Holfort *et al.*, 1998). Because anthropogenic carbon tends to build up more rapidly in the northward-flowing surface and upper thermocline water, the net effect of the anthropogenic transient is to reduce the present day net southward meridonal flux by about 0.1–0.2 Gt C yr^{-1}.

An additional and often neglected component of the meridional CO_2 transport arises because there is a net southward mass flux in the basin required to balance the net atmospheric water vapour flux from the Atlantic to the Pacific (see, for example, Wijffles *et al.*, 1992). The mass flux, which can be estimated from flow through the Bering Strait modified by the net freshwater input over the Arctic and North Atlantic, can be thought of as a zonal average barotropic flux and contributes another approximate 0.6 Gt C yr^{-1} to the southward transport (Holfort *et al.*, 1998). Some confusion has arisen in the literature because this throughflow is often not included in section estimates or in many numerical models.

Several of the other, smaller fluxes (riverine input, sediment burial, net DOC re-mineralisation) must be included to balance the basin carbon budget at the ± 0.1 Gt C yr^{-1} level. Globally, rivers deliver about 0.8 Gt C yr^{-1} of DOC and DIC to the ocean (Siegenthaler & Sarmiento, 1993), or about 0.1–0.2 Gt C yr^{-1} for the North Atlantic. Most of the riverine carbon, however, is re-mineralised and released back to the atmosphere on the continental shelves, and the small remainder is generally assumed to be in approximate balance with the sedimentary organic and inorganic carbon burial flux over the basin. The magnitude and variability of the meridional DOC fluxes and net DOC convergence or divergence for the basin are still almost entirely unknown but may be addressed in the near future with the development of improved DOC analytical techniques (see, for example, Ducklow *et al.*, 1995).

Anthropogenic carbon uptake

Model calculations suggest that the ocean takes up at present about 40% of the fossil fuel carbon released to the atmosphere. The resulting anthropogenic signal in the surface water (*c.* 50 μmol kg^{-1}) decreases rapidly with depth and is small relative to the natural variations in background DIC concentration (1900–2200 μmol kg^{-1}). The high-quality DIC time series collected at Hydrostation S and BATS show an average increase of about 1.7 μmol kg^{-1} yr^{-1} in surface waters for the past decade (Bates *et al.*, 1996a), close to the expected equilibrium response.

Measurement errors for DIC have, until the recent introduction of reference DIC standards (DOE, 1994), limited efforts to directly estimate anthropogenic CO_2 uptake by monitoring the change in DIC concentrations over time. A number of more indirect methods have been introduced, each with its own weaknesses. However, together, they provide a relatively consistent picture for the removal of excess CO_2 by the North Atlantic (Wallace, 1995).

One approach, originally introduced by Brewer (1978) and Chen (1993), involves removing from the observed DIC distributions a pre-industrial DIC field computed with pre-formed nutrients, mixing relations, and Redfield ratios. Using the GEOSECS and TTO data sets, Gruber *et al.* (1996) calculate a North Atlantic (10°N–80°N) anthropogenic CO_2 burden of 20 ± 4 Gt C for the early 1980s.

Quay *et al.* (1992) proposed another method relying on the change over time of the DIC ^{13}C : ^{12}C ratio as a proxy for the penetration of anthropogenic CO_2 into the ocean. Sufficient high-quality $\delta^{13}C$ data are not yet available, however, for detailed regional budgets. Preliminary efforts to use correlations of anthropogenic CO_2 with the measured distributions of transient tracers such as tritium and the chlorofluorocarbons also show good promise, particularly for the main thermocline.

Anthropogenic CO_2 uptake is primarily limited by ocean transport and therefore can be calculated from 3-D general circulation models (GCM). A standard practice is to assume that the natural carbon system is in steady state and then compute the anthropogenic DIC as a perturbation field, either with or without explicit treatment of the variability in ocean chemistry due to the natural background. One predicted uptake estimate for the North Atlantic of 18.4 Gt C (or about 0.4 Gt C yr^{-1} in the 1980s; Sarmiento *et al.*, 1995) is similar in magnitude to that found by Gruber *et al.* (1996) but with a different spatial distribution. The current generation of numerical models depend perhaps too heavily on bomb-radiocarbon simulations for validating the circulation fields and contain a number of known deficiencies, particularly with the formation and outflow of North Atlantic Deep Water. Another important caveat is that the oceanic GCM-based estimates do not reflect any potential future modifications of the natural marine carbon cycle.

The biological pump

The strength of the biological carbon pump in the North Atlantic is difficult to assess directly from observations. This is partly due to our limited understanding of the controls on export flux from the surface. Despite a significant investment in floating sediment traps over the passed decade, fundamental questions remain over how to directly measure the particle flux leaving the euphotic zone. Other export components such as the DOC flux and zooplankton migration are also not well constrained, and efforts to close the local carbon budget for the BATS site identified a large carbon imbalance over the annual cycle (Michaels *et al.*, 1994).

Using CZCS ocean colour composites and primary production algorithms, Longhurst *et al.* (1995) estimate that the total primary productivity for the basin is about 6.3 Gt C yr^{-1}. Applying their suggested basin-wide *f* ratio of 0.1, the ratio of export production to total, yields a new production of about 0.6 Gt C yr^{-1}. Campbell & Aarup (1992) compute a slightly lower integrated new production rate of 0.56 Gt C yr^{-1} for the extratropical North Atlantic; their method accounts for only the production fuelled by the seasonal drawdown in surface nutrients and also relies on the CZCS data indirectly to compute the initial wintertime nutrient levels and the timing of the spring bloom. For comparison, the Princeton GCM ecosystem model (Sarmiento *et al.*, 1993) produces 0.9–1.3 Gt C yr^{-1} of new production for the North Atlantic, with an average *f* ratio of 0.42. The model *f* ratio is probably too high, related to excessive nutrient upwelling, and a reasonable first-order estimate is probably in the range of 0.6–0.9 Gt C yr^{-1}. Based on data from BATS and a few other sites where good DOC data exist, roughly about a third of this export production may be leaving the euphotic zone as semi-labile DOC rather than particulate matter (Ducklow *et al.*, 1995).

The bulk of the downward carbon flux is re-mineralised back to DIC in the seasonal and main thermoclines, the rates of which can be quantified by using vertical trap flux profiles, tracer-based techniques, and microbial activity rates. Excluding the possibility of net convergence or divergence of DOC, the integrated tracer-based values should provide good estimates for the overlying regional export flux, although many problems have arisen in practice because of the different time and space scales sampled by various techniques (Jenkins, 1995). Only about 10% of the surface export flux reaches the deep ocean, and almost all of this organic matter and most of the inorganic carbon are re-mineralised either in the water column or at the sediment–water interface. Neglecting the high depositional environments on the continental shelf, Jahnke (1996) estimates that about 0.06 Gt C yr^{-1} of organic carbon is oxidised at the sediment–water interface over the North Atlantic, with about two thirds of the flux along the continental slope and margin. A comparable magnitude of $CaCO_3$ dissolution is also occurring at the sediment–water interface (Archer, 1996). The burial rate of carbon (total) in the North Atlantic is small, about 0.01–0.03 Gt C yr^{-1}.

Based on the large recycling efficiencies and relatively tight coupling between carbon and nutrients in the marine biological carbon cycle, many have suggested that biology is unimportant for basin-scale budgets such as Fig. 12.3, at least on the time scale of the anthropogenic CO_2 signal. Although true for steady-state conditions, the biological pump is comparable in magnitude to both the carbon cycle of the pre-industrial thermohaline and the modern anthropogenic CO_2 uptake.

Substantial perturbations could arise either from changes in the physical circulation or from process that decouple carbon and nutrient cycling, examples of which include shifts in ecosystem structure and export Redfield ratios, changes in the balance of DOM to POM production, and variable re-mineralisation with depth. Considerable attention has focused recently on the role of iron and other micronutrients in limiting production in high-nutrient – low-chlorophyll (HNLC) regions such as the Equatorial Pacific, North Pacific and Southern Ocean (Coale *et al.*, 1996). Iron may also be an important control for nitrogen fixation in the oligotrophic subtropical gyres (Michaels *et al.*, 1996; Falkowski, 1997), although the effect of altering external nutrient supplies to the Atlantic carbon cycle remains unclear. Also relevant to the North Atlantic, shifts in ecosystem structure towards a coccolithophorid-dominated system with enhanced biocalcification could represent a potentially significant mechanism for altering sea-surface p_{CO_2} (Bates *et al.*, 1996*b*).

North Atlantic carbon budgets

Although carbon budgets can be constructed for the North Atlantic that balance to about 0.1 Gt C yr^{-1} (Fig. 12.3), a closer examination of the terms suggests that our present uncertainties are no better than about 0.2–0.3 Gt C yr^{-1}, or about at the edge of saying whether the pre-industrial North Atlantic was an important net carbon sink. For example, Sarmiento *et al.* (1995) find that their model-based air–sea CO_2 fluxes for the present North Atlantic agree in a reasonable fashion with the observational estimates of Takahashi *et al.* (1995) and that the current net uptake over the basin is comparable to the storage of anthropogenic carbon. Clearly, the attribution of most of the current net air–sea CO_2 flux to anthropogenic CO_2 leaves little room for a substantial pre-industrial CO_2 sink. In contrast, Holfort *et al.* (1998) suggest from their ocean carbon transport estimates that more than 50% of the anthropogenic CO_2 supply to the North Atlantic actually enters the ocean in the south and is advected northward in the upper limb of the thermohaline overturning circulation. Their interpretation of the measured ocean CO_2 transport divergence, then, is in support of a significant pre-industrial North Atlantic sink.

The North Atlantic surface water Δp_{CO_2} data could support Keeling *et al.*'s (1996) estimate for northern hemisphere ocean uptake (1.0–1.3 Gt C yr^{-1}) if the upper value of 0.7 Gt C yr^{-1} from Takahashi *et al.* (1995) is used (about half anthropogenic and half thermohaline) and if about 0.3–0.5 Gt C yr^{-1} is taken up in the Pacific. The current range of estimates for the Pacific varies from zero to about this level (see, for example, Stephens *et al.*, 1995). The much larger ^{13}C derived estimate of Ciais *et al.* (1995) is difficult to reconcile with the current ocean data.

We should note, however, that significant care should be taken in comparing published estimates of regional carbon budgets for both the atmosphere and the ocean because of the common mismatch of domain boundaries and time scales. For example, the carbon uptake partitioning derived from the relatively short time series of atmospheric $\delta^{13}C$ and $O_2 : N_2$ shows substantial inter-annual variability and may diverge greatly in any particular year from longer-term (i.e. decadal) model- or tracer-based ocean uptake values. As another example, the tropical ocean is generally a net source of CO_2 to the atmosphere, and extending the southern boundary of the Takahashi *et al.* (1995) air–sea flux calculation to include the tropics cuts the net maximum carbon uptake to 0.5 Gt C yr^{-1}. The atmospheric gradient methods, however, are rather insensitive to tropical fluxes, and the extratropical uptake rate is probably a more appropriate comparison measure for the net hemispheric uptake values.

Future work

Solving the question of the North Atlantic CO_2 sink will require a co-ordinated field and modelling programme demanding a variety of new and novel

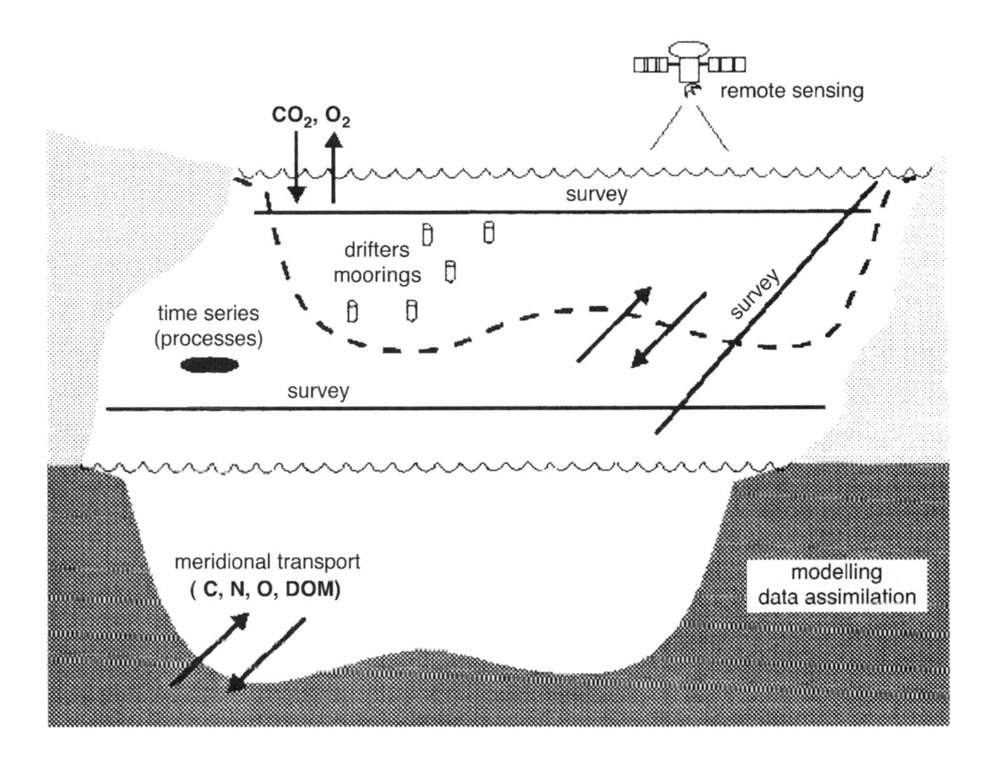

Figure 12.4 *A schematic of a proposed coordinated sampling and modelling programme for quantifying a basin-scale carbon budget.*

approaches (Fig. 12.4). The major elements of such a programme would include efforts to quantify the net air–sea CO_2 flux by drifters, moorings, ships of opportunity and satellite data, zonal hydrographic lines to constrain the net divergence of inorganic and organic carbon, and time-series stations and repeat survey lines to diagnose temporal changes in the inorganic carbon inventory. Process studies are also crucial if we are to understand mechanisms and the potential for biological feedbacks on carbon uptake. Three areas that show particular promise are automated chemical measurement systems, the next generation of biogeochemical data-assimilation models, and improved atmospheric inversion techniques.

At the root of much of the present uncertainty in the North Atlantic carbon budget is the scarcity of data, particularly at high latitudes during the winter. The desired temporal and spatial data scales are incompatible with traditional oceanographic shipboard sampling, and alternatives based on the remote measurement of chemical species from volunteer observing ships (VOS), moorings, drifters and floats are required. For example, Friederich *et al.* (1995) demonstrated a capability from tests on a mooring in Monterey Bay for measuring the air–sea p_{CO_2} difference with 1–2 μatm accuracy hourly for up to several months. *In situ* colorimetric-based p_{CO_2} sensors have been developed and are undergoing field tests (DeGrandpre *et al.*, 1995).

Ongoing improvements in three-dimensional numerical models offer an alternative, complementary approach, and recent results from the Princeton carbon cycle model (Sarmiento *et al.*, 1995), which includes a complete carbonate system and simple phosphate-based biological model, are generally consistent with both the net air–sea and the meridional CO_2 flux estimates. Future modelling directions will include emphasis on more sophisticated biogeochemistry (see, for example, Six & Maier-Reimer, 1996) and on data assimilation, with the ability to directly incorporate near-real-time, basin-scale satellite data sets such as wind speed, sea surface height, SST and ocean colour.

Finally, efforts are under way to generate improved surface CO_2 flux patterns by inverting atmospheric data from the extensive NOAA–CMDL CO_2 monitor programme using atmospheric transport codes (Tans *et al.*, 1996). Properly constraining the meridional ocean carbon transports by basin will directly contribute to these atmospheric calculations.

The WOCE and JGOFS programmes have led to a dramatic increase in the amount of high-quality carbonate system data available for the ocean, and we have for the first time sufficient field data to actually calculate ocean carbon budget estimates, which were previously almost entirely the purview of numerical modellers. Further progress towards resolving the North Atlantic carbon budget will come only through the joint analysis of observational and model-based studies.

References

Archer, D. (1996). A data-driven model of the global calcite lysocline. *Global Biogeochemical Cycles*, **10**, 511–26.

Bates, N. R., Michaels, A. F. & Knap, A. H. (1996*a*). Seasonal and interannual variability of oceanic carbon dioxide species at the U.S. JGOFS Bermuda Atlantic Time-series Study (BATS) site. *Deep-Sea Research*, Part II, **43**, 347–83.

Bates, N. R., Michaels, A. F. & Knap, A. H. (1996*b*). Alkalinity changes in the Sargasso Sea: geochemical evidence of calcification? *Marine Chemistry*, **51**, 347–58.

Brewer, P. G. (1978). Direct observation of the oceanic CO_2 increase. *Geophysical Research Letters*, **5**, 997–1000.

Brewer, P. G., Sarmiento, J. L & Smethie, W. M. Jr (1985). The Transient Tracers in the Ocean (TTO) program: The North Atlantic Study, 1981; The Tropical Atlantic Study, 1983. *Journal of Geophysical Research*, **90**, 6903–5.

Brewer, P. G., Goyet, C. & Dyrssen, D. (1989). Carbon dioxide transport by ocean currents at 25N latitude in the Atlantic Ocean. *Science*, **246**, 477–9.

Broecker, W. S. & Peng, T.-H. (1992). Interhemispheric transport of carbon dioxide by ocean circulation. *Nature*, **356**, 587–9.

Campbell, J. W. & Aarup, T. (1992). New production in the North Atlantic derived from seasonal patterns of surface chlorophyll. *Deep-Sea Research*, **39**, 1669–94.

Chen, C.-T. A. (1993). The oceanic anthropogenic CO_2 sink. *Chemosphere*, **27**, 1041–64.

Chipman, D. W., Marra, J. & Takahashi, T. (1993). Primary production at 47°N and 20°W in the North Atlantic Ocean: A comparison between ^{14}C incubation method and the mixed layer carbon budget. *Deep-Sea Research*, **40**, 151–69.

Ciais, P., Tans, P. P., White, J. W. C., Trolier, M., Fancey, R. J., Berry, J. A., Randall, D. R., Sellers, P. J., Collatz, J. G. & Schimel, D. S. (1995). Partitioning of ocean and land uptake of CO_2 as inferred by $\delta^{13}C$ measurements from the NOAA Climate Monitoring and Diagnostics Laboratory Global Air Sampling Network. *Journal of Geophysical Research*, **100**, 5051–70.

Coale, K. H., Johnson, K. S., Fitzwater, S. E., Gordon, R. M., Tanner, S., Chavez, F. P., Ferioli, L., Sakamoto, C., Rogers, P., Millero, F., Steinberg, P., Nightingale, P., Cooper, D., Cochlan, W. P., Landry, M. R., Constantinou, J., Rollwagen, G., Trasvina, A. & Kudela, R. (1996). A massive phytoplankton bloom induced by an ecosystem-scale iron fertilization experiment in the equatorial Pacific Ocean. *Nature*, **383**, 495–501.

DeGrandpre, M. D., Hammar, T. R., Smith, S. P. & Sayles, F. L. (1995). *In situ* measurements of seawater p_{CO_2}. *Limnology and Oceanography*, **40**, 969–75.

DOE (1994). *Handbook of Methods for the Analysis of the Various Parameters of the Carbon Dioxide System in Sea Water*. Version 2, (ed. A. G. Dickson & C. Goyet). DOE, ORNL/CDIAC-74. Oak Ridge, TN: CDIAC.

Ducklow, H. W. & Harris, R. P. (1993). Introduction to the JGOFS North Atlantic Bloom Experiment. *Deep-Sea Research*, Part II, **40**, 1–8.

Ducklow, H. W., Carlson, C. A., Bates, N. R., Knap, A. H. & Michaels, A. F. (1995). Dissolved organic carbon as a component of the biological pump in the North Atlantic Ocean. *Philosophical Transactions of the Royal Society*, B **348**, 161–7.

Eglinton, G., Elderfield, H., Whitfield, M. & le B. Williams, P. J. (eds.) (1995). The role of the North Atlantic in the global carbon cycle. *Philosophical Transactions of the Royal Society*, B **348**, 121–264.

Falkowski, P. G. (1997). Evolution of the nitrogen cycle and its influence on the biological sequestration of CO_2 in the ocean. *Nature*, **387**, 272–5.

Friederich, G. E., Brewer, P. G., Herlien, R. & Chavez, F. P. (1995). Measurement of sea surface partial pressure of CO_2 from a moored buoy. *Deep-Sea Research*, Part I, **42**, 1175–86.

Gruber, N., Sarmiento, J. L. & Stocker, T. F. (1996). An improved method for detecting anthropogenic CO_2 in the oceans. *Global Biogeochemical Cycles*, **10**, 809–37.

Holfort, J., Johnson, K. M., Schneider, B., Siedler, G. & Wallace, D. W. R. (1998). Meridional CO_2 transport of dissolved inorganic carbon in the South Atlantic Ocean. *Global Biogeochemical Cycles*, **12**, 479–99.

Jahnke, R. A. (1996). The global ocean flux of particulate organic carbon: areal distribution and magnitude. *Global Biogeochemical Cycles*, **10**, 71–88.

Jenkins, W. J. (1995). Tracer based inferences of new and export primary productivity in the ocean. *Reviews of Geophysics*, Supplement, 1263–9.

Keeling, C. D., Piper, S. C. & Heimann, M. (1989). A three-dimensional model of atmospheric CO_2 transport based on observed winds: 4. Mean annual gradients and interannual variations. In *Aspects of Climate Variability in the Pacific and Western Americas*, Geophysical Monograph 55, ed. D. H. Peterson, pp. 305–63. Washington, DC: American Geophysical Union.

Keeling, R. F. & Peng, T.-H. (1995). Transport of heat, CO_2 and O_2 by the Atlantic's thermohaline circulation. *Philosophical Transactions of the Royal Society*, B **348**, 133–42.

Keeling, R. F. & Shertz, S. R. (1992). Seasonal and interannual variations in atmospheric oxygen and implications for the global carbon cycle. *Nature*, **358**, 723–7.

Keeling, R. F., Piper, S. C. & Heimann, M. (1996). Global and hemispheric CO_2 sinks deduced from changes in atmospheric O_2 concentration. *Nature*, **381**, 218–21.

Lefevre, N. (1995). A first step towards a reference Δp_{CO_2} map for the North Atlantic. IGBP-DIS Working Paper No. 11.

Longhurst, A. R., Caverhill, C. & Platt, T. (1995). An estimate of global primary production in the ocean from satellite radiometer data. *Journal of Plankton Research*, **17**, 1245–71.

Manabe, S. & Stouffer, R. J. (1993). Century-scale effects of increased CO_2 on the ocean-atmosphere system. *Nature*, **364**, 215–18.

Martel, F. & Wunsch, C. (1993). The North Atlantic circulation in the early 1980s – an estimate from inversion of a finite-difference model. *Journal of Physical Oceanography*, **23**, 898–924.

Michaels, A. F. & Knap, A. H. (1996). Overview of the U.S. JGOFS Bermuda Atlantic Time-series Study and the Hydrostation S program. *Deep-Sea Research*, Part II, **43**, 157–98.

Michaels, A. F., Bates, N. R., Buesseler, K. O., Carlson, C. A. & Knap, A. H. (1994). Carbon cycle imbalances in the Sargasso Sea. *Nature*, **372**, 537–40.

Michaels, A. F., Olson, D., Sarmiento, J. L., Ammerman, J. W., Fanning, K., Jahnke, R., Knap, A. H., Lipschultz, F. & Prospero, J. M. (1996). Inputs, losses and transformations of nitrogen and phosphorus in the pelagic North Atlantic Ocean. *Biogeochemistry*, **35**, 181–226.

Quay, P. D., Tilbrook, B. & Wong, C. S. (1992). Oceanic uptake of fossil fuel CO_2: Carbon-13 evidence. *Science*, **256**, 74–9.

Robbins, P. E. (1994). Direct observations of the meridional transport of total inorganic carbon in the South Atlantic ocean. *American Geophysical Union EOS Transactions*, **75**, Supplement, 161.

Sarmiento, J. L. & Le Quere, C. (1996). Oceanic carbon dioxide uptake in a model of century-scale global warming. *Science*, **274**, 1346–50.

Sarmiento, J. L., Slater, R. D., Fasham, M. J. R., Ducklow, H. W., Toggweiler, J. R. & Evans, G. T. (1993). A seasonal 3–dimensional ecosystem model of nitrogen cycling in the North Atlantic Euphotic Zone. *Global Biogeochemical Cycles*, **7**, 417–50.

Sarmiento, J. L., Murnane, R. & Le Quere, C. (1995). Air-sea CO_2 transfer and the carbon budget of the North Atlantic. *Philosophical Transactions of the Royal Society*, B **348**, 211–19.

Schimel, D., Enting, I. G., Heimann, M., Wigely, T. M. L., Raynaud, D., Alves, D. & Siegenthaler, U. (1995). CO_2 and the carbon cycle. In *Climate Change 1994*, IPCC, ed. J. T. Houghton, L. G. Meira Filho, J. Bruce, H. Lee, B. A. Callander, E. Haites, N. Harris & K. Maskell, pp. 35–71. Cambridge: Cambridge University Press.

Schlosser, P., Boenisch, G., Rhein, M. & Bayer, R. (1991). Reduction of deepwater formation in the Greenland Sea during the 1980s: evidence from tracer data. *Science*, **251**, 1054–6.

Siegenthaler, U. & Sarmiento, J. L. (1993). Atmospheric carbon dioxide and the ocean. *Nature*, **365**, 119–25.

Six, K. D. & Maier-Reimer, E. (1996). Effects of plankton dynamics on seasonal carbon fluxes in an ocean general circulation model. *Global Biogeochemical Cycles*, **10**, 559–83.

Stephens, M. P., Samuels, G., Olson, D. B., Fine, R. A. & Takahashi, T. (1995). Sea-air flux of CO_2 in the North Pacific using shipboard and satellite data. *Journal of Geophysical Research*, **100**, 13571–83.

Takahashi, T., Chipman, D. W. & Volk, T. (1993). Seasonal variation of CO_2 and nutrients in the high-latitude surface oceans: A comparative study. *Global Biogeochemical Cycles*, **7**, 843–78.

Takahashi, T., Takahashi, T. & Sutherland, S. C. (1995). An assessment of the role of the North Atlantic as a CO_2 sink. *Philosophical Transactions of the Royal Society*, B **348**, 148–60.

Tans, P. P., Fung, I. Y. & Takahashi, T. (1990). Observational constraints on the global atmospheric CO_2 budget. *Science*, **247**, 1431–8.

Tans, P. P., Bakwin, P. S. & Guenther, D. W. (1996). A feasible Global Carbon Cycle Observing System: a plan to decipher today's carbon cycle based on observations. *Global Change Biology*, **2**, 309–18.

Volk, T. & Hoffert, M. I. (1985). Ocean carbon pumps: analysis of relative strengths and efficiencies in ocean-driven circulation atmospheric CO_2 changes. In *The Carbon Cycle and Atmospheric CO_2: Natural Variations Archean to Present*, ed. E. T. Sundquist & W. S. Broecker, pp. 99–110. Washington, DC: American Geophysical Union.

Wallace, D. W. R. (1995). Monitoring global ocean carbon inventories. Ocean Observing System Development Panel, Texas A&M University, College Station, TX, 54 pp.

Wallace, D. W. R. & Lazier, J. R. N. (1988). Anthropogenic chlorofluoromethanes in newly formed Labrador Sea Water. *Nature*, **332**, 61–3.

Wanninkhof, R. (1992). Relationship between wind speed and gas exchange over the ocean. *Journal of Geophysical Research*, **97**, 7373–82.

Watson, A. J., Robinson, C., Robinson, J. E., le B. Williams, P. J. & Fasham, M. J. R. (1991). Spatial variability in the sink for atmospheric carbon dioxide in the North Atlantic. *Nature*, **350**, 50–3.

Watson, A. J., Nightingale, P. D. & Cooper, D. J. (1995). Modelling atmosphere-ocean CO_2 transfer. *Philosophical Transactions of the Royal Society*, B **348**, 125–32.

Wijffles, S. E., Schmitt, R. W., Bryden, H. L. & Stigebrandt, A. (1992). Transport of freshwater by the oceans. *Journal of Physical Oceanography*, **22**, 155–62.

13 Temporal studies of biogeochemical dynamics in oligotrophic oceans

A. F. Michaels, D. M. Karl and A. H. Knap

Keywords: time-series stations, carbon dynamics, nitrogen fixation, dissolved organic carbon, dissolved organic nitrogen, carbon dioxide, nitrate, phosphate

Introduction

The Joint Global Ocean Flux Study (JGOFS) includes three major field components: time-series studies, process studies and a global survey. In the United States, the time-series component is represented by two open-ocean stations, one near Bermuda (Bermuda Atlantic Time-series Study; BATS) and one near Hawaii (Hawaii Ocean Time-series; HOT). The goal of the research at both sites is to achieve an improved understanding of the time-varying fluxes of carbon and other biogenic elements within the ocean and the exchanges with the atmosphere, usually divided into three specific science questions and one technology-related objective:

to understand the seasonal and interannual variations in ocean physics, chemistry and biology;
to understand the relationships between the external physical forcings and the biological rate processes;
to understand the processes that determine sea-surface p_{CO_2}, including thermodynamics, particle export and gas exchange; and
to provide a test bed for the introduction and validation of new oceanographic instruments and technologies.

These goals are supplemented by more specific research objectives that are related to regional biogeochemical dynamics at each site and, to some extent, the interests and backgrounds of the scientists who are involved. These more specific goals include the need to understand the cycling of nitrogen as a control on the carbon cycle, the role of dissolved organic materials in both the elemental cycles and the net exports from the surface waters, and the role of community structure in affecting the cycling of elements. More recently, nitrogen fixation has been under increased study at both sites as this process appears to be crucial to understanding the seasonal and inter-annual patterns in the cycling of carbon, nitrogen and phosphorus in the sea.

Organisation of the HOT and BATS programmes

The HOT and BATS programmes were initiated in October 1988 (Michaels & Knap, 1996; Karl & Lukas, 1996). The stations are funded primarily through research grants from the US National Science Foundation. Although time-series research is closely related to monitoring and the JGOFS stations have some monitoring role, the explicit goal of the US JGOFS time-series research is to obtain an understanding of basic ocean processes. The monitoring function is related to the documentation of change over the time scale of the programmes, partly in the hope that putative climate changes or the incremental changes in biogeochemistry associated with ocean feedbacks will become apparent.

Many of the initial results of the research from these two programs have already been published. In this chapter, we summarise only a few of the observations and trends from both stations in the context of a comparison between the two sites. Although both stations are nominally located in low-biomass, low-nutrient oceanic gyres, they differ in latitude and seasonality and consequently in the rates and mechanisms of biogeochemical cycling. For the details of each specific study, we cite the appropriate literature and concentrate on the comparison between sites.

The HOT programme includes two separate components, the JGOFS programme with a focus on biogeochemistry and a World Ocean Circulation Experiment (WOCE) component that concentrates on the physical oceanography of the region. At BATS, all of the measurements were initially included in the JGOFS programme. Both sites also received 'seed' funding from the US National Ocean and Atmosphere Administration (NOAA) to initiate measurements of dissolved inorganic carbon (DIC) species at a very high level of precision and accuracy. The US National Aeronautics and Space Administration (NASA) has funded bio-optics and satellite oceanography programmes at Bermuda. Both programmes have also benefited from institutional support. For example, the Hawaii programme received significant matching funds from the state of Hawaii, and the Bermuda Biological Station for Research funded the purchase of a dedicated research vessel and other improvements to the physical infrastructure that were required to allow the BATS programme to be successful.

The combination of funding and focused research goals is a necessary but not sufficient condition for the operation of a time-series station. These stations also

Table 13.1. *Common measurements made at the BATS and HOT sites (each site also makes many additional measurements that are not performed at the other)*

Parameter	Depth range (m)	Technique/instrument
Continuous electronic measurements		
Temperature	0–4200	Thermistor on SeaBird SBE-911 plus CTD
Salinity	0–4200	Conductivity sensor on SeaBird SBE-9/11 plus CTD
Depth	0–4200	Digiquartz pressure sensor on SeaBird SBE-9/11 plus CTD
Dissolved oxygen	0–4200	SeaBird polarographic oxygen electrode
Fluorescence	0–500	SeaTech fluorimeter
Meteorological data	sfc	Mooring, ship and land stations
Surface p_{CO_2}, T, S, fluorometry	sfc	Underway system on some research vessels
Discrete measurements from Niskin bottles or nets		
Salinity	0–4200	Conductivity on guideline Autosal 8400A
Oxygen	0–4200	Winkler Titration, automated endpoint detection
Total CO_2	0–250	Automated colorimetric analysis
Alkalinity	0–250	High precision titration
p_{CO_2}	sfc, 0–250	Underway p_{CO_2} analysis, extraction
Nitrate	0–4200	CFA colorimetric
Nitrite	0–4200	CFA colorimetric
Phosphate	0–4200	CFA colorimetric
Silicate	0–4200	CFA colorimetric
Dissolved organic carbon	0–4200	High-temperature, catalytic oxidation
Dissolved organic nitrogen	0–4200	UV oxidation
Dissolved organic phosphorus	0–4200	UV and chemical oxidation
Particulate carbon	0–4200	High-temperature combustion, CHN analyser
Particulate nitrogen	0–4200	High-temperature combustion, CHN analyser
Particulate phosphorus	0–300	Chemical oxidation
Fluorimetric chlorophyll *a*	0–250	Acetone extraction, Turner fluorimeter
Phytoplankton pigments	0–250	HPLC, resolves 19 pigments
Bacteria	0–3000	Fluoroescence microscopy or flow cytometry
Zooplankton biomass	0–200	Day and night tows with 200 μm mesh net
Rate measurements		
Primary production	0–140	Trace-metal clean, *in situ* incubation, ^{14}C uptake
Bacterial activity	0–140	Thymidine and/or leucine incorporation
Particle fluxes	150,200,300	Free-drifting cylindrical traps (MultiPITs)
Mass		Gravimetric analysis
Particulate carbon flux		Manual swimmer removal, acidification, CHN
Particulate nitrogen flux		Manual swimmer removal, CHN analysis
Particulate phosphorus flux		Manual swimmer removal, chemical oxidation

require a staff of dedicated and well-trained scientists and other support personnel. Both stations have been extremely fortunate in attracting exceptional technical staffs who have been able to maintain a high level of data quality in the face of occasionally trying circumstances and the inevitable doldrums that come from repeated sampling of the same system. As these stations have matured, it has become apparent to the principal investigators that the human infrastructure is one of the most challenging components of running and sustaining a time-series station of this complexity. We are most grateful for the dedication of our technical staffs and both we and the entire scientific community who will also use these data owe them a debt of gratitude for their collective efforts.

The sampling at these stations is nominally monthly. The more striking seasonality at the Bermuda station dictated that the sampling frequency at this site be increased to bi-weekly in the spring. The Bermuda station is located approximately 80 km southeast of Bermuda near the site of a long-term (since 1976) sediment trap mooring started by Werner Deuser (WHOI) and now operated by Maureen Conte (WHOI). The BATS site is approximately twice as far offshore as the Hydrostation S station, the site of one of the longest continuous ocean time-series studies. The bi-weekly sampling at Hydrostation S began in 1954 and continues to this day. The Hawaii station is approximately 100 km north of Oahu. The Hawaii sampling also includes regular sampling at a near-shore station at the beginning of each cruise. Both stations measure a sub-set of the core JGOFS parameters (Table 13.1) and have spent considerable effort to ensure that the data quality is high and comparable, both to each other and to the standards of the broader oceanographic community. Details of the analytical and quality control procedures are available in the published papers (Michaels *et al.*, 1994*b*, 1996; Karl & Lukas, 1996).

Seasonal and inter-annual ecosystem patterns

The basic seasonal and inter-annual patterns at each of the sites have been well described elsewhere (Karl & Michaels, 1996). For the purposes of this chapter, we concentrate on a direct comparison of the general patterns in the basic biogeochemistry between the two oceanic time-series stations. The BATS site is approximately 8 degrees of latitude further north (*c.* 500 miles) than the HOT site and this difference is quite apparent in the strength of the seasonality in the thermal structure. The Bermuda site is characterised by a nearly 10 °C seasonal variation in surface temperature and fluctuations in mixed-layer depth between a few metres in the summer and over 250 m in some winters (Fig. 13.1). The seasonal variation in SST is much smaller at the HOT programme station

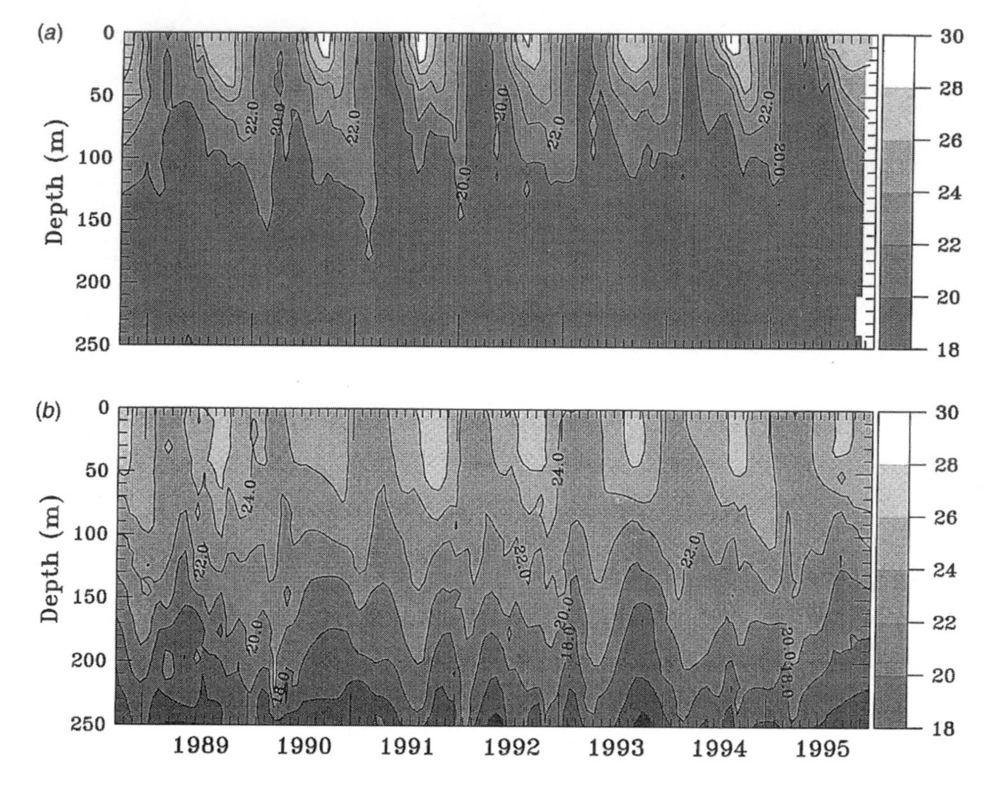

Figure 13.1 *Contour plot of temperature (°C) at BATS (*a*) and HOT (*b*), October 1988 – January 1996.*

ALOHA and the mixed layer rarely extends deeper than 100 m (Fig. 13.1). Bermuda is also characterised by a layer of near-isothermal waters between 300 and 600 m, the so-called 18° water that is formed to the north each winter and subducts below the Sargasso Sea. At Hawaii, the stratification is strong and fairly continuous throughout the upper 1000 m.

This sharp difference in the mixing and stratification seasonality is subsequently reflected in both the nutrient dynamics and biological productivity. The deep mixing near Bermuda intrudes into the nitricline and brings nitrate to the surface during most winters (Fig. 13.2). This nitrate injection lasts from a few days to a few months and it is rapidly consumed as the upper ocean is re-stratified in early spring. The integrated stock of nitrate that is introduced into the euphotic zone and subsequently consumed is up to 0.1–0.2 mol N m^{-2} in the BATS time series (1988–1995) and is inferred to reach 0.5–0.6 mol N m^{-2} (Michaels *et al.*, 1994*b*) in some of the years with the deepest mixed layers in the Hydrostation S time series (1954–1995).

Before the HOT programme, few data were available to document temporal or spatial variability in the inorganic nutrient concentrations in surface waters of

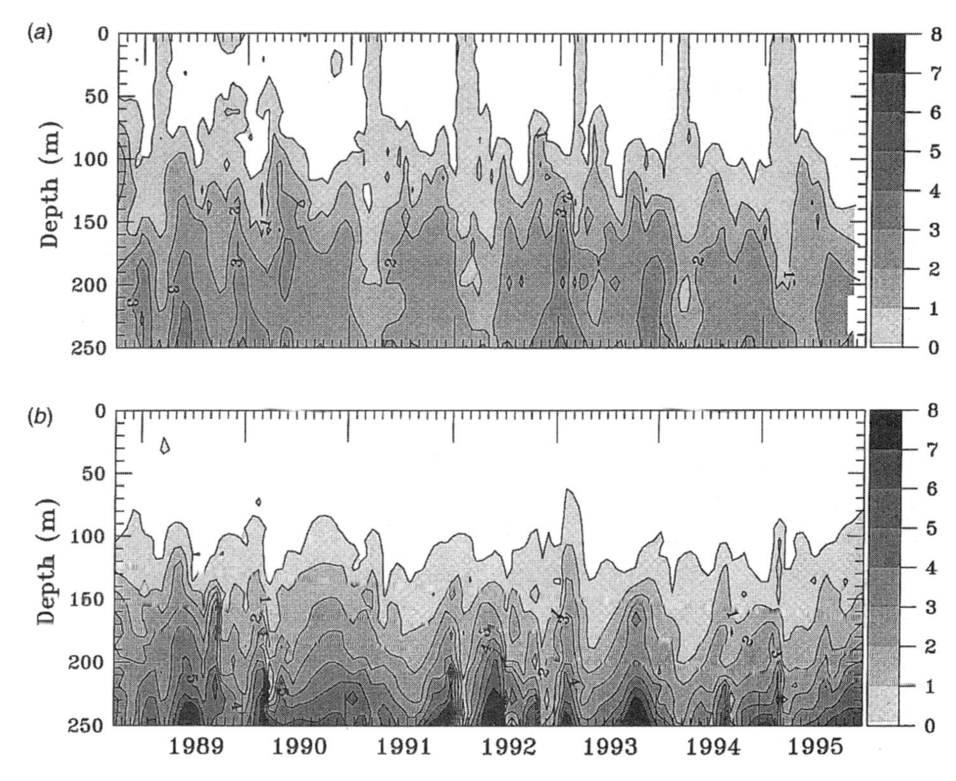

Figure 13.2 *Contour plot of nitrate (μmol kg⁻¹) at BATS (a) and HOT (b), October 1988 – January 1996.*

the North Pacific gyre. The majority of published surface water NO_3^- concentrations were below the detection limit for the standard automated nutrient analyses ($\leq$ 50 nmol kg^{-1}; see, for example, Hayward, 1987). A more detailed analysis based on high-precision, high-sensitivity nutrient measurements at Station ALOHA is now beginning to document a dynamic nutrient field in the oligotrophic North Pacific. The time-series data set can be summarised as follows: (1) NO_3^- is always detectable in surface waters (0–50 m) but the concentration rarely exceeds 10 nmol kg^{-1}; (2) NO_3^- concentrations exhibit a temporally variable increase with depth between 50 and 100 m with typical vertical NO_3^- gradients of 20–40 μmol N m^{-4}; and (3) the top of the nitricline ranges from 75 to 175 m. Furthermore, at least six events have been encountered during the approximately 300 days of at-sea observations (Fig. 13.3). These nutrient intrusions probably supply the majority of the NO_3^- used for new production at Station ALOHA. Even if one excludes these large mixing events from consideration, the month-to-month variability in total 0–100 m depth-integrated NO_3^- exceeds an order of magnitude (e.g. < 0.05–0.96 mmol N m^{-2}, excluding the six events in excess of 1 mmol N m^{-2}). This substantial

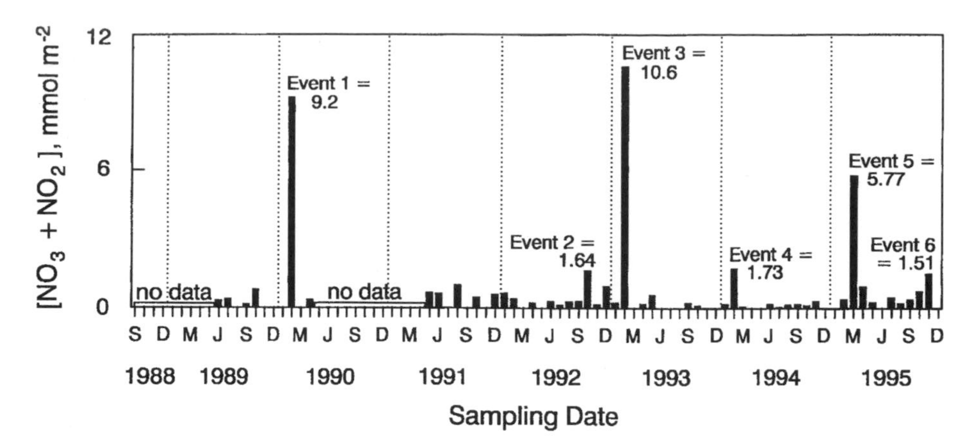

Figure 13.3 *Temporal variability in the depth-integrated inventories of nitrate plus nitrite concentrations (mmol m^{-2}) measured at Station ALOHA. The four 'events' shown are believed to be manifestations of mixing and/or stirring processes. With the exclusion of these perturbation events, the background mean and standard deviation values for the upper water column (0–100 m) inventories are 0.474(± 0.48) mmol m^{-2}, respectively (n = 41).*

variation is still two orders of magnitude smaller than the spring nutrient intrusions at BATS. The role of eddies and other mesoscale features in the temporal variability of NO_3^- at HOT is currently unknown, but is suspected to be substantial.

The deep mixing at Bermuda does not result in surface soluble reactive phosphorus (SRP) concentrations that exceed the detection limits of the analysis (0.05 μmol kg^{-1}) (Fig. 13.4). Nitrate: SRP ratios in the upper 1000 m at Bermuda occur at ratios that are significantly above the nominal Redfield ratio of 16 mol: 1 mol (Michaels *et al.*, 1994*b*), in part because SRP is depleted relative to nitrate to a depth of approximately 150 m. This SRP depletion is in a depth range (100–150 m) usually thought of as a zone of high net re-mineralisation. In addition, the deep mixing in spring probably introduces less SRP than would be required to support Redfield ratio growth. Yet the nitrate still disappears within a few weeks of stratification. This nutrient-ratio anomaly is not understood, but it may reflect either the variability in nutrient stoichiometries and processes that are possible in nature or the yet unmeasured role of dissolved organic nutrients.

Soluble reactive phosphorus (SRP) is also low but temporally variable in the surface waters near Station ALOHA throughout the year (Karl & Tien, 1997). During the first 6-year period of observation, upper water column (0–100 m) SRP concentrations varied from more than 150 nmol kg^{-1} to less than 10 nmol kg^{-1}, and generally conformed to one of two distinct concentration–depth patterns: 'Type I' profiles were characterised by uniform SRP

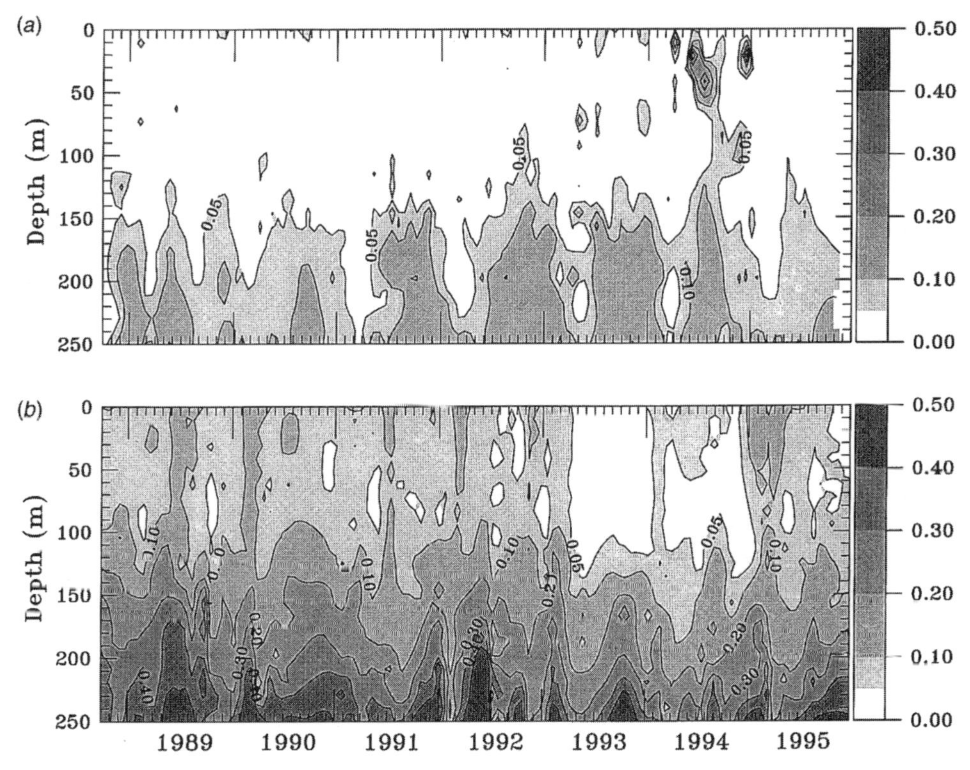

Figure 13.4 *Contour plot of soluble reactive phosphorus (SRP) (μmol kg^{-1}) at BATS* (a) *and HOT* (b), *October 1988 – January 1996.*

concentrations with depth (concentration gradients less than 0.05 μmol m^{-4} SRP) and 'Type II' profiles were characterised by distinct near-surface (0–30 m) SRP concentration maxima. The Type I profiles were further divided into low (Type I-L) or high (Type I-H) categories based on whether the average SRP concentration was less than 60 or greater than 60 nmol kg^{-1}, respectively (Fig. 13.5). Three independent, but not mutually exclusive, models are presented to explain these time-varying SRP concentrations: (1) convective mixing; (2) atmospheric deposition; and (3) upward P flux. The upward P flux model, including both passive (upward particle flux) and active (biological transport) processes, is the most consistent model for the available data set (Karl & Tien, 1997).

We have also observed a systematic decrease in the SRP pool inventory (0–100 m) from an initial value of 10 mmol P m^{-2} to 5 mmol P m^{-2} during the 6-year observation period (Fig. 13.6). This loss of bioavailable P from the surface waters is consistent with the hypothesised role of N_2 fixation as a major source of new N for the oligotrophic North Pacific gyre during the period 1988–1994 (see subsequent section).

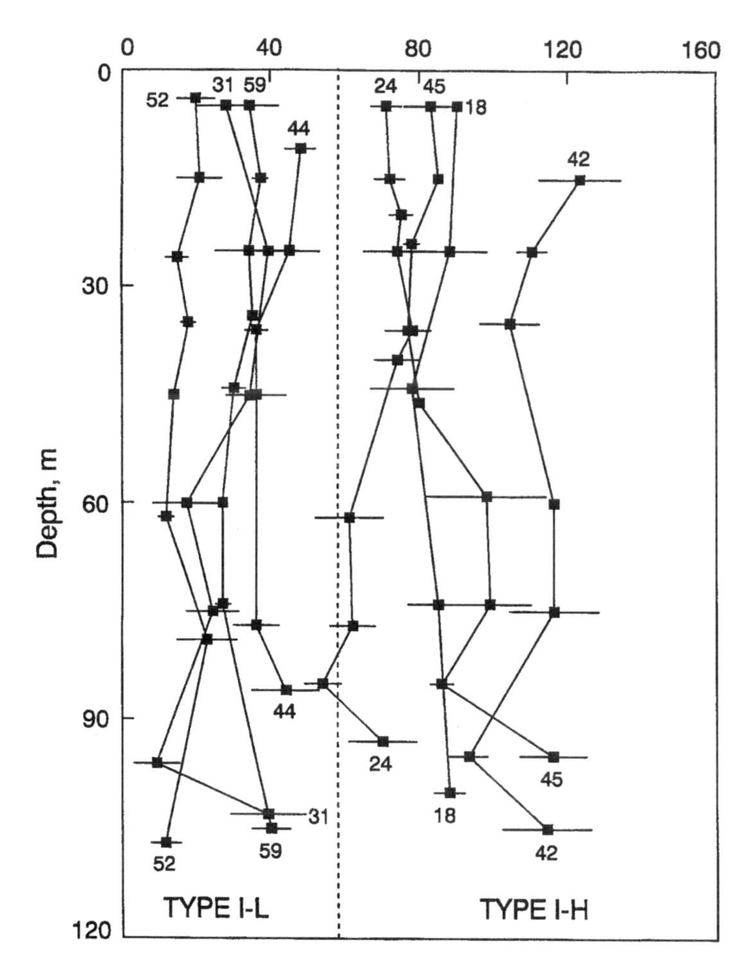

Figure 13.5 *Selected SRP-M (SRP by MAGIC procedure) concentrations with depth at Station ALOHA, showing two different profile patterns:* (a) *Type-I profiles displaying a relatively constant SRP concentration with depth; and* (b) *Type-II profiles displaying near-surface enrichments of SRP relative to mid-euphotic zone concentrations. The number above and below each profile indicates the HOT cruise on which each data set was collected. The data shown are mean values,* ± *1 standard deviation (n = 3).*

The seasonality of deep mixing at Bermuda induces seasonal fluctuations in phytoplankton biomass and productivity that are tightly phased with the timing of nutrient addition, whereas in Hawaii the seasonality is less pronounced (Figs. 13.7, 13.8). Some of the correlation between nutrient addition and plankton blooms may also be due to the aliasing of the bi-weekly to monthly cruises sampling eddies with different histories and bloom development. However,

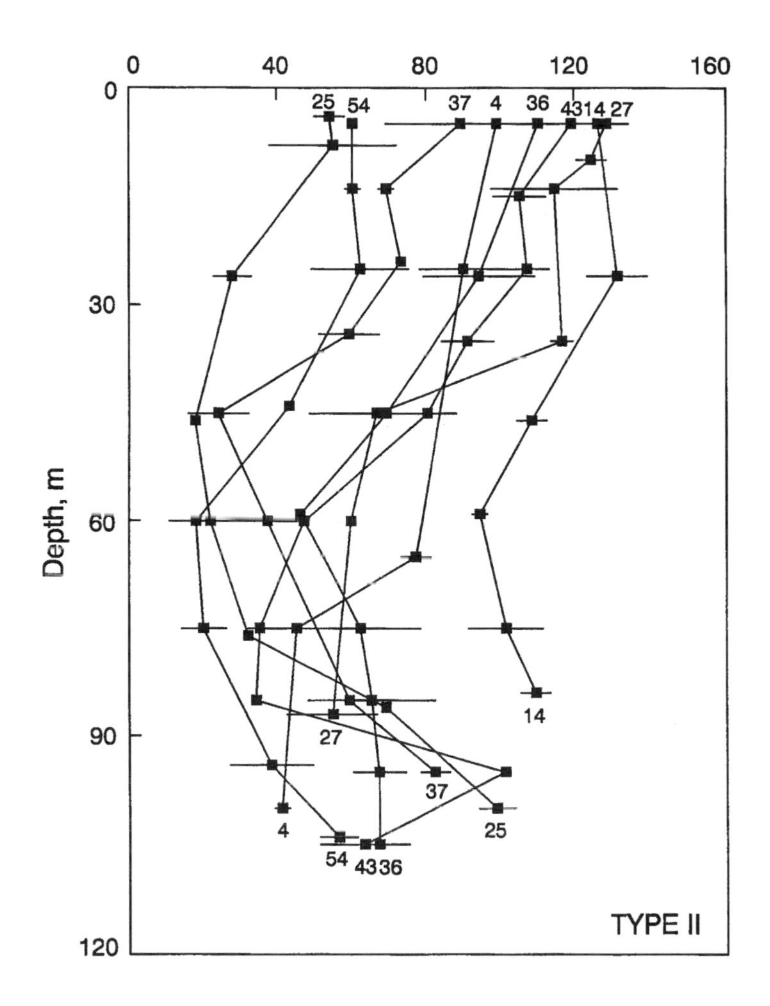

there is a reasonable relationship between the depth of the winter mixed layers and the peak values of spring production at BATS, with the exception of a single deep mixed layer in spring 1994 (Fig. 13.9). This anomalous mixed layer was warm and nutrient-free for its depth and was obviously due to the sampling of a large warm eddy.

Carbon dynamics and secular p_{CO_2} increases

Both stations show a seasonality in the total DIC at the surface, with the BATS station having a range of 60 μmol kg^{-1} and a range at HOT of 20 μmol kg^{-1}. This

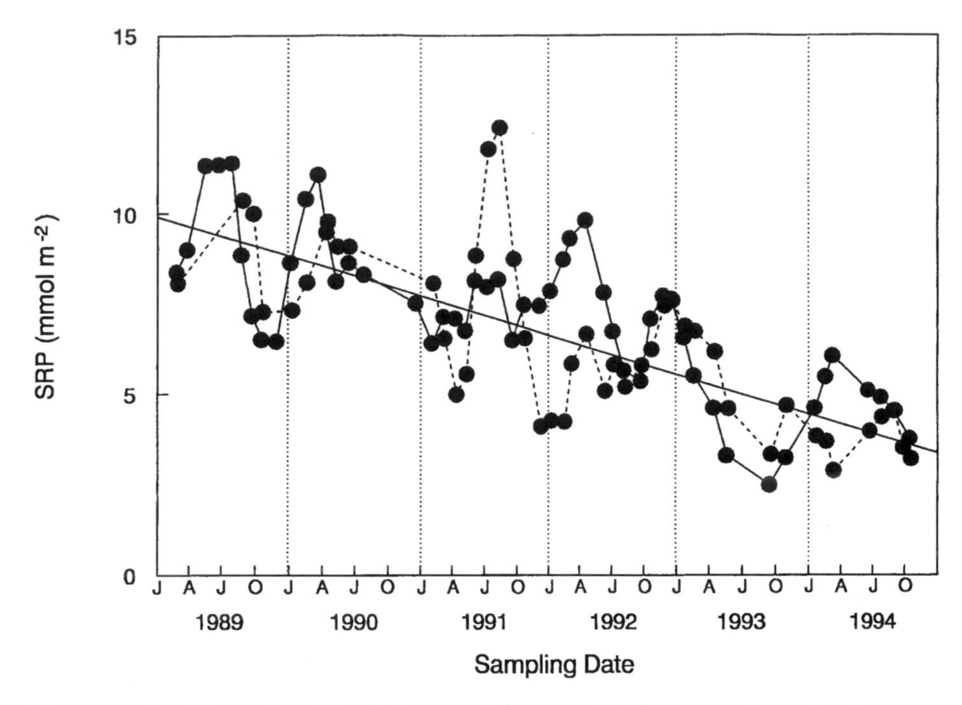

Figure 13.6 *Time series of depth-integrated (0–100 m) SRP inventories at Station ALOHA for the period 1989–94. The data points are the 3-point running mean values for either SRP-AA (SRP measured by autoanalyser, solid line) or SRP-M (SRP measured by MAGIC, dashed line). The model I linear regression fit for the SRP-M data set is: SRP (mmol m^{-2}) = 9.95 − [0.003 × (days since 1 Jan 1989)].*

seasonality is due to a combination of mixing, gas exchange and export. The deep mixing in winter at BATS brings DIC-rich waters to the surface. A small amount of this DIC is removed during the spring bloom as the 0.2–1.0 μmol kg^{-1} of nitrate removal results in 1–7 μmol kg^{-1} of DIC removal, assuming Redfield C: N stoichiometry.

The seasonal fluctuations in DIC and temperature result in a seasonal pattern of p_{CO_2} at the surface (Bates *et al.*, 1996) and thus of gas exchange. Despite the higher DIC in the winter, both stations are sinks for CO_2 at this time because the cooler waters lower p_{CO_2} (Bates *et al.*, 1996; Winn *et al.*, 1994). In the summertime, p_{CO_2} is elevated by the warmer temperatures causing an outgassing. On an annual basis, both sites are weak sinks of the order of 0.5–1.0 mol m^{-2} yr^{-1} (Bates *et al.*, 1996; Winn *et al.*, 1994). At HOT, the seasonal fluctuations in gas exchange are comparable to the net changes in DIC in the upper ocean, whereas at BATS the seasonal gas exchange is not large enough to explain the seasonal fluctuations in DIC in the upper 150 m (Michaels *et al.*, 1994*a*). If fact, BATS has an anomalous summertime

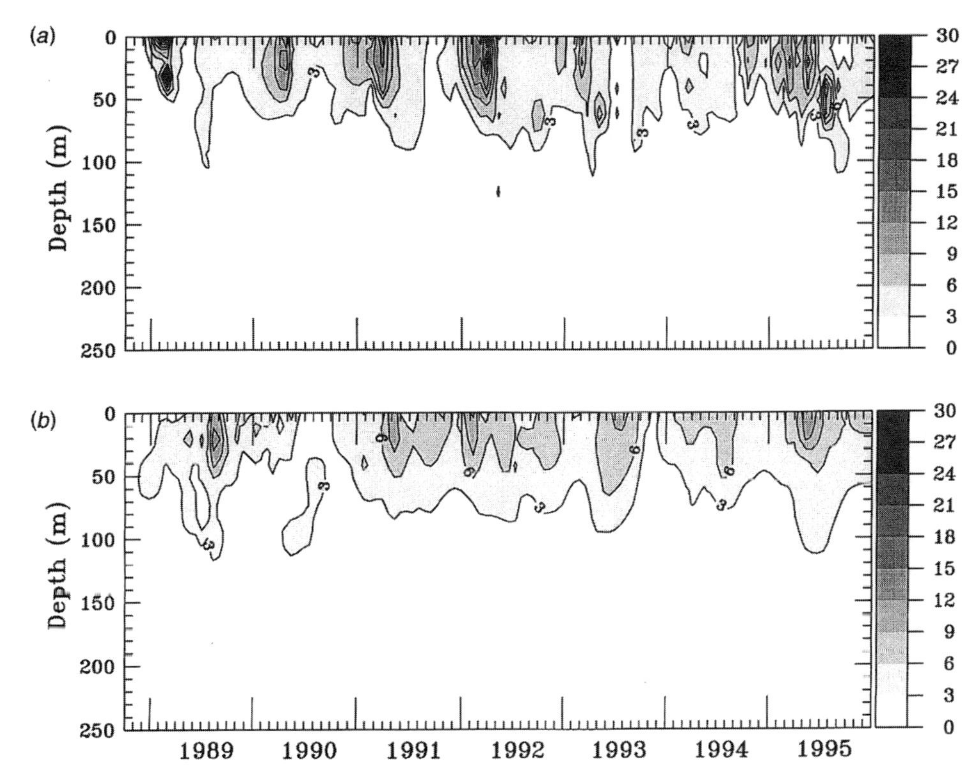

Figure 13.7 *Contour plot of primary productivity (mg C m^{-3} d^{-1}) at BATS (*a*) and HOT (*b*), October 1988 – January 1996.*

drawdown of DIC of the order of 3 mol m^{-2} that cannot be explained by the balance of fluxes as currently measured (Michaels *et al.*, 1994*a*).

Inter-annual variability in primary production and particle export

Both HOT and BATS stations show inter-annual fluctuations in production and flux that might be related to the large-scale climate system. At HOT, there is a measurable increase in productivity, a decrease in particle flux and an apparent alternation from nitrogen limitation to phosphate limitation in the transition from the La Niña conditions of 1989–1991 to the El Niño condition of 1991–93 (Karl *et al.*, 1995). At BATS, mixed layers were deeper during the 1991–93 El Niño period and this pattern also seems to occur in the 42 year record at Hydrostation S (Michaels *et al.*, 1996). As these deep winter-mixed layers are correlated with the intensity of the spring blooms (Fig. 13.9), there is a pattern comparable to HOT with an increased production during ENSO; however, at BATS, this is limited to the spring. Both stations share the decrease

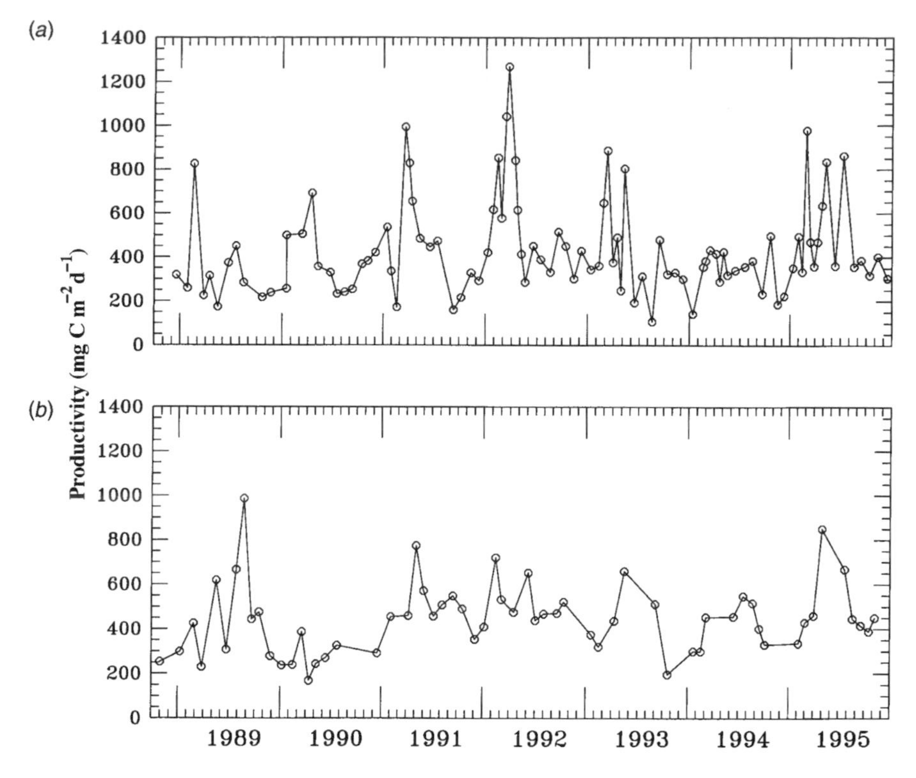

Figure 13.8 *Time series of integrated primary productivity at BATS (0–140 m integral) (a) and HOT (0–200 m integral) (b), October 1988 – January 1996.*

in the annual particle flux during 1991–93. Although there are a variety of speculations about the reasons for these ENSO-correlated patterns in the biogeochemistry, we still do not have full explanations. However, it is precisely in the identification of links between climate and biogeochemistry that we should look for the feedback mechanisms that may allow ocean biogeochemistry to have an effect on climate.

Nitrogen fixation

Nitrogen fixation has emerged at both stations as an important process for understanding the coupled nitrogen and carbon cycles. There are distinct differences in how each station has approached this problem. At HOT, nitrogen fixation is inferred from an examination of the biology and nutrient dynamics of the euphotic zone (Karl *et al.*, 1997). At Bermuda, the evidence comes from an investigation of the nutrient dynamics of the mesopelagic and a study of the anomalous nitrate : phosphate ratios at these depths. In both cases, this process seems quantitatively important and may explain some of the seasonal and

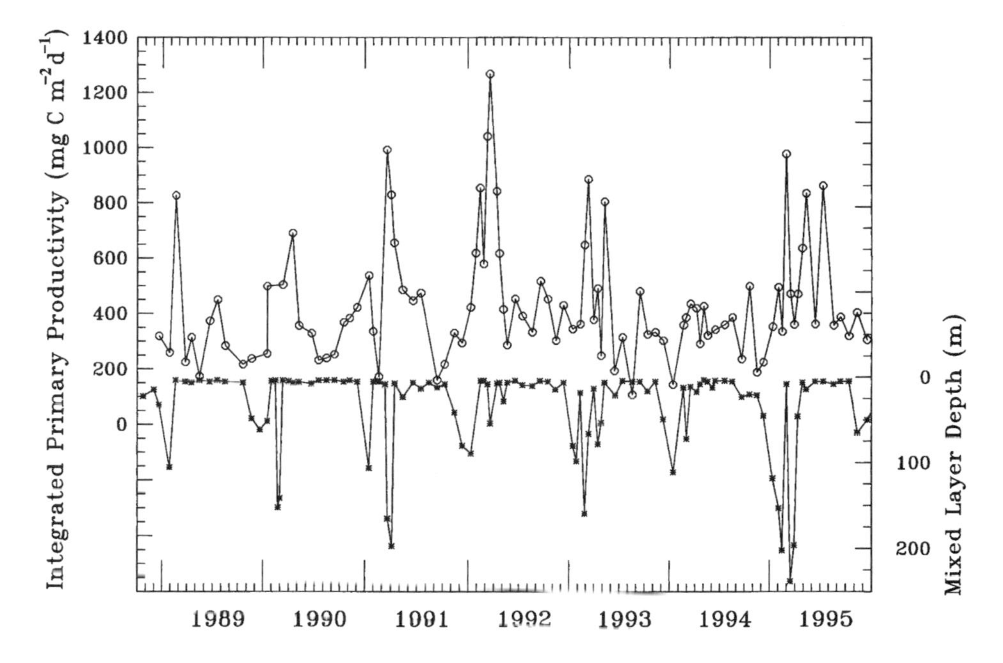

Figure 13.9 *Time series of integrated primary productivity at BATS (upper line) and the mixed layer depth on each cruise (lower line).*

interannual patterns in nutrient dynamics, primary production and particle export. However, nitrogen fixation is rarely included in models of the upper ocean and the data from HOT and BATS indicate that this may be an important oversight that needs to be reconciled.

Nitrogen fixation at HOT

Several independent lines of evidence from Station ALOHA suggest that nitrogen fixation is an important source of new nitrogen for the pelagic ecosystem of the subtropical North Pacific Ocean. The evidence is derived from independent measurements and data syntheses, including: (1) *Trichodesmium* population abundances and estimates of their potential rates of biological nitrogen fixation; (2) measurements of *Trichodesmium* nitrogenase activity (acetylene reduction assay) and extrapolations to nitrogen fixation rates; (3) assessment of the molar N: P stoichiometries of near-surface dissolved and particulate matter pools; (4) application of a one-dimensional N and P mass balance model; (5) observations of a secular decrease of 50% in the soluble reactive phosphorus (SRP) pool in the upper water column (0–100 m) and corresponding increase in the soluble non-reactive P (SNP) pool; and (6) seasonal variations in the natural ^{15}N isotopic abundances of particulate matter exported to the deep sea (Karl *et al.*, 1995, 1997). Considered separately, none of these data sets provides conclusive evidence for the quantitative role of nitrogen fixation. However, when viewed together and in the full context of the other HOT programme time-series observations, a conclusion regarding the significance of nitrogen fixation in the subtropical North Pacific Ocean is inescapable.

Excess nitrate and nitrogen fixation at BATS

In the Sargasso Sea, the ratio of nitrate to phosphate in the upper 1000 m exceeds the nominal Redfield ratio of 16. The waters above 1000 m are ventilated on time scales of decades or less (Sarmiento, 1983; Jenkins, 1988; Sarmiento *et al.*, 1990) and in the areas where these isopycnal surfaces outcrop, the concentrations of both nitrate and phosphate are often near zero. At steady state, the anomalous nutrient ratios on these isopycnal surfaces in the regions away from the ventilation areas must be sustained by a continual production of nutrients at these higher ratios to sustain the gradients. The rate of production of this excess nitrate has been estimated by using the well-studied resident times of these layers and the observed isopycnal gradients of the nutrient ratios (Michaels *et al.*, 1996).

If the organisms that compose the bulk of the organic biomass in sinking material have N : P elemental concentrations near the Redfield ratio of 16 : 1, the re-mineralisation of this biomass will lead to the production of 16 mol of nitrate for every 1 mol of phosphate produced. If organisms such as nitrogen-fixing cyanobacteria create biomass that is richer in nitrogen (more correctly, phosphate-poor) this will yield the re-mineralisation of more than 16 mol of nitrate for every 1 mol of phosphate. Conversely, denitrification would remove nitrate with respect to phosphate and decrease the measured *in situ* ratio. We define a parameter called 'Excess Nitrate' as the concentration of nitrate in excess (or deficit) of that expected from the re-mineralisation of phosphate at Redfield stoichiometries:

$$N^* = \text{Excess nitrate} = [NO_3^-] - 16 \times [PO_4^{3-}],$$

where $[NO_3^-]$ and $[PO_4^{3-}]$ are the concentrations of nitrate and phosphate, respectively. In a mean profile of Excess Nitrate at the BATS site, or in GEOSECS and TTO data for the central Sargasso Sea, N^* is elevated throughout the upper 800 m. If we compare the excess nitrate at BATS or the TTO stations with that from the regions where each density surface outcrops during the winter ventilation, we can determine an excess nitrate gradient (Fig. 13.10). This gradient can be compared to the ventilation time scale of each isopycnal surface to calculate the rate of supply of excess nitrate that would be required to maintain the gradient (using techniques described by Sarmiento, 1983). This calculation is similar to the approach commonly used to estimate oxygen utilisation rates (OUR) in the same region (Sarmiento *et al.*, 1990).

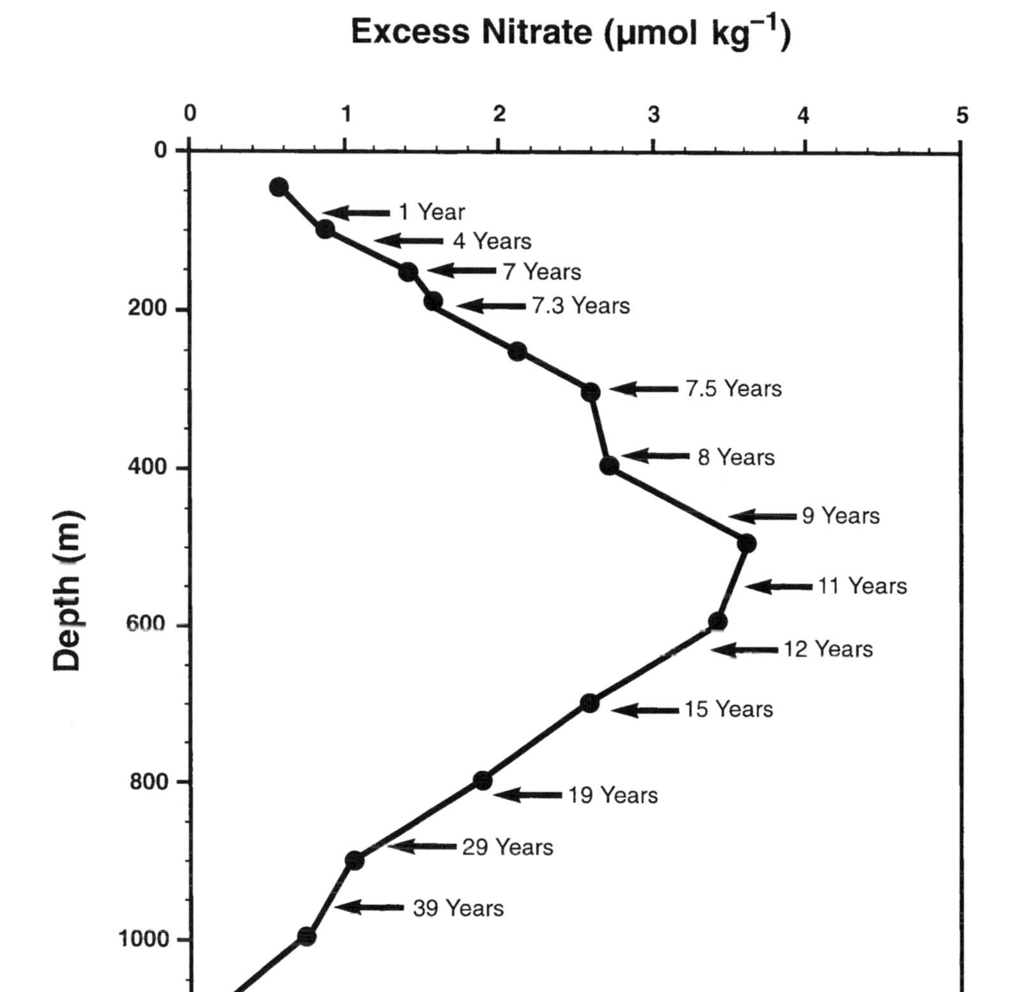

Figure 13.10 *Excess nitrate gradient at BATS. Excess nitrate is calculated as described in the text. The gradient is calculated by subtracting the value at the outcrop of the isopycnal surface from the value at BATS (see text). Ages indicate the ventilation residence time of that isopycnal surface.*

From these calculations, Michaels *et al.* (1996) estimated that $3.7–6.4 \times 10^{12}$ mol of 'excess' nitrate are created in the Sargasso Sea each year. The range in the estimates is largely a function of the different residence time estimates for each density surface and less a function of the uncertainty in the magnitude of the horizontal N^* gradient. In a different and more detailed analysis of N^* conducted by Gruber & Sarmiento (1996), they find a lower rate of excess nitrogen creation (2×10^{12} mol yr^{-1}). There are two possible sources of the nitrate excess in the main thermocline: nitrogen fixation and/or differential fluxes or lability in dissolved organic nitrogen compared with dissolved organic phosphorus. Unfortunately, neither process is well constrained by direct measurement. However, in a discussion of the available data, it appears that nitrogen fixation and the creation and export of high N:P organic biomass is the more likely explanation for this pattern.

Nitrogen fixation as a source of the excess nitrate may resolve some of the unanswered questions about the anomalous carbon drawdown in the Sargasso Sea (see above). Each year, between April and December, approximately 3 mol m^{-2} of carbon disappear from the surface waters at BATS (Michaels *et al.*, 1994*b*) in the absence of significant nitrate. If nitrate was mixed up from below, carbon would also be imported, as the deeper water has elevated DIC. Nitrogen fixation would cause the uptake of dissolved CO_2 as the new reduced nitrogen is converted into biomass. This may occur at the typical Redfield ratio of 6.6 mol C: 1 mol N, or probably at a much higher ratio as the most conspicuous diazotroph, *Trichodesmium* sp., uses a carbohydrate ballasting process to regulate buoyancy. Using C:N ratios of both 6.6 and 15, we estimate the total carbon uptake in the surface waters overlying the Sargasso Sea as 24–96 $\times 10^{12}$ mol C yr^{-1} (0.3–1.2 $\times 10^{15}$15 g C yr^{-1}). In the seasonal environment of BATS, this would probably be confined to the summer–autumn period where oligotrophy is conducive to nitrogen fixation, but could occur year-round to the south. Approximately 50–80% of this carbon would be sequestered for multi-year time scales, but little for time scales of longer than a decade. The area over which this nitrogen fixation probably occurs is approximately 20×10^{12} m^2, which gives an annual carbon fixation rate of 1.2–4.8 mol C m^{-2} yr^{-1}. This rate is very similar to the observed summertime carbon drawdown at the BATS site (Michaels *et al.*, 1994*a,b*).

Trichodesmium sp. and most other marine diazotrophs require elevated cell quotas of iron, approximately 100–1000-fold higher than those required by organisms that utilise nitrate (Hutchins, 1989; Rueter *et al.*, 1992; Elardo, 1994). The iron supply to the Sargasso Sea is primarily through the deposition of Saharan dust (Prospero *et al.*, 1996). Model estimates of total dust deposition to the Sargasso Sea between 10 and 40°N latitude (excluding the areas near the African coast) are approximately 44×10^{12} g dry mass yr^{-1} (Prospero *et al.*, 1996). Assuming 3.5% iron by mass, dust would supply 2.8×10^{10} mol Fe yr^{-1}. *Trichodesmium* sp. typically has an N:Fe ratio of the order of 1000 mol N: 1 mol Fe (range = 200–3000; Hutchins, 1989; Rueter *et al.*, 1992; Elardo, 1994). The export of particulate biomass produced by nitrogen fixing organisms would probably transport the intra-cellular iron to depth as well. Thus, the total iron deposition could support an annual excess nitrate production rate from nitrogen fixation of as much as 28×10^{12} mol N yr^{-1}. Much of the atmospheric iron is insoluble and thus unavailable to nitrogen-fixing phytoplankton. The iron supply would support the estimated excess nitrate production if *Trichodesmium* sp. were able to utilise about 13–23% of the iron supply. Interestingly, the relatively close match between the estimated nitrogen fixation rate and the rate that is potentially supported by the iron deposition may indicate that Fe supply limits the total amount of nitrogen fixation that can occur. Fe inputs are concentrated in the summer, a good temporal match to the timing of the DIC drawdown. There is a large decadal variation in dust deposition as a result of the large expansion of the Sahara in the early 1970s. If dust supply controls this process, we may thus expect annual and decadal fluctuations in nitrogen fixation that could drive 0.3–1.2 Gt C yr^{-1} fluctuations in the northern hemisphere atmospheric CO_2 stocks. However, the main thermocline ventilates on time scales of 7–30 years. Thus, nitrogen fixation in the Sargasso Sea cannot result in the sequestration of large amounts of CO_2 on the longer time scales associated with the anthropogenic signal in the atmosphere.

Surprises, assumptions and the inhererent value of long-term time series

As described above, it is apparent that both time-series studies have allowed an unparalleled look into the biogeochemistry of the two regions. These studies are neither perfect nor complete, but they do present the most intensive, time-dependent examination of our understanding of elemental cycling in the open oceans. Because open-ocean systems represent half of the surface area of the world, this understanding can be important on a global scale if it can be extrapolated to this scale. At the same time, the BATS and HOT research programmes present a mechanism to examine some of the assumptions we use as we create models to do much of the spatial and temporal extrapolation.

Our current conceptual model for the cycling of carbon in oligotrophic systems assumes that nitrogen limits plant growth and that the cycling of carbon can be linked to the nitrogen cycle by the Redfield stoichiometry. This is expressed in models by a simplified version of the new production hypothesis (Dugdale & Goering, 1967) that concentrates on the dynamics of nitrate and ammonia as a reflection of new and regenerated processes, respectively (Fig. 13.11*a*). The carbon cycle is then reflected as a simple multiplication of the nitrogen fluxes by 6.6 (mol : mol) (Fig. 13.11*b*). This is computationally tractable and, in many cases, seems to represent the dynamics of the drawdown

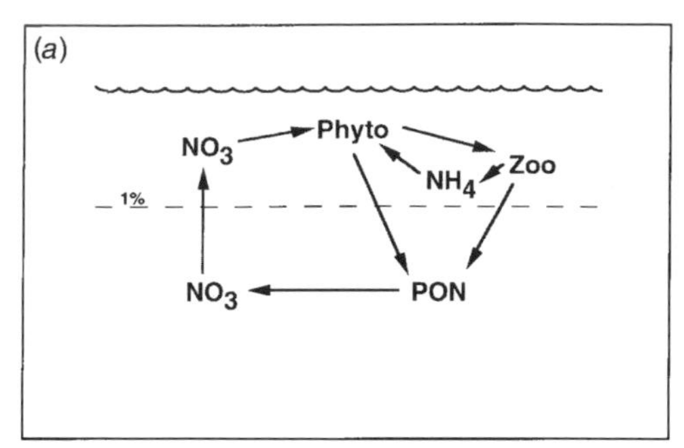

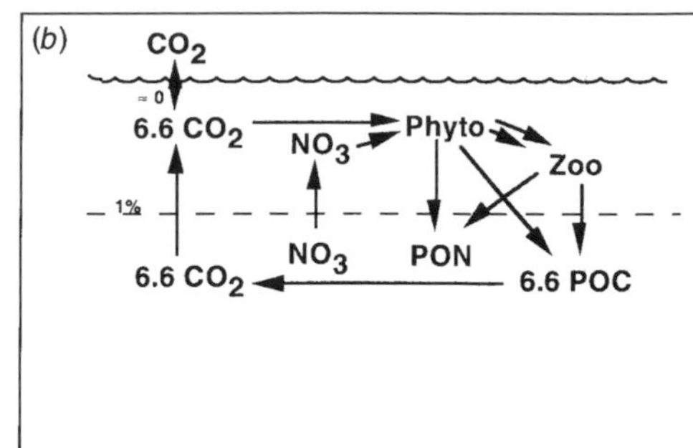

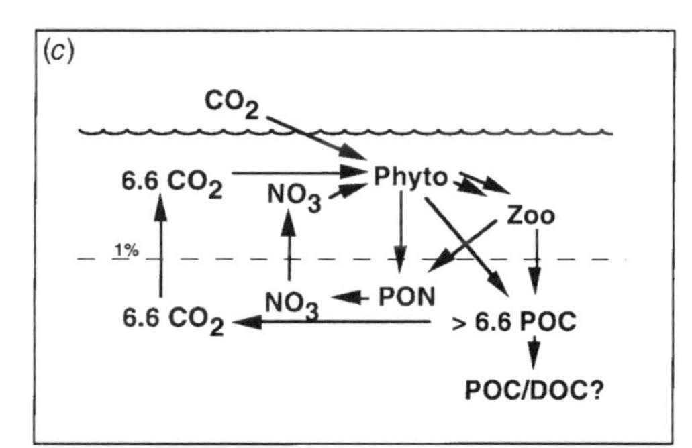

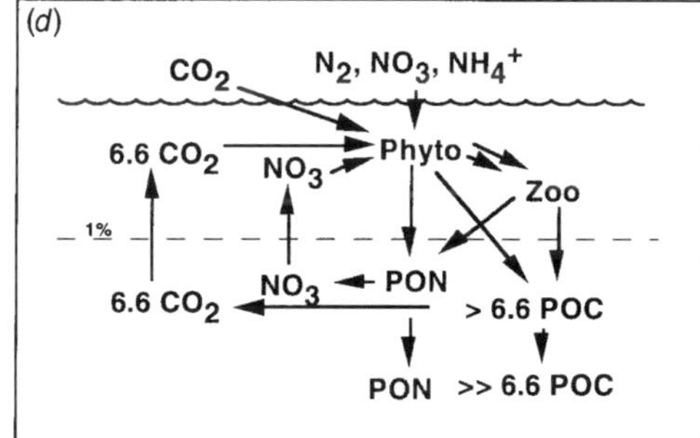

Figure 13.11 *Conceptual diagrams of the nitrogen and carbon cycle. (*a*) Simple nitrogen cycle as often found in numerical models. (*b*) Simple carbon cycle as driven by the nitrate dynamics multiplied by the Redfield C: N ratio of 6.6 (mol: mol). (*c*) Conceptual change in carbon fluxes if the C: N ratio of export exceeds the Redfield ratio. (*d*) Conceptual change in carbon fluxes if the nitrogen cycle includes an air–sea exchange component through nitrogen fixation.*

of a spring bloom or upwelling in highly productive areas. However, the assumption that this behaviour can be abstracted globally has an important and perhaps un-intended consequence.

The simple assumptions in Fig. 13.11*a,b* basically require that biological processes are irrelevant in the global carbon dynamics. In a world that operates with these simple assumptions, all of the DIC that is required to fuel phytoplankton growth is introduced when the nutrients are upwelled or mixed into the euphotic zone. Because the simple N cycle has no gas-exchange component, a simple carbon system that is linearly linked to nitrogen will have no net gas exchange either. Most of the global biogeochemical models of the ocean are represented by these dynamics. Thus, it is not surprising that they have trouble articulating the role of biological processes in the uptake of CO_2 by the ocean. Either these simple assumptions are true and ocean biology is minimally important (in which case we can probably invest our global change effort elsewhere), or we have to study the biology of the ocean from a different perspective.

The important processes for the net role of biology in the air–sea carbon balance will be those deviations from the simple assumptions that allow for a net exchange of carbon with the atmosphere. Many of these have been described, but are not usually the focus of research. If the Redfield ratio assumption is relaxed, the carbon dynamics become more interesting (Fig. 13.11*c*, high C:N export). An increased C:N ratio of export compared with uptake has been described in many studies and is seen at both BATS and HOT. Dissolved organic carbon and nitrogen may be even more out of Redfield balance. If this dynamic is included in the link between the simple nitrogen dynamics and the resultant carbon dynamics, there is a net transfer of carbon from the atmosphere to the deep sea. This basically increases the length scale for the depth of re-mineralisation of carbon compared with nitrogen. On millennial time scales, this enriched deep carbon will ultimately ventilate back to the surface. However, on inter-annual, decadal and centennial time scales, it means that a change in the rate of biological activity (and export in particular) can have a net effect on the atmospheric carbon concentration.

The other obvious modification of these assumptions is the inclusion of a gas phase or air–sea exchange in the nitrogen cycle. This will then require a gas exchange for carbon at some appropriate stoichiometry. Nitrogen deposition and nitrogen fixation are two processes where nitrogen is introduced from the atmosphere or from a dissolved gas phase that is ultimately equilibrated with the atmosphere (Fig. 13.11*d*). On a global scale, N deposition may play a modest role (Galloway *et al.*, 1995). As described above, nitrogen fixation may be very important in the carbon dynamics on the inter-annual scales of climate variability and on the centennial scales of global change.

The key point is that the examination of the simple nitrogen dynamics and rate processes in the upper ocean that come from the basic paradigm (new production, export) may be less important than understanding how the deviations in our assumptions affect the links between elements. One of the most important lessons from the two time-series stations is that an intensive, multi-year study of each of these systems has revealed processes that are at odds with the assumptions of the simplest versions of our basic paradigms. When we include these new processes and changed assumptions, we come up with air–sea carbon dynamics that differ from the patterns in our simple models. Biology may be more important for global processes when it breaks the rules, and we have to look for these unique patterns to understand their impacts.

Conclusion

The two US JGOFS time-series stations have proved a valuable component of the overall JGOFS efforts. They have provided insights into the inter-annual variability in ocean biogeochemistry and into the processes that control this variability. As the two stations pass their tenth year of operation, longer-period patterns are also emerging. The coupling of the detailed observations at these two locations with the longer, but more limited time-series observations at the Hydrostation S (Atlantic) and Climax (Pacific) stations strongly indicate that these ecosystems vary on decadal time scales and provide potential mechanisms for feedbacks between ocean biogeochemistry and climate. JGOFS is scheduled to end within the next few years. The current plans include continuation of both US stations at least through the end of US JGOFS and hopefully for decades to come. The utility of this approach, coupled with the strong commitments to rapid and complete public release of the time-series data, make these stations a resource for the entire oceanographic community that should continue to provide insight and opportunities for a long time to come.

References

Bates, N. R., Michaels, A. F., & Knap, A. H. (1996). Seasonal and interannual variability of oceanic carbon dioxide species at the U.S. JGOFS Bermuda Atlantic Time-series Study (BATS) site. *Deep-Sea Research*, Part II, **43**, 347–83.

Dugdale, R. C. & Goering, J. J. (1967). Uptake of new and regenerated forms of nitrogen in primary productivity. *Limnology and Oceanography*, **12**, 196–206.

Elardo, K. M. (1994). Changes in proteins associated with nitrogen fixation and iron nutrition in the marine cyanobacterium *Trichodesmium*. M.S. Thesis, Portland State University.

Galloway, J. N., Schlesinger, W. H., Levy, H. II, Michaels, A. F. & Schnoor, J. L. (1995). Nitrogen fixation: Anthropogenic enhancement-environmental response. *Global Biogeochemical Cycles*, **9**, 235–52.

Gruber, N. & Sarmiento, J. L. (1996). Global patterns of marine nitrogen fixation and denitrification. *Global Biogeochemical Cycles*, **11**, 235–66.

Hayward, T. L. (1987). The nutrient distribution and primary production in the central North Pacific. *Deep-Sea Research*, **34**, 1593–627.

Hutchins, D. A. (1989). Nitrogen and iron interactions in filamentous cyanobacteria. M.S. Thesis, Portland State University.

Jenkins, W. J. (1988). The use of anthropogenic tritium and helium-3 to study subtropical gyre ventilation and circulation. *Philosophical Transactions of the Royal Society of London*, A **325**, 43–61.

Karl, D. M. & Lukas, R. (1996). The Hawaii Ocean Time-series (HOT) program: Background, rationale and field implementation. *Deep-Sea Research*, Part II, **43**, 129–56.

Karl, D. M. & Tien, G. (1997). Temporal variability in dissolved phosphorus concentrations in the subtropical North Pacific Ocean. *Marine Chemistry*, **56**, 77–96.

Karl, D. M., Letelier, R., Hebel, D., Tupas, L., Dore, J., Christian, J. & Winn, C. (1995). Ecosystem changes in the North Pacific subtropical gyre attributed to the 1991–92 El Niño. *Nature*, **373**, 230–4.

Karl, D., Letelier, R., Tupas, L., Dore, J.,

Christian, J. & Hebel, D. (1997). The role of nitrogen fixation in the biogeochemical cycling in the subtropical North Pacific Ocean. *Nature*, **388**, 533–8.

Karl, D. M. & Michaels, A. F. (eds.) (1996). Topical Studies in Oceanography. *Deep-Sea Research*, Part II, vol. 43 (no. 2–3). Oxford: Elsevier Science.

Michaels, A. F. & Knap, A. H. (1996). Overview of the U.S. JGOFS Bermuda Atlantic Time-series Study and the Hydrostation S program. *Deep-Sea Research*, Part II, **43**, 157–98.

Michaels, A. F., Bates, N. R., Buesseler, K. O., Carlson, C. A. & Knap, A. H. (1994*a*). Carbon system imbalances in the Sargasso Sea. *Nature*, **372**, 537–40.

Michaels, A. F., Knap, A. H., Dow, R. L., Gundersen, K., Johnson, R. J., Sorensen, J., Close, A., Knauer, G. A., Lohrenz, S. E., Asper, V. A., Tuel, M. & Bidigare, R. R. (1994*b*). Seasonal patterns of ocean biogeochemistry at the U.S. JGOFS Bermuda Atlantic Time-series Study site. *Deep-Sea Research*, **41**, 1013–38.

Michaels, A. F., Olson, D., Sarmiento, J., Ammerman, J., Fanning, K., Jahnke, R., Knap, A. H., Lipschultz, F. & Prospero, J. (1996). Inputs, losses and transformations of nitrogen and phosphorus in the pelagic North Atlantic Ocean. *Biogeochemistry*, **35**, 181–226.

Prospero, J. M., Barrett, K., Church, T., Dentener, F., Duce, R. A., Galloway, J. N., Levy, H. II, Moody, J. & Quinn, P. (1996). Atmospheric deposition of nutrients to the North Atlantic Basin. In *Nitrogen Cycling in the North Atlantic Ocean and Its Watersheds*, ed. R. Howarth, pp. 27–74. Dordrecht: Kluwer Academic Publishers.

Rueter, J. G., Hutchins, D. A., Smith, R. W. & Unsworth, N. L. (1992). Iron nutrition of *Trichodesmium*. In *Marine Pelagic Cyanobacteria: Trichodesmium and Other Diazotrophs*, ed. E. J. Carpenter, D. G. Capone & J. G. Rueter, pp. 289–306. Dordrecht: Kluwer Academic Publishers.

Sarmiento, J. L. (1983). A tritium box model of the North Atlantic thermocline. *Journal of Physical Oceanography*, **13**, 1269–74.

Sarmiento, J. L., Thiele, G., Key, R. M. & Moore, W. S. (1990). Oxygen and nitrate new production and remineralization in the North Atlantic subtropical gyre. *Journal of Geophysical Research*, **95(C10)**, 18303–15.

Winn, C. D., Mackenzie, F. T., Carrillo, C. J., Sabine, C. L. & Karl, D. M. (1994). Air-sea carbon dioxide exchange in the North Pacific subtropical gyre: Implications for the global carbon budget. *Global Biogeochemical Cycles*, **8**, 157–63.

PART FOUR

GLOBAL-SCALE ANALYSIS AND INTEGRATION

14 Advances in ecosystem modelling within JGOFS

M. J. R. Fasham and G. T. Evans

Keywords: ocean carbon models, biogeochemical processes, dissolved and particulate organic carbon, North Atlantic Ocean

Introduction

The goals of JGOFS are to understand the biological, physical and chemical processes controlling the carbon cycle within the ocean and to predict how this cycle might alter with global climate change. It is impossible to sample the ocean with sufficient temporal and spatial resolution to quantify all the biogeochemical cycles on a global scale and so these goals can only be achieved with the aid of models (SCOR, 1990). The observational programme must be focused on providing the necessary information about the essential processes involved, so that good models of these processes can be developed (SCOR, 1992). This is the key justification for the rolling programme of world-wide JGOFS process studies, and the JGOFS data sets will become a fruitful source of data for testing marine biogeochemical models in the same way that the GEOSECS data were for marine chemical models.

The North Atlantic Bloom Experiment in 1989 (NABE89) was the first JGOFS process study and so most of the JGOFS models published to date (1996) have attempted to simulate these data by using either mixed-layer or one-dimensional (1–D) models. The published models will be reviewed and then methods for fitting nonlinear models to observations will be discussed. These methods enable us to evaluate the fit of a model to observations in a more objective way than before and to compare the fit of alternative models. As an example, two models will be tested for their ability to model the NABE89 observations at 47°N 20°W. Finally, some observations on future priorities for JGOFS modelling will be made.

A review of NABE89 ecosystem models

The first attempt to model some of the NABE89 results was made by Taylor *et al.* (1991), who developed a phytoplankton–nutrient model in which zooplankton grazing of the phytoplankton was specified by a loss function that varied seasonally with temperature. The model was two-layered, consisting of a mixed layer of varying depth (specified from climatological data) below which was a thermocline layer of constant depth (40 m). The carbon chemistry was modelled with the standard chemical equations, and the effects of biological processes on dissolved inorganic carbon (DIC) and alkalinity were modelled using fixed Redfield ratios. The seasonal cycles of phytoplankton chlorophyll, nitrate, DIC and p_{CO_2} were predicted for the equator and latitudes 35°N, 47°N and 60°N. The model showed reasonably good agreement with the observations at 47°N and 60°N obtained on the UK JGOFS cruise in late May–June (see also Taylor *et al.* (1992)). Taylor & Stephens (1993) later showed that including diurnal changes in the mixed layer due to convective mixing produced a better fit to the observations of surface p_{CO_2} and O_2 obtained during a Lagrangian experiment at 60°N 20°W.

Fasham & Evans (1995, described below) and Taylor *et al.* (1993) have used more complex ecosystem models. The latter model is one of the most complex marine ecosystem models developed to date, consisting of four phytoplankton groups (diatoms, dinoflagellates, phytoflagellates and picoplankton), two zooplankton groups (heteroflagellates and microzooplankton), plus bacteria. The mesozooplankton grazing was parameterised, as in Taylor's previous models, by a seasonally varying loss function. The abiotic compartments were detritus, DOC, nitrate and silicate. The model was run at four latitudes (47°N, 52°N, 56°N and 60°N) and compared with biomass data for phytoplankton and zooplankton groups as well as with nutrient data collected on the UK and US NABE cruises. The model reproduced well the observed phytoplankton succession at 47°N and 60°N and the latitudinal variability observed on the transect between these two sites. Taylor also concluded that 'there is no need to invoke limitation by a trace nutrient, such as iron, to reproduce the events in the NE Atlantic during 1989'.

Ducklow (1994) investigated the ability of the 7-compartment model of Fasham *et al.* (1990) to reproduce the spring bacterial results obtained during NABE at 47°N. He obtained a good fit to bacterial biomass but found that bacterial production was under-estimated by a factor of at least two. This could be due either to inadequacies in the bacterial sub-model or to incorrect values of the conversion factors used to convert production estimates to nitrogen units. Some support for the latter reason was provided by the modelled ratios of bacterial to primary production, which were within the range observed during the cruises. Ducklow also compared the performance of this model with a more complex 9-compartment size-structured model (Ducklow & Fasham, 1991) in fitting the bacterial data obtained at the US JGOFS Bermuda time-series station. There is no obvious spring bacterial bloom at Bermuda and only the size-structured model reproduced this. Ducklow concludes that the compartment of fast-growing protozoan bacterivore is needed to track the increase in bacterial production in spring thereby preventing a bacterial bloom.

One-dimensional ecosystem models of NABE89 have mainly focused on the development of the spring bloom rather than on a whole annual cycle. Marra & Ho (1993) used the Price–Weller–Pinkel thermocline model (Price *et al.*, 1986) and a PZN ecosystem model to simulate the development of the spring bloom during a 2 week period (days 115–128) at 47°N 20°W in 1989. The cell quota model was used to describe phytoplankton growth, and zooplankton grazing was parameterised using the Ivlev expression as modified for adaptive feeding by Mayzaud & Poulet (1978). Model closure consisted of a linear zooplankton mortality term. A reasonably good fit was obtained to the observed temporal change in the temperature, nitrate and chlorophyll distributions within the surface 60 m. Zooplankton biomass data were available only as integrated values over 0–100 m but these agreed well with the equivalent variable calculated from the model.

Stramska & Dickey (1993, 1994) have modelled the development of the 1989 spring bloom at the site of the Marine Light in the Mixed Layer (MLML) bio-optical buoy (59°29'N 20°50'W) using the Mellor–Yamada model of thermocline development (Mellor & Yamada, 1974). Phytoplankton primary production was modelled by using the Kiefer–Mitchell model (Kiefer & Mitchell, 1983), nutrient limitation was not considered (nitrate and silicate levels remained high throughout the period being modelled), and other phytoplankton losses were parameterised by a linear loss term. As pointed out by the authors, such a simple one-component model could not be used to model an annual cycle but, over a period of a few weeks (days 100–162), the model did reproduce the temporal and spatial development of temperature and chlorophyll as measured by the instrumented buoy. One very interesting result of the model was that it reproduced the observed initiation of the bloom at a time before significant temperature stratification had developed. The authors stressed that such a result would not have been predicted by any model that assumed the mixed layer to be uniform and mixing actively at all times.

McGillicuddy *et al.* (1995*a,b*) have used a five-compartment model (phytoplankton, zooplankton, nitrate, ammonium, and 'exportable' nitrogen) to describe the development of the spring bloom at 47°N 20°W during NABE. The first of their papers used a 1-D physical model and they obtained a good fit to the observation time series as long as the mesoscale spatial variability of the initial conditions was taken into account. In their second paper the same ecosystem model was embedded in a 3-D physical model and it was shown that interactions between eddies can produce intense vertical motions. The consequent nutrient fluxes into the euphotic zone could substantially enhance production when surface nutrients have reached limiting concentrations in late spring.

More recently, Doney *et al.* (1996) have used a four-component ecosystem model (including phytoplankton photo-adaptation) in a 1-D non-local turbulent mixing model of the seasonal thermocline to model the BATS time series at Bermuda. The model reproduced the basic features of the seasonal cycle such as mixed-layer depth, nutrients, the spring bloom, and the deep chlorophyll maximum, but was less successful at simulating the seasonal variation of primary production, especially during the summer.

Most of the models described have only attempted to model the spring – early summer period. Although this period had the best data coverage for model testing, the aims of JGOFS can only be realised by models capable of modelling an annual cycle; ultimately we need to be able to predict the annual particle flux to the deep ocean and its relationship to surface primary productivity. We believe that such models will as a minimum require the zooplankton to be modelled explicitly and so some of the models described above would not be suitable for this task.

The use of optimisation methods for fitting models to data

In recent years a number of fairly simple ecosystem models have been developed that attempt to model the annual cycle of biological production in the ocean mixed layer (Evans & Parslow, 1985; Frost, 1987; Fasham *et al.*, 1990; Steele & Henderson, 1992). Such models have been used to simulate the seasonal observations from time-series stations such as Bermuda Station 'S' (Fasham *et al.*, 1990; Steele & Henderson, 1993) and Ocean Weather Stations I (Fasham, 1993) and P (Frost, 1993; Fasham, 1995) with reasonable success. After the initial sense of achievement has worn off, it has become clear to the modelling community that we now have to be more objective in asking such questions as 'does a model give a good fit to the observations?' and 'does model A fit the observations better than model B?' The first of these questions can be approached by first defining a measure of misfit between the model predictions and the observations, and a number of possible measures exist such as sums of squares of deviations or likelihood functions. Although this is a step in the right direction, there is still the problem of deciding whether the value of misfit obtained by a model is good or bad. This complex statistical problem has not yet, to our knowledge, been solved. One possible approach would be to compare the fit with that obtained by some 'random' or 'ecology-free' model that had the same degrees of freedom as the ecosystem model.

Once a measure of misfit has been defined then the second question could in principle be answered by comparing the misfit values for the two different models. However, for this approach to be useful we must have determined the 'best' parameter values for each model, a far from simple task. All these models require a large number of parameters (10–40) to specify the ecological processes, and estimating suitable values for these parameters can be extremely difficult. Some parameters, such as phytoplankton and bacterial growth rates, or zooplankton grazing and excretion rates, can in principle be measured at sea by means of suitable experiments. However, even if this is achieved the modeller often requires some time average of the parameters rather than values for just the few times of the year when cruises took place. Some other parameters, such as natural mortality rates or detrital re-mineralisation are, at present, very difficult to measure at all. The approach of most modellers to this problem has been to use experimentally determined parameter values if available and then adjust some, or all, of the remaining parameters until the model gives good agreement with the observation set. However, with a large parameter set it is often difficult to get the model to give a good fit to all the observations with such a hit-and-miss method. When this happens, one is uncertain if this is due to basic inadequacies in the model, or whether the correct part of the parameter space has been missed. What is obviously needed is a more rational and automated approach and computing capacity is now commonly great enough that standard methods of non-linear optimisation can be applied to estimating this many parameters. Some results using one such technique to compare the fit of two models to the NABE89 data set at 47°N 20°W is described below.

The optimisation method

This method is described more fully in Fasham & Evans (1995) and by G. T. Evans (personal communication). We define misfit as the sum of squared deviations between model predictions and observed values. It was further assumed that the variance increases as the square root of the actual value (a compromise between constant absolute and constant relative variance) and so the misfit measure T_{obs} was defined as

$$T_{\mathrm{obs}} = \Sigma(\sqrt{X_{\mathrm{obs}_i}} - \sqrt{X_{\mathrm{mod}_i}})^2,$$

where X_{obs_i} is the ith observation and X_{mod_i} is the modelled value of that observation type at the time of the ith observation. Note that each observation type (e.g. phytoplankton chlorophyll, primary production) should be linked with a modelled state variable or inter-compartment flow. However, it is not necessary to have observations for all the state variables, although the ability to validate the model would be improved if this were the case.

We also have prior ideas about what the parameter values should be: both intrinsic bounds (fractions lie between 0 and 1; most physiological rate parameters must be non-negative) and a target value that we would choose in the absence of data, but ignore if the data strongly suggest another value. These ideas are incorporated in a function depending on three terms: T, the suggested or target value of the parameter, and U and L, its upper and lower bounds. If p is a trial parameter value with weight w, then the function is defined as

$$\begin{aligned} P(p) &= w(T-p)/(p-L) \qquad L < p \leq T \\ &= w(p-T)/(U-p) \qquad T \leq p < U. \end{aligned}$$

The significance of the weight w is that if the trial parameter is half-way between the target and the bound (lower or upper) then this has the same contribution to the total misfit as one model–observation pair for which $\sqrt{\text{model}} - \sqrt{\text{observation}} = w$. A value of $w = 0.1$ was chosen for most parameters, or 0.03 for parameters for which little *a priori* information was available. The quantity $T_{par} = \Sigma(P(p_k))^2$ is added to T_{obs} to give a combined penalty function for the minimisation algorithm.

Once we have chosen a function to minimise, wandering through parameter space in search of a minimum is a standard, if lengthy, procedure. The only small complication is that we do not use methods that rely on gradients of the function. We have found numerical differentiation of numerical solutions of systems of differential equations, where automatic step size control changes the points at which the solution is evaluated, to be unreliable. We therefore used Powell's conjugate direction method (Press *et al.*, 1992), which does not require gradients.

Fitting different models to the NABE89 observations

The observation set

The techniques described above were applied to an observation set derived from measurements made during the 1989 JGOFS North Atlantic Bloom Experiment at 47°N 20°W (Ducklow & Harris, 1993; Lochte *et al.*, 1993). Most of the data were collected between 24 April and 31 May on two US cruises on *Atlantis II* and one German cruise on *Meteor*. The *Meteor* cruise covered the period while *Atlantis II* was in port, thus ensuring a complete coverage of the spring bloom period. Some data were also obtained during July from British cruises on *Discovery*, and during August from the Netherlands cruise on *Tyro*. To provide data for comparison with the model, all values obtained on a given day within the mixed layer were averaged. There were many observations of phytoplankton chlorophyll, primary production and nitrate but fewer of ammonium, bacteria and microzooplankton (Fig. 14.1). All data have been converted to units of mmol N m^{-3} (or mmol N m^{-3} d^{-1} for primary production) using standard conversion constants (Fasham & Evans, 1995). During the spring bloom the nitrate concentration declined from 7 mmol m^{-3} to less than 0.5 mmol m^{-3}, whereas the phytoplankton concentrations increased from less than 0.5 mmol m^{-3} to peak values of *c.* 3 mmol m^{-3} (Fig. 14.1*a,b*). During July and August nitrate concentrations remained low and phytoplankton concentrations declined, reaching values less than 0.5 mmol m^{-3}. The data set for primary production was more limited (Fig. 14.1*f*). Lochte *et al.* (1993) suggested that the UK and German ^{14}C incubations during the spring bloom period may have been affected by metal contamination and so to ensure internal consistency only the US data have been used for this period. For the July–August period ^{14}C measurements made on *Discovery* and *Tyro* were used.

A good set of bacterial count data was obtained on the two *Atlantis* cruises (Fig. 14.1*d*) and showed a strong increase in biomass associated with the spring bloom. Very few ammonium measurements were made and only observations made on *Atlantis II* have been used (Fig. 14.1*e*).

According to Burkill *et al.* (1993) and Verity *et al.* (1993), most of the grazing of phytoplankton, at least during the spring bloom, was due to microzooplankton. Obtaining good estimates of microzooplankton biomass is time-consuming and only six estimates were available from *Atlantis* cruise 119–5 (Verity *et al.*, 1993) and one each from *Discovery* cruises 183 and 184 (Fig. 14.1*c*). Note that, unlike the values used in Fasham & Evans (1995), the US microzooplankton biomass values now include estimates of the biomass of copepod nauplii.

A closer analysis of the spring bloom shows that it consists of two separate peaks (Fig. 14.2*a*). As the *Meteor* and *Atlantis* were not occupying exactly the same position during their sampling, this double peak could perhaps be attributed to spatial variability. However, the nitrate data showed a good continuity between all three cruises, suggesting that the double phytoplankton peak is a real temporal event, although there was an increase in nitrate concentration between days 140 and 141 that could have been due to mesoscale variability. The silicate was almost totally utilised during the period of the first peak suggesting that this bloom consisted mainly of diatoms (Fig. 14.2*b,c*). This is supported by phytoplankton species counts made on the *Meteor* cruise (Lochte *et al.*, 1993). The main contribution to the second chlorophyll peak came from small flagellates (Sieracki *et al.*, 1993; Taylor *et al.*, 1993).

The resulting data set represents a combination of temporal changes and the effects of horizontal spatial variability. The spatial distribution of eddies in the region was well mapped (Robinson *et al.*, 1993) and biological variability associated with these eddies was observed (Lochte *et al.*, 1993; McGillicuddy *et al.*, 1995*a*). However, for the purposes of this analysis any effect of spatial

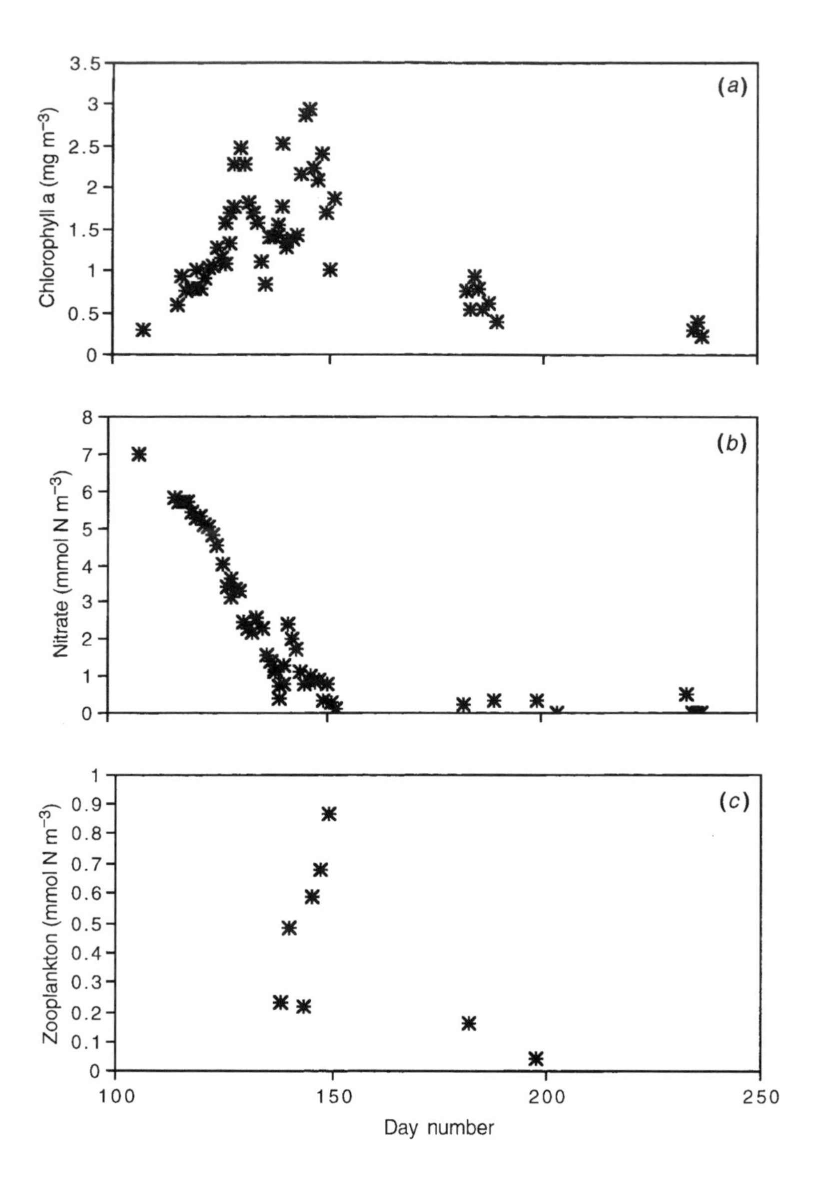

Figure 14.1 *The mixed layer average concentration of various quantities observed during the 1989 North Atlantic Bloom Experiment at 47°N 20°W.* (a) *Phytoplankton chlorophyll;* (b) *nitrate;* (c) *microzooplankton biomass;* (d) *heterotrophic bacterial biomass;* (e) *ammonium; (f) primary production. Units are mmol N m^{-3} except phytoplankton chlorophyll (mg m^{-3}) and primary production (mmol N m^{-1} d^{-1}).*

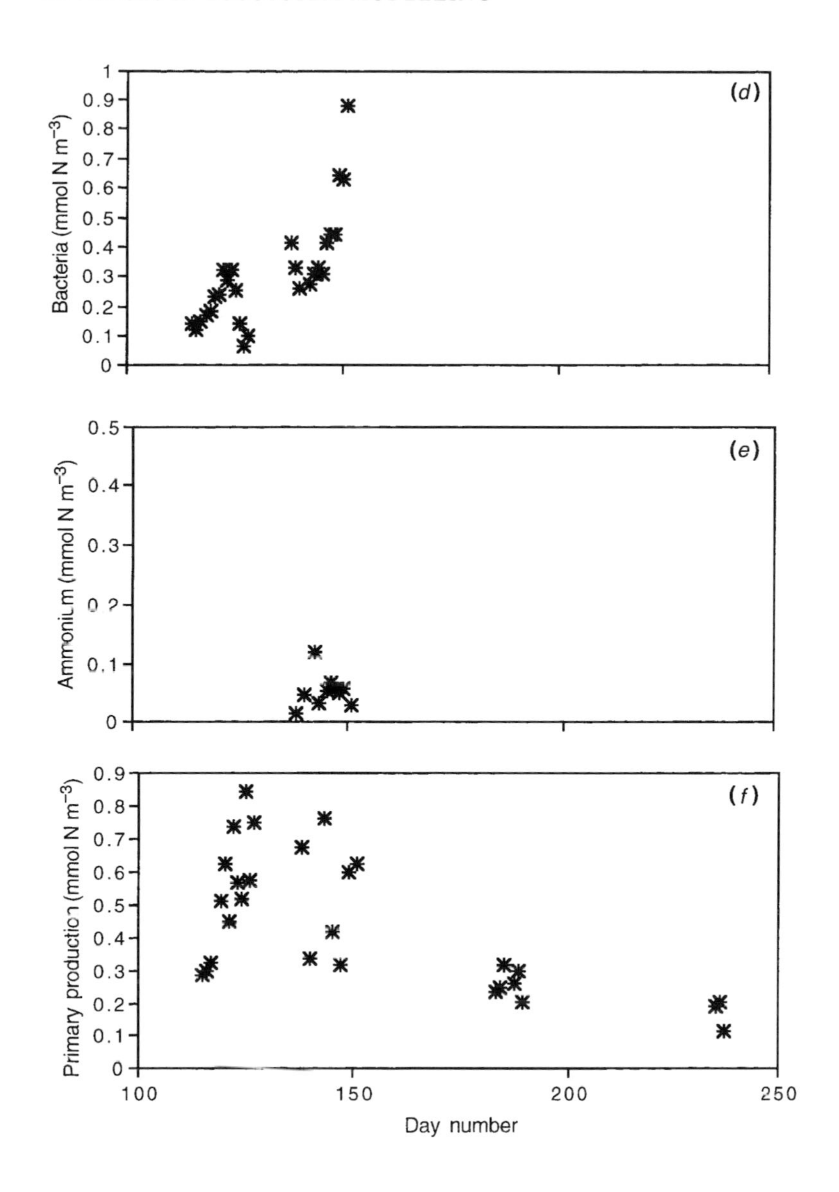
(d)
Bacteria (mmol N m^{-3})
0
0.1
0.2
0.3
0.4
0.5
0.6
0.7
0.8
0.9
1
(e)
Ammonium (mmol N m^{-3})
0
0.1
0.2
0.3
0.4
0.5
(f)
Primary production (mmol N m^{-3})
0
0.1
0.2
0.3
0.4
0.5
0.6
0.7
0.8
0.9
100
150
200
250
Day number

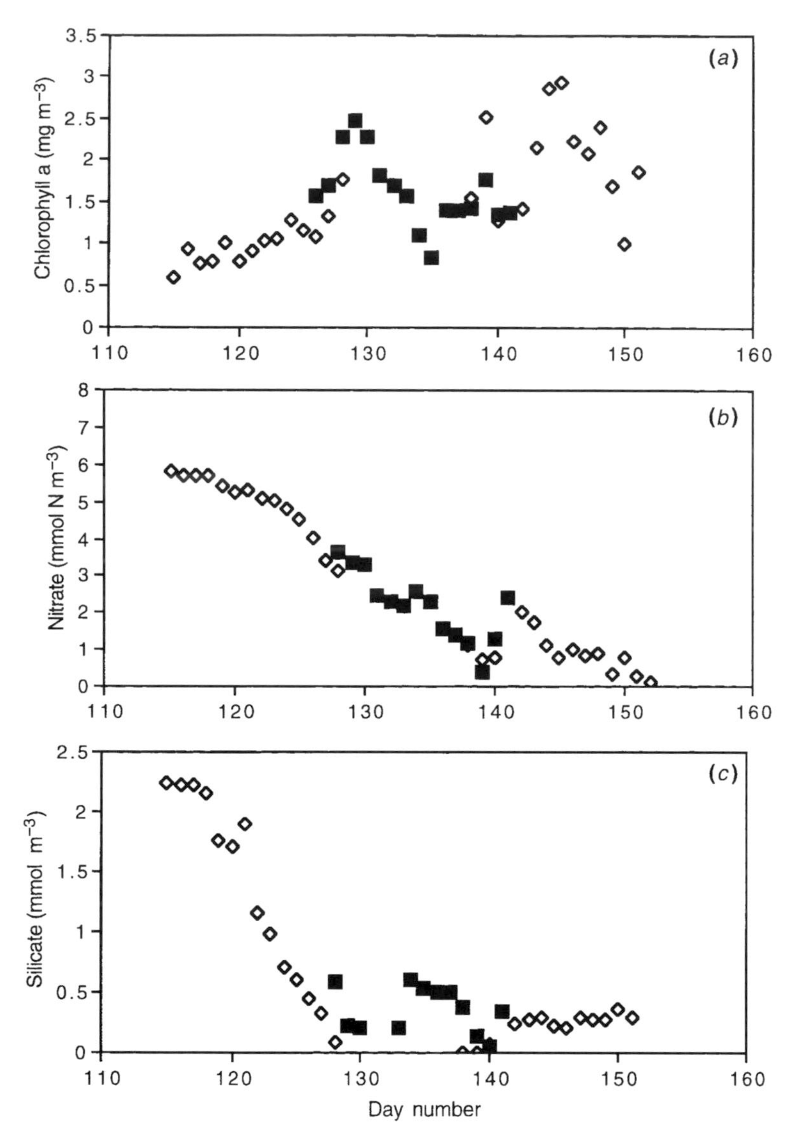

Figure 14.2 *Phytoplankton chlorophyll (*a*), nitrate (*b*) and silicate (*c*) observations made during the spring bloom period at 47°N 20°W (squares are* Meteor *observations and diamonds are* Atlantis II *observations).*

variability was treated as noise superimposed on a strong seasonal cycle representing the average seasonal change in the ecosystem dynamics.

The ecosystem models

We investigated two ecosystem models, both based on the mixed-layer nitrogen model of Fasham, Ducklow & McKelvie (1990; hereafter referred to as FDM).

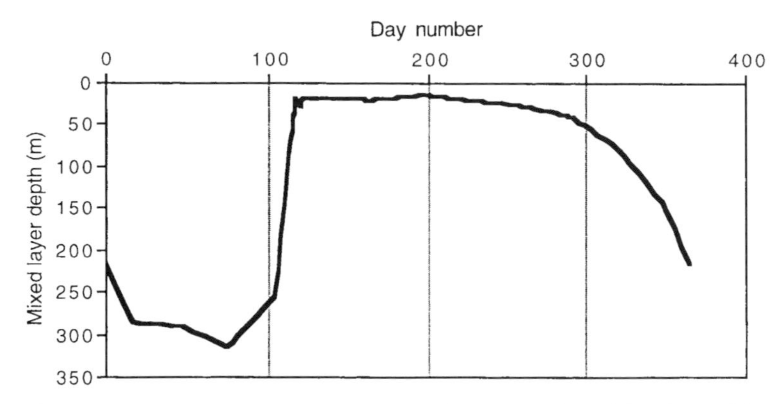

Figure 14.3 *The seasonal cycle of mixed layer depth used for the model runs.*

The equations governing the FDM model are given in the original paper and only a brief summary of the model structure will be given here.

Seasonal changes in mixed-layer depth are not explicitly modelled but are assumed to be known from observations. At 47°N 20°W climatic mixed-layer depths (Levitus, 1982) were supplemented by ship's observation for the period of the spring thermocline development (Fig. 14.3). The deepening of the mixed layer throughout the late autumn and winter entrains nitrate from below the mixed layer that fuels the new primary production. The amount of nitrate entrained, and therefore the nitrate concentration at the start of the spring bloom, depends on the assumed vertical nitrate gradient below the mixed layer (this is specified by the equation $N = a + bz$, where z is depth and a, b are parameters of the model) . Mixing across the pycnocline can also be effected by such processes as breaking internal waves, diurnal convective mixing, intermittent storm events; these processes are all parameterised by a constant mixing velocity. The other physical forcing function is the solar radiation, parameterised by the standard astronomical formulae (Brock, 1981), and a model for the atmospheric transmittance of clouds (Evans & Parslow, 1985) using a constant fractional cloudiness. Light transmittance through the water column was modelled by Beer's law with a water attenuation coefficient and phytoplankton self-shading coefficient.

The two ecosystem models used were as follows.

*Model-1 (Fig. 14.4*a*)*

This is the standard FDM model having the following compartments: phytoplankton, zooplankton, bacteria, labile dissolved organic nitrogen (DON), ammonium, nitrate and detritus. The zooplankton in this model are the sort that obey a differential equation, whose grazing is immediately used to produce more zooplankton, typically protozoa. The microzooplankton of the

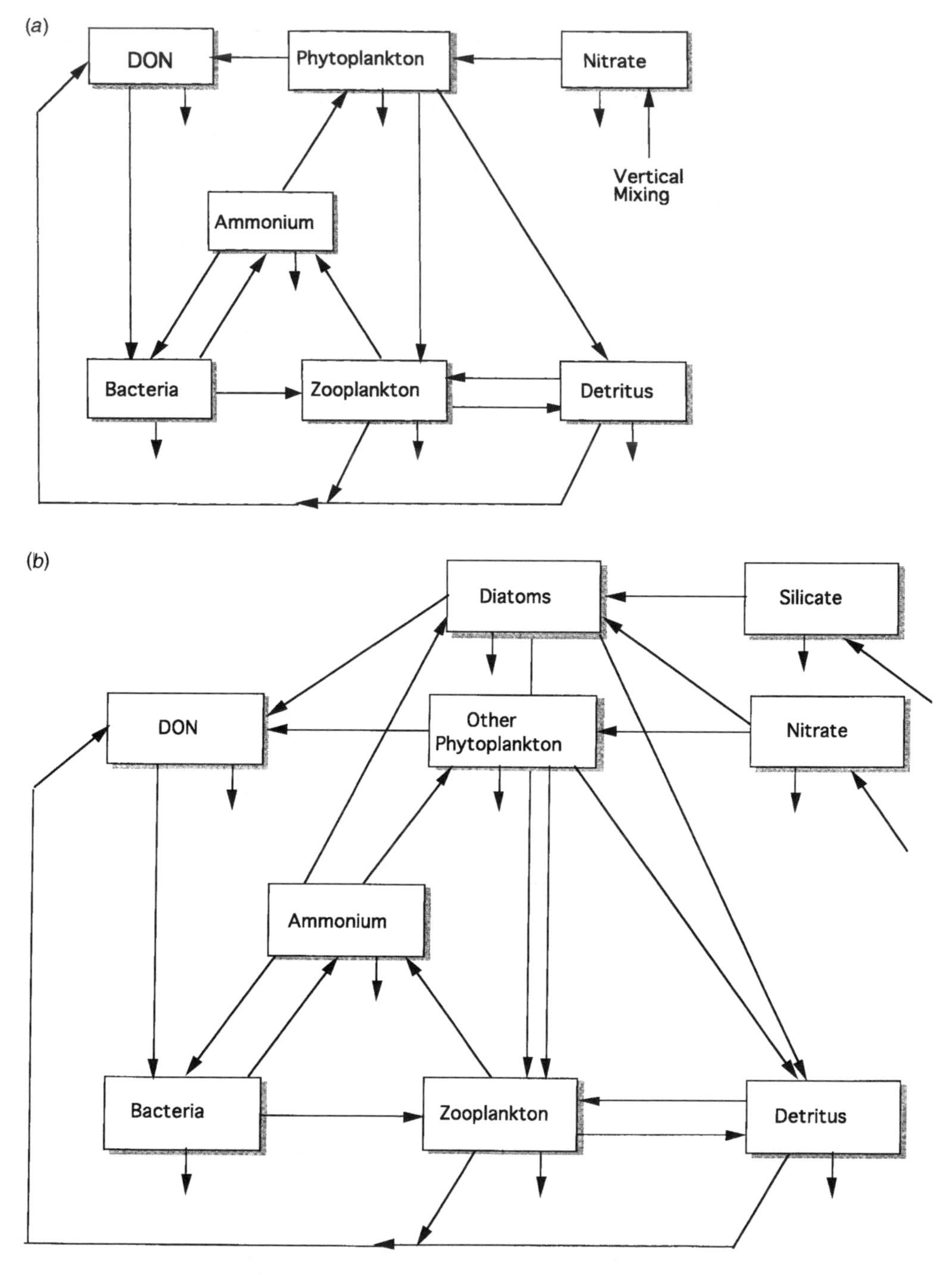

Figure 14.4 *Model flow diagrams for (a) Model-1 and (b) Model-2.*

observations include other organisms such as copepod nauplii, which can have a similar effect as far as grazing phytoplankton is concerned, but whose own growth may in fact follow different rules. There are two differences from the published FDM model. Firstly, the zooplankton mortality is parameterised by a quadratic function of zooplankton biomass rather than by a linear function (see below for details); and secondly, the sub-mixed layer nitrate concentration is treated as a linear function of depth. The number of model parameters was 29.

*Model-2 (Fig. 14.4*b*)*

In this model the phytoplankton have been split into diatoms and non-diatom phytoplankton and an extra nutrient, silicate, has been added. The number of model parameters was 40.

The functional relationship between phytoplankton growth rate and light and the interaction between ammonium and nitrate uptake was described in FDM. In model-2 the diatom growth rate σ was assumed to be jointly limited by nitrogen and silicate in the form of a multiplicative Michaelis–Menten function thus:

$$\sigma = J((N_n e^{-\psi \mathrm{dNr}})/(k_{1\mathrm{d}} + N_n) + N_r/(\mathrm{k}_{2\mathrm{d}} + N_r))(S/(k_S + S)),$$

where J is the nutrient-saturated growth rate averaged over the day and the mixed layer (Evans & Parslow, 1985), and N_n, N_r, and S are the concentrations of nitrate, ammonium and silicate respectively. Other symbols are defined in Table 14.1. Silicate uptake by diatoms was assumed to be in constant ratio to the total nitrogen uptake (ammonium plus nitrate); the value of this ratio was determined by the optimisation. Both types of phytoplankton could exude a fixed fraction of their primary production to labile DON.

Both models have only one zooplankton compartment that feeds on phytoplankton, bacteria and detritus (and in model-2, two types of phytoplankton). Feeding preferences for the prey of zooplankton refer to the case of equal prey concentrations. In general zooplankton ingest relatively more of abundant prey and this was parameterised by using a non-linear switching term that also acts as a stabilising property of the model equations (FDM). Losses due to faecal pellet egestion were parameterised by an assimilation efficiency, and excretion was assumed to be a linear function of biomass.

The parameterisation of the zooplankton mortality is not trivial, as it represents the effect of unmodelled higher predators. Steele & Henderson (1992) have shown that the mathematical form of this closure term can have a large influence on the dynamics and they favoured a mortality function that was a quadratic function of zooplankton biomass. Recently, Fasham (1995) has provided further support for the quadratic form and it was used in both models.

Table 14.1. *The model parameter sets: definitions with the values obtained from the two model solutions*

The symbols apply to model-1 as described in Fasham *et al.* (1990). For model-2, the term phytoplankton should be interpreted as non-diatom phytoplankton.

Parameter	Symbol	Units	Model-1	Model-2
Fractional cloudiness	C	—	0.75	0.75
Cross-thermocline mixing rate	m	$m\,d^{-1}$	0.1	0.1
Surface nitrate concentration	N_s	$mmol\,m^{-3}$	4.276	4.276
Nitrate depth gradient	N_g	$mmol\,m^{-4}$	0.0265	0.0265
Water attenuation coefficient	k_w	m^{-1}	0.05	0.05
Phytoplankton self-shading coefficient	k_c	$m^2\,(mmol\,N)^{-1}$	0.0853	0.068
Phytoplankton maximum growth rate	V_p	d^{-1}	2.02	1.27
Diatom maximum growth rate	—	d^{-1}	—	0.88
Initial slope of *P–I* curve	α	$(W\,m^2)^{-1}\,d^{-1}$	0.164	0.222
Initial slope of *P–I* curve (diatoms)	—	$(W\,m^2)^{-1}\,d^{-1}$	—	0.157
Half-saturation constant for nitrate uptake	k_1	$mmol\,m^{-3}$	0.83	0.80
Half-saturation constant for nitrate uptake (diatoms)	k_{1d}	$mmol\,m^{-3}$	—	0.28
Half-saturation constant for ammonium uptake	k_2	$mmol\,m^{-3}$	0.73	0.75
Half-saturation constant for ammonium uptake (diatoms)	k_{2d}	$mmol\,m^{-3}$	—	0.74
Nitrate uptake ammonium inhibition parameter	ψ	$mmol\,N\,m^{-3})^{-1}$	3.53	3.42
Nitrate uptake ammonium inhibition parameter (diatoms)	ψ_d	$mmol\,N\,m^{-3})^{-1}$	—	1.49
Half-saturation constant for silicate limitation (diatoms)	k_s	$mmol\,N\,m^{-3}$	—	0.83
Phytoplankton-specific mortality rate	μ_1	d^{-1}	0.044	0.027

Diatom-specific mortality rate	—	d^{-1}	—	0.035
Phytoplankton LDON exudation fraction	γ	—	0.13	0.18
Diatom LDON exudation fraction	—	—	—	0.03
Ratio of silicate uptake to ($NO_3 + NH_4$) uptake	—	—	—	0.88
Zooplankton maximum ingestion rate	g	d^{-1}	1.27	1.04
Zooplankton ingestion half-saturation constant	k_3	$mmol\ m^{-3}$	1.03	0.99
Zooplankton assimilation efficiency	β	—	0.82	0.81
Zooplankton excretion rate	μ_2	d^{-1}	0.13	0.13
Fraction of zooplankton excretion going to ammonium	ε	—	0.87	0.85
Zooplankton mortality parameter	μ_5	$(mmol\ m^{-3}\ d)^{-1}$	0.41	0.25
Fraction of zooplankton losses going export	Ω	—	0.16	0.31
Zooplankton feeding preference for phytoplankton	p_1	—	0.19	0.29
Zooplankton feeding preference for bacteria	p_2	—	0.47	0.59
Zooplankton feeding preference for detritus	p_3	—	0.34	0.06
Zooplankton feeding preference for diatoms	—	—	—	0.06
Bacterial maximum uptake rate	V_b	d^{-1}	2.06	0.54
Bacterial excretion rate	μ_3	d^{-1}	0.22	0.02
Bacterial half-saturation constant for uptake	k_4	$mmol\ m^{-3}$	0.89	1.35
Ratio of NH_4 uptake : LDON uptake for bacteria	η	—	0.16	0.21
Detrital sinking rate	V	$m\ d^{-1}$	4.47	3.55
Detrital breakdown rate	μ_4	d^{-1}	0.015	0.049
Surface silicate concentration	—	$mmol\ m^{-3}$	—	3.97
Silicate depth gradient	—	$mmol\ m^{-4}$	—	0.006

A fraction of the zooplankton mortality flux is exported as fast-sinking detritus from the mixed layer; the remainder is assumed to be recycled to ammonium within the mixed layer.

The bacteria are modelled exactly as in FDM: the growth is determined by a Michaelis–Menten function of DON and then ammonium uptake is calculated as a constant ratio of DON uptake. Bacteria can also excrete ammonium, making it possible for them to act as either net users or re-mineralisers of nitrogen.

In both models, detritus is derived from faecal pellets and dead phytoplankton. However, detritus can also be recycled within the mixed layer by two mechanisms, namely its re-ingestion by zooplankton and its breakdown into DON and subsequent uptake by bacteria. The fate of the ungrazed detritus is determined by two parameters, the detrital sinking rate and the breakdown rate of detritus to DON.

In order to make a fair comparison between the two models, the three physically determined parameters (cloudiness, cross thermocline mixing rate, light attenuation coefficient due to water) were fixed (see Table 14.1). In addition, the sub-mixed layer nitrate gradient determined by using the model-1 optimisation was used for model-2. The target parameter values were taken from FDM. The model equations were solved with a variable time-step 4.5 order Runge–Kutta algorithm (Press *et al.*, 1992) and were run for 2 years to achieve a repeating annual cycle. The results from the third year of the simulation were used to optimise the parameters.

Fit of the models to the observation set

Model-1

The optimal fit of model-1 to the observations is shown in Fig. 14.5 (solid lines) and the optimised parameter values are given in Table 14.1. The parameter values and model simulations were very similar to those obtained previously by Fasham & Evans (1995; model-1).

The nitrate observations were modelled very successfully (Fig. 14.5*b*, thick line). This is not surprising given that the spring nitrate decline was so well covered by the observations and that the nitrate values have the highest magnitude range and so the greatest reduction in the misfit can be achieved by fitting these observations closely. The fact that the optimisation method usually gives good fits to the nitrate observations is very satisfactory as it ensures that there is a good temporal match between model and observations for the main factor controlling the spring bloom. However, the model over-estimated the few ammonium observations (Fig. 14.5*e*).

The initial spring increase in phytoplankton is modelled well (Fig. 14.5*a*) but the double peak structure is not; the optimisation averages between the two blooms and the inter-bloom minimum to give a lower modelled peak values at about the time of this minimum. This failing might not seem surprising from a model that has just one phytoplankton compartment, but such models have been shown to produce oscillatory solutions with two or more peaks at a latitude of 60°N (Fasham, 1993). The fact that the optimisation did not produce such a solution for the lower latitude of 47°N is in agreement with the theoretical predictions of Ryabchenko *et al.* (1997).

The initial bacterial bloom is also well modelled but the later bloom values are under-estimated (Fig. 14.5*d*). As a large fraction of the nitrogen for bacterial growth comes from phytoplankton exudation, then this under-estimation is consistent with the under-estimation of the peak values of the phytoplankton bloom.

Because of the limited data set, it is difficult to say whether the model gives a good representation of microzooplankton or not (Fig. 14.5*c*). However, it can be said that the inclusion of the copepod nauplii biomass into the spring microzooplankton biomass means that the modelled biomass is closer to the observations than when these nauplii values were excluded (cf. the model-1 results in Fasham & Evans, 1995).

The predicted primary production values are well within the range of observations during the spring bloom, although the July values are slightly under-estimated. It should be remembered that some of the day-to-day variability in primary production is due to daily variations in cloud cover (Martin *et al.*, 1993) that the model, with its fixed annual cloudiness, could not reproduce. It is therefore better to compare the primary production averaged over a given time. Martin *et al.* (1993) estimated that the average primary production within the top 35 m between 24 April and 1 June was 86 mmol $C\,m^{-2}\,d^{-1}$ or 13 mmol $N\,m^{-2}\,d^{-1}$. The modelled production within the mixed layer for the same period was 12.6 mmol $N\,m^{-2}\,d^{-1}$, an excellent agreement with the data. However, this comparison is not exact because the model mixed-layer depth was shallower than 35 m for most of this period (Fig. 14.3). The modelled *f* ratios (Fig. 14.5*g*) declined from a winter value of *c.* 0.7 to a minimum of *c.* 0.1 around day 165. After day 200, the *f* ratio increases again, owing to upward mixing of nitrate as the mixed layer deepens (Fig. 14.3).

Model-2

In order to determine a parameter set for model-2, additional observations of silicate concentration (Fig. 14.6*b*) were used and the phytoplankton chlorophyll observations were partitioned between diatoms and other phytoplankton. This was done by calculating a linear relation between the fraction of phytoplankton biomass attributed to diatoms and time using the observations made by M.

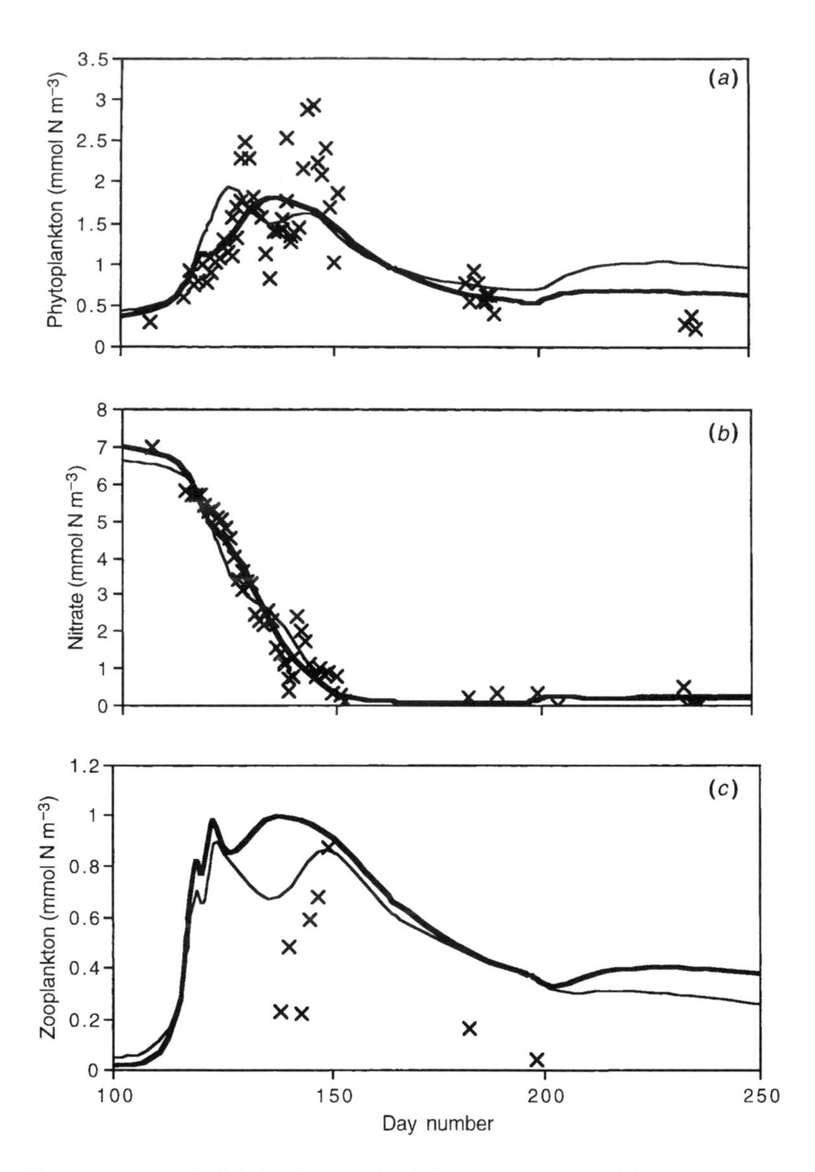

Figure 14.5 *Model simulations (model-1: thick line, model-2: thin line) compared with observations (crosses) for:* (a) *phytoplankton,* (b) *nitrate,* (c) *microzooplankton,* (d) *bacteria,* (e) *ammonium,* (f) *total primary production,* (g) *f ratio.*

Decker on the *Meteor* cruise (as shown in Fig. 8c of Lochte *et al.*, 1993); an extrapolation of this equation was used for the two *Atlantis II* cruises.

The optimised fits of the model to the data are shown in Figs. 14.5 (thin lines) and 14.6 and model parameter values are given in Table 14.1. The model-2 parameter values for zooplankton and non-diatom phytoplankton are generally similar to those for model-1, except for phytoplankton maximum growth rate

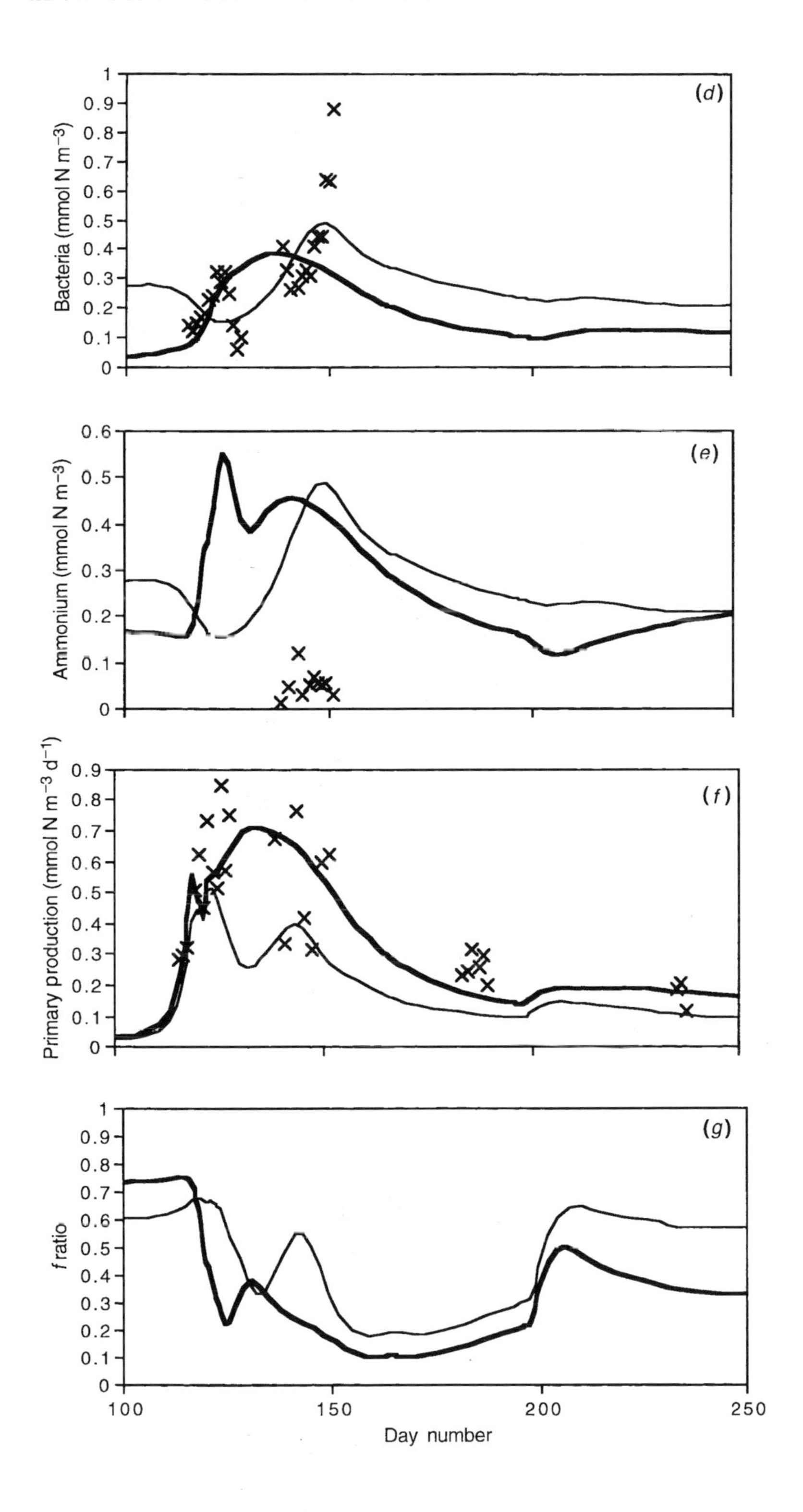

1
0.9
0.8
0.7
0.6
0.5
0.4
0.3
0.2
0.1
0
Bacteria (mmol N m^{-3})
(d)
0.6
0.5
0.4
0.3
0.2
0.1
0
Ammonium (mmol N m^{-3})
(e)
0.9
0.8
0.7
0.6
0.5
0.4
0.3
0.2
0.1
0
Primary production (mmol N m^{-3} d^{-1})
(f)
1
0.9
0.8
0.7
0.6
0.5
0.4
0.3
0.2
0.1
0
f ratio
(g)
100
150
200
250
Day number

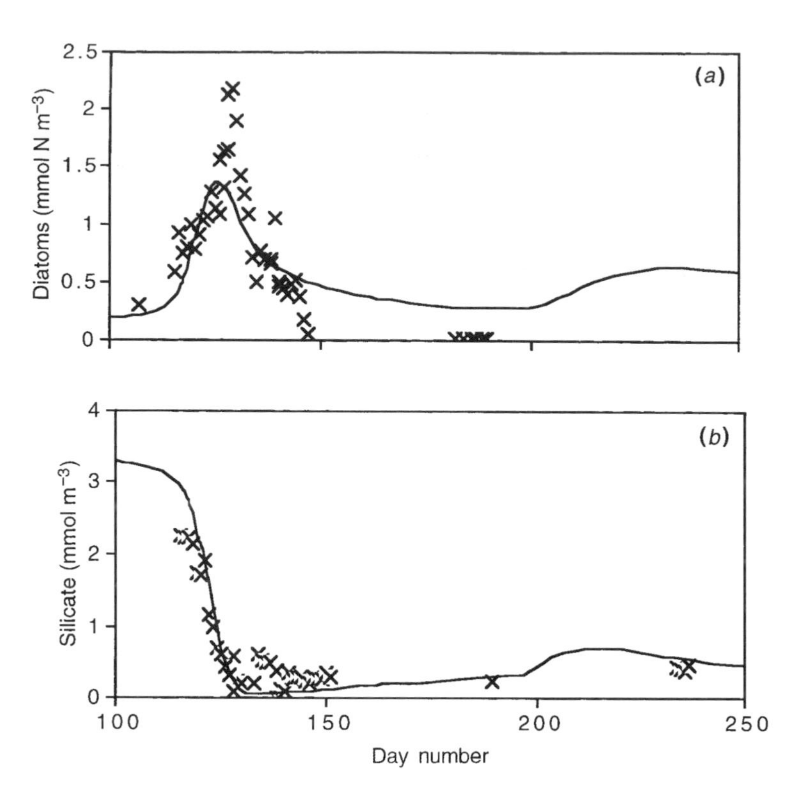

Figure 14.6 *Model-2 simulations (lines) compared with observations (crosses) for:* (a) *diatoms,* (b) *silicate.*

(V_p), which has a much lower value in model-2. The bacterial parameters for the two models are quite different and the parameters for diatoms show some significant differences from those for non-diatoms. The half-saturation constant for nitrate and the ammonium inhibition parameter are much lower for diatoms, giving them a higher affinity for nitrate. Note also that the maximum growth rate of the diatoms is also low compared with the model-1 value for the undifferentiated phytoplankton.

It had been hoped that the inclusion of two phytoplankton compartments would enable the double peak of phytoplankton observations to be simulated. Although this was found to be the case (Fig. 14.5*a*), the fit to the observations was not very good and the peak values were still under-estimated. The first of these peaks was due to diatoms (Fig. 14.6*a*) and the model gave a good fit to the silicate observations (Fig. 14.6*b*). The modelled zooplankton and ammonium were similar to model-1 (Figs. 14.5*c* and 14.5*e*). However, as might be expected from the differences in bacterial parameters, the simulated bacteria blooms for the two models were very different (Fig. 14.5*d*).

The model-2 primary production (Fig. 14.5*f*) showed a double peak structure but the values were lower than for model-1; the average primary production for the period 24 April – 1 June was only 60% of the model-1 estimate. Part of the problem was the lower V_p values already referred to, but a comparison of the f ratios between the two models shows that there was also less regenerated production during the model-2 spring bloom (Fig. 14.5*g*).

Comparison of the model-1 results with other observations

A model optimised to a given data set might of course be ecologically unrealistic in other ways. Indeed, bearing in mind that we are using extremely simplified models to simulate a highly complex multi-species ecosystem, such an outcome is quite possible. One check on this problem is to compare various stocks and flux observations, which were not used for optimisation, with equivalent values derived from the model. It would obviously be impossible for the models used here to simulate accurately all the measurements made during the JGOFS surveys, but if the key standing stocks and fluxes of relevance to the nitrogen cycle can be simulated then the JGOFS goal of modelling the global biological pump will have been considerably advanced. Because the values of model-1 primary production showed better agreement with the observations than model-2, the analysis will be restricted to model-1.

A comparison of modelled and observed particulate organic nitrogen concentration (PON) has especial value as it does not depend on any assumptions about conversion factors from chlorophyll or bacterial cell numbers to nitrogen units, and also includes detritus for which we have no independent observations. Two methods were used to calculate model PON. The first summed phytoplankton, bacteria and detritus concentrations and the second also included microzooplankton, thereby giving lower and upper bounds for PON. The observed PON values (Fig. 14.7) mainly fall below the lower modelled PON values during the spring bloom, but even so the predicted concentrations of organic matter are within 20–25% of the observed values. There is thus no indication of gross errors in the factors used to convert biomass to nitrogen units.

We have seen that, for both models, the simulated microzooplankton biomass was higher than the observed values. Another independent check on the validity of the zooplankton component would have been to compare modelled and observed zooplankton grazing rates. Microzooplankton grazing rates were measured in 1989 using the dilution technique and fractionation experiments by Verity *et al.* (1993), and using the dilution technique alone by Burkill *et al.* (1993). Such experiments give estimates of the phytoplankton specific growth and grazing rates. However, there may be some problems in comparing these values with the model results because the dilution experiments apparently overestimated the natural *in situ* phytoplankton growth rates. For example, the

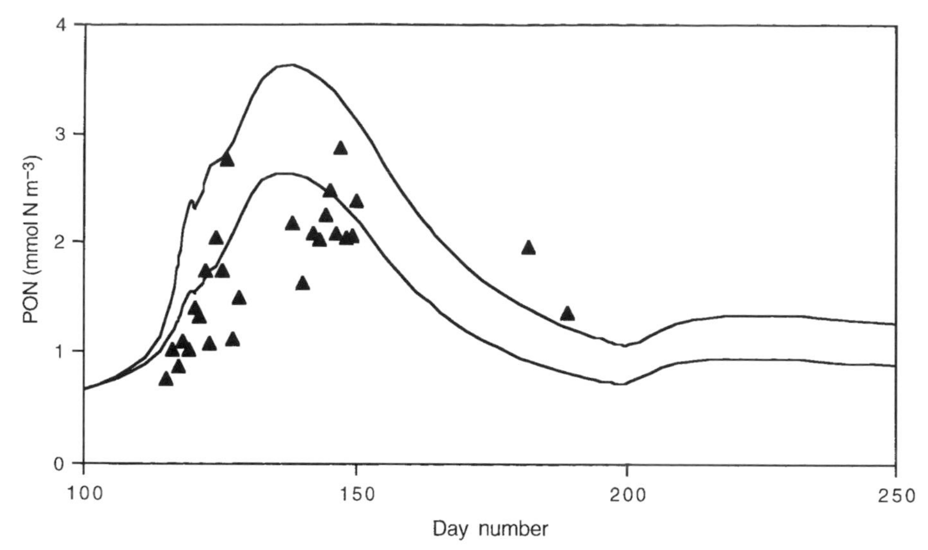

Figure 14.7 *Comparison of observed PON values with the model-1 results. The lower line is the sum of modelled bacteria, phytoplankton and detritus concentrations. The upper line also includes the zooplankton biomass.*

average phytoplankton growth rate between days 138 and 149 calculated from the experiments of Verity *et al.* (1993) was $0.69 \pm 0.20\ d^{-1}$, whereas the growth rate calculated by dividing observed primary production by phytoplankton biomass was $0.26 \pm 0.10\ d^{-1}$. The average modelled growth rate over the same period was $0.38 \pm 0.01\ d^{-1}$, in reasonable agreement with the latter observations.

One way round this problem is to compare the observed and modelled estimates of the fraction of phytoplankton production grazed by the zooplankton. The average experimentally measured values of this quantity was 79% between days 138 and 149 (Verity *et al.*, 1993) and *c.* 100% between days 182 and 189 (Burkill *et al.*, 1993). The predicted mean values were 77% for the first period and 73% for the second (Fig. 14.8*a*). Using this limited data, the model grazing rates are reasonable, although more observations will be needed before we can feel confident with the model predictions.

An interesting feature of the model results is that, during the winter, zooplankton grazing accounts for only a small proportion of the primary production (Fig. 14.8), implying that natural mortality must be the main loss term for the phytoplankton. Other predictions are that the largest food source for the microzooplankton in the winter is detritus and that bacteria contribute less than 10% of the food supply during the post-bloom and summer period (Fig. 14.8*b*).

During the period that bacterial production estimates were made (*Atlantis II* leg 2), the simulated bacterial production, both total (Fig. 14.9) and per unit bacterial biomass, was less than the observed values. In order to convert these

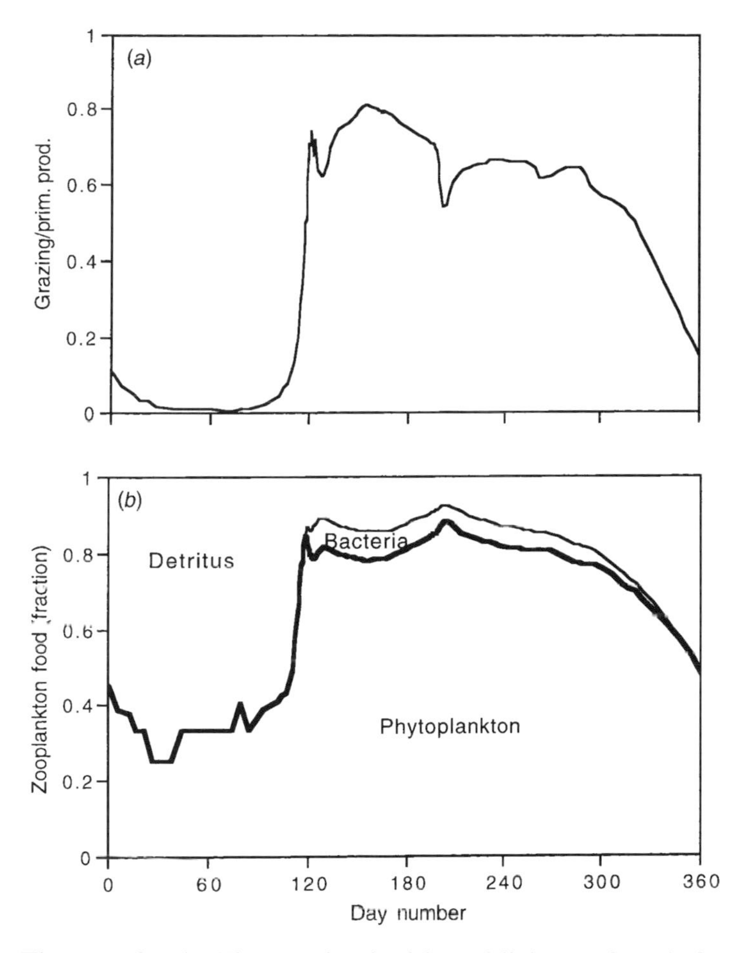

Figure 14.8 (a) *The annual cycle of the modelled ratio of zooplankton grazing to primary production.* (b) *The annual cycle of the fraction of zooplankton food derived from phytoplankton, bacteria and detritus.*

observations to nitrogen units various average conversion factors had to be assumed (Ducklow *et al.*, 1993) and the observations could be brought into better agreement with the model results by using the lowest measured value of the ratio of cells produced to the uptake of tritiated thymidine, rather than the average value. During the period from day 138 to 151, the observed mean growth rate (using the mean conversion factors) was $0.32 \pm 0.22\ d^{-1}$ whereas the modelled value was $0.13 \pm 0.01\ d^{-1}$. Kirchman *et al.* (1994) estimated that bacterial production accounted for 22–39% of the total ammonium uptake during this period, whereas the model value was 7%. Note that Ducklow (1994) also found that a model based on FDM under-estimated bacterial production; some possible reasons for this discrepancy will be discussed below.

One of the most important fluxes that any JGOFS model should be able to

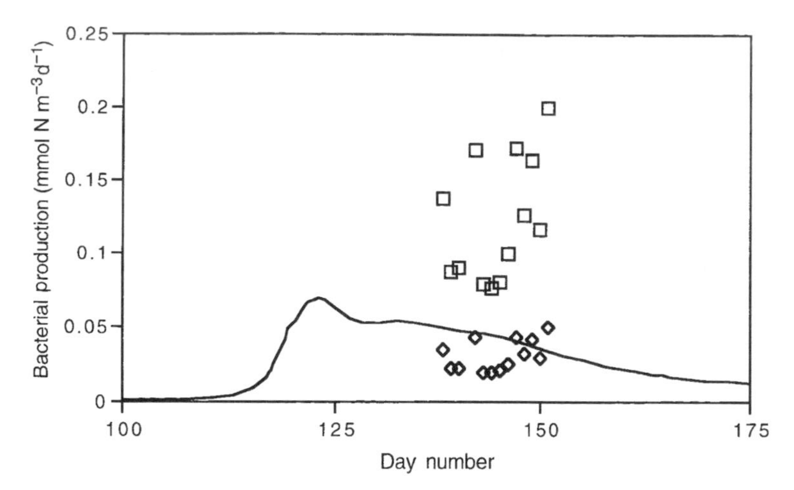

Figure 14.9 *Comparison of the observed bacterial production (mmol N m^{-3} d^{-1}) with model-1 simulations. The upper points (squares) use an average value for the conversion of thymidine uptake to bacterial nitrogen uptake and the lower points (diamonds) use the lower limit of values.*

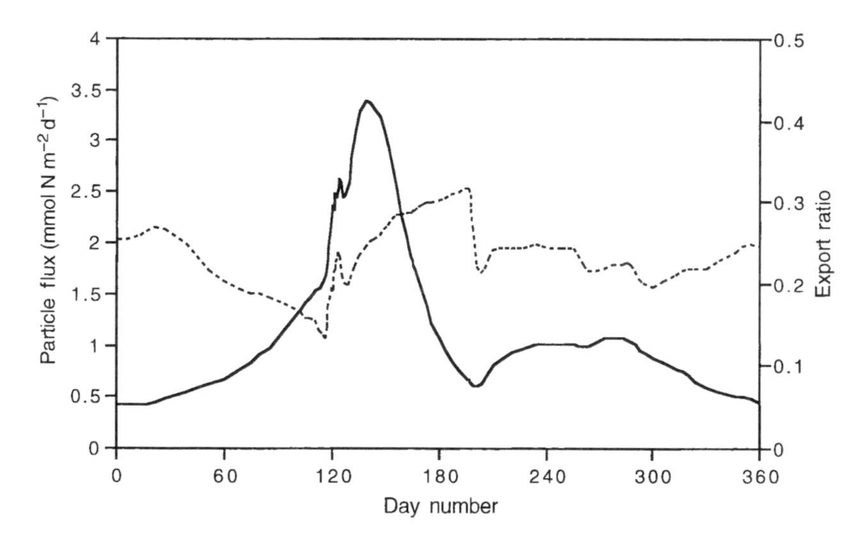

Figure 14.10 *The annual cycle of modelled particle flux (mmol N m^{-3} d^{-1}) from the mixed layer (solid line) and the export ratio (dashed line).*

predict is the particle flux out of the euphotic zone. Martin *et al.* (1993) estimated that the average *e* ratio (the ratio of export flux to primary production) at a depth of 35 m between days 114 and 151 was 0.30; the equivalent modelled value was 0.20 ± 0.04 (Fig. 14.10). More directly, the observed and predicted particle fluxes out of the mixed layer can be compared for the spring bloom period. Martin *et al.* (1993) used free-floating sediment traps at a number of depths to derive a particle flux – depth relation for the period between days 114 and 151. This was used to extrapolate to a depth of 35 m to give an average PON flux of 6.8 mol N m^{-2} d^{-1}, i.e. an average over the 35 m depth of 0.19 mol N m^{-3} d^{-1}. The equivalent model estimate was 0.11 ± 0.04 mol N m^{-3} d^{-1}. Both these results show that the model under-estimates the particle flux from the mixed layer, at least during the spring bloom period.

Another interesting result from the model is that, although the particle flux varies by a factor of seven (Fig. 14.10) over the annual cycle, the *e* ratio is much less variable (mean = 0.23 ± 0.04). If this proves to be a general result of such models then it may greatly simplify the calculation of particle fluxes from satellite-derived estimates of primary production.

Discussion

Assessment of the model results

We have shown how non-linear optimisation can be used to minimise the misfit between a model and a given set of observations. The speed with which such fits can be obtained makes it possible to compare the predictive ability of a large range of models. There are now a number of groups using similar methods to fit models to data (Matear, 1995; Hurtt & Armstrong, 1996; Lawson *et al.*, 1996; Prunet *et al.*, 1996; Harmon & Challenor, 1997) and it is likely that these exercises will bring about an improvement in the predictive ability of marine ecosystem models. However, some words of caution should be added concerning the problems of global minima and parameter independence. Any method that attempts to find the minimum of a function in multi-dimensional space is always open to the danger of finding a local rather than a true global minimum. It is often claimed that the simulated annealing technique has a very good probability of finding the global minimum but it uses much more computer time. It would be helpful if some comparisons were made between simulated annealing and other faster methods.

Matear (1995) fitted models to time-series observations at OWS Papa using simulated annealing and found a high degree of correlation among the parameters. He concluded that, with the limited data sets presently available, it is usually impossible to obtain a unique model solution. Obviously, better data sets would alleviate this problem but another approach may be to simultaneously fit data sets from a number of separate time series; G. Hurtt & R. Armstrong (personal communication) are presently attempting this.

Judged by its ability to model the NABE data at 47°N 20°W, the seven-compartment model as modified from Fasham *et al.* (1990) was reasonably successful. The model simulated the overall development and decline of the phytoplankton spring bloom and the associated nitrate decline (Fig. 14.5*a*,*b*). It also yielded values of PON and primary production that were close to the observed values, but under-estimated particle flux. The ability to model export flux correctly is obviously a prime requirement of a JGOFS model. The model was less successful in its representation of zooplankton and bacteria. Although the modelled zooplankton grazing (measured as a fraction of primary production) was similar to that measured by shipboard experiments, the model biomass was higher than most of the observations of microzooplankton biomass. Bearing in mind the paucity of microzooplankton data, especially for the early stage of the bloom, and the fact that mesozooplankton will also contribute to phytoplankton grazing, it may however be too hasty to condemn the zooplankton model. One obvious development of model-1 would be to explicitly include two size classes of zooplankton (see, for example, Fasham, 1995). The uncertainties concerning zooplankton modelling highlight the need for zooplankton biomass data to be obtained at the same temporal frequency as phytoplankton and nutrients.

Although the early stages of the spring bacterial bloom were simulated accurately by model-1, the peak values were greatly under-estimated, as was the bacterial production and growth rate. The optimised value of the half-saturation coefficient for bacterial uptake of DON (k_4) was 0.89, which, for the modelled DON values during the spring bloom of 0.15 mmole N m^{-3}, means that the bacterial growth rate was only 14% of its maximum value. This large value of k_4 may arise because no semi-labile DON observations were used for the optimisation and so there were no effective data to constrain the value of k_4. This situation might be improved if bacterial production is included in the optimisation data set.

The second model described included silicate nutrient and partitioned phytoplankton chlorophyll between diatoms and other phytoplankton. It reproduced the observed double peak structure during the spring phytoplankton bloom but still under-estimated the peak values. This problem requires further study but may be caused by the model having only one zooplankton compartment grazing both classes of phytoplankton, rather than a separate zooplankton for each class. Despite these problems the model proved capable of realistically modelling the spring declines of both nitrate and silicate.

Model-2 also under-estimated the primary production compared with model-1 and produced very different parameter values for the bacteria (Table 14.1). The reasons for this are unclear but presumably relate to the problems of indeterminacy in parameter values for models with large numbers of parameters. This will be investigated in the future by calculating the variances and covariances of the optimised parameters (Matear, 1995) for the two models.

The future for JGOFS ecosystems models

Finally, we discuss some possible future directions for marine ecosystem models within the JGOFS context. First, it will be necessary to take into account the concentration of iron in order to model production in areas where iron is likely to be limiting to phytoplankton growth. The first iron experiment carried out in the equatorial Pacific (Martin *et al.*, 1994; Watson *et al.*, 1994) gave some surprising results that could possibly be explained by the effects of zooplankton grazing. The results from the second experiment (Coale *et al.*, 1996) were less ambiguous and showed a pronounced phytoplankton bloom after iron fertilisation. However, the effects of any proposed continuous iron fertilisation would have to take into account the grazing response of the zooplankton. This again shows the importance of obtaining good zooplankton data when carrying out any such experiments. From the modelling point of view, before we can attempt to model the effects of iron there is a need for more physiological experiments to determine more precisely how iron limitation affects the rates of photosynthesis and nutrient uptake.

The second problem raised by the JGOFS studies concerns the role of dissolved organic matter (DOM). The two models described above only attempted to model labile DOM that is directly utilised by bacteria and even that presented problems with obtaining suitable observation sets. However, there is increasing evidence (Carlson *et al.*, 1994; Ducklow *et al.*, 1995) that a fraction of DOM, having a residence time of months rather than days and referred to as semi-labile DOM, may play a significant role in the export of carbon and nitrogen from the euphotic zone. Some modelling work with 3-D models (Bacastow & Maier-Reimer, 1991; Najjar *et al.*, 1992) showed that allowing a large fraction of export carbon to be in this form gave better fits to the meridional nutrient distributions. Modelling the production of semi-labile DOM in the upper ocean is not easy, as it can be produced by a variety of processes. However, some work in progress (T. R. Anderson, personal communications) has shown that it is possible to model these processes and, if the semi-labile DOM is assumed to break down to labile DOM with a time scale of a few months, such a model can explain some observed time series of DON and DOC as well as that of the more standard biological variables.

Finally, there is the trend towards increased model complexity, as better data sets become available. We have already mentioned the work of Taylor *et al.* (1993) and the diatom model described herein is a further contribution. However, such models require larger numbers of parameters and may therefore fall foul of the parameter uniqueness problem discussed above. Hurtt & Armstrong (1996) tried to circumvent this problem by using a size-structured model with allometrically determined parameters (see also Moloney & Field (1991)). There may also be a restraint on complexity provided by the need to develop models of the biological pump for 3-D coupled ocean–atmosphere climate models; the computational requirements of such models limits the number of compartments that can be used. The challenge for JGOFS modellers may therefore be to develop simple models that, although not perfect for any given time series of upper-ocean data, still give good simulations of global data sets of particle flux to the deep ocean. In this way, we may achieve our ambition of modelling the global biological CO_2 pump and understanding its role in future climate change scenarios.

References

Bacastow, R. & Maier-Reimer, E. (1991). Dissolved organic carbon in modeling oceanic new production. *Global Biogeochemical Cycles*, **5**, 71–85.

Brock, T. D. (1981). Calculating solar radiation for ecological studies. *Ecological Modelling*, **14**, 1–19.

Burkill, P. H., Edwards, E. S., John, A. W. G. & Sleigh, M. A. (1993). Microzooplankton and their herbivorous activity in the northeastern Atlantic Ocean. *Deep-Sea Research*, Part II, **40**, 479–94.

Carlson, C. A., Ducklow, H. W. & Michaels, A. F. (1994). Annual flux of dissolved organic carbon from the euphotic zone in the northwestern Sargasso Sea. *Nature*, **371**, 405–8.

Coale, K. H., Johnson, K. S. *et al.* (1996). A massive phytoplankton bloom induced by an ecosystem-scale iron fertilisation experiment in the equatorial Pacific Ocean. *Nature*, **383**, 495–501.

Doney, S. C., Glover, D. M. & Najjar, R. G. (1996). A new coupled one-dimensional biological-physical model for the upper ocean: applications to the JGOFS Bermuda Atlantic Time-series Study (BATS) site. *Deep-Sea Research*, Part II, **43**, 591–624.

Ducklow, H. W. (1994). Modeling the microbial food web. *Microbial Ecology*, **28**, 303–20.

Ducklow, H. W. & Fasham, M. J. R. (1991). Bacteria in the greenhouse: modeling the role of oceanic plankton in the global carbon cycle. In *New Concepts in Environmental Microbiology*, ed. R. Mitchell, pp. 1–31. New York: Wiley–Liss.

Ducklow, H. W. & Harris, R. P. (1993). Introduction to the JGOFS North Atlantic Bloom Experiment. *Deep-Sea Research*, Part II, **40**, 1–8.

Ducklow, H. W., Kirchman, D. L., Quinby, H. L., Carlson, C. A. & Dam, H. G. (1993). Stocks and dynamics of bacterioplankton carbon during the spring bloom in the eastern North Atlantic Ocean. *Deep-Sea Research*, Part II, **40**, 245–64.

Ducklow, H. W., Carlson, C. A., Bates, N. R., Knap, A. H., Michaels, A. F. & Takahashi, T. (1995). Dissolved organic carbon as a component of the biological pump in the North Atlantic ocean. *Philosophical Transactions of the Royal Society of London*, B **348**, 161–7.

Evans, G. T. & Parslow, J. S. (1985). A model of annual plankton cycles. *Biological Oceanography*, **3**, 327–47

Fasham, M. J. R. (1993). Modelling the marine biota. In *The Global Carbon Cycle*, ed. M. Heimann, pp. 457–504. Heidelberg: Springer-Verlag.

Fasham, M. J. R. (1995). Variations in the seasonal cycle of biological production in subarctic oceans: A model sensitivity analysis. *Deep-Sea Research*, **42**, 1111–49.

Fasham, M. J. R. & Evans, G. T. (1995). The use of optimisation techniques to model marine ecosystem dynamics at the JGOFS station at 47°N 20°W. *Philosophical Transactions of the Royal Society of London*, B **348**, 203–9.

Fasham, M. J. R., Ducklow, H. W. & McKelvie, S. M. (1990). A nitrogen-based model of plankton dynamics in the oceanic mixed layer. *Journal of Marine Research*, **48**, 591–639.

Frost, B. W. (1987). Grazing control of phytoplankton stock in the subarctic Pacific: a model assessing the role of mesozooplankton, particularly the large calanoid copepods, *Neocalanus* spp. *Marine Ecology Progress Series*, **39**, 49–68.

Frost, B. W. (1993). A modelling study of processes regulating plankton standing stock and production in the open subarctic Pacific. *Progress in Oceanography*, **32**, 17–57.

Harmon, R. & Challenor, P. (1997). A Markov chain Monte Carlo method for estimation and assimilation into models. *Ecological Modelling*, **101**, 41–59.

Hurtt, G. C. & Armstrong, R. A. (1996). A pelagic ecosystem model calibrated with BATS data. *Deep-Sea Research*, Part II, **43**, 653–83.

Kiefer, D. A. & Mitchell, B. G. (1983). A simple steady-state description of phytoplankton growth based on absorption cross section and quantum efficiency. *Limnology and Oceanography*, **28**, 770–5.

Kirchman, D. L., Ducklow, H. W., McCarthy, J. J. & Garside, C. (1994). Biomass and nitrogen uptake by heterotrophic bacteria during the spring phytoplankton bloom in the North Atlantic Ocean. *Deep-Sea Research*, **41**, 879–95.

Lawson, L. M., Hofmann, E. E. & Spitz, Y. H. (1996). Time series sampling and data assimilation in a simple marine ecosystem model. *Deep-Sea Research*, Part II, **43**, 625–52.

Lochte, K., Ducklow, H. W., Fasham, M. J. R. & Stienen, C. (1993). Plankton succession and carbon cycling at 47°N 20°W during the North Atlantic Bloom Experiment. *Deep-Sea Research*, Part II, **40**, 91–114.

Levitus, S. (1982). *Climatological Atlas of the World Ocean*. NOAA Professional Paper 13. Washington, DC: US Government Printing Office.

McGillicuddy, D. J. Jr, McCarthy, J. J. & Robinson, A. R. (1995*a*). Coupled physical and biological modeling of the spring bloom in the North Atlantic (I): model formulation and one-dimensional bloom processes. *Deep-Sea Research*, **42**, 1313–57.

McGillicuddy, D. J. Jr, Robinson, A. R. & McCarthy, J. J. (1995*b*). Coupled physical and biological modeling of the spring bloom in the North Atlantic (II): three-dimensional bloom and post-bloom processes. *Deep-Sea Research*, **42**, 1359–98.

Marra, J. & Ho, C. (1993). Initiation of the spring bloom in the northeast Atlantic (47°N, 20°W): a numerical simulation. *Deep-Sea Research*, Part II, **40**, 55–74.

Martin, J. H., Fitzwater, S. E., Gordon, R. M., Hunter, C. N. & Tanner, S. J. (1993). Iron, primary production and carbon-nitrogen flux studies during the North Atlantic Bloom Experiment. *Deep-Sea Research*, Part II, **40**, 115–34

Martin, J. H., Coale, K. H. *et al.* (1994). Testing the iron hypothesis in ecosystems of the equatorial Pacific Ocean. *Nature*, **371**, 123–9.

Matear, R. J. (1995). Parameter optimisation and analysis of ecosystem models using simulated annealing: A case study at Station P. *Journal of Marine Research*, **53**, 571–607.

Mayzaud, P. & Poulet, S. A. (1978). The importance of the time factor in the response of zooplankton to varying concentrations of naturally occurring particulate matter. *Limnology and Oceanography*, **23**, 1144–54.

Mellor, G. L. & Yamada, T. (1974). A hierarchy of turbulence closure models for planetary boundary layers. *Journal of Atmospheric Science*, **31**, 1791–806.

Moloney, C. L. & Field, J. G. (1991). The size-based dynamics of plankton food webs. I. A simulation model of carbon and nitrogen flows. *Journal of Plankton Research*, **13**, 1003–38.

Najjar, R., Sarmiento, J. L. & Toggweiler, J. R. (1992). Downward transport and fate of organic matter in the ocean: simulations with a general circulation model. *Global Biogeochemical Cycles*, **6**, 45–76.

Press, W. H., Teukolsky, S. A., Vetterling, W. T. & Flannery, B. P. (1992). *Numerical recipes in C.* Cambridge:

Cambridge University Press.
Price, J., Weller, R. A. & Pinkel, J. (1986). Diurnal cycling: observations and models of the upper ocean response to diurnal heating, cooling, and wind mixing. *Journal of Geophysical Research*, **91**, 8411–27.
Prunet, P., Minster, J.-F., Ruiz-Pino, D. & Dadou, I. (1996). Assimilation of surface data in a one-dimensional physical-biogeochemical model of the surface ocean. 1. Method and preliminary results. *Global Biogeochemical Cycles*, **10**, 111–38.
Robinson, A.R., McGillicuddy, D. J., Calman, J., Ducklow, H. W., Fasham, M. J. R., Hoge, F. E., Leslie, W. G., McCarthy, J. J., Podewski, S., Porter, D. L., Saure, G. & Yoder, J. A. (1993). Mesoscale and upper ocean variabilities during the 1989 JGOFS bloom study. *Deep-Sea Research*, Part II, **40**, 9–36.
Ryabchenko, V. A., Fasham, M. J. R., Kagan, B. A. & Popova, E. E. (1997). What causes short-term oscillations in ecosystem models of the ocean mixed layer? *Journal of Marine Systems*, **13**, 33–50.
SCOR (1990). *The Joint Global Ocean Flux Study (JGOFS) Science Plan.* JGOFS Report No 5. Halifax, Canada: Scientific Committee on Oceanic Research.
SCOR (1992). *JGOFS Implementation Plan.* JGOFS Report No 9. Halifax, Canada: Scientific Committee on Oceanic Research.
Sieracki, M. E., Verity, P. G. & Stoecker, D. K. (1993). Plankton community response during the 1989 North Atlantic spring bloom. *Deep-Sea Research*, Part II, **40**, 213–26.
Steele, J. H. & Henderson, E. W. (1992). The role of predation in plankton models. *Journal of Plankton Research*, **14**, 157–72.
Steele, J. H. & Henderson, E. W. (1993). The significance of interannual variability. In *Towards a Model of Ocean Biogeochemical Processes,* ed. G. T. Evans & M. J. R. Fasham, pp. 237–60. Heidelberg: Springer-Verlag.
Stramska, M. & Dickey, T. D. (1993). Phytoplankton bloom and the vertical thermal structure of the upper ocean. *Journal of Marine Research*, **51**, 819–42.
Stramska, M. & Dickey, T. D. (1994). Modeling phytoplankton dynamics in the northeast Atlantic during the initiation of the spring bloom. *Journal of Geophysical Research*, **99**, 10241–53.
Taylor, A. H. & Stephens, J. A. (1993). Diurnal variations of convective mixing and the spring bloom of phytoplankton. *Deep-Sea Research*, Part II, **40**, 389–408.
Taylor, A. H., Watson, A. J., Ainsworth, M., Robertson, J. E. & Turner, D. R. (1991). A modelling investigation of the role of phytoplankton in the balance of carbon at the surface of the North Atlantic. *Global Biogeochemical Cycles*, **5**, 151–71.
Taylor, A. H., Watson, A. J. & Robertson, J. E. (1992). The influence of the spring phytoplankton bloom on carbon dioxide and oxygen concentrations in the surface waters of the northeastern Atlantic during 1989. *Deep-Sea Research*, **39**, 137–52.
Taylor, A. H., Harbour, D. S., Harris, R. P., Burkill, P. H. & Edwards, E. S. (1993). Seasonal succession in the pelagic ecosystem of the North Atlantic and the utilization of nitrogen. *Journal of Plankton Research*, **15**, 875–91.
Verity, P. G., Stoecker, D. K., Sieracki, M. E. & Nelson, J. R. (1993). Grazing, growth and mortality of microzooplankton during the 1989 North Atlantic spring bloom at 47°N, 18°W. *Deep-Sea Research*, **40**, 1793–814.
Watson, A. J., Law, C. S., van Scoy, K. A., Millero, F. J., Yao, W., Friederich, G. E., Liddicoat, M. I., Wanninkhof, R. H., Barber, R. T. & Coale, K. H. (1994). Minimal effect of iron fertilisation on sea-surface carbon dioxide concentrations. *Nature*, **371**, 143–5.

15 Remote sensing of primary production in the ocean: promise and fulfilment

T. Platt, S. Sathyendranath and A. Longhurst

Keywords: remote sensing, ocean colour, primary production, biogeochemical provinces, pigment profile, photosynthesis parameters, local algorithm, basin scale modelling

Introduction

Remote sensing of ocean colour affords us our only window into the synoptic state of the pelagic ecosystem and is likely to remain the only such option into the near future. Estimation of primary production from remotely sensed ocean colour is a problem in two parts: first, the construction of a local algorithm; and second, the development of a protocol for extrapolation. Good local algorithms exist, but their implementation requires that certain parameters be specified. Protocols for extrapolation have to include procedures for the assignment of these parameters. One suitable approach is based on partition of the ocean into a suite of domains and provinces within which the physical forcing, and the algal response to it, are distinct. This approach is still in its infancy, but is best developed for the North Atlantic. Using this method, and using the accumulated data from oceanographic expeditions, leads to an estimate for the annual primary production of the North Atlantic at the basin scale. Direct validation of the result is not possible in the absence of an independent calculation, but the potential errors involved may be assessed.

Motivation for the use of satellite data on ocean colour

There are at least five reasons why remotely sensed data on ocean colour are useful in global biogeochemical studies. One is to provide a synoptic field of chlorophyll concentration to compare against the calculated field derived from coupled, ocean–ecosystem models, or to use for initialising such models (Wroblewski *et al.*, 1988; Ishizaka & Hofmann, 1993; Yentsch, 1993). Another is as a basis for the computation of regional-, basin-, and global-scale estimates of marine primary production (Platt & Sathyendranath, 1988; Platt *et al.*, 1991*a*; Morel & André, 1991; Campbell & Aarup 1992; Sathyendranath *et al.*, 1995; Longhurst *et al.*, 1995; Antoine *et al.*, 1996). A third is as a generic vehicle for the extrapolation to large horizontal scale of small numbers of discrete observations, made from ships, of various eco-physiological rates and pools (Platt *et al.* 1991*b*). A fourth is in the elaboration of a typology of the seasonal cycles in the pelagic ecosystem for various oceanic regimes or provinces (Longhurst, 1995). Finally, ocean-colour data are useful in assessing the influence of the pelagic microbiota on the physics of the mixed layer (Sathyendranath *et al.*, 1991*a*) or *vice versa* (Obata *et al.*, 1996).

Each of these applications will require that we emphasise particular aspects of the ocean-colour database, but, at the same time, there are features common to all applications. In this chapter, we shall stress mainly the estimation of primary production at large scale as an example of how the ocean-colour data can be used.

Computation of primary production from remotely sensed data is an application of algal physiology. Not all the information required for this computation is accessible to remote sensing. We must combine the data from satellites with data collected by ship: however, the incompatibility of scales in time and space between the two kinds of data is a major obstacle. We begin by outlining the salient properties of ocean-colour data.

The ocean-colour database

Until recently, ocean-colour data taken by satellites existed only for the period 1978–1986. Within this window, the coverage of the global ocean provided by the Coastal Zone Colour Scanner (CZCS) was uneven in time and space: partly because the sensor was not always activated, and partly because useful data cannot be recorded through clouds (a limitation with both systematic and random components, and therefore a potential source of bias in some applications). Because the satellite flew in a polar orbit, the frequency of coverage varied with latitude. However, the large swath width of the sensor ensured that every point on the surface of the earth could be sampled at least once per day, cloud-free conditions permitting (Robinson, 1985).

The sensor recorded ocean-leaving radiance in several wavebands, and the ratio of two of these (the blue–green ratio) can be calibrated against phytoplankton concentration in the surface layer (Gordon & Morel, 1983). The signal detected by the sensor emanates not just from the surface proper: it contains (progressively weaker) contributions from deeper horizons (Gordon & McCluney, 1975). Contributions from horizons deeper than one optical depth (inverse of the diffuse attenuation coefficient) are generally agreed to be negligible (Gordon & Clark, 1980).

The data were collected at a spatial resolution of better than 1 km at the sea surface. Often, however, in the interest of making the files more compact, the data are reported at a 4 km resolution (Feldman *et al.*, 1989). For the time resolution, one should distinguish between instantaneous images constructed for comparison with data collected by ship on a particular day, and 'averages' constructed from serial images. These are averages only in a special sense, and they are usually biased. For example, in regions subject to frequent cloud cover, individual pixels in the average may be represented by only one or a small number of observations, or they may even be represented by no observations.

Between 1986 and 1996, no synoptic data on ocean colour were available to science. However, beginning in 1996, the situation began to improve considerably. In March 1996, a German sensor (MOS) was launched on an Indian satellite (IRS-P3). In August 1996, French (POLDER) and Japanese (OCTS) satellites were launched on a Japanese satellite (MIDORI); these sensors functioned until June 1997. In August 1997, the USA launched SeaWiFS, which, at the time of writing (September 1997), has just begun to transmit useful data. Further ocean-colour missions are planned by various national space agencies. Access to ocean-colour data from now on should therefore be excellent, and with moderate luck we can expect an unbroken data stream into the indefinite future.

The database collected from ships

The relevant data accumulated by observations from ships consist of vertical profiles of pigment concentration, and measurements of the characteristics of the photosynthesis–light curve. Vertical profiles of *in situ* primary production can also be useful for checking and tuning the local algorithm. The spatial distribution of entries in these archives is very uneven. Of the quantities mentioned, the pigment profile has the richest archive, primary production the next richest, and the photosynthesis parameters the least rich of all. For example, in an archive we compiled for the North Atlantic (defined, for the present purpose, to be the area from 89°W to 1°E between 10°S and 70°N), we found some 6280 pigment profiles that met the required standards of quality, but only 1862 sets of photosynthesis parameters (Sathyendranath *et al.*, 1995). The spatial distribution of the latter was heavily biased towards the western side of the basin (mainly a consequence of their having all been measured by the same research group).

Sporadic coverage, in both space and time, is the most salient characteristic of the ship archive. The existing coverage, or lack of it, is a useful guide to where future sea sampling might be conducted.

The local algorithm

When the ocean-colour data are used to estimate primary production, the usual goal is to calculate the daily production of the ocean water column, $P_{Z,T}$. The problem has two principal parts: calculation of $P_{Z,T}$ at a particular time and place, where it is assumed that all the necessary information is given (the local algorithm); and then the extrapolation of the results of the local algorithm to yield a figure representative of a larger region (Platt & Sathyendranath, 1988). We deal first with the local algorithm.

Various procedures exist for calculation of primary production at a given location, all having broadly similar requirements for the information that must be supplied (Platt & Sathyendranath, 1993).

The first requirement is to know the irradiance at the sea surface in the photosynthetically-active waveband (400–700 nm), $I_0(t)$, a function of the time t. This can be calculated by a standard astronomical method on the assumption that the sky is free of clouds. The effect of clouds can be allowed for by using information collected by remote sensing (Bishop & Rossow, 1991), or, failing that, using climatological data on clouds. The loss of irradiance by reflection at the sea surface depends on sea state, which is also accessible to remote sensing. In the strictly local context, the algorithm for primary production may be forced with irradiance measured on the spot.

The next requirement is the vertical distribution of phytoplankton pigment, important as the principal determinant of the penetration of visible light through the sea and of the absorption of photons in the photosynthesis process. This profile cannot be derived directly by remote sensing. The most convenient way to supply the information is in the form of the parameters of a standardised vertical profile, such as the shifted Gaussian (Platt *et al.*, 1988).

Finally, we need to know the parameters of the photosynthesis–light curve, the initial slope, α^B, and the production at light saturation or assimilation number P^B_m, where the superscripts indicate normalisation to pigment biomass B.

For the calculations of primary production presented below, we use the procedure of Platt & Sathyendranath (1988) in its most recent version, as presented in Sathyendranath *et al.* (1995). This method is usually applied with the photosynthesis parameters independent of depth, but since the integration over depth is done numerically, their possible variation with depth is easily accounted for, provided the necessary information is available.

Extrapolation to large horizontal scale

Calculations at large horizontal scales require that we implement the local algorithm at many field points (pixels) in the region of interest. The question immediately arises as to how the supplementary information, that is the parameters of the photosynthesis–light curve and of the pigment profile, should be supplied at each and every pixel. If these properties were invariant (constant and spatially uniform) over the region, this would be a trivial problem. Experience persuades us that things are not so simple.

In this context, two main lines of attack are open. One is to derive the required parameters as functions of environmental variables, such as temperature, accessible to remote sensing (Method I). The other is to assign the parameters based on a data archive, taking into account location and season (Method II).

In principle, Method I would be the preferred method: it would be more objective, it would rely less on the skill and judgement of the user; and it would give unambiguous parameter estimates at every field point, with some idea of the variance associated with those estimates. The difficulty with following this approach is that reliable estimator functions, as required by the method, do not exist. They certainly do not exist as universal laws.

Lacking the necessary estimator functions, we can consider Method II, the assignment of parameters based on prior knowledge of their magnitudes at a given location in a particular season. This approach is based on the belief that the shape of the pigment profile and of the photosynthesis–light curve are, in general, more stable properties than the pigment biomass at the surface (the only biological property whose value can be updated continuously by remote sensing). The difficulties in following this approach are that the existing databases, especially that on photosynthesis parameters, are modest in size; that the existing observations are unevenly distributed in space and time; and that therefore some protocol has to be erected by which the archived data can be translated into the parameter field at a given time. Notice that, because we are unable to deduce the estimator functions directly from first principles, Method I would also rely on the use of a data archive: to the extent that the database is limited, both methods will be compromised.

Inherent weaknesses in both methods (lack of universally valid estimator functions in Method I and sparse, unevenly distributed archives in Method II) impose a requirement for partition of the region of interest to facilitate the large-scale calculations. In the next section, we consider how such a partition could be carried out.

Dynamic biogeochemical provinces

The conceptual basis for partitioning the ocean (Platt & Sathyendranath, 1988; Mueller & Lange, 1989; Longhurst, 1998) is as follows. We assume that in the pelagic ecosystem, the rates of important ecophysiological and biogeochemical processes (in particular the photosynthetic rate) are under physical control. Physical forcing regulates the environmental conditions that will determine the species assemblage flourishing at a given place and time; the magnitude of the nutrient flux into the photic zone and the manner in which it is supplied; the rate of vertical mixing and therefore the rate of photoadaptation; the stratification in the water column and therefore the vertical distribution of photosynthetic pigments. We then partition the ocean by delineating those areas that share a common physical forcing, in so far as the forcing can be expected to control the rates of the processes just mentioned.

A complication is that the nature and intensity of physical forcing is known to vary seasonally, and between years, in many parts of the ocean. Therefore, it is essential that the partition scheme allow the partition boundaries to vary with time. This is one meaning of the qualifier 'dynamic' in the name of the method: dynamic biogeochemical provinces. The other connotation is that the parameters themselves may be allowed to vary from the default (constant) values in response to local forcing, provided that we know how to quantify their response.

The tendency of the boundaries to move with time requires that we conceive of the provinces at two distinct levels: first in the Platonic or ideal sense (a pair of provinces with definable and dissimilar properties will exist adjacent to each other thus creating a boundary of a known, general character); and second in the particular sense (at a stated time, the said boundary is believed to follow a stated trajectory). The question then arises of how the realised boundaries may be delineated in routine applications. Because we intend to draw the boundaries at large geographical scale, the most useful criteria to use will be those that can be implemented from remotely sensed data: sea-surface temperature, sea-surface elevation, wind stress and ocean colour itself. This means that the characteristics by which we distinguish one province from another must be translatable into objective rules that can be applied to data on one or more of these properties, collected by satellites, such that any pixel in the ocean-colour field can be assigned unambiguously to a particular province. Such is the long-range goal of this approach. Until now, however, the appropriate rules have not been formulated.

Once the provinces are established, what role do we expect them to play? They form a template upon which the parameters of the photosynthesis–light curve and of the vertical distribution of pigments can be assigned. In our view, some such template is essential no matter what protocol (archive; empirical function of environmental variables; rational function of environmental variables based on first principles) is used to assign parameters. This is because, at present, no universal rules exist to derive these parameters from environmental properties. At best, we have rules that are useful only in a limited region, so that at large scale a series of such rules would have to be applied in a piecewise manner.

The provinces, then, depict the underlying oceanographic structure of an area, represented as an outline map on which the necessary parameters can be filled in as assigned according to one of several possible procedures. The simplest, or default, procedure is to assume that for a given season, whose extent may be defined to suit the particular requirements of the province, the parameters that we need are constant inside the boundaries of the province: the preferred way to assign them is on the basis of archived data for the same region and season (Method II). In the North Atlantic, as elsewhere, the weakness with this approach is the small size of the archive and its uneven coverage.

Observe that Methods I and II are not mutually exclusive. It may be that, for a given province and season, particular information is available about one or more of the parameters such that the default procedure can be improved for this or these parameters. Thus, we are led to consider a Method III, in which the default procedure (Method II) would be used where no alternative was available, but particular rules, province-specific or otherwise (Method I), would be applied wherever and whenever possible. Method III would be 'adaptive', in the sense that the word is used in computational mathematics, selecting the algorithms to assign parameters according to circumstances.

A specific example of a rule that could be used in Method III is available for the North Atlantic. It is known both from observations and from theoretical work that for a broad swath of this ocean in the summer, the depth of the chlorophyll maximum shallows to the north at a rate of 3.5 m per degree of latitude (Strass & Woods, 1991). This is a robust result. It can be easily codified into an objective rule, and it makes sense to use it, no matter what procedure is followed for the assignment of the rest of the parameters. In other words, it is in no way essential for all the parameters to be assigned by the default procedure, nor indeed by any single procedure: the preferred approach is to use the best information available for any parameter in the given circumstances. The example just given is region-specific: it would make little sense to use the same rule in the North Pacific.

Another example is the use of surface temperature data to estimate properties, such as surface nitrate concentration, useful for the calculation of new production (Dugdale *et al.*, 1989; Sathyendranath *et al.*, 1991*b*). Although it has a basis that is easily understandable (upwelling water tends to be both colder and richer in nutrients than the water it replaces), and although it has been applied with success in a variety of oceanographic regimes, this approach lacks a detailed theoretical foundation. Its implementation in a particular region is based strictly on observation, but within this limitation it is easily codified into a rule: qualitatively, the rule would be the same everywhere – a regression of nitrate on temperature – but the coefficients in the regression would vary from province to province.

Ideally, we would prefer rules that were grounded in first principles and therefore free of region-specificity. At present, no such rules exist, but this does not mean that we should not continue to seek them. Again, it is worth stressing that we are not required to assign all parameters by the same procedure, and a universal, first-principle rule for even a single parameter would constitute a major advance.

At this point, it may be useful to summarise the concept of the biogeochemical provinces. They partition the ocean into a suite of regions having similar physical forcing with respect to the properties important for algal growth, and thus form a rational template, spatial and temporal, for the assignment of the parameters necessary for calculation of water-column primary production. Such a partition appears to be essential whether the parameters are assigned by empirical rules, by rules derived from theory, or based on a data archive. In practice, a combination of approaches may be used to assign parameters (Method III).

Against this background, what progress has been made in the partition of the North Atlantic? Longhurst (1995, 1998) and Longhurst *et al.* (1995) recognise four primary algal domains, each with a distinctive coupling between physical forcing and algal ecology (Sathyendranath *et al.*, 1995). The *Polar Domain* is characterised by the presence in spring and summer of a stable, low-salinity layer at the surface consequent upon ice-melt. In such conditions, development of phytoplankton blooms is under strict control of irradiance (intensity and daylength). The *West-wind Domain* covers the mid-latitudes, where vertical mixing by storms competes with stabilisation by solar heating to determine the structure of the upper water column. Here, the critical-depth model of Sverdrup is a useful basis for interpretation of phytoplankton dynamics as a response to local forcing. In contrast, the *Trade-wind Domain* is seen as a domain where the principal forcing of algal dynamics may be remote, rather than local, a consequence of geostrophic adjustment to the large-scale wind field. Finally, the *Coastal-boundary Domain* refers to the continental margins where coastal winds and local topography exert a strong influence on the local circulation. It includes the upwelling regions.

These four domains are sufficiently general to be useful for any ocean. Indeed, they have already provided the basis for a calculation of global-scale

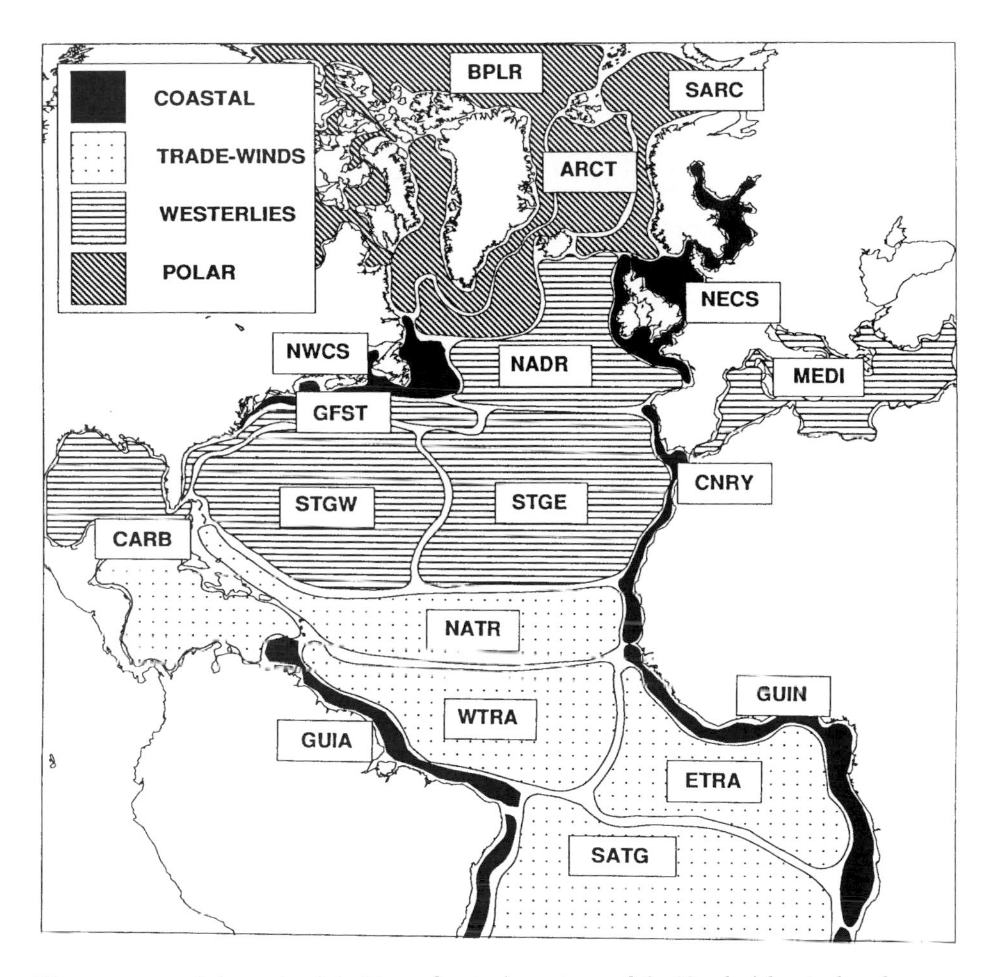

Figure 15.1 *Schematic of the biogeochemical provinces of the North Atlantic, based on Longhurst (1998) and Sathyendranath* et al. *(1995).*

primary production (Longhurst *et al.*, 1995). If we want to consider the particular properties of the North Atlantic, however, a further partition is required: this is the partition that yields the biogeochemical provinces themselves. Thus, in the Polar Domain, we identify the Boreal–Polar, Arctic, and Subarctic Provinces; in the West-wind Domain we distinguish the Gulf Stream, N Atlantic Drift, Subtropical Gyre (East and West) and Mediterranean Provinces; in the Trade-wind Domain we recognise the Tropical Gyre, the W Tropical Atlantic, the E Tropical Atlantic, the S Subtropical Gyre, and the Caribbean and Gulf Provinces; and in the Coastal-boundary Domain we separate the NW Atlantic, the NE Atlantic, the E Atlantic, the Guinea, and the Guiana Provinces. This gives some 18 provinces (Fig. 15.1).

Implementation in the North Atlantic

We have used the suite of biogeochemical provinces outlined in the previous section, and the relevant database assembled from observations at sea, to calculate annual primary production in the North Atlantic, given the satellite archive of ocean colour (Sathyendranath *et al.*, 1995). The steps involved in the implementation are as follows.

An archive of chlorophyll profiles was built, with data compiled from the work of a number of institutes over several decades. Each profile was checked for quality, and then fitted to a shifted Gaussian profile. After rejecting some 1665 profiles on the grounds of poor quality, we judged a further 6280 profiles to give acceptable fits. The fitted profiles were partitioned by province and by season, and average shapes of the chlorophyll profiles were established for each case (note that the archived chlorophyll profiles were used only to establish the *shape* of the profile, but not the *magnitude*). Similarly, an archive of some 1862 *P–I* parameters was used to establish mean parameters by season and by algal domain (the data set was insufficient for partition into provinces). In other words, the parameter assignment was made primarily according to Method II. A tentative step towards Method III was made for the Trade-wind Domain, where the photosynthesis parameters were assigned based on surface light and nitrate availability.

Cloud-cover data were used with an atmospheric transmission model (Bird, 1984) to compute the light available at sea level, on a monthly basis, for every 1° grid in the N Atlantic. Monthly-average values of satellite-derived chlorophyll data for 1979, also gridded by one degree, were combined with information on the shape of the chlorophyll profile for the season and the province (from archived data), to obtain the absolute profile at each grid point. A spectral, light-transmission model that uses the chlorophyll profile as input was then used to compute the light available at depth, and thus the primary production at that depth. The computations were carried out for some 12 time intervals over the day, and for every 0.5 m in the euphotic zone, and summed over depth and over time to obtain the daily, water-column production at that pixel.

The computations were repeated for each pixel for each month of the year, and summed over the year, and over the pixels assigned to each province, to obtain annual production per province. According to these calculations for the North Atlantic Basin (10°S to 70°N; 89°W to 1°E; roughly 15% of the total area of the global ocean), the primary production in 1979 was 10.5 Gt C yr^{-1}. The West-wind Domain contributed 2.3 Gt C yr^{-1} to the total and the Polar Domain, 2.2 Gt C yr^{-1}. The Trade-wind Domain was responsible for 3.1 Gt C yr^{-1}, and the Coastal Domain for 2.9 Gt C yr^{-1}. We do not insist that these figures are correct in any absolute sense. They merely represent the present state of our ability to combine the satellite and the ship archives to draw conclusions about the annual primary production in the North Atlantic.

Because the parameter assignment was based on seasonal averages, and because the calculations used monthly-averaged fields of pigments and clouds, the resolution of the primary production estimates cannot be better than monthly. The estimated fields of primary production for each month are shown in Fig. 15.2 (colour plate).

Problems with validation of results

Having made an estimate of basin-scale primary production for the North Atlantic, we can ask whether it is possible to validate the result, which is to test it against the result of a similar calculation made according to an independent method. Unfortunately, the answer is no. There are no independent methods available. There are two principal reasons for this. First, properly done, the remote sensing method for primary production uses all the available data from direct primary production measurements made by ships. It subsumes the ship database, and in addition uses the rich satellite archive as a vehicle for extrapolating to times and places where direct measurements are lacking. No data remain unused and therefore independent. Further, previous estimates of basin-, or global-, scale primary production have, necessarily, combined data from many different years as a device to offset, partly, the incompleteness of the database. Remote sensing is the only method that affords (albeit in the medium to long term) a realistic chance of estimating annual primary production for a particular year, and therefore of detecting inter-annual variations.

One method that might be thought of for validation is to use the results as the basis for predicting the pigment fields at some future time. The difficulty with applying this approach is that we would need information on the loss terms for phytoplankton, which are not accessible to remote sensing, and for which the data archive is considerably more sparse than that for the growth terms. The flow field would also be required, but this is a more feasible objective than the loss terms, whether approached by modelling or by remote sensing.

Another major obstacle to validation of estimates of primary production by remote sensing relates to comparisons with results from indirect or bulk-property methods. An example of these methods is that based on apparent oxygen utilisation rate. Such methods usually have a long intrinsic time scale as well as large spatial scales. One of the characteristics of primary production measurements is that long-time-scale methods do not estimate the same component of primary production as those with short temporal scales (Platt *et al.*, 1989, 1992; Platt & Sathyendranath, 1993). Specifically, bulk-property methods estimate new production, whereas the ship database of short-term incubations refers to gross production. The fundamental incompatibility of scales between these two classes of method implies a limitation on our ability to compare their results. At the very least, the results of the one class of method would have to be extrapolated before they could be compared with the results of the other, and the robustness or otherwise of the extrapolation procedure itself would interfere with the attempt to validate.

A basin-scale estimate of primary production using remotely sensed data is unusual in that it refers to large spatial scales, but is based on physiological parameters that are valid at short temporal scales, being derived from short-term incubations using a radioactive tracer for the inorganic carbon assimilated. The same would be true of basin-scale estimates made using the ship database alone, but any such calculation that overlooks the potential of the ocean-colour data as a tool to aid interpolation of sparse observations should be regarded as of only limited value.

New production

One way to circumvent the incompatibility between scales and components of primary production is to attempt to estimate new production directly by remote sensing. The basic procedure is first to estimate total primary production, as before, and then to multiply the result by the ratio of new production to total production, as estimated by remote sensing or by any other method (the calculation should be a weighted integration over time and space). To date, this has not been done at the basin scale in the Atlantic or in any other ocean. However, localised calculations have been made, for example on Georges Bank (Sathyendranath *et al.*, 1991*a*) and off North Africa (Dugdale *et al.*, 1989). The pigment fields derived from satellite have been used as a guide to the calculation of new production in the North Atlantic, from information on nitrate distribution (Campbell & Aarup, 1992).

A crude way to establish a lower bound for new production at the basin scale might be based on the observation that any primary production below the mixed layer is very likely to be new production. Stating this another way, any production in the deep chlorophyll maximum is probably new production. The difference between daily production calculated for a uniform water column (surface value of pigment maintained throughout) and that for a non-uniform column is one way to estimate primary production in the chlorophyll maximum, and therefore the lower limit of new production. The result of such an admittedly approximate calculation for the North Atlantic is 1.0 Gt C m^{-2} yr^{-1}, implying an *f* ratio (equal to the ratio of new production to total production) of at least 0.1. Any losses of organic material from the mixed layer would augment the estimate of new production and of the implied *f* ratio.

Approximate though it may be, this calculation emphasises the value of including information on the shape of the pigment profile in the estimation of primary production. It might be argued that, because the estimate increases by only about 10% when the chlorophyll maximum is accounted for, the effort involved in assembling the database on pigment profiles and controlling it for quality is not worthwhile. However, we maintain that the database on pigment profiles, and the estimates of primary production made from them, contain useful information whose value should not be overlooked. Moreover, the error is systematic rather than random, and will not disappear through mutual cancellation when the estimates of primary production are averaged over many pixels.

Error analysis

Although it is not possible to validate the basin-scale estimates in the normal sense, it is possible to assess the errors inherent in the calculation. To do this, we consider each element of the calculation separately.

Following Platt *et al.* (1988), we suppose that the precision of the surface irradiance is within 10% (Gautier & Katsaros, 1984), and that that of the pigment concentration derived from ocean colour is within 35% (Gordon *et al.*, 1983). The error on the photosynthesis parameters will contain a component arising from the measurement itself, and one arising from the aggregation of data within domains. The first component is well established (5% for P^B_m, 20% for α^B, Platt *et al.*, 1980). The second can be estimated from the parameter archives (Sathyendranath *et al.*, 1995) as the standard error relative to the mean: it is about 7% for both α^B and P^B_m. Platt *et al.* (1988) found the compounded error on the local algorithm, as used in the context of remote sensing, to be *c.* 42%, with the error in biomass retrieval dominating. If we now include the further 7% error associated with aggregating the photosynthesis parameters into domains, we can estimate the resultant error to be *c.* 50%.

With respect to the errors associated with defining the shape of the pigment profile, a set of data from three provinces of the Westerlies Domain is available. These data were collected using an undulating body equipped with a profiling fluorometer calibrated for chlorophyll retrieval. The data were averaged over segments 5 km long of transects up to *c.* 1000 km long, run during September–October 1992, and fitted to the standard, shifted-Gaussian shape. Of interest is the stability of the profile parameters within a province. In the notation of Platt *et al.* (1988), the appropriate properties to consider are z_m and

σ, respectively the depth and thickness of the chlorophyll maximum, and ρ, the ratio of the Gaussian amplitude at z_m to the total pigment at z_m. Using the standard error divided by the mean for an estimate of relative error as before, the results for ρ were 0.8% in the Eastern Subtropical Gyre province (244 profiles); 0.7% in the Western Subtropical Gyre province (187 profiles); and 0.7% in the Gulf Stream province (47 profiles). For z_m, the error averaged 1.7%, and could be reduced by roughly one third if the data were normalised to photic depth. For σ, the variation was 2.8% in the Gyre and 3.4% in the Gulf Stream: normalising to photic depth did not reduce these figures in either case.

The observed stability in ρ supports our basic approach to specification of the pigment profile, where the shape is fixed by the profile parameters but the magnitude is scaled to the satellite signal. Further, we have seen that, at the basin scale, the maximum systematic error that could be associated with vertical structure in the pigment field (if it were ignored) is about 10%. In this context, the variability observed in z_m and σ must be considered insignificant at this scale.

Ocean colour data as a generalised extrapolation tool

In addressing the use of remotely sensed data on ocean colour as a vehicle for the extrapolation of sparse observations on primary production, we should not forget the broader use as a tool for the extrapolation of eco-physiological rates in general. The basis for this universal application is the widespread use in modern theoretical work on biological oceanography of biomass-specific variables. For example, in summing the effects of the various loss processes to balance against phytoplankton growth in accounting for the incidence of algal blooms, it is expedient to normalise all the terms, including those representing the effect of zooplankton, to phytoplankton biomass (Platt *et al.*, 1991*b*). This is certainly a mathematical convenience. If biomass-normalised rates are measured at a few stations, they can be extrapolated more easily if the synoptic field of biomass is given. Of course, the limitations imposed by the colour scanner's inability to collect an unweighted image of the *vertical distribution* of pigment biomass remain: the requirement to have independent data on the vertical profile of biomass still exists.

Another sense in which the ocean-colour data serve as a general vehicle for the extrapolation and synthesis function is that they provide a yardstick of comparison for that other method capable of delivering synoptic fields: three-dimensional modelling. One long-term goal of both remote sensing and modelling must be to make comparisons in real time between the fields generated by the two methods. An important intermediate step, and one that will continue to be useful in the future, is to apply the ocean-colour results as data to be assimilated, intermittently or continuously, into the three-dimensional models, as a device to prevent the models diverging indefinitely from reality.

A final example of the use of remotely sensed data in the extrapolation and synthesis function of marine biogeochemistry is that it can provide the oceanographic context for detailed ship studies carried out in a limited area. The importance of this application would be difficult to over-estimate.

Discussion

It is no exaggeration to state that the ocean-colour data provided by the Coastal Zone Colour Scanner revolutionised biological oceanography. We can expect similar rewards, and superior performance, from the new generation of ocean-colour sensors. The ability to see a synoptic image of the pelagic ecosystem, even if only a two-dimensional one, is of the highest value. But for biogeochemical applications, the greatest reward will come when the biomass fields can be turned, in a routine and reliable way, into fields of primary production.

As an ancillary to the estimation of basin-scale primary production in the ocean, the remote-sensing method has been developed furthest in the North Atlantic. The reason is that, in this ocean, the archive of collateral data, especially data on the photosynthesis parameters, is larger than that for any other ocean. If the remote-sensing method is judged and found wanting in the North Atlantic, it must be found wanting in every other ocean, highly localised applications always excepted.

Although the archive of complementary data is richest in the North Atlantic, one should not conclude that it is adequate for the job. The historical information on vertical profiles of pigments is the most complete: one would like to see a more even distribution over regions and seasons. But the archive of photosynthesis parameters should certainly be improved, whether we intend to assign parameters by Method I, II or III. Development of empirical algorithms for prescription of photosynthesis parameters is as strongly dependent on the existence of a good database as is the direct assignment by region and season. The construction of prescriptive algorithms from first principles merits further study, but robust results are unlikely to appear overnight.

Whatever method is used for the assignment of parameters, it seems inevitable that it will be applied within the context of some scheme of algal domains and biogeochemical provinces. Considerable progress has been made to this end in the North Atlantic. However, the paucity of the data on photosynthesis parameters precludes exploitation of this work to the full: it was judged feasible to assign them only at the level of the algal domain, rather than at the finer level of the biogeochemical provinces.

We believe that the theory available for calculation of primary production, given information on the irradiance, the pigment profile and the photosynthesis parameters, is adequate for the job. It is certainly not the major obstacle to progress. Where effort needs to be invested is in the translation of knowledge about the provinces into rules that can then be incorporated into an adaptive computational scheme. The algorithms to assign parameters should respond to the input fields in an 'intelligent' (in the sense that the word is used in cybernetics) way, rather than a passive one. To date, the protocols are only rudimentary in this respect.

Ocean-colour data have many weaknesses. Nevertheless, they remain our only windows into the synoptic state of the pelagic ecosystem. Learning to extract from them the information they contain about the ocean carbon cycle, an exercise in applied plant physiology, must continue to be a priority activity in biological oceanography. Increasing reliance on remotely sensed data will not displace the requirement for oceanographic vessels, but it will provide strong indicators of where, when and how they should be deployed to best advantage. Our ability to interpret remotely sensed data on ocean colour will improve as more relevant sampling is carried out from ships, and thus more generalisations uncovered that can be coded into rules for inclusion in the adaptive algorithms on which interpretation will be based.

Compared with the data that can be collected by ship, the data collected by remote sensing are of lower precision and lack information on vertical structure. However, they make up in sample size what they lack in precision. For example, a ship might be able to sample 50 points on the perimeter of a 100 km box in one day. At a local resolution of 1 km, the satellite will provide 10^4 simultaneous observations inside the box. (Generally speaking, sampling variance is inversely proportional to the number of observations.) Furthermore, remote sensing will show how conditions inside the box relate to those outside, even as far as the edges of the ocean. In addition, the observations can be repeated on the next and subsequent days, for as long as required. Finally, sea-surface temperature, and other physical properties, can be collected by remote sensing on the same time and space scales, opening the possibility of a multidisciplinary, synoptic approach to marine ecosystem. Remote sensing gives us the possibility to examine the spatial gradients of primary production, their seasonal variation and their fluctuations between years. It is the best option we have to establish the long-term history of the pelagic ecosystem in any region.

Acknowledgements

18 119–32.

An earlier version of this paper was published in *Phil. Trans. R. Soc. Lond.* B (1995) **348**, 191–202. The present, slightly modified version is published with the permission of The Royal Society, for which the authors are grateful. This work would not have been possible but for a great number of people whose painstaking measurements of chlorophyll profiles were made available to us. The sources of all the data used are listed in Table 1 of Sathyendranath *et al.* (1995) and we sincerely thank those responsible. We also thank those scientists at NASA, especially Gene Feldman and Norman Kuring, who processed and made available the satellite data. The work presented in this paper was supported by the European Space Agency, with additional support from the Office of Naval Research; the National Aeronautics and Space Administration; the Department of Fisheries and Oceans, Canada; the Natural Sciences and Engineering Research Council; and the Department of National Defence, Canada. This research was carried out as part of the Canadian contribution to the Joint Global Ocean Flux Study. We thank Carla Caverhill, Cathy Porter and George White for their help with the data analysis.

References

Antoine, D., André, J.-M. & Morel, A. (1996). Oceanic primary production 2. Estimation at global scale from satellite (Coastal Zone Color Scanner) chlorophyll. *Global Biogeochemical Cycles*, **10**, 57–69.

Bird, R. E. (1984). A simple, solar spectral model for direct-normal and diffuse horizontal irradiance. *Solar Energy*, **32**, 461–71.

Bishop, J. K. B. & Rossow, W. B. (1991). Spatial and temporal variability of global surface solar irradiance. *Journal of Geophysical Research*, **96**, 16 839–58.

Campbell, J. & Aarup, T. (1992). New production in the North Atlantic derived from seasonal patterns of surface chlorophyll. *Deep-Sea Research*, **39**, 1669–94.

Dugdale, R. C., Morel, A., Bricaud, A. & Wilkerson, F. P. (1989). Modeling new production in upwelling centers: a case study of modeling new production from remotely sensed temperature and color. *Journal of Geophysical Research*, **94**,

Feldman, G., Kuring, N., Ng, C., Esaias, W., McClain, C. R., Elrod, J., Maynard, N., Endres, D., Evans, R., Brown, J., Walsh, S., Carle, M. & Podesta, G. (1989). Ocean colour. Availability of the global data set. *EOS Transactions, American Geophysical Union*, **70**, 634–5.

Gautier, C. & Katsaros, K. B. (1984). Insolation during STREX. 1. Comparisons between surface measurements and satellite estimates. *Journal of Geophysical Research*, **89**, 11 779–88.

Gordon, H. R. & Clark, D. K. (1980). Atmospheric effects in the remote sensing of phytoplankton pigments. *Boundary-Layer Meteorology*, **18**, 299–313.

Gordon, H. R. & McCluney, W. R. (1975). Estimation of the depth of sunlight penetration in the sea for remote sensing. *Applied Optics*, **14**, 413–16.

Gordon, H. R. & Morel, A. (1983). *Remote Assessment of Ocean Color for*

Interpretation of Satellite Visible Imagery. A Review. 114 pp. New York: Springer-Verlag.

Gordon, H. R., Clark, D. K., Brown, J. W., Brown, O. B., Evans, R. H. & Broenkow, W. W. (1983). Phytoplankton pigment concentrations in the Middle Atlantic Bight: comparison of ship determinations and CZCS estimates. *Applied Optics*, **22**, 20–36.

Ishizaka, J. & Hofmann, E. E. (1993). Coupling of ocean color data to physical-biological models. In *Ocean Colour: Theory and Applications in a Decade of CZCS Experience*, ed. V. Barale & P. M. Schlittenhardt, pp. 271–88. Dordrecht: Kluwer Academic Publishers.

Longhurst, A. (1995). Seasonal cycles of pelagic production and consumption. *Progress in Oceanography*, **36**, 77–167.

Longhurst, A., Sathyendranath, S., Platt, T. & Caverhill, C. (1995). An estimate of global primary production in the ocean from satellite radiometer data. *Journal of Plankton Research*, **17**, 1245–71.

Longhurst, A. R. (ed.) (1998). *Ecological Geography of the Sea*. San Diego: Academic Press. 398 pp.

Morel, A. & André, J.-M. (1991). Pigment distribution and primary production in the western Mediterranean as derived and modelled from Coastal Zone Color Scanner observations. *Journal of Geophysical Research*, **96**, 12 685–98.

Mueller, J. L. & Lange, R. E. (1989). Bio-optical provinces of the Northeast Pacific Ocean: A provisional analysis. *Limnology and Oceanography*, **34**, 1572–86.

Obata, A., Ishizaka, J. & Endoh, M. (1996). Global verification of critical depth theory for phytoplankton bloom with climatological *in situ* temperature and satellite ocean color data. *Journal of Geophysical Research*, **101**, 20 657–67.

Platt, T. & Sathyendranath, S. (1988). Oceanic primary production: Estimation by remote sensing at local and regional scales. *Science*, **241**, 1613–20.

Platt, T. & Sathyendranath, S. (1993). Estimators of primary production for interpretation of remotely sensed data on ocean color. *Journal of Geophysical Research*, **98**, 14 561–76.

Platt, T., Gallegos, C. L. & Harrison, W. G. (1980). Photoinhibition of photosynthesis in natural assemblages of marine phytoplankton. *Journal of Marine Research*, **38**, 687–701.

Platt, T., Sathyendranath, S., Caverhill, C. M. & Lewis, M. R. (1988). Ocean primary production and available light: Further algorithms for remote sensing. *Deep-Sea Research*, **35**, 855–79.

Platt, T., Harrison, W. G., Lewis, M. R., Li, W. K. W., Sathyendranath, S., Smith, R. E. & Vézina, A. F. (1989). Biological production of the oceans: The case for a consensus. *Marine Ecology Progress Series*, **52**, 77–88.

Platt, T., Caverhill, C. & Sathyendranath, S. (1991*a*). Basin-scale estimates of oceanic primary production by remote sensing The North Atlantic. *Journal of Geophysical Research*, **96**, 15 147–59.

Platt, T., Bird, D. & Sathyendranath, S. (1991*b*). Critical depth and marine primary production. *Proceedings of the Royal Society of London*, B **246**, 205–17.

Platt, T., Jauhari, P. & Sathyendranath, S. (1992). The importance and measurement of new production. In *Primary Productivity and Biogeochemical Cycles in the Sea*, ed. P. G. Falkowski & A. D. Woodhead, pp. 273–84. New York: Plenum Press.

Robinson, I. S. (1985). *Satellite Oceanography: An Introduction for Oceanographers and Remote-Sensing Scientists*. 455 pp. Chichester: Ellis Horwood Limited.

Sathyendranath, S., Gouveia, A. D., Shetye, S. R., Ravindran, P. & Platt, T. (1991*a*). Biological control of surface temperature in the Arabian Sea. *Nature*, **349**, 54–6.

Sathyendranath, S., Platt, T., Horne, E. P. W., Harrison, W. G., Ulloa, O., Outerbridge, R. & Hoepffner, N. (1991*b*). Estimation of new production in the ocean by compound remote sensing. *Nature*, **353**, 129–33.

Sathyendranath, S., Longhurst, A., Caverhill, C. M. & Platt, T. (1995). Regionally and seasonally differentiated primary production in the North Atlantic. *Deep-Sea Research*, **42**, 1773–802.

Strass, V. H. & Woods, J. D. (1991). New production in the summer revealed by the meridional slope of the deep chlorophyll maximum. *Deep-Sea Research*, 38, 35–56.

Wroblewski, J. S., Sarmiento, J. L. & Flierl, G. R. (1988). An ocean basin scale model of plankton dynamics in the North Atlantic. 1. Solutions for the climatological oceanographic conditions in May. *Global Biogeochemical Cycles*, **2**, 199–218.

Yentsch, C. S. (1993). CZCS: Its role in the study of the growth of oceanic phytoplankton. In *Ocean Colour: Theory and Applications in a Decade of CZCS Experience*, ed. V. Barale, & P. M. Schlittenhardt, pp. 17–32. Dordrecht: Kluwer Academic Publishers.

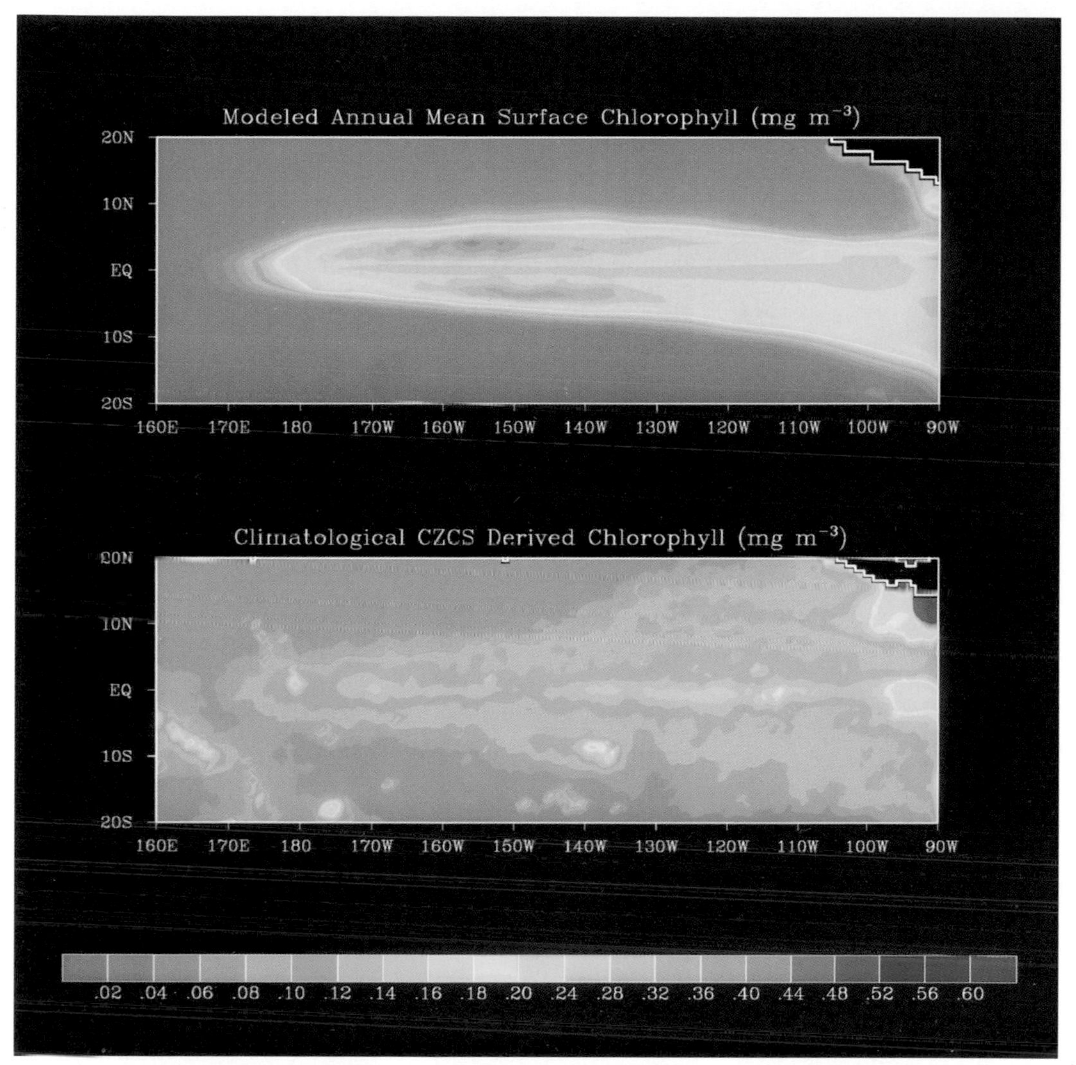

Figure 6.2 *Comparison of modelled annual mean (year six) surface chlorophyll (top panel) with the climatological chlorophyll derived from Coastal Zone Color Scanner (CZCS) data (below).*

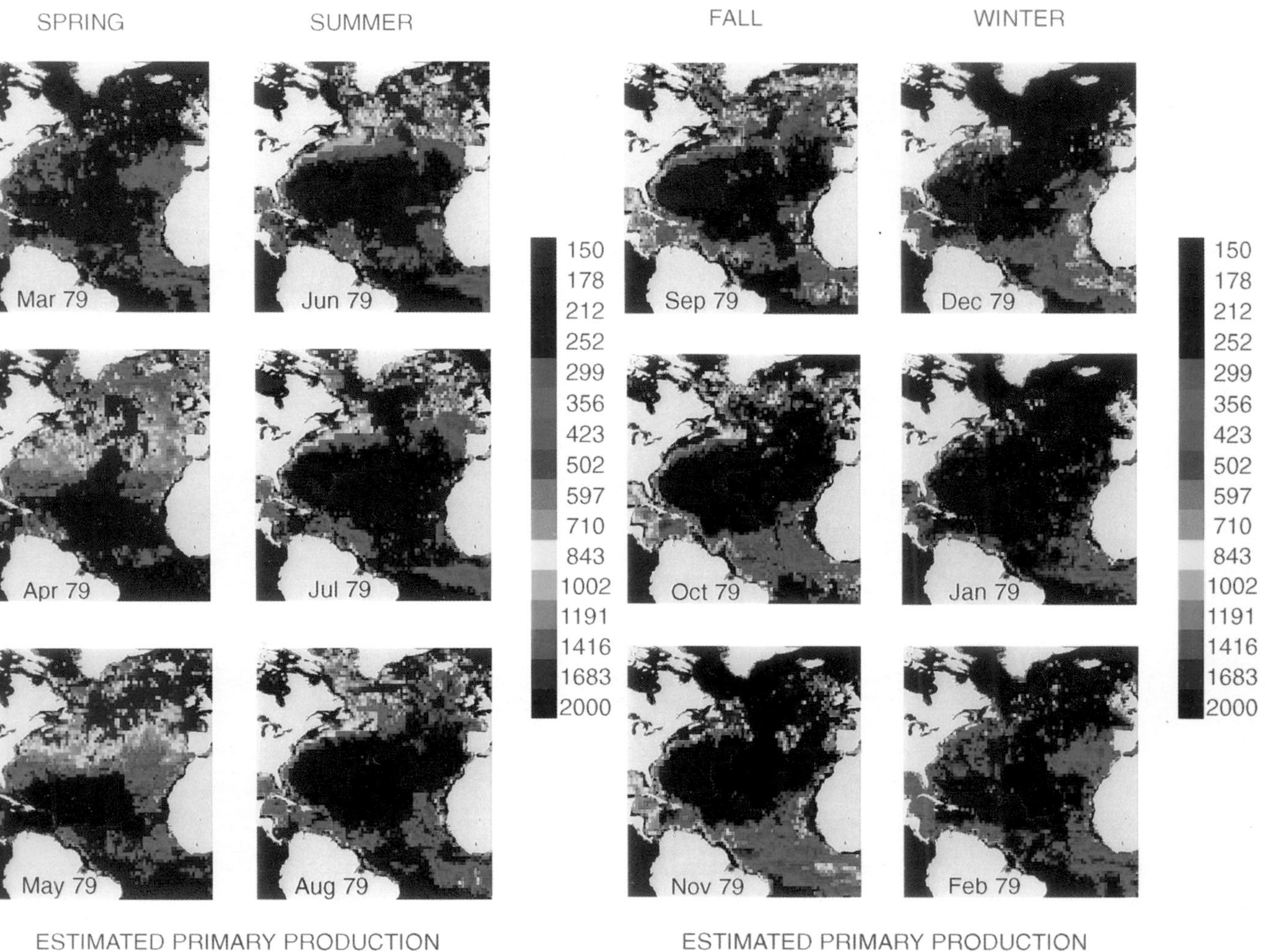

Figure 15.2 *Monthly averages of daily primary production estimated from ocean-colour data for 1979. The grey areas represent land; the black areas indicate lack of ocean-colour data. By virtue of the spatial resolution in the satellite data, the images show considerable detail. However, in using them it should be borne in mind that the parameters on which the computations are based are resolved only to the level of the algal domains (for the photosynthesis parameters) or the level of the biogeochemical provinces (in the case of the parameters describing the shape of the pigment profile). The resolution of the parameters will not improve until prescriptive functions become available for estimating their magnitudes at each pixel.*

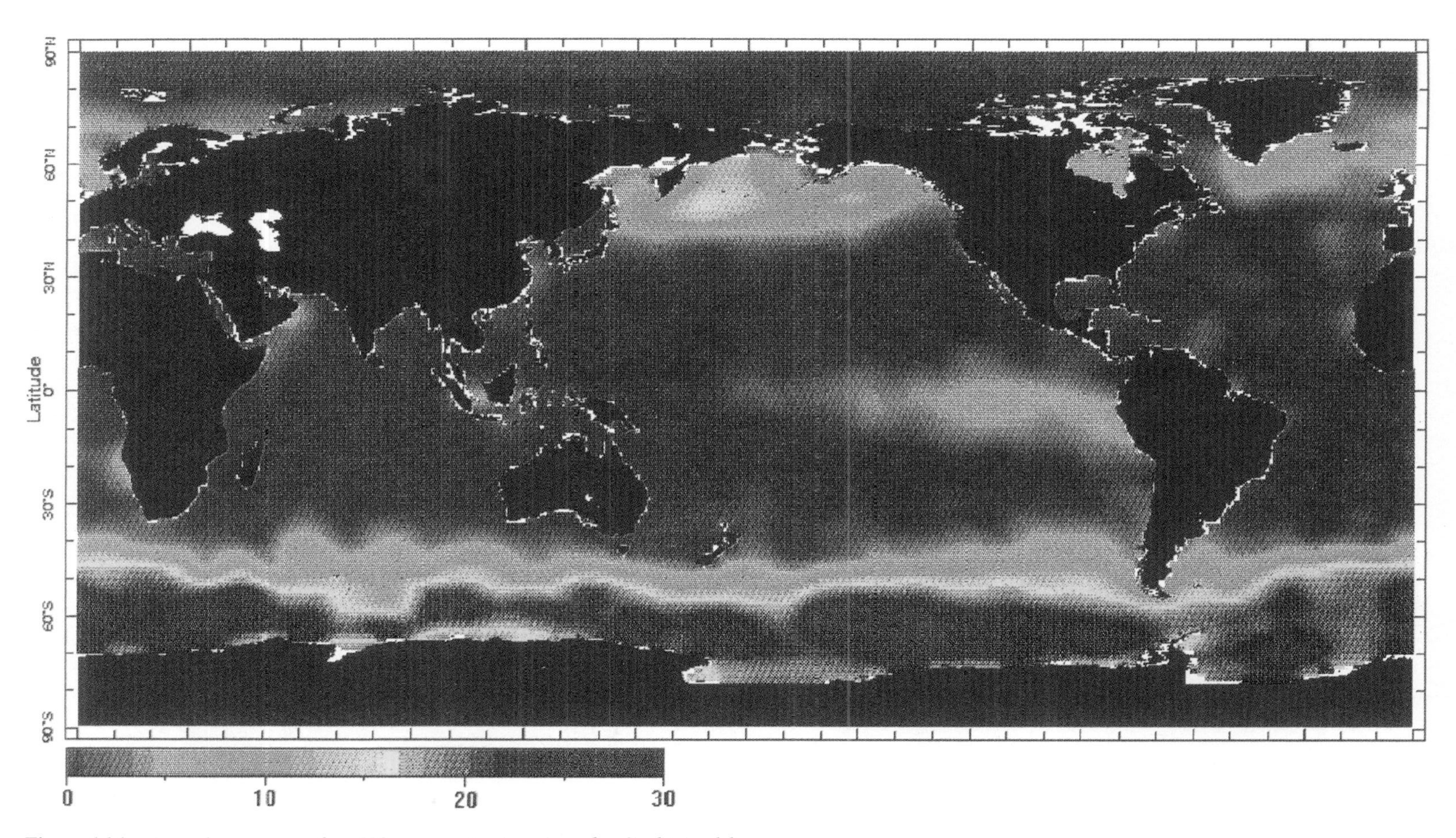

Figure 16.1 *Annual average surface NO_3 concentrations (mmol m^{-3}) obtained from the Levitus data set. Plot downloaded from the Lamont–Doherty Earth Observatory web page* (http://rainbow.ldeo.columbia.edu/).

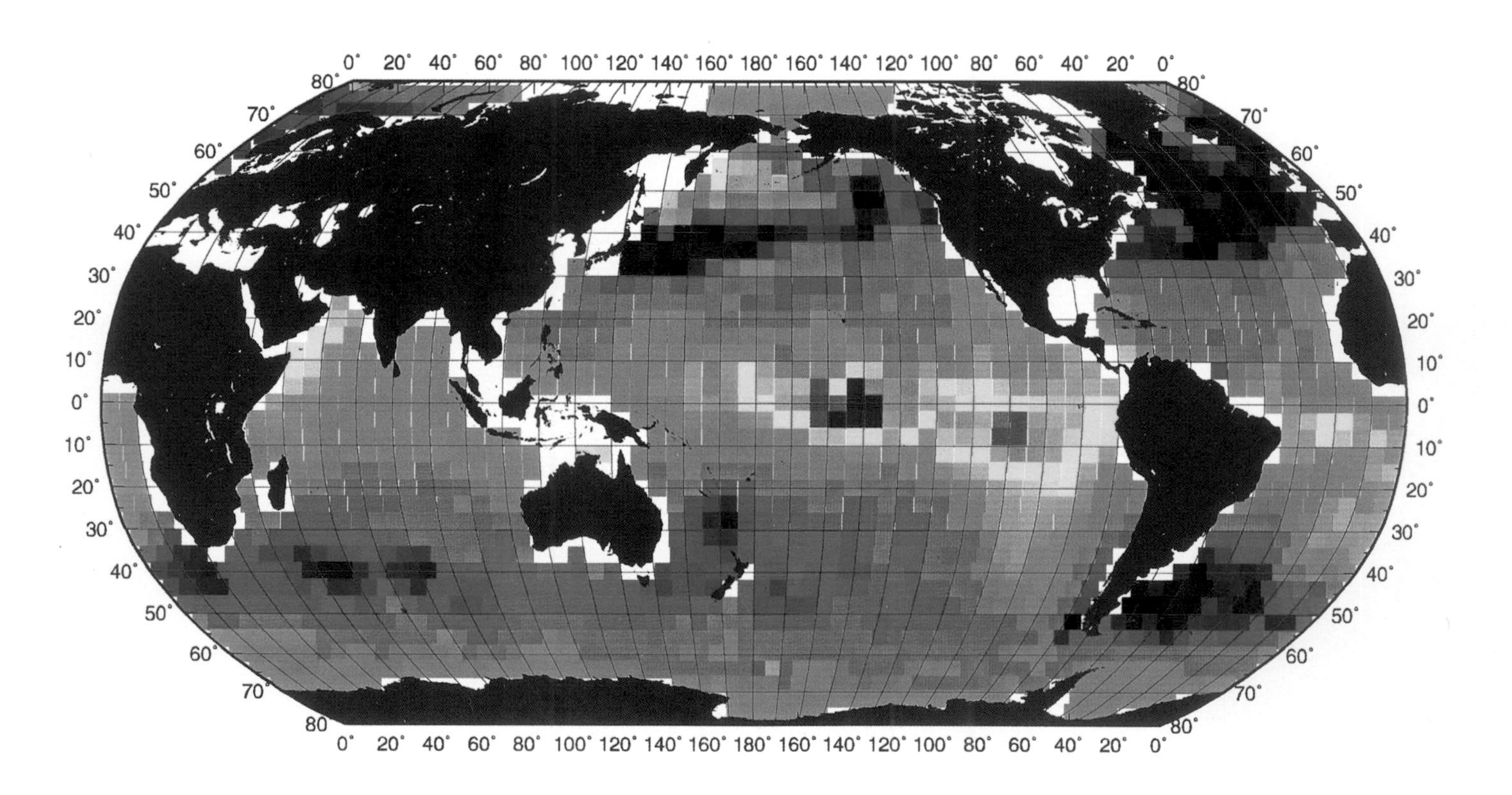

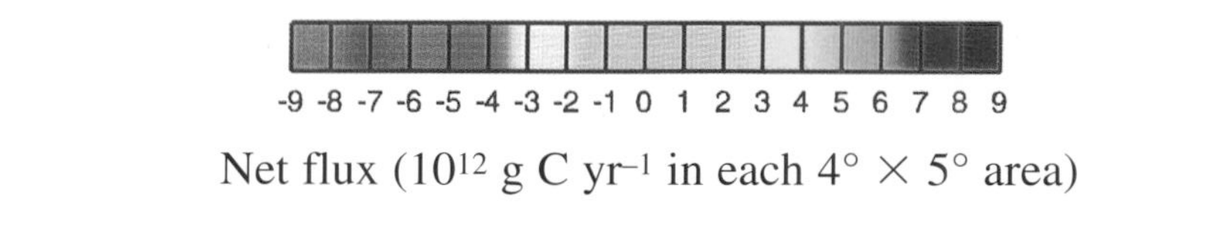

Figure 17.1 *Mean annual net CO_2 flux over the global oceans (in 10^{12} g C yr^{-1} for each pixel area) computed for 1990 using the gas transfer coefficient formulated by Wanninkhof (1992). The effect of full atmospheric CO_2 increase is assumed for normalising observed $^{P}CO_2$ values in high latitude areas to the reference year of 1990. Areas covered with ice are assumed to have zero sea–air CO_2 flux.* From Takahashi *et al.* (1997). *Reprinted with permission from* Proceedings of the National Academy of Sciences. *Copyright 1997 by the National Academy of Sciences, USA.*

PART FIVE

FUTURE CHALLENGES

16 Beyond JGOFS

K. L. Denman and M. A. Peña

Keywords: atmospheric CO_2 exchange, carbon and nitrogen export, 'biotic pump', re-mineralisation, Redfield ratio, ocean carbon models

Introduction

This chapter attempts to assess the progress towards greater understanding of the ocean carbon cycle made by JGOFS by the mid-1990s and to project, in the wider context of climate change, what scientific issues need to be addressed during the remaining years of JGOFS and what future scientific activities and programmes might be needed to complement and extend the findings of JGOFS. Two events stimulated the ideas presented in this paper. First, the JGOFS Scientific Symposium held in Villefranche-sur-mer during May 1995, part of a scientific review of the International Geosphere–Biosphere Programme (IGBP), provided the impetus for preparation of a group of reviews of JGOFS science (this volume). Second, we participated in the 1995 scientific assessment of climate change conducted by Working Group I of the Intergovernmental Panel on Climate Change (IPCC), specifically Chapter 10: Marine biotic responses to environmental change and feedbacks to climate (Denman *et al.*, 1996). This task required us to attempt an assessment of marine biogeochemical processes with potential for feedbacks to climate, covering topics far broader than the carbon-based research of JGOFS.

Where was JGOFS in the mid-1990s?

The state of our knowledge and understanding of ocean biogeochemical processes has progressed markedly in the decade since JGOFS was first being conceived and developed as an international programme. Early in the planning of JGOFS, the decision was made to focus on the natural ocean cycle of carbon and associated biogenic elements, (see, for example, JGOFS Science Plan, 1990), because CO_2 accounts for roughly half of the radiative greenhouse gas effect and because the ocean appears to accept a net input of CO_2 from the atmosphere equivalent to roughly one third to one half the input to the atmosphere from fossil-fuel burning and deforestation. The JGOFS strategy included (1) intensive process studies in key geographic areas, (2) a global survey using remote sensing, ship transects and long time series, (3) model studies for system identification, data assimilation, and prediction, (4) reconstruction of past climate from sediments, and (5) an international data archiving activity. In 1996, most process studies had completed at least one or all their planned intensive sampling periods; yet there was still no widely available colour sensor in orbit to aid in scaling up from the limited-area process studies to ocean basin and global scales.

Results from the various process studies have not yet allowed us to develop better budgets for the pool sizes and fluxes of carbon, especially in the upper ocean. Rather, the explosion in our knowledge of the ocean carbon cycle has raised our awareness of previously poorly known or unknown processes, but has yet to be assimilated into a revised paradigm for the ocean carbon cycle. It seems that at present the definitions of new, export, and recycled carbon and nutrients seem to be inadequate in an operational sense, water-column transports and transformations of carbon and related bionic elements have turned out to be extremely complex, and although modelling and synthesis must span from the microscope to the process study scale to the global scale, we are slow in developing the necessary parameterisations. In addition, new problems have been identified and tackled by JGOFS: most notably, trying to understand the factors controlling nutrient utilisation in high-nutrient – low-chlorophyll (HNLC) regions (see Fig. 16.1, colour plate) has became an area of intense interest and activity within JGOFS. Finally, other global marine programmes are under way – LOICZ and the Global Ocean Ecosystem Dynamics programme (GLOBEC) within IGBP, and CLIVAR within WCRP – that offer opportunities for collaboration.

The IPCC 1995 Scientific Assessment

Chapter 10 of the IPCC 1995 Scientific Assessment (Denman *et al.*, 1996) identified a number of critical mechanisms or processes through which marine biota (may) respond to environmental change in such a way as to develop feedbacks to the climate system. Several of these, which we will group under the theme that we term *operation of the biotic pump*, are processes that are part of or can influence the ocean carbon cycle, the central theme of JGOFS. [We use the term 'biotic pump' to include biological uptake and transport from the ocean surface layer to the ocean interior of carbon, both as organic molecules ($C_{organic}$) and as $CaCO_3$ ($C_{carbonate}$).] We summarise below the assessment of these processes in Chapter 10.

- Carbon is usually considered to be taken up by phytoplankton during primary production and released during re-mineralisation of organic matter in constant (atom to atom) proportions of the major nutrients $C_{organic}:N:P$, referred to as the Redfield ratios. If these ratios (or their vertical gradients) change in time in response to changes in ocean circulation or other properties, model simulations indicate that there is a large potential for the biotic pump to influence atmospheric CO_2 concentrations.

- $CaCO_3$ is fixed in the surface ocean by those marine algae and animals that have 'hard' parts of $CaCO_3$. This $CaCO_3$ also sinks out of the surface layer with the exported organic carbon, but each molecule of $CaCO_3$ removed is accompanied by creation of a molecule of CO_2 in the surface ocean, counteracting the removal of the organic carbon. Globally, the 'rain ratio' $C_{organic} : C_{carbonate}$ exported from the surface ocean is about 4 : 1, but a shift in phytoplankton species causing a shift to a ratio of 1 : 1 would cause the 'biotic pump' to have no net effect on surface p_{CO_2}. At present, we cannot quantify the probability or extent of such a shift occurring as a response to climate change.
- Biogeochemical processes occur principally in the top few hundred metres of the sea. The ocean margins play a major, but poorly assessed, role in oceanic biogeochemical cycling and are the burial sites of a substantial amount of organic carbon derived both from these oceanic processes and from terrestrial sources. A potential feedback may result from anthropogenic eutrophication of continental shelf areas. There, increased nutrient availability may stimulate the production of organic matter in the surface layer and its re-mineralisation at depth, reducing O_2 concentrations to levels that promote denitrification and thus the release of elemental N_2 and of N_2O, a potent greenhouse gas.
- New nutrients (including iron) coming from outside the ocean (both as a result of increased atmospheric deposition or coastal runoff from human activities) would increase organic carbon production, its export to the deep ocean, and the drawdown of atmospheric CO_2 (probably by less than 1 Gt C yr^{-1}). Increased export of carbon production would also result in increased re-mineralisation of organic matter at depth with an accompanying increase in production of N_2O, especially (as above) if low O_2 concentrations develop. N_2O is the third most important increasing greenhouse gas after CO_2 and CH_4, and the oceans now contribute about 20% of the total input of N_2O to the atmosphere.
- Iron in atmospheric dust may have contributed to the transition from glacial to interglacial periods. Lack of iron appears to limit phytoplankton growth in oceanic regions where macronutrients, particularly nitrogen, phosphorus and silica, are abundant yet phytoplankton concentrations are generally low (see Fig. 16.1). According to modelling studies (Sarmiento & Orr, 1991; Kurz & Maier-Reimer, 1993), intentional iron fertilisation of the Southern Ocean to promote phytoplankton growth and hence the drawdown of atmospheric CO_2 appears not to be viable.

The second theme of IPCC 1995 Chapter 10 was *the dimethyl sulphide (DMS) cycle* whereby DMS formed by (some) phytoplankton can outgas to the atmosphere and contribute to the formation of clouds. DMS is the principal volatile sulphur-containing compound in the sea and the major natural source of sulphur to the atmosphere. In the atmosphere, DMS is oxidised to produce aerosols, which promote the development of clouds and thus influence climate. The production of DMS is a function of the composition and concentration of the plankton. DMS is removed from seawater by physical, chemical and biological processes. In temperate and tropical waters, the rate of biological consumption can be more than ten times greater than atmospheric ventilation. A much better understanding of sulphur cycling in the surface ocean is required for quantitative assessment.

The final theme of Chapter 10 was concerned with potential damage to the marine food web from *solar ultra-violet (UV) radiation*. UV radiation has increased as a consequence of stratospheric ozone depletion. UV-B (280–320 nm) exposure depresses primary production and growth of phytoplankton. As species differ in their tolerance to UV-B, a shift in community structure favouring the more tolerant would be expected. UV-B radiation inhibits marine bacterial activity and has differing effects on dissolved organic carbon (DOC) derived from various sources. The effects of UV-B on marine biogeochemical processes must be taken into account for a realistic assessment of the role of these processes in environmental change.

These summaries from the IPCC 1995 report represent a scientific assessment of what we believe that we know about marine biotic processes that have realised or potentially important feedbacks to the climate system. We now need to identify gaps in our knowledge in those specific areas considered to be of high priority in the context of increasing our understanding of the processes controlling climate change. Particularly relevant to JGOFS was the recognition that external nutrients entering the ocean, generation of low O_2 conditions, and formation of gases such as N_2O, N_2 and CH_4 that can outgas to the atmosphere represent processes where the ocean cycles of O_2, C, N and P are not related through the Redfield ratios, as they are so often assumed to be.

Model sensitivities of atmospheric CO_2 to ocean processes

There is little doubt that the ocean carbon cycle has played an essential role in controlling the atmospheric concentration of CO_2 in the past, and will continue to do so into the future. The oceans contain about 40 000 Gt C in dissolved, particulate and living forms, more than 50 times that in the atmosphere and almost 20 times that contained in land biota, detritus and soils (Denman *et al.*, 1996). In addition, the biotic pump maintains a considerable vertical gradient in carbon in the ocean, with the surface-layer inorganic carbon concentration (and hence the partial pressure of CO_2) being much lower than in the absence of marine biota. Consider that the pre-industrial equilibrium atmospheric concentration of CO_2 was *c.* 280 ppm (v). Various model simulations (see, for example, Sarmiento & Toggweiler, 1984; Shaffer, 1993) indicate that a fully efficient biotic pump (capable of utilising all surface nitrate) would have resulted in an atmospheric CO_2 concentration of *c.* 160 ppm (v), and an abiotic ocean (with extinction of all marine production) would have led to an atmospheric CO_2 concentration of *c.* 450 ppm (v). Despite this wide range of potential impact on atmospheric CO_2 concentrations by changes in operation of the biotic pump, it is considered to have remained in approximate steady-state during the build-up of atmospheric CO_2 over the last century (Siegenthaler & Sarmiento, 1993).

Since the beginning of JGOFS, models of the ocean carbon cycle have become more realistic than the first-generation box models. In particular, two studies have carried out model sensitivity analyses to determine how much atmospheric CO_2 concentrations might change in response to changes in various aspects of the ocean carbon cycle. We summarise the results of these simulations below in an attempt to rank various processes within the ocean carbon cycle according to the sensitivity of atmospheric CO_2 concentrations to changes in these processes. We hope that by identifying in this manner the processes with potential for the greatest feedbacks to the climate system, we can then evaluate our understanding of these processes, now and possibly by the completion of JGOFS.

The first model, the HIgh Latitude exchange/interior Diffusion-Advection (HILDA) model (Shaffer, 1993, 1996; Shaffer & Sarmiento, 1995), consists of a warm surface box, a cold surface box, a high latitude well-mixed sub-surface box, and a low-latitude diffusive continuously stratified sub-surface box. A thermohaline circulation cell was applied to the model, and the various exchange coefficients were calibrated with observations of temperature and ^{14}C. The second model (Heinze *et al.*, 1991) consists of a three-dimensional ocean carbon cycle model with transports from an 11-layer $3.5^\circ \times 3.5^\circ$ version of the Hamburg large-scale geostrophic ocean general circulation model. The physical model was integrated up to quasi-steady-state (15 000 years), and each of the sensitivity runs was integrated for 20 000 years to a stationary state. Hence, the responses in this model (and in the steady-state Shaffer model) are representative of glacial–interglacial time scales, but most of those processes related to functioning of the biotic pump would still be operative on the decadal to millennial time scale relevant to climate change.

Table 16.1. *Sensitivity analysis of two-ocean carbon cycle models*
Changes in atmospheric CO_2 concentrations, Δp_{CO_2} (ppm (v)), are taken relative to a pre-industrial concentration of *c.* 280 ppm (v).

Process or parameter	Variation	Atmospheric Δp_{CO_2} (ppm)(v)
Redfield ratio, Carbon: nutrient	$2\times^{S}$	−126
	$+30\%^{H}$	−72
Re-mineralisation depth scales		
Carbon	$2\times^{S}$	−62
Both carbon and nutrient	$2\times^{S}$	−35
Nutrient	$2\times^{S}$	+20
Increase high latitude productivity	$2\times^{S}$	−59
	c. $3\times^{H}$	−32
High latitude mixing/ventilation		
Vertical exchange	$2\times^{S}$	+44
Ventilation (current amplitudes)	$0.5\times^{H}$	−26
Change 'rain ratio' $C_{organic}:C_{carbonate}$	$0.5\times^{S}$	+36
	$2\times^{H}$	−28
Surface solubility of inorganic carbon		
Sea-surface temperature of warm box	$+5\,°C^{S}$	+12
Sea-surface temperature of model	$-2\,°C^{S}$	−22

^{S}Shaffer model.
^{H}Heinze model.

We have summarised the results of the sensitivity studies in Table 16.1, grouping similar parameter changes where possible and ranking the parameters and processes from the largest to the smallest change. Whereas Shaffer has examined the sensitivity to doubling or halving certain processes, Heinze *et al.* have examined the sensitivity to changes in each parameter that might be expected to have occurred during glacial–interglacial transitions. The models are most sensitive to changes in the Redfield ratio of the uptake of C to the limiting nutrient in primary production. Increasing the number of atoms of carbon taken up for every molecule of nutrient by, say, 30% could reduce the (pre-industrial) atmospheric CO_2 concentration by 25%. The Shaffer model is next most sensitive to the depth scale for re-mineralisation; i.e. the depth above which some fraction, say half, of the organic carbon has been re-mineralised. If carbon were re-mineralised at twice the depth (on average) as the limiting nutrient, then the inorganic carbon would take proportionally longer to return to the sea surface than the limiting inorganic nutrient. As the limiting nutrient would reach the surface ocean before the corresponding carbon, it could remove through photosynthesis other carbon from the surface layer, and the atmospheric CO_2 concentration would be reduced by more than 20%. Doubling the re-mineralisation depth scale of both carbon and the limiting nutrient would reduce atmospheric CO_2 by about half as much, because more carbon would be stored in the ocean owing to the longer return route to the surface for both carbon and nutrient. The models are almost as sensitive to the doubling of high-latitude productivity as to the doubling of the re-mineralisation depth scales. Next, doubling $CaCO_3$ production (or halving the 'rain ratio' of $C_{organic}:C_{carbonate}$ from the surface ocean), as a result of a major shift in phytoplankton community composition towards coccolithophores, would increase the atmospheric CO_2 concentration by *c.* 10% (because every molecule of $CaCO_3$ fixed releases a molecule of CO_2, thereby counterbalancing the effect of formation of organic carbon), a change about half as large (but of opposite sign) as doubling the C re-mineralisation depth scale or doubling high-latitude productivity. Finally, changing the solubility of CO_2 by increasing sea-surface temperature by several degrees would increase atmospheric CO_2 by roughly half as much again as halving the rain ratio.

In summary, the models suggest that the ocean carbon system is sensitive, in terms of the effect on levels of atmospheric CO_2, to changes (in decreasing order) in the uptake Redfield ratio, the re-mineralisation depth scale for carbon relative to the limiting nutrient, high-latitude primary production, high-latitude mixing or ventilation, the rain ratio from the surface ocean, and finally sea-surface temperature. In the next section, we briefly review our knowledge of the ranges of variability of and factors controlling some of these processes.

Adjusting the biotic pump

In the context of climate change, we are interested in how changes in the biotic pump – ocean carbon cycle that occur in response to environmental change might provide feedbacks to the climate system. In the last section, we ranked several aspects of the biotic pump in terms of the sensitivity of modelled atmospheric CO_2 levels to their variation. Two questions need to be addressed. First, can we measure these critical variables well enough to detect change at inter-annual and longer time scales? Second, do we understand these aspects of the operation of the biotic pump well enough to predict possible feedbacks to climate?

The null paradigm: constant Redfield ratios

Carbon is usually considered to be taken up by phytoplankton during primary production and released during re-mineralisation of organic matter in constant proportion to the major nutrients, $C_{organic}:N:P = 106:16:1$, referred to as the

Redfield ratios (Redfield *et al.*, 1963). Furthermore, the rate of carbon uptake during primary production is considered to be controlled by the supply of the limiting nutrient into the sunlit upper ocean from the ocean interior and from external sources. Therefore, a constructive approach towards analysing the C, N, P and O cycles within the ocean might be to adopt as the null paradigm that these cycles are related by constant Redfield ratios. By removing the portions of the measured, calculated or modelled fluxes expected according to the null paradigm, we would be left with only the small but important departures from the null paradigm, which often are lost among the various sources of noise and uncertainty because numerically the largest fraction of the fluxes are related according to constant Redfield ratios.

Departures from Redfield ratios

If climate change causes changes in the input of limiting nutrients from land, the atmosphere, and/or the ocean interior to the upper ocean (and thus changes the production of export carbon), then the rate of sequestering of carbon by the biotic pump could change with possible feedbacks to the build-up of greenhouse gases in the atmosphere. Similarly, if climate change causes changes in the Redfield ratio of the utilisation of C to that of the limiting nutrient in primary production, then more or less carbon would be transferred by the biotic pump for each unit of limiting nutrient, again with possible implications for oceanic carbon sequestration. The assumption of constant Redfield ratios for utilisation by phytoplankton has been inferred from the general constancy of the ratios in the composition of phytoplankton (Slawyk *et al.*, 1978; Parsons *et al.*, 1984), although, in laboratory experiments, C: N composition ratios vary with growth rate and the inferred degree of nutrient limitation (see, for example, Sakshaug, 1977). For re-mineralisation, the ratios of change of dissolved inorganic forms (i.e. concentrations corrected for 'pre-formed' amounts) in the deep ocean generally conform to constant Redfield ratios (Anderson & Sarmiento, 1994).

Recent studies, however, have documented departures from Redfield ratios in phytoplankton composition and uptake. Phytoplankton C: N (and C: P) uptake ratios tend to be higher and more variable than composition ratios, a difference that could be due to the release by phytoplankton of dissolved organic matter (DOM) with a C: N ratio higher than the Redfield value. Slawyk *et al.* (1978) measured atomic C: N composition ratios in the range 4 to 10, but utilisation ratios (from uptake studies) in the range 4 to 45, with only one value significantly below the Redfield value of 6.6 ($n = 20$). In a review, McCarthy (1980) cited a number of investigations where the C: N uptake ratio generally differed from the Redfield value. Sambrotto *et al.* (1993) estimated a 'net community production' C: N ratio in several areas by measuring changes in nitrate and dissolved inorganic carbon (DIC) in the surface ocean layer, obtaining values in the range 9 to 12.

Contemporary studies comparing the magnitude of the export of particulate organic matter (POM) from the euphotic zone with the magnitude of new production (from ^{15}N-labelled compounds, as defined by Dugdale & Goering, 1967) suggest that suspended and sinking particles represent only a fraction of the new production. As much as 50% may be released as DOM (see, for example, Carlson *et al.*, 1994; Michaels *et al.*, 1994; Murray *et al.*, 1994; Williams, 1995). Usually, the DOM (and suspended POM) accumulates in the upper layer during the summer season. That fraction which has not yet been re-mineralised or broken down by photo-degradation within the euphotic zone may be exported to greater depths during autumn and winter through mixing, diffusion and advection. Williams (1995) reviewed the few studies where DOM and POM were both measured and concluded that usually the accumulated DOM pool was elevated in C relative to N. Ducklow (1995) cited C: N ratios for bulk DOM ranging from 15 to 38.

W. G. Harrison (personal communication) has deployed drifting sediment traps for periods from 2 to 9 days in the North Atlantic immediately below the euphotic layer (at depths ranging from 70 to 150 m), thereby capturing particles representing the sinking POM fraction of export production. Over a 7 year period at latitudes between 23 and 45°N, he observed particulate C: N ratios ranging between 6.2 and 11.7 with a mean of 9.1 ($n = 12$). Williams (1995) inferred an inorganic C: N utilisation ratio of 15 from seasonal increases in mixed-layer DOM and POM in the English Channel. However, Banse (1994) re-analysed previous observations of nitrate and DIC changes from some enclosure experiments and calculated C: N utilisation ratios below 6.6. The balance of evidence reviewed here suggests that phytoplankton C: N utilisation ratios exceed the Redfield value (6.6) in the order of 30%.

Re-mineralisation depth scales

If particulate N (or P) has a smaller re-mineralisation depth scale relative to C, then N (or P) is re-mineralised at shallower depths and the remaining sinking POM sampled throughout the water column should be fractionated towards increasing C: N (and C: P) ratios with increasing depth. Early observations by Bishop *et al.* (1977, 1980) from large-volume *in situ* pumps and Martin *et al.* (1987) from moored sediment traps support this hypothesis for both suspended and sinking fractions of POM. Contemporary observations from JGOFS studies are more compelling. In samples from sediment traps moored during the JGOFS 1989 North Atlantic Blooms Study, Honjo & Manganini (1993) observed C: N: P ratios of 154: 18: 1 (at 34°N, *c.* 4500 m depth) and 148: 18: 1 (at 48°N, *c.* 3700 m depth), the C: N ratios elevated by about 20% over Redfield values.

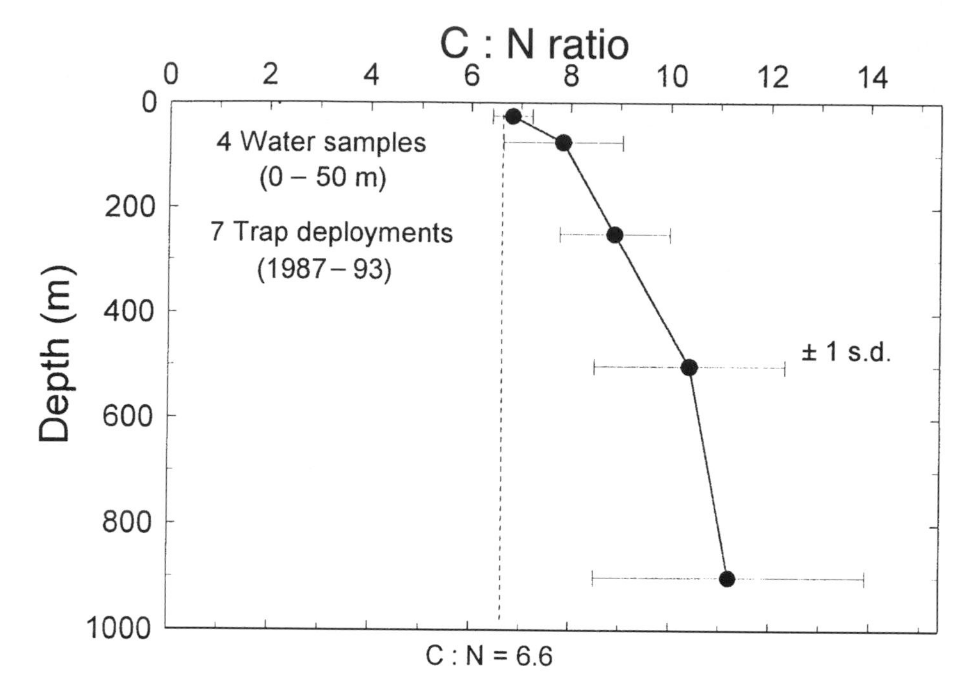

Figure 16.2 *C: N ratios in organic material collected by large-volume filtering (0–50 m) and by drifting sediment traps near Ocean Station Papa (50°N, 145°W). All deployments were for a week or less duration. Vertical dashed line indicates atomic Redfield C: N of 106: 16. (From C. S. Wong and F. Whitney, personal communication.)*

Because most of the re-mineralisation occurs in the top few hundred metres, drifting traps deployed for periods of a week or less may provide more relevant data with better vertical resolution. Off Bermuda, Lohrenz *et al.* (1992) found the C: N ratio in POM to increase from approximately the Redfield value at 150 m to about 10 (range 5–29) at 400 m. Off Hawaii, Karl *et al.* (1996) found the ratio to increase from a mean of about 8 at 150 m to about 12 at 500 m. At Ocean Station Papa (50°N, 145°W) in the subarctic Pacific Ocean, C. S. Wong & F. Whitney (personal communication) have found the C: N ratio of suspended particles in the upper 50 m to equal the Redfield value (6.6) but sinking particles caught in drifting traps yielded a C: N ratio increasing from 8 at 100 m to about 11 at 900 m (averages over 7 deployments; Fig. 16.2). All these observations taken together indicate a significantly larger re-mineralisation depth scale for particulate C relative to particulate N (or P).

Toggweiler (1989) and Najjar *et al.* (1992) have shown that general circulation models that include significant re-mineralisation of DOM (and suspended POM) produce more realistic vertical nutrient profiles than those that equate export production to sinking POM only. In particular, the inclusion in their models of only sinking POM generated excess re-mineralisation of nutrients at relatively shallow depths in equatorial upwelling regions, which they call 'nutrient trapping'. However, re-mineralisation depth scales for DOM (and suspended POM) are poorly known for two reasons. First, different fractions of DOM (not usually measured) have re-mineralisation time scales ranging from hours to weeks to years, referred to as labile, semi-labile and refractory fractions (see, for example, Carlson & Ducklow, 1995). Second, the vertical transport mechanisms (mixing, diffusion and advection) differ in dominance and magnitude both temporally and spatially. For example, one might expect the DOM re-mineralisation depth scale to be of the order of 100 m in the subarctic Pacific and several hundred metres in the North Atlantic, reflecting the large difference in maximum depth of surface-induced winter mixing. However, we cannot quantify with precision the physical processes transporting DOM and suspended POM downwards from the euphotic zone, especially subduction associated with mesoscale eddies and large scale oceanic fronts or transition zones. Finally, we know even less about the relative rates of re-mineralisation of C, N and P in DOM and suspended POM than in sinking POM.

We can infer some properties from models of re-mineralisation rates for POM and DOM taken together. Shaffer (1996) performed extensive parameter estimation and sensitivity studies with the HILDA model to achieve the best fit to global observations of physical variables, nutrients, carbon and oxygen. His best fit for the C: N: P ratios of new production is 105: 15: 1, close to the Redfield values. Best-fit re-mineralisation depth scales were 1370, 1110 and 1060 m for C, N and P. C: P re-mineralisation ratios (for material converted, as opposed to material not yet converted from the organic phase) were 81, 99, 122 and 150 at model depths of 100, 1000, 2000 and 3000 m. C: N ratios (assuming a constant N: P ratio) were 5.1, 6.2, 7.6 and 9.4 at the same depths. That is, P and N must be preferentially re-mineralised at shallower depths, leaving POM and DOM increasingly enriched in C with increasing depth, consistent with observations from the top 500 m which clearly indicate that POM settling in traps becomes enriched in C relative to N with increasing depth. Anderson & Sarmiento (1994) did find a more or less constant ratio for C: N: P of 117: 16: 1 (near the Redfield values) from an inverse analysis of dissolved inorganic nutrient concentrations (corrected for preformed nutrients) on 20 neutral surfaces over depths from 400 to 4000 m, but as with re-mineralisation, mechanisms supplying nutrients to the euphotic zone from below on time scales less than decades (e.g. upwelling, winter mixing) occur primarily in the top 500 m.

The effect on the surface-layer C concentration of such a partitioning of limiting nutrient re-mineralisation and recycling within the top 500 m, and carbon re-mineralisation and recycling primarily at depths below 500 m, is captured in the simple 3-layer model described by Evans & Fasham (1993). Exchange between the surface layer and the middle layer (here the layer between the surface layer and 500 m) would be vigorous because it is driven by surface-driven mixing and upwelling, whereas exchange between the middle and deep layers would be less vigorous because it occurs primarily through the thermohaline circulation. The (limiting) nutrient would thus be returned from the middle layer to the surface layer much faster than the carbon that it fixed during photosynthesis, which would have to be transported from the deep layer to the middle layer and then to the surface layer. The limiting nutrient returned to the surface layer could fix through photosynthesis more surface layer carbon, some of which would sink out of the surface layer before a significant fraction of the original carbon had returned to the surface layer. Surface C concentrations, and hence p_{CO_2} would be reduced relative to an ocean in which both utilisation by phytoplankton and re-mineralisation occurred according to constant Redfield ratios.

What controls the rain ratio?

For every C atom fixed into an organic molecule by photosynthesis, a molecule of CO_2 is removed from the surface ocean layer. For every C atom fixed into $CaCO_3$, 'hard' body parts in mainly coccoliths, foraminifera and pteropods, one molecule of CO_2 is released into the surface ocean according to the calcification equation:

$$Ca^{2+} + 2HCO_3^- \rightarrow CaCO_3 + CO_2 + H_2O,$$

thereby cancelling the uptake of one molecule of CO_2 through creation of organic carbon by photosynthesis. The composition ratio $C_{organic} : C_{carbonate}$ of particles sinking out of the euphotic zone, the 'rain ratio', represents the net effect on removal of CO_2 from the surface ocean. If the rain ratio reaches 1 : 1, then the biotic pump releases as much CO_2 to the surface layer through $CaCO_3$ formation as it removes through formation of organic carbon molecules.

According to Broecker & Peng (1982), the average $C_{organic} : C_{carbonate}$ ratio in the upper ocean is about 4: 1, because phytoplankton communities are usually dominated by organisms that do not build $CaCO_3$ shells. Tsunogai & Noriki (1991) tabulated most of the deep-ocean sediment-trap records up to 1989 and found a median ratio of about 4: 1 (they plotted the inverse) at about 500 m, their shallowest depth, although the observations themselves span a wide range (from just greater than 1 to about 20). As expected because of re-mineralisation of organic carbon at shallower depths, the median ratio decreased to about 1.6: 1 at 5000 m, although some observations showed a reduction in carbonates at the greatest depths, as would be expected owing to dissolution below the lysocline. Generally, the highest rain ratios were for latitudes greater than 40° and the lowest were for latitudes less than 35°, corresponding most likely to high and low proportions of diatoms, with silica structures rather than carbonate structures. From the analysis of dissolved properties, Anderson & Sarmiento (1994) calculated a $C_{organic}:C_{carbonate}$ 'utilisation ratio' for the Indian Ocean decreasing from about 7 at 400 m to 2 at 1000 m to about 1 at 4000 m (and for the Pacific Ocean decreasing from 5 to 4 to 1 over the same depth range), confirming that there is some dissolution of $CaCO_3$ throughout the water column. It is likely that the rain ratio out of the euphotic zone is largely controlled by the dominant species fixing carbon through photosynthesis. Robertson *et al.* (1994) documented $C_{organic}:C_{carbonate}$ production ratios approaching 1 : 1 in the North Atlantic during a bloom of the coccolithophore *Emiliania huxleyi*. They estimated that the effect of such high calcification was to reduce the air–sea gradient in p_{CO_2} by *c.* 15 μatm from that expected during a similar bloom of low-carbonate-forming organisms.

We do not understand well the factors that control phytoplankton community composition, or those that cause changes. However, recent observations with long-term moored sediment traps demonstrate that multi-year trends in community composition occur. Deuser *et al.* (1995) observed a 50% reduction over 14 years in the ratio of the sinking fluxes of bionic silica (opal) to $CaCO_3$ in deep moored traps near Bermuda, indicating a shift away from diatoms towards coccolithophorids. The wind speed also increased by *c.* 25 % over the same period, suggesting variation in the region's climate. Fischer *et al.* (1996) observed approximately 100% increases in the $C_{organic}$: opal and $C_{organic}:C_{carbonate}$ ratios between 1988 and 1991 in deep moored traps off West Africa, primarily due to total fluxes decreasing while $C_{organic}$ fluxes remained unchanged. Sea-surface temperatures (estimated from $\delta^{18}O$ ratios in the samples) decreased by about 3 °C, with a possible increase in average wind speed. In an attempt to understand the factors that favour a bloom of carbonate-forming organisms, Westbroek *et al.* (1993) have recommended a research strategy for further study and quantification of the 'carbonate pump' (and the related DMS cycle) based on laboratory and modelling studies of the coccolithophore *Emiliania huxleyi*, which they argue is both an adequate representative of all calcifying marine organisms and accessible for laboratory experimentation.

Increasing productivity through fertilisation of HNLC regions

Martin (1990) proposed that because nutrients in the Southern Ocean remain at high non-limiting concentrations (see Fig. 16.1, colour plate), probably because of the lack of the trace metal iron, we should consider the possibility of increasing the uptake by the ocean of atmospheric CO_2 by fertilising the Southern Ocean with iron. A number of studies have addressed this possibility in the laboratory and in the ocean, with mixed results (see, for example, Kolber *et al.*, 1994; Martin *et al.*, 1994; Watson *et al.*, 1994; Cullen, 1995). However, model simulations (see, for example, Sarmiento & Orr, 1991; Kurz & Maier-Reimer, 1993) of the effects of perpetual fertilisation of the Southern Ocean to utilise all the available nutrients suggest that the reduction of the expected build-up of CO_2 in the atmosphere would be minimal. Basically, after roughly a decade from the start of fertilisation, there would be a continuing reduction of *c.* 50 ppm (v) in the increasing atmospheric concentration; that is, instead of the atmospheric concentration of CO_2 increasing from about 350 ppm (v) in 1990 to *c.* 900 ppm (v) by 2100, it would increase from about 300 ppm (v) to *c.* 850 ppm (v).

In the same way as we have used model sensitivity studies to rank processes affecting operation of the biotic pump, we can use model studies to suggest gaps in our understanding of iron fertilisation. The three-dimensional model of Sarmiento & Orr (1991) developed very low oxygen concentrations around a depth of 1500 m, resulting from increased oxygen utilisation through re-mineralisation of the increased rain of organic particles. A box model in fact predicted the generation of anoxic conditions (see, for example, Peng & Broecker, 1991). Fuhrman & Capone (1991) reviewed possible biogeochemical effects of widespread fertilisation: the species composition of the Southern Ocean might change in such a way as to alter the production of DMS or $CaCO_3$, both with feedbacks to climate change; widespread anoxia would lead to increased denitrification, i.e. production of N_2O and N_2 gases, some of which would escape to the atmosphere upon transport into the surface layer; and significant loss of nitrogen from the ocean would eventually tend to counteract the increased production of organic carbon by fertilisation.

Decoupling of C, N, P and O cycles

Different sources and sinks

The mounting evidence that particulate organic C, N and P re-mineralise at different rates would be enough to decouple the oceanic C, N, P and O cycles from simple conversion by fixed Redfield ratios. However, in addition these elements have different sources and sinks in the ocean.

As sinks of N, during re-mineralisation, about one atom of N is converted to N_2O for every 10^3 atoms of N converted to NO_3 (Broecker & Peng, 1982). Some of this N_2O gas eventually reaches the upper ocean and escapes to the atmosphere, a loss of both N and O, but not of C and P. Under oxygen-deficient conditions, oxygen needed in re-mineralisation is removed from NO_3 and N_2O, forming N_2 gas (denitrification), some of which also eventually escapes to the atmosphere.

As sources of N, there is significant atmospheric deposition of bioavailable NO_3 (see, for example, Galloway *et al.*, 1995), and fixation of atmospheric N_2 by the cyanobacterium *Oscillatoria* (*Trichodesmium*) (see, for example, Michaels *et al.*, Chapter 13, this volume). For example, during the 1991–92 warming event in the North Pacific, Karl *et al.* (1995) observed near Hawaii an increase in the proportion of *Trichodesmium* in the phytoplankton community sufficient to shift the planktonic ecosystem from N-limitation to P-limitation. O_2 consumption also takes place in sea-floor sediments, and especially in near-shore environments close to the sediment–water interface where N_2O, N_2 and CH_4 can be produced and diffused into the near-bottom waters.

The oceanic N budget: is it balanced?

Traditionally, ocean biogeochemical models use P rather than N as the nutrient, under the argument that there is a fixed inventory of oceanic P on time scales shorter than about 10^4 years. However, a more pragmatic reason is that such models do not have to deal with losses of N through denitrification and the addition of N by nitrogen fixation (Shaffer, 1996). Codispoti (1989), however, has reviewed the arguments for P versus N as the nutrient limiting new and export production. He concluded that N is most likely to be the limiting nutrient unless open-ocean N fixation is much greater than published estimates. Are the additional sources and sinks for N that operate on time scales less than those for P (principally runoff and sedimentation) in balance, or might we expect amplification of imbalances in response to environmental change sufficient to cause feedback to the climate?

Codispoti & Christensen (1985), Christensen *et al.* (1987) and Christensen (1994) estimated that losses (primarily from denitrification) are nearly twice the inputs in the ocean N cycle. Christensen (1994) estimated a net loss of N from the global oceans of 76 Tg N yr^{-1} (total losses 166 not 176, in his Table 1, J. P. Christensen, personal communication), of which 129 Tg N yr^{-1} results from denitrification. Converting with the standard Redfield ratio, 76 Tg N yr^{-1} is equivalent to a lost potential C-fixation of 0.4 Gt C yr^{-1}. Compared with a total oceanic N inventory of *c.* 1.1×10^6 Tg N (Codispoti, 1989), a loss of 76 Tg N yr^{-1} represents a turnover time for N in the ocean of *c.* 15 000 yr. Ganeshram *et al.* (1995) adopted a higher range for the present net loss of 85–115 Tg N yr^{-1}, but Michaels *et al.* (Chapter 13, this volume) estimate higher rates of N fixation, which would reduce the imbalance considerably.

From sediment cores, Ganeshram *et al.* (1995) and Altabet *et al.* (1995) found evidence of reduced water-column denitrification during glacial periods in the Eastern Tropical Pacific and the Arabian Sea, respectively. Ganeshram *et al.* argue that enhanced water-column denitrification in these areas during interglacial periods (such as the present period) would enhance the previously estimated 25% decrease in oceanic N inventory during interglacial periods, due to increased water column and sediment denitrification, because of a more intense oxygen minimum and much greater continental shelf area. Human activities in the coastal zone that increase the biological oxygen demand and thereby increase denitrification, and intentional fertilisation of the Southern Ocean that would increase re-mineralisation and denitrification, both have the potential to reduce the oceanic N inventory at a greater rate than at present. We do not know whether such a change would affect operation of the biotic pump, or on what time scales the effect would be manifest.

Components of future ocean biogeochemical studies

This book represents an overview of JGOFS scientific activities, achievements and outstanding problems in the mid-1990s. In this chapter, we have attempted to view JGOFS from the larger perspective of necessary global change science, gained through participation in the 1995 IPCC Scientific Assessment of Climate Change. Now we make a largely arbitrary attempt (1) to recommend the major scientific issues that should be emphasised during the remaining few years of JGOFS, (2) to identify future weaknesses in understanding that we believe will remain at the end of JGOFS as currently envisioned, and finally (3) to outline the structure of a global scientific study to follow JGOFS that we feel would best address the gaps in our scientific knowledge of biogeochemical cycles in the ocean required in the context of understanding climate change.

Remaining JGOFS issues

During the remaining data analysis, synthesis and modelling activities of JGOFS, we believe that the observations should be analysed and interpreted with special attention paid to the following themes:

- resolving gross imbalances in attempts at upper-ocean carbon/nitrogen budgets, i.e. new production $\neq$ export production (measured as sinking particles);
- organic–inorganic transformations, especially formation, transport and re-mineralisation of different fractions of DOM as well as POM;
- the effect on nutrient utilisation of fertilisation with the trace element iron;
- careful assessment of the sequestration of C on continental shelves and slopes, and the exchange of C, N and P with the open ocean;
- connected multi-disciplinary processes (putting the process studies in a better physical context);
- integrating from the process studies to the annual cycle (and longer) and to global scales.

Addressing this final theme has been made especially difficult by the absence of widely available high-quality remotely sensed data from a satellite-borne colour sensor during the observational phase of JGOFS. From the initial planning, it was clearly recognised that satellite colour imagery would be integral to the process of 'scaling up' from intensely sampled short-term regional process studies and time-series stations to ocean basin spatial scales and multi-year time scales. The absence of satellite colour imagery placed unexpected pressure on models to provide the linkages between short-term regional studies and ocean-basin, multi-year scales.

Future weaknesses in our understanding

To maintain scientific focus, JGOFS emphasises the carbon cycle in the open ocean, but important scientific issues that cut across the boundaries of JGOFS (and most other international studies) are becoming increasingly obvious. We project that after JGOFS finishes, the following scientific areas will need increased effort:

atmosphere–ocean exchanges of particulates and biogenic gases;
coupling and decoupling of C, N, P, S, Si and O cycles in the ocean;
connections between the oceanic margins and the open ocean in the C, N, P, S, Si and O cycles (collaboration with LOICZ), especially because the impact of human activities is and will continue to be greatest in the coastal zone;
photochemical processes in the upper ocean and increased UV-B, especially effects on DOM and DMS;
parameterising biogeochemical processes in models (what are the best formulations and how do the parameters vary in time and space);
detecting annual and inter-annual variation from longer-term trends.

We feel that these scientific activities can best be tackled with greater emphasis on the upper ocean and on quantifying exchanges with the lower atmosphere. These activities would involve greater participation by upper ocean physicists, and atmospheric chemists and physicists, as outlined below.

A Surface Ocean–Lower Atmosphere Study (SOLAS)

We propose a multi-disciplinary ocean biogeochemical study with emphasis on the surface ocean and the lower atmosphere, to be planned and implemented within IGBP with close linkages to WCRP. We envision the following three themes or foci.

Focus 1: Multiple Ocean Cycles study. A study to follow JGOFS focusing on the coupling–decoupling of the C, N, P, S, Si and O cycles in the ocean. The prevailing paradigm, that they are coupled by the (constant) Redfield ratios, has many exceptions due to sources and sinks of individual elements, e.g. denitrification in anoxic waters, surface deposition of inorganic N species, and short-term uptake by phytoplankton of C, N and P at non-Redfield ratios.

Focus 2: Multidisciplinary Ocean–Atmosphere Transfers study. A study concentrating on fluxes of gases, heat, water vapour, and relevant substances (deposition of nitrogen, sulphur, iron compounds to the sea surface) across the air–sea interface.

Focus 3: Coupled Boundary Layers study. A study to develop a coupled mixed-layer model of the planetary boundary layer and the upper ocean mixing layer, for implementation into coupled atmosphere–ocean general circulation models. Development of biogeochemical modules for embedding in the coupled physical models would be of mutual interest to IGBP and WCRP.

Focus 1, a Multiple Ocean Cycles study, would be a natural successor to JGOFS, and would complement the IGBP programmes LOICZ and GLOBEC (the Global Ocean Ecosystem Dynamics study), and the WCRP programme CLIVAR. Focus 2, a Multidisciplinary Ocean–Atmosphere Transfers study, would be conducted jointly with IGAC. Focus 3, a Coupled Boundary Layers study, should be planned and implemented jointly with the CLIVAR Upper Ocean Panel. SOLAS would replace GOEZS (the Global Ocean Euphotic Zone Study) in the planning of IGBP, since we believe that a Surface Ocean–Lower Atmosphere Study as described here better addresses the contemporary scientific requirements of IGBP.

The breadth and scope of the proposed activities are daunting. Nevertheless, we believe that it is possible to constrain the range of activities (1) by careful examination of sensitivity studies of the rapidly improving generations of coupled models of ocean–atmosphere–land biogeochemical processes, and (2) by attempts (such as IPCC Climate Change 1995) to identify the state of our knowledge (and ignorance) of the scientific issues most relevant in the context of understanding climate change.

Acknowledgements

We thank C. S. Wong and F. Whitney of the Institute of Ocean Sciences and W. G. Harrison of the Bedford Institute of Oceanography for use of their unpublished data, S. Calvert, G. Evans and two anonymous reviewers for their careful and constructive comments on the original manuscript, our many colleagues whose studies have stimulated our thinking, and the JGOFS Scientific Steering Committee for providing a forum for us to express our ideas. The Canadian Department of Fisheries and Oceans Ocean Climate Greenplan and the Natural Sciences and Engineering Research Council JGOFS programmes supported our activities leading to this paper.

References

Altabet, M. A., Francois, R., Murray, D. W. & Prell, W. L. (1995). Climate-related variations in denitrification in the Arabian Sea from sediment 15N/14N ratios. *Nature*, **373**, 506–9.

Anderson, L. A. & Sarmiento, J. L. (1994). Redfield ratios of remineralization determined by nutrient data analysis. *Global Biogeochemical Cycles*, **8**, 65–80.

Banse, K. (1994). Uptake of inorganic carbon and nitrate by marine plankton and the Redfield ratio. *Global Biogeochemical Cycles*, **8**, 81–4.

Bishop, J. K., Collier, R. W., Ketten, D. R. & Edmond, J. M. (1980). The chemistry, biology and vertical flux of particulate matter from the upper 1500 m of the Panama Basin. *Deep-Sea Research*, **27**, 615–40.

Bishop, J. K., Edmond, J. M., Ketten, D. R., Bacon, M. P. & Silker, W. B. (1977). The chemistry, biology and vertical flux of particulate matter from the upper 400 m of the equatorial Atlantic Ocean. *Deep-Sea Research*, **24**, 511–48.

Broecker, W. S. & Peng, T.-H. (1982). *Tracers in the Sea*. Lamont-Doherty Geological Observatory, Columbia University: ELDEGEO Press.

Carlson, C. A., Ducklow, H. W. & Michaels, A. F. (1994). Annual flux of dissolved organic carbon from the euphotic zone in the northwestern Sargasso Sea. *Nature*, **371**, 405–8.

Carlson, C. A. & Ducklow, H. W. (1995). Dissolved organic carbon in the upper ocean of the central equatorial Pacific Ocean, 1992: Daily and finescale vertical variations. *Deep-Sea Research*, Part II, **42**, 639–56.

Christensen, J. P. (1994). Carbon export from continental shelves, denitrification and atmospheric carbon dioxide. *Continental Shelf Research*, **14**, 547–76.

Christensen, J. P., Murray, J. W., Devol, A. H. & Codispoti, L. A. (1987). Denitrification in continental shelf sediments has major impact on the oceanic nitrogen budget. *Global Biogeochemical Cycles*, **1**, 97–116.

Codispoti, L. A. (1989). Phosphorus vs. nitrogen limitation of new and export production. In *Productivity of the Ocean: Present and Past*, ed. W. H. Berger, V. S. Smetacek & G. Wefer, pp. 377–94. Chichester: John Wiley.

Codispoti, L. A. & Christensen, J. P. (1985). Nitrification, denitrification and nitrous oxide cycling in the eastern tropical South Pacific Ocean. *Marine Chemistry*, **16**, 277–300.

Cullen, J. J. (1995). Status of the iron hypothesis after the Open-Ocean Enrichment Experiment. *Limnology and Oceanography*, **40**, 1336–43.

Denman, K. L., Hofmann, E. E. & Marchant, H. (1996). Marine biotic responses to environmental change and feedbacks to climate. In *Climate Change 1995*, ed. J. T. Houghton, L. G. Meira Filho, B. A. Callander, N. Harris, A. Kattenberg & K. Maskell, pp. 483–516. Intergovernmental Panel on Climate Change. Cambridge: Cambridge University Press.

Deuser, W. G., Jickells, T. D., King, P. & Commeau, J. A. (1995). Decadal and annual changes in biogenic opal and carbonate fluxes to the deep Sargasso Sea. *Deep-Sea Research*, Part I, **42**, 1923–32.

Ducklow, H. W. (1995). Ocean biogeochemical fluxes: New production and export of organic matter from the upper ocean. *Reviews of Geophysics*, **33**, Supplement, 1271–6.

Dugdale, R. C. & Goering, J. J. (1967). Uptake of new and regenerated forms of nitrogen in primary productivity. *Limnology and Oceanography*, **12**, 196–206.

Evans, G. T. & Fasham, M. J. (1993). Themes in modelling ocean biogeochemical processes. In *Towards a Model of Ocean Biogeochemical Processes*, ed. G. T. Evans & M. J. Fasham, pp. 1–19. Berlin: Springer-Verlag.

Fischer, G., Donner, B., Ratmeyer, V., Davenport, R. & Wefer, G. (1996). Distinct year-to-year particle flux variations off Cape Blanc during 1988–1991: Relation to $\delta^{18}O$-deduced sea-surface temperatures and trade winds. *Journal of Marine Research*, **54**, 73–98.

Fuhrman, J. A. & Capone, D. G. (1991). Possible biogeochemical consequences of ocean fertilization. *Limnology and Oceanography*, **36**, 1951–9.

Galloway, J. N., Schlesinger, W. H., Levy, H. II, Michaels, A. & Schnoor, J. L. (1995). Nitrogen fixation: Anthropogenic enhancement-environmental response. *Global Biogeochemical Cycles*, **9**, 235–52.

Ganeshram, R. S., Pedersen, T. F., Calvert, S. E. & Murray, J. W. (1995). Large changes in oceanic nutrient inventories from glacial to interglacial periods. *Nature*, **376**, 755–8.

Heinze, C., Maier-Reimer, E. & Winn, K. (1991). Glacial p_{CO_2} reduction by the world ocean: experiments with the Hamburg carbon cycle model. *Paleoceanography*, **6**, 395–430.

Honjo, S. & Manganini, S. J. (1993). Annual biogenic particle fluxes to the interior of the North Atlantic Ocean; studied at 34°N 21°W and 48°N 21°W. *Deep-Sea Research*, Part II, **40**, 587–607.

Houghton, J. T., Meira Filho, L. G., Callander, B. A., Harris, N., Kattenberg, A. & Maskell, K. (eds.) (1996). *Climate Change 1995*. Intergovernmental Panel on Climate Change. Cambridge: Cambridge University Press.

JGOFS (1990). *Science Plan*. Report No. 5. Baltimore: Scientific Committee on Oceanic Research, International Council of Scientific Unions.

Karl, D. M., Christian, J. R., Dore, J. E., Hebel, D. V., Letelier, R. M., Tupas, L. M. & Winn, C. D. (1996). Seasonal and interannual variability in primary production and particle flux at station ALOHA. *Deep-Sea Research*, Part II, **43**, 539–68.

Karl, D. M., Letelier, R., Hebel, D., Tupas, L., Dore, J., Christian, J. & Winn, C. (1995). Ecosystem changes in the North Pacific subtropical gyre attributed to the 1991–92 El Niño. *Nature*, **373**, 230–4.

Kolber, Z. S., Barber, R. T., Coale, K. H., Fitzwater, S. E., Greene, R. M., Johnson, K. S., Lindley, S. & Falkowski, P. G. (1994). Iron limitation of phytoplankton photosynthesis in the equatorial Pacific Ocean. *Nature*, **371**, 145–9.

Kurz, K. D. & Maier-Reimer, E. (1993). Iron fertilisation of the austral ocean – the Hamburg model assessment. *Global Biogeochemical Cycles*, **7**, 229–44.

Lohrenz, S. E., Knauer, G. A., Asper, V. L., Tuel, M., Michaels, A. F. & Knap, A. H. (1992). Seasonal variability in primary production and particle flux in the northwestern Sargasso Sea: U.S. JGOFS Bermuda Atlantic Time-series Study. *Deep-Sea Research*, **39**, 1373–91.

Martin, J. H. (1990). Glacial-interglacial CO_2 change: the iron hypothesis. *Palaeoceanography*, **5**, 1–13.

Martin, J. H., Coale, K. H., Johnson, K. S., Fitzwater, S. E., Gordon, R. M., Tanner, S. J., Hunter, C. N., Elrod, V. A., Nowicki, J. L., Coley, T. L., Barber, R. T., Lindley, S., Watson, A. J., Van Scoy, K., Law, C. S., Liddicoat, M. I., Ling, R., Stanton, T., Stockel, J., Collins, C., Anderson, A., Bidigare, R., Ondrusek, M., Latasa, M., Millero, F. J., Lee, K., Yao, W., Zhang, J. Z., Friederich, G., Sakamoto, C., Chavez, F., Buck, K., Kolber, Z., Greene, R., Falkowski, P., Chisholm, S. W., Hoge, F., Swift, R., Yungel, J., Turner, S., Nightingale, P., Hatton, A., Liss, P. & Tindale, N. W. (1994). Testing the iron hypothesis in ecosystems of the equatorial Pacific Ocean. *Nature*, **371**, 123–9.

Martin, J. H., Knauer, G. A., Karl, D. M. & Broenkow, W. W. (1987). VERTEX: carbon cycling in the northeast Pacific. *Deep-Sea Research*, **34**, 267–85.

McCarthy, J. J. (1980). Nitrogen. In *The Physiological Ecology of Phytoplankton*, ed. I. Morris, pp. 191–233. Oxford:

Blackwell Scientific.
Michaels, A. F., Bates, N. R., Buesseler, K. O., Carlson, C. A. & Knap, A. H. (1994). Carbon-cycle imbalances in the Sargasso Sea. *Nature*, **372**, 537–40.
Murray, J. W., Barber, R. T., Roman, M. R., Bacon, M. P. & Feely, R. A. (1994). Physical and biological controls on carbon cycling in the Equatorial Pacific. *Science*, **266**, 58–65.
Najjar, R. G., Sarmiento, J. L. & Toggweiler, J. R. (1992). Downward transport and fate of organic matter in the ocean: simulations with a general circulation model. *Global Biogeochemical Cycles*, **6**, 45–76.
Parsons, T. R., Takahashi, M. & Hargrave, B. (1984). *Biological Oceanographic Processes*. Oxford: Pergamon.
Peng, T.-H. & Broecker, W. S. (1991). Factors limiting the reduction of atmospheric CO_2 by iron fertilisation. *Limnology and Oceanography*, **36**, 1919–27.
Redfield, A. C., Ketchum, B. H. & Richards, F. A. (1963). The influence of organisms on the composition of sea water. In *The Sea*, vol. 2, ed. M. N. Hill, pp. 26–77. New York: Wiley Interscience.
Robertson, J. E., Robinson, C., Turner, D. R., Holligan, P., Watson, A. J., Boyd, P., Fernandez, E. & Finch, M. (1994). The impact of a coccolithophore bloom on oceanic carbon uptake in the northeast Atlantic during summer 1991. *Deep-Sea Research*, **41**, 297–314.
Sakshaug, E. (1977). Limiting nutrients and maximum growth rates for diatoms in Narragansett Bay. *Journal of Experimental Marine Biology and Ecology*, **28**, 109–23.
Sambrotto, R. N., Savidge, G., Robinson, C., Boyd, P., Takahashi, T., Karl, D. M., Langdon, C., Chipman, D., Marra, J. & Codispoti, L. (1993). Elevated consumption of carbon relative to nitrogen in the surface ocean. *Nature*, **363**, 248–50.
Sarmiento, J. L. & Orr, J. C. (1991). Three-dimensional simulations of the impact of Southern Ocean nutrient depletion on atmospheric CO_2 and ocean chemistry. *Limnology and Oceanography*, **36**, 1928–50.
Sarmiento, J. L. & Toggweiler, J. R. (1984). A new model for the role of the oceans in determining atmospheric p_{CO_2}. *Nature*, **308**, 621–4.
Shaffer, G. (1993). Effects of the marine biota on global carbon cycling. In *The Global Carbon Cycle*, ed. M. Heimann, pp. 431–55. Berlin: Springer-Verlag.
Shaffer, G. (1996). Biogeochemical cycling in the global ocean 2. New production, Redfield ratios, and remineralization in the organic pump. *Journal of Geophysical Research*, **101**, 3723–45.
Shaffer, G. & Sarmiento, J. L. (1995). Biogeochemical cycling in the global ocean 1. A new, analytical model with continuous vertical resolution and high-latitude dynamics. *Journal of Geophysical Research*, **100**, 2659–72.
Siegenthaler, U. & Sarmiento, J. L. (1993). Atmospheric carbon dioxide and the ocean. *Science*, **365**, 119–25.
Slawyk, G., Collos, Y., Minas, M. & Grall, J.-R. (1978). On the relationship between carbon-to-nitrogen composition ratios of the particulate matter and growth rate of marine phytoplankton from the northwest African upwelling area. *Journal of Experimental Marine Biology and Ecology*, **33**, 119–31.
Toggweiler, J. R. (1989). Is the downward dissolved organic matter (DOM) flux important in carbon transport? In *Productivity of the Ocean: Present and Past*, ed. W. H. Berger, V. S. Smetacek & G. Wefer, pp. 65–85. New York: J. Wiley and Sons.
Tsunogai, S. & Noriki, S. (1991). Particulate fluxes of carbonate and organic carbon in the ocean. Is the marine biological activity working as a sink of the atmospheric carbon? *Tellus*, **43B**, 256–66.
Watson, A. J., Law, C. S., Van Scoy, K. A., Millero, F. J., Yao, W., Friederich, G. E., Liddicoat, M. I., Wanninkhof, R. H., Barber, R. T. & Coale, K. H. (1994). Minimal effect of iron fertilization on sea-surface carbon dioxide concentrations. *Nature*, **371**, 143–9.
Westbroek, P., Brown, C. W., van Bleijswijk, J., Brownlee, C., Brummer, G. J., Conte,

M., Egge, J., Fernández, E., Jordan, R., Knappersbusch, M., Stefels, J., Veldhuis, M., van der Wal, P. & Young, J. (1993). A model system approach to biological climate forcing: the example of *Emiliania huxleyi*. *Global Planetary Change*, **8**, 27–46.

Williams, P. J. le B. (1995). Evidence for the seasonal accumulation of carbon-rich dissolved organic material, its scale in comparison with changes in particulate material and the consequential effect on net C/N assimilation ratios. *Marine Chemistry*, **51**, 17–29.

PART SIX

CONCLUSION

17 Some conclusions and highlights of JGOFS mid-project achievements

J. G. Field, H. W. Ducklow and R. B. Hanson

Keywords: carbon sink, CO_2 drawdown, ocean colour satellites, biological and physical pumps, export flux, iron fertilisation, high-nutrient – low-chlorophyll, marginal ice zone, monsoon, dissolved organic carbon, JGOFS data sets, ecological three-dimensional models

Introduction

This volume provides a cross-section of JGOFS and related research during the first half of its expected life. It would take many volumes to adequately cover all JGOFS findings to date and indeed the official list of reviewed journal publications already exceeds 1500. There have also been several journal issues devoted to JGOFS; those of *Deep-Sea Research* and *Philosophical Transactions of the Royal Society* spring to mind.

JGOFS was planned as a co-operative international venture between marine scientists of different countries, each with their own national programmes and funding systems. But more importantly, it was planned and implemented by joint efforts of previously different communities of marine scientists: chemical, physical, biological and geological oceanographers. In this it has succeeded admirably and created a new community of marine biogeochemical scientists with new insights into the role of the oceans in the global carbon cycle. We will highlight some of the achievements of JGOFS below. It should be clearly understood that it is impossible to separate the achievements of officially funded JGOFS research from the related research inspired by JGOFS, and the questions raised and challenged by individual scientists inside and outside the programme; hence the phrasing of 'JGOFS and related research'.

JGOFS was planned as a series of complementary studies, which can be broadly categorised as intensive process and time-series studies on the one hand, and extensive surveys to give global and synoptic coverage on the other. The latter includes the CO_2 survey performed co-operatively with the WOCE

Hydrographic Programme, and also remote sensing of sea-surface temperatures, ocean colour and altimetry in particular. Delays in launching the second generation of ocean colour sensors have been a setback for JGOFS, because many observations were planned in the context of real-time ocean colour images, which were not available. Nevertheless, improvisation allowed successful completion of the main open-ocean process studies reported in this volume.

Extensive studies

One of the major achievements of JGOFS has been to publish a suite of protocols to set the minimum standards for the basic measurements made as part of JGOFS (Knap *et al.*, 1994, reprinted in 1996). Thus, comparable basic measurements were made globally, including the global CO_2 survey. This made possible, for example, the observational estimates of oceanic CO_2 fluxes (Takahashi *et al.*, 1997), which identified regions of CO_2 drawdown and outgassing (Fig. 17.1, colour plate). It also includes the finding that the Southern Ocean is a net sink for CO_2, whereas it had been believed to be a source. The region remains under-sampled, especially in winter, and more work is needed in this complex region. Overall, the observations show that the world ocean is a net sink for about 1.4 Gt C, compared with pre-industrial estimates that the oceans were a source for about 0.5 Gt C. Thus, the changed flux into the ocean is observed to be some 2 Gt C, agreeing with model estimates (see, for example, Sarmiento *et al.*, 1992; Siegenthaler & Sarmiento, 1993).

Without JGOFS there would have been much less incentive to launch ocean colour satellites. Possibly JGOFS failed in not exerting more pressure for timely launching of SeaWiFS. Nevertheless, the satellite sensor is now operational and new improved algorithms are being developed to estimate phytoplankton pigment densities in both oceanic and coastal waters, heralding a new era in biological oceanography. In anticipation of these developments, the concept of biogeochemical provinces (Platt *et al.*, Chapter 15, this volume) has been developed to provide greatly improved global estimates of primary production, which are considerably higher than most previous estimates.

Intensive studies

Each of the intensive process study regions was chosen for a particular reason, and the chapters of this volume show that important new insights have been gained into biogeochemical processes in every case.

North Atlantic Ocean

The North Atlantic Bloom Experiment was the first JGOFS process study and showed clearly the role of spring phytoplankton blooms in drawing down CO_2 as the 'green carpet unrolled northwards', taking weeks for the ocean and atmosphere to be restored to CO_2 equilibrium: the 'biological pump' in action. Coccolithophore blooms were also shown to be important, both in removing $CaCO_3$ from the euphotic zone and also in releasing CO_2 into solution and altering the 'rain ratio' (see Denman & Peña, Chapter 16, this volume). These and many other studies showed that the activities of the biological pump are strongly influenced by physical processes at the mesoscale, with much activity at fronts and eddies. A weakness in the present generation of global GCM carbon models is that this kind of intense biological activity, which often results in sedimentation, has to be represented by parameterisation, if it is represented at all.

Pacific Ocean

The equatorial Pacific was studied intensively by JGOFS because of the strong inter-annual variation of the El Niño – Southern Oscillation. Fortune smiled on JGOFS in that major cruises took place at the height of the 1992 El Niño, and afterwards, providing strong contrast in physical conditions and the ecosystem response. As expected, the equatorial Pacific was shown to be a net source of CO_2 to the atmosphere in the upwelling region. The source term was reduced during the El Niño conditions (Murray *et al.*, 1994, 1996). The carbon system was also shown to be strongly influenced by equatorial waves, which drive mesoscale patterns of biomass accumulation (Barber *et al.*, 1996; Foley *et al.*, 1997). The export flux from the euphotic zone was estimated by sediment traps and by the thorium isotope technique (Jannasch *et al.*, 1996; Murray *et al.*, 1996; Gardner, Chapter 8, this volume). The combined use of sediment-trap carbon and thorium measurements has also been undertaken at JGOFS time-series stations, providing a valuable check on export estimates (see Gardner, Chapter 8, this volume).

The iron-fertilisation controversy has been explored by the biogeochemical community using innovative field experiments in which a patch of carefully fertilised water was marked by an inert tracer (SF_6) and followed for days to weeks (Coale *et al.*, 1996). The first experiment demonstrated some stimulation of biological productivity but the fertilised patch of water was subducted beneath another water mass and no CO_2 drawdown was observed. The second experiment was much clearer and, to the surprise of many, demonstrated increased primary production and CO_2 drawdown. This helps explain the paradox of the high-nitrogen – low-chlorophyll (HNLC) region in the equatorial Pacific, (see Chapters 4, 5, and 6, this volume). The interacting roles of iron, grazing and physical processes were examined in detail in the JGOFS Equatorial Pacific Study (Landry *et al.*, 1997).

Southern Ocean

The Southern Ocean programme was built around several national Antarctic research programmes. Before JGOFS, it was the most under-sampled of the world oceans, especially in winter. This remains the case, but great advances have been made and the vast region has been shown to be very complex, that is, it is a mosaic of sources and sinks for CO_2, spatially, seasonally, and inter-annually. Overall it has been shown to be a net sink for CO_2 (Takahashi *et al.*, 1997). There has been a greatly improved understanding of biological activity at the subtropical convergence, the polar front and the marginal ice zone, again demonstrating the importance of physical processes in driving biogeochemistry (Bathmann *et al.*, Chapter 10, this volume). The enigma of Southern Ocean HNLC regions remains, that is, are these regions caused by iron-limitation, deep mixing, grazing or other factors, probably in combination?

Indian Ocean

The Arabian Sea was chosen for intensive study because of the strong seasonality of the reversing monsoon winds, which drive seasonal upwelling along most of the western side of the basin. The results of process studies are still being synthesised, but already demonstrate the close coupling of physical and biological processes with feedback from plankton to light and heat absorption (Smith, 1998; Sathyendranath & Platt, Chapter 9, this volume). The importance of the microbial loop and dissolved organic matter in closing the carbon budget is emerging here, as in many other JGOFS studies. During the lifetime of JGOFS, it has become generally accepted that the microbial loop provides a relatively constant-spinning background of production, respiration and nutrient regeneration in the small size fractions of plankton, whereas blooms result from development of the large size fractions under favourable conditions (Joint *et al.*, 1993).

Time series

The findings of the two major JGOFS time-series stations off Bermuda (BATS) and Hawaii (HOT) have been synthesised (Michaels *et al.*, Chapter 13, this volume). The other two JGOFS time-series sites started later, with a less complete suite of physical and biogeochemical measurements: ESTOC off the Canary Islands in the North Atlantic (Llinás *et al.*, 1997) and KERFIX off Kerguelen Island in the Indian Ocean sector of the Southern Ocean. The time-series stations have proved to be one of the outstanding successes of

JGOFS, because so many biogeochemical processes are intense events operating at short time scales. One of the most important results in JGOFS has been the classification of dissolved organic carbon into labile and refractory portions, the latter forming part of the export flux from the mixed layer when it deepens by internal- or surface-wave action (see Doney *et al.*, Michaels *et al.* and Fasham & Evans, Chapters 12, 13 and 14, respectively, this volume; Carlson *et al.*, 1994). Indeed the role of dissolved organic carbon has been found to be an important component in all the JGOFS Process Studies.

A fact that has not been emphasised in previous chapters, and was not explicit in the aims of the time-series stations when they were planned, is the crucial role the time-series observations have played in validating and calibrating regional models (Doney *et al.*, 1996; Doney *et al.*, Chapter 12, this volume). It is through models that we express our understanding of processes and this understanding has to be consistent with observations made at the appropriate scales. The value of time-series observations increases with the length of the time series, and with the range of observations made there is tremendous added value. Clearly there is a large cost of time and effort involved, and there need to be good reasons for continuing to measure each parameter. The time-series observations have been far from perfect, but without them our understanding of the coupled physical–biogeochemical ocean systems would have been much less complete than it is today.

Data, modelling and synthesis

JGOFS has been criticised for its decentralised data management policy, having data scattered in different data centres around the world. There were times when one wished for a more centralised system, but biogeochemical data are much more varied and complex than physical data and practically demand a distributed system. The Internet has developed during the lifetime of JGOFS and now provides access to JGOFS data sets and the principal investigators through the JGOFS home page (http://ads.smr.uib.no/jgofs/jgofs.htm). Many JGOFS data sets are now available on CD-ROMs, making them available to all investigators.

JGOFS modelling has deliberately developed in parallel along several different lines, each with its own objectives. The models range from simple geochemical box models of the global ocean and detailed time-dependent ecological models driven by physical processes in the vertical dimension only to fully coupled three-dimensional models including biology. Considerable advances have been made in coupling simple ecological models of the biological pump to regional and global three-dimensional general circulation models (GCMs) (see Fasham & Evans, Chapter 14, this volume). In co-operation with its sister project GAIM (Global Analysis Interpretation and Modelling),

JGOFS is entering into an Ocean Carbon-Cycle Model Inter-comparison Project (OCMIP) (http://gaim.unh.edu/ocean.html) to improve and compare coupled three-dimensional global general circulation models. The inclusion of explicit biology in many of the models tested in the OCMIP Phase II comparison would not have happened without the impetus provided by JGOFS. Dynamic biogeochemical province models have also been developed to synthesise basin-scale and global primary production from remotely sensed data coupled with ship observations (Platt *et al.*, Chapter 15, this volume). These have all been exciting developments, which have advanced JGOFS towards synthesis.

As JGOFS moves into its synthesis phase from 1998 to 2004, there are many outstanding questions. As always, scientific research reveals as many new questions and gaps in understanding as it provides answers. Denman & Peña (Chapter 16, this volume) pose some of the longer-term questions. Others include the following examples.

What is the magnitude of carbon fluxes across continental margins? (See Liu *et al.*, Chapter 7, this volume.)

To what extent does bottom- and deep-water formation vary inter-annually, how is this likely to be affected by global change, and how are its chemical properties influenced by biogeochemical processes? (The physical pump.)

What is the role of intermediate water formation and upwelling in the carbon cycle? (This is being addressed in the new North Pacific JGOFS study.)

Can models be developed fast enough to include data assimilation into coupled ecological three-dimensional models within the lifetime of JGOFS?

These are all challenging and important questions, requiring great effort as the results of JGOFS continue to come from the newer observational studies and as synthesis activities stimulate new hypotheses, as well as conclusions.

References

Barber, R. T., Sanderson, M. P., Lindley, S. T., Chai, F., Newton, J., Trees, C. C., Foley, D. G. & Chavez, F. P. (1996). Primary productivity and its regulation in the Equatorial Pacific during and following the 1991–92 El Niño. *Deep-Sea Research*, Part II, **43**, 933–69.

Carlson, C. A., Ducklow, H. W. & Michaels, A. F. (1994). Annual flux of dissolved organic carbon from the euphotic zone in the northwestern Sargasso Sea. *Nature* **37**, 405–8.

Coale, K. H., Johnson, K. S., Fitzwater, S. E., Gordon, R. M., Tanner, S., Chavez, F. P., Ferioli, L., Sakamoto, C., Rogers, P., Millero, F., Steinberg, P., Nightingale, P., Cooper, D., Cochlan, W. P., Landry, M. R., Constantinou, J., Rollwagen, G., Trasvina, A. & Kudela, R. (1996). A massive phytoplankton bloom

induced by an ecosystem-scale iron fertilization experiment in the equatorial Pacific Ocean. *Nature*, **383**, 495–501.
Doney, S. C., Glover, D. M. & Najjar, R. G. (1996). A new coupled, one-dimensional biological-physical model for the upper ocean: applications to the JGOFS Bermuda Atlantic Time Series (BATS) site. *Deep-Sea Research*, Part II, **43**, 591–624.
Foley, D. G., Dickey, T. D., McPhaden, M. J., Bidigare, R. R., Lewis, M. R., Barber, R. T., Lindley, S. T., Garside, C., Manov, D. V. & McNeil, J. D. (1997), Longwaves and primary productivity variations in the Equatorial Pacific at 0 deg., 140 deg W. *Deep-Sea Research*, Part II, **44**, 1801–26.
Jannasch, H. W., Honeyman, B. D & Murray, J. W. (1996). Marine scavenging: The relative importance of mass transfer and reaction rates. *Limnology and Oceanography*, **41**, 82–8.
Joint, I., Pomroy, A., Savidge, G. & Boyd, P. (1993). Size-fractionated primary productivity in the northeast Atlantic in May–July 1989. *Deep-Sea Research*, Part II, **40**, 423–40.
Knap, A. A., Michaels, A., Close, A., Ducklow, H. W. & Dickson, A. (eds.) (1996). *Protocols for the Joint Global Ocean Flux Study (JGOFS) Core Measurements*. JGOFS Report No. 19, vi + 170 pp. (Reprint of the IOC Manuals and Guides No. 29, UNESCO 1994.)
Landry, M. R., Barber, R. T., Bidigare, R. R., Chai, F., Coale, K. H., Dam, H. G., Lewis, M. R., Lindley, S. T., McCarthy, J. J., Roman, M. R., Stoecker, D. K., Verity, P. G. & White, J. R. (1997). Iron and grazing constraints on primary production in the central equatorial Pacific: An EqPac synthesis. *Limnology and Oceanography*, **42**, 405–18.
Llinás, O., Rodríguez de León, A., Siedler, G. & Wefer, G. (eds.) (1997). European Station for Time Series in the Ocean Canary Islands (ESTOC) data report. Canarian Institute of Marine Sciences. Technical Reports No. 3.
Murray, J. W., Barber, R. T., Roman, M. R., Bacon, M. P. & Feely, R. A. (1994). Physical and biological controls on carbon cycling in the equatorial Pacific. *Science*, **266**, 58–65.
Murray, J. W., Young, J., Newton, J., Dunne, J., Chapin, T., Paul, B. & McCarthy, J. J. (1996). Export flux of particulate organic carbon from the central equatorial Pacific determined using a combined drifting trap Th^{234} approach. *Deep-Sea Research*, Part II, **43**, 1095–132.
Sarmiento, J. L., Orr, J. C. & Siegenthaler, U. (1992) A perturbation simulation of CO_2 uptake in an ocean general-circulation model. *Journal of Geophysical Research*, **97**, 3621–45.
Siegenthaler, U. & Sarmiento, J. L. (1993). Atmospheric carbon-dioxide and the ocean. *Nature*, **365**, 119–25.
Smith, S. L. (1998). The 1994–1996 Arabian Sea Expedition: Oceanic response to monsoonal forcing, part I. *Deep-sea Research*, Part II, **45**, 1905–2501.
Takahashi, T., Feely, R. A., Weiss, R. F., Wanninkhof, R. H., Chipman, D. W., Sutherland, S. C. & Takahashi, T. T. (1997). Global air-sea flux of CO_2: An estimate based on measurements of sea-air p_{CO_2} difference. *Proceedings of the National Academy of Sciences, U.S.A.*, **94**, 8292–9.
Wanninkhof, R. (1992). Relationship between wind-speed and gas-exchange over the ocean. *Journal of Geophysical Research*, **97**, 7373–82.

Index